D0778510

CLIMATE AND HISTORY

EDITED BY

T. M. L. WIGLEY M. J. INGRAM AND G. FARMER

CLIMATE AND HISTORY

Studies in past climates and their impact on Man

CAMBRIDGE UNIVERSITY PRESS

Cambridge

London New York New Rochelle

Melbourne Sydney

Published by the Press Syndicate of the University of Cambridge
The Pitt Building, Trumpington Street, Cambridge CB2 1RP
32 East 57th Street, New York, NY 10022, USA
296 Beaconsfield Parade, Middle Park, Melbourne 3206, Australia

First published 1981

Printed in Malta by Interprint Limited

British Library cataloguing in publication data

Climate and history.
1. Climate changes – Congresses
2. Man – Influence on nature – Congresses
I. Wigley, T. M. L. II. Ingram, M. J.
III. Farmer, G.
551.6 QC981.8.C5 80-42275
ISBN 0 521 23902 8

CONTENTS

CONTRIBUTORS

Mr J. L. Anderson, School of Economics, La Trobe University, Bundoora, Victoria, Australia 3083

Dr W. R. Baron, History Department, University of Maine, Orono, Maine 04469, USA

Dr H. J. B. Birks, Botany School, University of Cambridge, Downing Street, Cambridge CB2 3EA, UK

Professor H. W. Borns, Institute for Quaternary Studies, University of Maine, Orono, Maine 04469, USA

Professor M. J. Bowden, Graduate School of Geography, Clark University, Worcester, Massachusetts 01610, USA

Ms A. E. Bridges, History Department, University of Maine, Orono, Maine 04469, USA

Mr G. Farmer, Climatic Research Unit, University of East Anglia, Norwich NR4 7TJ, UK

Professor H. Flohn, Meteorologisches Institut der Universität Bonn, Auf dem Hügel 20, D-5300 Bonn 1, Federal Republic of Germany

Professor H. C. Fritts, Laboratory of Tree-Ring Research, University of Arizona, Tucson, Arizona 85721, USA

Dr G. A. Gordon, Laboratory of Tree-Ring Research, University of Arizona, Tucson, Arizona 85721, USA

Dr H. A. Gould, Graduate School of Geography, Clark University, Worcester, Massachusetts 01610, USA

Dr J. Gray, Department of Physics, University of Alberta, Edmonton, Alberta, Canada T6G 2JG

Dr M. J. Ingram,* Department of Modern History, The Queen's University of Belfast, Belfast BT7 1NN, UK

* Formerly of the Climatic Research Unit, University of East Anglia, Norwich, UK.

Dr D. L. Johnson, Graduate School of Geography, Clark University, Worcester, Massachusetts 01610, USA

Dr R. W. Kates, Graduate School of Geography, Clark University, Worcester, Massachusetts 01610, USA

Dr P. A. Kay, Graduate School of Geography, Clark University, Worcester, Massachusetts 01610, USA

Professor H. H. Lamb, Climatic Research Unit, University of East Anglia, Norwich NR4 7TJ, UK

Mr G. R. Lofgren, Laboratory of Tree-Ring Research, University of Arizona, Tucson, Arizona 85721, USA

Dr R. McGhee, Archaeological Survey of Canada, National Museum of Man, Ottawa, Ontario, Canada K1A OM8

Dr T. H. McGovern, Department of Anthropology, Hunter College, Colombia University, New York, NY 10021, USA

Dr A. MacKay, Department of History, University of Edinburgh, 5 Buccleuch Place, Edinburgh, EH8 9JX, UK

Dr D. A. Mooley, Indian Institute of Tropical Meteorology, Ramdurg House, University Road, Puna 411005, India

Dr S. E. Nicholson, Graduate School of Geography, Clark University, Worcester, Massachusetts 01610, USA

Dr G. B. Pant, Indian Institute of Tropical Meteorology, Ramdurg House, University Road, Puna 411005, India

Dr M. L. Parry, Department of Geography, University of Birmingham, P.O. Box 363, Birmingham B15 2TT, UK

Dr C. Pfister, Geographisches Institut der Universität Bern, Hallerstrasse 12, CH-3012 Bern, Switzerland

Professor S. C. Porter, Quaternary Research Center, University of Washington, Seattle, Washington 98195, USA

Mr W. E. Riebsame, Graduate School of Geography, Clark University, Worcester, Massachusetts 01610, USA

Dr B. D. Shaw, Classical Studies, University of Lethbridge, 4401 University Drive, Lethbridge, Alberta, Canada T1K 3M4

Professor D. C. Smith, History Department, University of Maine, Orono, Maine, 04469, USA

Dr D. M. G. Sutherland, History Department, Brock University, St Catherines, Ontario, Canada L2S 3A1

Mr D. J. Underhill,* 34 Nancy Street, Bolton, Ontario, Canada

* Formerly of the Climatic Research Unit, University of East Anglia, Norwich, UK.

Professor Wang Shao-wu, Head, Section of Meteorology, Department of Geophysics, Peking University, People's Republic of China

Dr R. A. Warrick, National Center for Atmospheric Research, P.O. Box 3000, Boulder, Colorado 80303, USA

Dr D. Weiner, Graduate School of Geography, Clark University, Worcester, Massachusetts 01610, USA

Dr T. M. L. Wigley, Climatic Research Unit, University of East Anglia, Norwich NR4 7TJ, UK

Dr Zhao Zong-ci, c/o Wang Shao-wu, Head, Section of Meteorology, Department of Geophysics, Peking University, People's Republic of China

PREFACE

This book developed out of a conference on the subject of Climate and History, held at the University of East Anglia in July 1979. The Conference was very much the brainchild of Professor Hubert H. Lamb, founder and first Director of the Climate Research Unit whose persistent commitment to the study of past climates is well known.

Anyone who studies historical sources for evidence of past climatic change cannot fail to consider the question of the impact of climate and weather on human history; the records abound with suggestive passages detailing the effects of weather fluctuations on harvests and other economic and social activities. However, only a little reflection is required to recognise that the study of the inter-relationships between climate and human affairs is a matter of great complexity. To make real progress requires the fruitful co-operation of historians, archaeologists, and exponents of the many scientific disciplines which bear on climate and climatic reconstruction. The purpose of the Conference was to bring together members of all the disciplines interested in the possible interactions between climate and history.

The development of the Conference, and indeed the Climatic Research Unit's involvement in historical climatology owes much to the support of the Rockefeller Foundation. Their generous grants in support of a programme of climatic reconstruction using documentary sources provided the context in which the idea of an international conference on climate and history developed. In a real sense, this book is a product of the Rockefeller Foundation's commitment to climatic research.

Apart from the Rockefeller Foundation, many other bodies sponsored the Conference. They include the Ford Foundation, National Science Foundation (USA), United Nations Environment Programme, World Meteorological Organisation, British Council, Anglia Television, The Lord Mayor and Norwich City Council, Royal Historical Society, Past and Present Society, Prehistoric Society, American Meteorological Society, Association Internationale d'Histoire Économique, and the University of East Anglia. We offer our sincere thanks to these organisations for their support.

In the event, the Conference was attended by some 250 individuals from over 30 countries, including China, India, Kenya, Peru, Australia, New Zealand, the United States, Canada, and many European countries. Over 65 papers were given. The fact that so many were of high quality posed an invidious problem when it came to choosing papers for publication in this volume. The essential criterion was to select contributions which could together form a well-structured book and represent a variety of approaches to the study of climate and history.

Many individuals contributed to the success of the Conference and to the work involved in preparing this volume. In particular we should like to thank: Ms S. Boland, Ms B. Harris, Dr A. Hassell Smith, Mr N. J. Huckstep, Dr P. M. Kelly, Professor H. H. Lamb, Ms J. M. Lough, Dr R. McGhee, Dr T. H. McGovern, Mr V. F. G. Morgan, Dr H. T. Mörth, Dr R. Mortimer, Ms S. Napleton, Ms A. E. J. Ogilvie, Dr J. P. Palutikof, Dr M. L. Parry, Dr J. Pilcher, Dr R. Pitman, Dr A. M. Swain and Mr D. J. Underhill.

T. M. L. Wigley
M. J. Ingram
G. Farmer

Norwich
October 1980

PART I

INTRODUCTION

1

Past climates and their impact on Man: a review

M. J. INGRAM, G. FARMER AND
T. M. L. WIGLEY

Abstract

The study of the interactions between climate and history embraces a number of distinct aspects: climate reconstruction; the identification and measurement of the impact of climate on past societies; the adaptation of societies to climatic stress; and human perceptions of climate and climatic change. In this review we discuss all of these aspects.

We begin with a critical evaluation of the methods that may be used to unravel the climatic changes of the last 5000 years, on the basis of various types of proxy data (glacial evidence, pollen, tree rings, stable isotopes and archaeological data), historical documents and, for recent centuries, meteorological instruments. Many uncertainties are involved in the use and interpretation of individual types of proxy and historical data. Nevertheless, a reasonably detailed record of past changes in climate can be gained by use of a variety of sources. We discuss this record with emphasis on the Medieval Warm Epoch and the Little Ice Age, and with particular reference to the question of changes in variability.

In reviewing the impact of climate on Man we first discuss prevailing schools of thought and then present a detailed methodology for future studies with a critical analysis of approaches used in the past. From a consideration of these aspects, and of a number of case studies, we conclude that the complex interrelationships between various parts of society make the detection of climatic impact extremely difficult, while in many historical situations climatic factors may be dismissed as of negligible importance. Although the reality of the primary impact of climate on the biosphere (crops, animals, etc.) cannot be disputed, higher order impacts on prices, social unrest, politics, etc. are so complicated by other factors that it is almost impossible to construct models which are amenable to rigorous statistical testing. These difficulties hold on all time scales, but are particularly acute when one is considering the possible impact of long-term climatic change. In certain circumstances, marginal societies being the best example, simpler models of impact can be constructed. Furthermore, these are the very situations where impact is expected to be most readily detectable. But even here, in the most detailed studies made to date, there are methodological difficulties associated with the quality of climatic and societal data which must be used to test the models, and more fundamental problems in that alternative explanatory hypotheses can be put forward. In spite of the lack of totally convincing case studies demonstrating the existence and magnitude of climatic impact, sufficient evidence exists to demand further well-structured investigations. One of the obfuscating variables in determining impact is the adaptability of society to climatic stress. Some societies have shown a considerable ability to adapt to both minor and major stresses by diffusing the impacts either internally or externally into a wider social network. Other societies, in contrast, seem to have shown less ability to adapt. Lessons to be learnt from further case studies of adaptability (or lack of adaptability) could have considerable application to modern-day problems of climatic impact in developing societies. The ability to adapt depends in part on whether and how society perceives climatic change; and the question of perception is interesting in other ways, too. Little work has been done on this aspect and it would appear to be a fruitful avenue for further research.

Introduction

As the theme of 'climate and history' invites a variety of approaches, the contributions to this volume have been chosen to represent as many of these approaches as possible, in order to provide information for a wide range of readers from a variety of different disciplines. The purpose of this review is to provide a synoptic view of the different approaches to the main theme, to review the findings of the separate contributions (and, to a limited extent, of previous writers in the field), and to identify areas where further research is necessary and likely to be fruitful. To facilitate this we will consider separately four main aspects: climate reconstruction; the identification and measurement of impact; adaptation; and perception. One of our aims is to show how different researchers, approaching the study of climate in the past with a multitude of different questions and preconceptions, and with the aid of a wide range of skills and techniques, may learn valuable lessons from each other and engage fruitfully in joint endeavours. 'Climate and history' as a field of study is located at the point of intersection of many different disciplines, and progress in the field demands interdisciplinary cooperation.

One approach is simply to study the history of climate itself, to attempt to reconstruct the pattern of climatic changes and fluctuations over past centuries and millennia. This approach is particularly associated with scientific disciplines but has also attracted some historians, most notably Le Roy Ladurie (e.g. 1972). The numerous techniques now available for reconstructing past climates – the strengths and weaknesses, possibilities and limitations, and the major conclusions about the course of climatic change which can be derived from these techniques – are discussed in the second part of this book and the second section of this review.

Historians, historical geographers and archaeologists are by definition concerned with analysing the course of economic, social and political change in past time. The impact of climatic fluctuation and change has long been recognised as one of the factors that demands consideration, though there is and always has been dispute about just how much attention needs to be paid to this variable. Moreover, the question of the past impact of climate is of interest to other disciplines. Many of the environmental scientists engaged in climatic reconstruction are not motivated simply by a spirit of scientific inquiry, but are deeply committed to Man's welfare and believe that their research can be of practical benefit. One of the possible applications of an improved understanding of the climates of the past and of the present is to predict the likely range of climatic variation in the future. The utility of such predictions (assuming them to be feasible) depends on an understanding of the ways in which climatic variation affects human activities. Present-day studies can contribute to this understanding, but there is an obvious incentive to extend the field of investigation to include the past. The fact that past societies (especially those more remote in time) differed markedly from those of the modern world makes it unrealistic to draw simple parallels. Nevertheless, it is reasonable to suppose that some useful lessons might be learnt from past events, especially perhaps from the study of the impact of climatic variations and conditions which have not been experienced in recent times. Accordingly, politicians and planners have begun to show a lively interest in identifying and measuring the effect of climatic fluctuations and changes on past societies.

Thus researchers from a variety of disciplines and with a variety of aims and preconceptions share some interest in the question of assessing the possible impact of climate on past societies. It is this theme which provides the focus for many of the contributions to this book, and will provide much of the substance of this review. It should be recognised at once, however, that the theme of climatic impact on human society can itself be approached from more than one direction. Until recently the subject was mostly conceived in terms of investigating the possible impact of climatic variations, whether on long or short time scales, with little reference to the possible range of human responses. Many of the contributions to this volume, discussed in the third section of this review, fit into this pattern. However, recognition and a growing understanding of the complexity of the inter-relationships between climate and human affairs have prompted some scholars, including a number of the contributors to this book, to reorientate their aims. They have begun to shift their attention to the study of the processes of human *adaptation* (or lack of adaptation) to climatic or other stimuli. This approach, which is markedly anti-determinist, is reviewed in the fourth section.

One factor relevant to the question of adaptation is how far, and with what degree of accuracy climatic changes were actually *perceived* by contemporaries; and the same questions are also of importance to those scientists and historians who have tried to use contemporary descriptions of weather and related phenomena in reconstructing the record of past climates. However, the perception of climate is a matter of much wider scope and interest: yet another approach to the study of 'climate and history', covered here in the final major section of this review. Here the focus shifts from the concrete impact of climate on socio-economic structures to the changing place which climate and weather have occupied in Man's mental world down to the present. This theme is relevant not only to understanding the thought processes of our more remote ancestors but also to comprehending why climate is currently of such great interest to scientists and others, and indeed why this very book came to be written. But before embarking on such intimate speculations, we must return to the beginning and review the available methods and data for reconstructing the climatic record and the information which may be derived from them.

Reconstructing the climate of the past five thousand years

When constructing a record of past climate for a particular region it is not sufficient simply to group together any information from previously published sources. There are very important questions concerning data accuracy, consistency and reliability that must be considered before any synthesis can be performed. These problems are discussed in the various chapters of this book. Here we will present an outline of the available data, give some guidance as to the reliability and usefulness of the different data types, and discuss some of the major climatic episodes of the last five millennia.

The data sources

Instrumental data The late seventeenth century was the starting point for the era of quantitative, instrumental meteorology. Although the thermometer was invented by Galileo around 1590, it was many years before standardised instruments appeared, which yielded data that are useful today, and still longer before instrument exposure was standardised. The longest homogeneous temperature series is that developed by

Manley (1974, and earlier papers) for central England, but even this careful reconstruction is subject to some uncertainty for perhaps the first 40 years of record. Pressure readings go back to 1654, some 11 years after the barometer was invented by Torricelli, with daily observations being made in Italy between 1654 and 1670, and English data being published as early as 1663 in the *Philosophical Transactions of the Royal Society*. For pressure, instrument exposure is less critical, although early records do have to be corrected for temperature. The earliest pressure maps based on instrumental data are those of Lamb & Johnson (1966), who give monthly mean maps for Januarys and Julys back to 1750, and of Kington (1981, and earlier papers) who has constructed daily maps for the period 1781–86 covering the European – eastern North Atlantic sector. Rainfall measurements were apparently made as early as the first century AD (Hellman, 1908) and rain gauges were used in Korea as early as the 1440s (although none of these records survives). Rain gauges were used in England in the 1670s, but again there are exposure problems, and the longest homogeneous series is that for Kew back to 1697 (Wales-Smith, 1971, 1973, 1980). As with rainfall, wind directions were also measured in Roman times. The earliest surviving systematic observations, however, are those of the astronomer Tycho Brahe for the period 1582–97 (La Cour, 1876). Accurate wind speed measurements are very much more recent, although early ships' logs recorded wind force qualitatively. The longest evaporation series is that derived for Kew by Wales-Smith (1973), and this has been used to construct a record of soil moisture deficits back to 1698 by Wigley & Atkinson (1977).

The overall impression of the instrumental period is that many data exist, decreasing in amount and in geographical coverage as we move back in time. However, the length of the instrumental era is short when considered in the context of the wide range of time scales on which climate changes. There is, therefore, a great need to extend the quantitative record back into the historical and prehistorical periods, and to do this pre-twentieth century instrumental data need to be supplemented with data from other sources.

Documentary data Certain documentary sources can provide data that are reliable, easily interpreted and accurately dated. These historical records include: annals and chronicles; records of public administration and government; private estate records (such as manorial account rolls); maritime and commercial records (including ships' logs); personal papers (diaries, correspondence, etc.); and scientific and proto-scientific writings (including journals of non-instrumental weather observations). All these sources have been used and extensively studied by historians, but many are still untapped as sources of climatological information.

These sources are of varying quality, and a rigorous historical methodology is required in order to reject unsuitable and inadequate material, and to extract the maximum amount of information from what remains. Although historians have developed such a methodology, historical climatologists have, in general, been uncritical of their sources and there is, as a consequence, considerable uncertainty in past reconstructions of climate based on historical sources. Many such reconstructions have been based on earlier collections of historical climate information generally referred to as 'compilations'. Unfortunately, many of the compilations used by historical climatologists (for example, those of Hennig (1904), Weikinn (1958) and, to a lesser extent, Britton (1937)) contain unreliable material; although some excellent compilations do exist (e.g. Gottschalk, 1971, 1975, 1977). Bell & Ogilvie (1978), in their review of compilations as climate data sources, give a number of examples of their unreliability. Further information on the techniques used for determining source reliability and on the many other pitfalls of using documentary historical sources is given by Ingram, Underhill & Wigley (1978) and by Ingram, Underhill & Farmer (this volume).

Documentary sources can be used to construct spatial pictures of the weather at a particular time, or to construct time series of indices of temperature or precipitation at a particular place. Much of the pioneering work has been performed by Lamb (summarised in Lamb, 1977, and this volume). The earliest daily weather maps are those for the summer of the Spanish Armada, 1588 (Douglas, Lamb & Loader, 1978; Douglas & Lamb, 1979). Seasonal maps can be constructed for extreme seasons in even earlier times. Examples given by Lamb (1977) and Ingram *et al.* (1978, and this volume) show seasonal reconstructions of circulation patterns. The longest continuous series of maps is that showing rainfall patterns over China from 1470 to 1979 (Wang Shao-wu & Zhao Zong-ci, this volume).

Documentary sources can also be used to construct time series of climatological data. The basic methods, which were developed many years ago, have recently been formalised on the basis of content analysis techniques (Moodie & Catchpole, 1975, 1976). Such time series may take two forms. First, quantitative data on certain phenomena may be correlated with meteorological parameters such as temperature or rainfall to produce quantitative estimates of these variables with well-defined error limits. Examples are the extension of the record of winter temperatures for the Netherlands (e.g. Labrijn, 1946) back to 1634 using data on the freezing of canals (de Vries, 1977; van den Dool, Krijnen & Schuurmans, 1978), and the use of days of snowfall as a winter temperature indicator by Flohn (1949). Secondly, qualitative data on the frequency of extreme months or seasons may be used to construct indices of temperature or rainfall. These may then be correlated with instrumental records to produce quantitative temperature or rainfall estimates; but such estimates are subject to uncertainties of two types, arising in the construction of the original indices, and in the statistical extension from index to temperature or rainfall. Examples are given by Lamb (1977, and this volume). Le Roy Ladurie (1972), Alexandre (1976a, b, 1977) and Wang Shao-wu & Zhao Zong-ci (this volume). More sophisticated indices (based on both documentary and proxy data) have been calculated by Pfister (this volume). Le Roy Ladurie and Pfister have made use of phenological data, in particular vine harvest dates. Further information on this valuable documentary proxy climate data source is given by Le Roy Ladurie & Baulant (1980).

One of the main problems with these time series is that some of the earlier examples were based on unreliable historical compilations. Comparisons between these and more reliably based indices from nearby regions give some idea of the influence of unreliable data (discussed by Lamb, this volume). Such comparisons serve to emphasise the importance of applying the meticulous methodology of Alexandre (1976a, b, 1977) to other regions.

Proxy data The term 'proxy' is used to denote any material that provides an indirect measure of climate. Proxy data therefore embrace documentary evidence of crop yields, harvest dates and glacier movements (see Pfister, this volume), as well as tree rings, varves, glaciers and snow lines, insect remains, pollen remains, and many other sources (summarised in Lamb, this volume, Table 11.1). We are concerned here with those that are capable of giving information for the last five thousand years or less. From Fig. 4.1 of Birks (this volume) it can be seen that some proxy data overlap with the documentary, and even instrumental, data periods, providing the opportunity for both combined and comparative approaches.

There are three main problems in using proxy data: those of dating, lag and response time (i.e., sensitivity to fluctuations on various time scales), and meteorological interpretation. We will briefly consider each in turn.

The main dating methods are stratigraphic (in particular, counting of annual layers or varves), dendrochronological, and isotopic (mainly carbon-14). For reconstructing the climate of the last five millennia three of the most promising proxy data sources are tree rings, pollen deposits from varved lakes, and ice cores. Here the dating may be precise to the year. With carbon-14 the dating precision is much less: counting errors alone give 1σ limits of around 50 years or more, implying 95 per cent confidence limits of ± 100 years, while additional uncertainties arise in converting from carbon-14 to calendar date. This latter uncertainty is particularly important because of the large fluctuations which have occurred in atmospheric carbon-14 (Stuiver & Quay, 1980; Suess, 1980).

The second problem with proxy data arises from the fact that the magnitude of response depends on the time scale of climatic change: many proxy data are relatively insensitive to rapid fluctuations and only give a record of long-term changes. Tree-line changes, for example, are at best only sensitive to climatic change on a 50-year time scale (e.g. Karlén, 1976), and may be far less responsive than this (e.g. La Marche, 1973). Pollen profiles are rarely interpretable on time scales of less than a century, comparable with the dating accuracy of most of this material: although recent work on varved lake deposits in North America by Bernabo (1977) and Swain (1978) has shown that much more detailed information can be obtained, possibly down to the 25-year time scale (Bernabo, personal communication). Related to the frequency-dependence of response is the problem of lags in response. Many proxy variables do not respond instantly to a climatic change, and (as is the case of glaciers and tree lines) the associated lag may differ, depending on the direction of change.

Interpretation of proxy data in terms of meteorological variables is the third problem associated with this data group. Tree rings provide a good illustration of this problem since they respond to both temperature and precipitation to varying degrees at different times of the year (Fritts, 1976); although with a careful choice of site trees can be located which respond primarily to a single variable (La Marche, 1974). Glaciers and pollen data also respond to both temperature and precipitation (see Porter, this volume, and Birks, this volume).

Common to all these proxy variables is the fact that only a part of their variations can be explained by climatic changes, so that they give us only a blurred image of these changes. It is frequently tempting to accept these messages literally and to ignore their uncertainties. This is a dangerous procedure, however, especially when one is examining the possibility of climatic impact on Man, and the uncertainties in the record of past climates must always be borne in mind.

Because there are so many different proxy data types, each with its inherent problems of dating, interpretation and response time, it is not feasible to discuss them all here. These details are covered by individual contributions to this volume. Lamb provides a general introduction. Glaciers and snow lines receive detailed treatment by Porter and are used as evidence in the chapters by Pfister and Lamb. Fritts, Lofgren & Gordon provide a review of climate reconstruction using tree rings. (Recent work on tree-ring density measurements (used by Pfister) is described by Schweingruber, Bräker & Schär (1979) and Schweingruber *et al.* (1978).) A concise introduction to the use of pollen data as a climate proxy is given by Birks; and Gray concentrates on the climate of the last 5000 years in his review of isotope analysis. The use (and abuse) of archaeological data as a proxy climate indicator is reviewed by McGhee, and a more detailed case study is given by McGovern. Agricultural data are discussed in a number of chapters – in particular those of Anderson, Parry, and Bowden *et al.*

The climate of the last five thousand years

Our main concerns in this chapter are with the methods of climate reconstruction and with the impacts of climatic change and fluctuation on society. In the latter, only detailed local climate information will suffice for rigorous analysis of climate–society links. We have here (echoing the more detailed discussions in other chapters of this book) stressed the uncertainties in our knowledge of past climate. The following review of the climate of the last 5000 years is therefore far from complete. Rather, it is biased towards illustrating more general methodological aspects. We will consider separately three different periods, defined with reference to the major type of evidence (instrumental/documentary/proxy) available for each period, and slant the discussion towards the particular climatic questions which each data type is capable of elucidating.

The instrumental period Although any local or regional impact will be determined by local or regional climatic conditions, it is still of interest to consider recent large-scale trends in climate since they illustrate the complexity of short-term climatic change. Perhaps the simplest indicator is the hemispheric or global mean annual surface temperature. Fig. 1.1 shows northern hemisphere temperature changes since 1880 (see also Lamb, this volume, Fig. 11.1). From about 1925 onward these data cover a sufficiently large fraction of the hemisphere to provide a good estimate of hemispheric changes. One of the most striking features of Fig. 1.1 is the large interannual variability, so large in fact that it masks all but the strongest longer-term trends. Thus, although a cooling trend is evident between about 1940 and 1965, no statistically significant trend is evident in recent years (although subjective interpretation might point to a slight warming).

This hemispheric curve, however, cannot be considered representative of regional variations; nor can one assume that mean annual trends apply to individual seasons. On annual to decadal time scales climatic changes are characterised by marked spatial inhomogeneity. This is particularly true of rainfall changes. Similar spatial inhomogeneity most probably extends to the century time scale at least (Williams, Wigley & Kelly, 1981; see also Fritts *et al.*, this volume).

The year-to-year variability so evident in Fig. 1.1 is also a feature of regional and local climate. It has been suggested that the impact of climate on society may be related more to changes in variability than to

Fig. 1.1. Northern hemisphere mean annual surface temperature variations in °C: deviations from the 1946–60 mean (from Jones & Wigley, 1980). The dashed curve shows the variations filtered using a nine-term binomial filter.

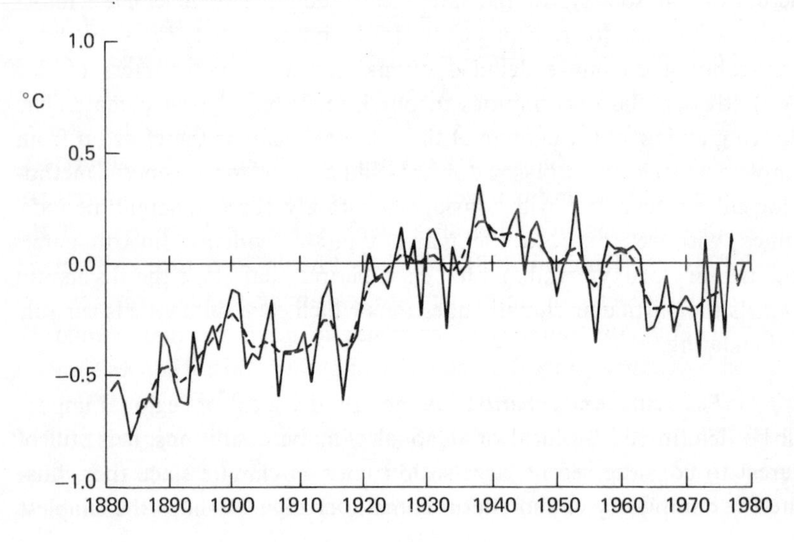

changes in the average conditions (see the chapters by Pfister, Flohn and Parry). Parry, for example, points out that the potential impact of two successive bad years is much higher than twice the impact of one bad year – and a change in variability may increase the probability of successive bad years without changing the average climate. Changes in variability may occur either independently or associated with changes in the mean. Bryson (see Bryson & Murray, 1977), for example, has suggested that the climate of middle latitudes is more variable in cooler periods. Pfister (this volume) presents evidence that may support this idea in Fig. 8.3 which shows that, in the late sixteenth to early seventeenth centuries, there was an increase in the frequency of extreme months which roughly paralleled the cooling trend of this period.

This leads us to two controversial questions: have there been any statistically significant changes in variability; and do changes in variability accompany changes in the mean? Such questions can only be answered by reference to good instrumental data. Ratcliffe, Weller & Collison (1978) have made a detailed analysis of recent instrumental data for the British Isles and find no evidence for increased variability in recent (and generally cooler) years, and van Loon & Williams (1978) (see also comments by Court (1979) and by Duchon & Koscielny (1979)) have found no link between changes in mean temperature and changes in interannual variability in the United States. Van Loon and Williams do, however, show considerable changes in year-to-year variability in recent decades, with increases in the western United States occurring at the same time as decreases in the eastern United States.

Lamb and some other climatologists disagree with some of these conclusions, but the problem is largely semantic. For example, there is no doubt that some individual decades are more variable than others. However, one would expect changes in variability to occur by chance, and those who argue that there have been no changes in variability are, in fact, arguing that any changes that have occurred are no more than random fluctuations. An interesting case is Kington's analysis of the period 1781–85 (Kington, 1981) which is known as a time of frequent extremes. However, for most climate parameters (temperature, precipitation, etc.), the number of extremes is no more than would be expected by chance. Only the frequency of cyclonic disturbances shows a statistically significant perturbation.

Both Flohn (this volume) and Mooley & Pant (this volume) present evidence that supports the contention that observed changes in variability are random. Other information on changes in variability in historical times (Lamb, this volume; Pfister, this volume) has not been subjected to any rigorous statistical analysis. Pfister's evidence, which might support the idea that the Little Ice Age was a time of more frequent extremes, does not necessarily imply a change in variability, but can be due to changes in the mean (Sawyer, 1980). If an extreme is perceived when the monthly mean temperature falls below a certain level then a depression of the long-term average temperature will be evidenced by more extremes without any change in variability. It is precisely this sort of relationship which Lamb (1977, and this volume) has used to interpret his decadal indices in terms of long-term climatic change.

As far as the impact of climate is concerned the *cause* of, for example, two successive bad years is only of secondary importance. Such juxtapositions of extreme events *must* occur occasionally by chance alone. Nevertheless, the probability may well be increased by changes in the mean and/or by changes in variability, and so to this extent it is important to understand how (and whether) such changes have occurred. Whether or not the Little Ice Age was a time of more variable climate is still an open question. Recent instrumental analyses show no evidence for a link between cooler climates and greater variability, but the Little Ice Age was a unique period which has no recent analogue.

The documentary period The instrumental period covers only the last 100 years or so, except for isolated regions (mainly in Europe) where records extend back into the eighteenth and seventeenth centuries. Prior to the nineteenth century, then, documentary (largely qualitative) evidence is the main direct source of climate information. The documentary period extends back only a few centuries in many parts of the world, but in Europe and China documentary evidence is available for about 1000 and 3000 years respectively. However, for the reasons explained above, one must be careful in using climatic interpretations based on these data.

As an illustration, perhaps the best known historically based time series are the decadal winter severity and summer wetness indices derived by Lamb (1977, and this volume) for three regions in Europe. Lamb's indices go back to AD 1100, but there is doubt about their reliability on the decadal time scale prior to about AD 1500. Lamb (this volume) has compared his indices with those of Alexandre (1976a, b, 1977) and further comparisons between these works and Pfister's data (this volume) are shown in Table 1.1. The correlations are generally low, but no attempt has been made to test their statistical significance. In doing so,

Table 1.1. *A comparison of decadal climate indices for Europe. The figures shown are correlation coefficients with the sample size shown in brackets. The Lamb and Alexandre indices are for regions around latitude 52° N at the longitudes shown, while Pfister's indices are for Switzerland, approximately 47° N, 8° E. Data from Lamb (1977), Alexandre (1976a, b) and Pfister (this volume).*

	Lamb (12° E)	Alexandre (5° E)	Pfister (8° E)
Summer Wetness			
Lamb (0° E)	0.51 (44[b])	0.41 (44[b])	0.27 (28[e])
Lamb (12° E)		0.28 (44[b])	0.16 (28[e])
Alexandre (5° E)			0.77 (7[d])
Winter Severity			
Lamb (0° E)	0.54 (24[a])	0.22 (24[a])	0.27 (28[e])
Lamb (12° E)		0.37 (24[a])	0.17 (28[e])
Alexandre (5° E)			—[c]

[a] 1100s–1390s (some decades missed because of insufficient data).
[b] 1100s–1610s (some decades missed because of insufficient data).
[c] No overlap period.
[d] 1550s–1610s.
[e] 1550s–1820s.

due allowance would have to be made for autocorrelation (as outlined by Mitchell *et al.*, 1966). Correlations between recent instrumental data from the four regions are generally higher than those in Table 1.1. It is interesting that some of the highest correlations in Table 1.1 are between the two indices derived by Lamb, even though these regions are the most widely separated geographically. In spite of uncertainties in these decadal indices, they are almost certainly reliable indicators of the most severe decades, even in medieval times. Decades like the wet decade of the 1310s, and the cold decade of the 1430s show up strikingly.

There are, however, difficulties in using such decadal indices to derive long-term (century time scale) trends in climate. In Lamb's estimated curve showing long-term temperature trends (see Lamb, this volume, Fig. 11.2) the preferred values prior to AD 1400 are the 'analyst's own opinion'. This is based on various proxy indicators and lies outside the confidence limits of estimates based solely on the documentary data of single seasons which form the basis of decadal indices. The fact is that our knowledge of long-term climate trends is based almost entirely on proxy data (including documentary proxy data). The main value of historical documentary data lies, therefore, in its illumination of short-term (seasons, months or shorter) climatic events, and on this time scale there are a wealth of untapped sources not only in Europe, but in many other parts of the world. The work of Wang Shao-wu & Zhao Zong-ci (this volume) for China, and Smith, Borns, Baron & Bridges (this volume) for the United States is only the tip of the iceberg.

The proxy data period Our knowledge of past climates depends heavily on proxy data even as late as the sixteenth and seventeenth centuries. On short time scales, however, only tree rings offer the chance of climate reconstructions which might match the detail and dating certainty of documentary historical data. To review in detail the available proxy climate data for the period 3000 BC to the present is plainly beyond the scope of this chapter. Prior to about AD 500 the data allow only the broadest of generalisations to be made: time resolution is relatively poor and only scattered sites are represented, often in locations far from civilisations contemporary with the data. Only the record for the last 1500 years can be covered with any degree of confidence, and even for this period there are considerable uncertainties.

The Medieval Warm Epoch and the Little Ice Age On the basis of the foregoing discussion we will restrict our description of past climates to two of the most noticeable features of the last 1500 years; the Medieval Warm Epoch and the Little Ice Age.

The Medieval Warm Epoch (MWE) is generally assigned to some interval within the period AD 1000–1400 (see Lamb, this volume, Fig. 11.2). A review of the available well-dated proxy data by Williams *et al.* (1981b) shows, however, considerable differences in timing from place to place, and, in general, numerous shorter-term (50-year time scale) fluctuations. For example, data from Scandinavia (e.g. Denton & Karlén, 1973; Karlén, 1976; Karlén & Denton, 1976; Karlén, 1979) on changes in tree-line and glacial advances imply that warm conditions prevailed from at least as early as AD 600 until around AD 1200. Ice core data from Crête in Greenland (Dansgaard *et al.*, 1975) show a warm period beginning in AD 700 with an oscillating decline in temperatures between AD 900 and 1400 in broad agreement with Fredskild's (1973) pollen data which imply warm conditions in southern Greenland between AD 600 and 1100. Pollen data from numerous sites in northern Canada (Nichols, 1975) and in Michigan (Bernabo, 1977) point to maximum warmth between about AD 700 and 1300. These data contrast with Alaskan glacial and tree-line fluctuations (e.g. Denton & Karlén, 1973, 1977) which indicate that conditions were warm between AD 500 and 800 and colder thereafter. In China and Japan (Chu Ko-chen, 1973) the MWE is scarcely evident, and severe winters were recorded during this period. What is most apparent when one examines these data in detail is that the MWE was a phenomenon largely confined to the European – North Atlantic sector, and even here there were significant asynchronous changes in climate on short and medium time scales. Even in Europe there is evidence that Alpine glaciers were advanced as far as at any time in the last 3000 years during the twelfth century (Röthlisberger, 1976), at odds with the general conception of the MWE.

From AD 1200 to 1600 the climate of those parts of the globe that were warm seems to have cooled significantly, and by the seventeenth century it appears that conditions colder than today were almost ubiquitous. Significant warming did not begin until the nineteenth century, and the period from around AD 1400 to 1800 is commonly referred to as the Little Ice Age. Conditions during the Little Ice Age were, however, far from stable, and there were complex spatial patterns of warming and cooling throughout the period. For example, in Europe at least, there appears to have been a noticeable warming of climate around AD 1500 (Lamb, 1977: 221) which may well have been common to the North Atlantic region in general, and which also shows up in La Marche's tree-ring record from the White Mountains in California (La Marche, 1974). There is strong evidence that the Little Ice Age also affected the southern hemisphere. Glacial advances occurred at a number of places (Mercer, 1968, 1972; Wardle, 1973; Denton & Karlén, 1973), implying cooler summer temperatures. The climax of this cool period may have been slightly earlier in the southern hemisphere than in the north.

We should stress, however, that during the MWE and the Little Ice Age, the climate was not continuously warm or cold, but continued to experience the 'natural' year-to-year fluctuations evident in Fig. 1.1. Thus, the seventeenth century, a time of frequent severe winters in many parts of the world, also experienced a number of hot summers (such as that before the Great Fire of London in 1666). It is the juxtaposition of these opposing extremes that creates an impression of greater variability during the Little Ice Age. The data preclude a rigorous statistical verification of this impression at present, but further analysis could give considerable insight into the mechanisms of climatic change. As noted above, changes in variability are at least as important in determining the impact of climate on Man as changes in the mean.

Identifying and measuring the impact of climate on past societies

Schools of thought

Although most historians, historical geographers and archaeologists have long recognised the possible importance of the impact of climate on human affairs, the literature reveals much disagreement about how seriously the possibility should be taken. The majority of historians have been content largely to ignore it. It is generally accepted that short-term (intra-annual, annual and inter-annual) variations in climate and weather, having an immediate effect on harvests and other economic activities, are relevant to short-term economic fluctuations. But long-term climatic influences have been commonly regarded as of little or no historical interest, on the basis of one or more of the following assumptions. First, that climate has been essentially stable in historical times. Second, that the magnitude of past long-term climatic shifts, though striking from a scientific point of view, has been too small to warrant their consideration as significant variables in the processes of economic and social change. Third, that lack of detailed information on past weather and climates and imperfect understanding of the complex processes of climate – history interactions preclude any serious study of the subject.

However, for a long time a small number of historians, and larger numbers of archaeologists and natural and environmental scientists, have been convinced of the importance of climate as a major independent variable affecting the development of human societies. The most extreme exponents may be labelled 'climatic determinists', of whom Ellsworth Huntington (1907, 1915) is the best known. Although the term 'climatic determinism' does not necessarily imply a belief that the whole course of history is explicable in terms of climate (Pearson, 1978), it certainly implies that climatic factors have been among the most important influences on the development of civilisations. Such views are still held today, for example by Chappell (1970) and in a less extreme form by Bryson & Murray (1977), Lamb and others. For example, Lamb (1969: 1209) has asserted that 'climatic history must be central to our understanding of human history'.

Others have been more cautious, eschewing grand generalisations about the significance of climate in world history, yet strongly urging the importance of climatic factors in particular areas and periods. Among historians they include Utterström (1955), Braudel (1949, 1973), and more recently Parker & Smith (1978) and Parker (1979); but these contributions, though interesting, are marred by methodological weaknesses (see, for example, the extensive criticisms of Utterström by Le Roy Ladurie, 1972: 8–11). In recent years a new breed of historians interested in climate–history interactions has emerged. These scholars, of whom Pfister (1975, 1978, and this volume), de Vries (1977, 1980), Post (1977) and Parry (1978, and this volume) are the most outstanding, admit the possible importance of climatic variations on human affairs both in the long and short term, make serious attempts to investigate the possibility with the aid of all the available methodological resources of history, economics and climatology, but demand the highest standards of proof. Their work has transferred the study of climate–history impact onto a higher plain of methodological rigour. The standards that they have set are discussed in the next subsection.

Methodology

Philosophical assumptions A philosophical issue basic to the study of the impact of climatic variations on human societies is on whom the onus of proof should lie. (A similar problem has recently been exposed by Pittock (1980) with reference to the controversy over what influence solar variations may have on weather and climate.) It is plain that climatic variations must have had *some* influence on economies and societies. Hence, some writers feel justified in accepting poorly substantiated 'demonstrations' of the importance of climatic effects in particular cases, in effect casting the onus of disproof on the sceptics. But the real issue is not whether climate has had an influence, but just what that influence was and whether it was of any real significance. The importance of climatic influences cannot be assumed. It is totally inadmissible to demand either explicitly or implicitly that the burden of proof should lie with those who doubt the importance of climatic effects, or to use methods of 'proof' which essentially rest on an implicit assumption of the social and economic significance of climatic variations. An example of such an unacceptable method of proof is the demonstration of a space–time coincidence between climatic variations and adverse or beneficial economic, social or political developments. Unfortunately, this method recurs again and again in the literature; it is time that it was universally recognised that, as a method of proof, it is valueless (cf. Parry, this volume).

The problem of data One searching question which anyone attempting to investigate possible climate–history links in particular cases must ask is whether the available data are sufficient to justify the attempt. The nature and extent of the available data on past climate and weather have already been discussed. However, it must be emphasised here that the kinds and quantity of data which may satisfy the purposes of some scientists seeking only to extend the climate record back in time may be quite inadequate for the purpose of demonstrating climate–history interactions. As will be seen later, the latter purpose often demands very detailed information about individual seasons, months, or even weeks, and such data are unfortunately absent or sparse for large regions of the world until the very recent past.

Data on the complexities of human economic and social activities are also highly variable both in quantity and quality in different locations and at different periods; and, in general, the further back into the past one penetrates, the fewer the available data. It is impossible to summarise the extent of the available information on the varieties of human activity most likely to have been affected by climate. However, a few examples taken from western Europe, a region whose history over the last thousand years or so is relatively well documented, will illustrate the problems (for further information on European economic data, see Wilson & Parker (1977)).

Grain production has been an activity of major economic importance in Europe down to the present day; and over large areas and for long periods, a considerable proportion of cereal production was controlled by large-scale secular and religious landholders whose activities are comparatively well recorded. Much information on grain production accordingly survives. Yet for the purposes of the economic historian it suffers from many imperfections. For example, while grain *price* data (a complex compound of many variables) are relatively abundant, information which bears more directly on agricultural productivity and production is 'generally exiguous, representative of only specific locations, and often difficult to interpret' (Anderson, this volume: 351). As another example, demographic data, in many ways fundamental to an understanding of social change, are distressingly sparse and imprecise. In most parts of Europe before the sixteenth or seventeenth centuries, the available information suffices only to indicate, with considerable margins of uncertainty, gross trends in population. Thereafter, the evidence begins to multiply, but demographic data only become really abundant and satisfactory in the nineteenth century (Wrigley & Schofield, 1981).

For localities outside Europe the problems are, in many cases, even greater. This is not to say that *no* data adequate to investigate climate–history links are available, but it is as essential to be aware of the gaps and problems in the record of human activities as it is to recognise the imperfections in the climatic evidence. It must be accepted that, for many parts of the globe over long periods, the data currently or potentially available on those human activities likely to show a discernible climatic impact are so sparse or poor as to vitiate any attempt to demonstrate it.

Preliminary models All attempts to identify and measure climate–history interactions invariably rest on models of the processes involved; and a preliminary model is necessary to identify, in general terms, how climate may be expected to affect human history. A simple model of a type which, explicitly or implicitly, underlies the majority of studies of climatic impact in the past is shown in Fig. 1.2. Variations in the atmospheric circulation manifest themselves as meteorological, hydrological and oceanographical phenomena. These may have a direct (first-order) effect on biophysical processes important to Man, including crop and animal survival and growth, marine and other aquatic life, and the activity of micro-organisms capable of causing disease in plants, animals and Man. In addition, they may have effects on aspects of the purely physical environment of importance to human societies, for example, causing rivers to freeze or flood. These direct biological and physical effects may have economic or social (second-order) significance, affecting food and raw material supplies derived from arable or animal husbandry, human health, the performance of machines (such as industrial and food processing mills driven by wind and water), transport

Fig. 1.2. First- to *n*th-order effects of climate on the environment and society. As the specific impact becomes more removed from the 'cause', progressively more interactions intervene to disguise and modify the link.

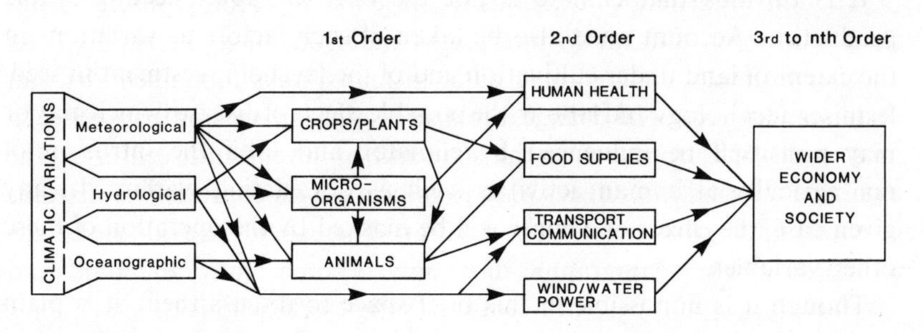

and communications, and military and naval operations. Depending on their magnitude, these effects may ramify into the wider economy and society (third to *n*th order). For example, food shortages may (in association with other factors) lead to rebellions, which in turn may help to undermine political systems (Tilly, Tilly & Tilly, 1975).

The complexity of the links between climate and human affairs Such a conceptual model is, in itself, of little value. It serves only to focus attention on the indirect and complex nature of the links between climate and the majority of economic and social phenomena, and to aid the identification of data relevant to the testing of climatic impact. The fundamental limitations of the model are that it does not include the many other variables which, apart from climate, affect the various types of human activity; nor does it specify in detail which climate variables are critical for each activity.

These complexities, which are considerable, may be illustrated by reference to the effects of weather on arable husbandry, perhaps the most obvious link between climate and human activities. Different crops are sensitive to different climatic factors (e.g. Slicher van Bath, 1977: 57–66). Moreover, their response to variations in temperature or other variables is often complex and markedly non-linear (Thompson, 1975; McQuigg, 1975; Starr & Kostrow, 1978). In any particular study of the effects of climate on agriculture, it is necessary to establish precisely in what respects the relevant crops are sensitive to climate. This is often possible by reference to the findings of modern agronomists on the effects of different climatic conditions on crop performances in the present, though obviously it is necessary to take account of the fact that crop varieties in use in the past were often less resilient and took longer to mature than those normally employed today (for an example of the use of this method to establish the climatic conditions critical for high-altitude oats cultivation, see Parry, this volume: 325).

It is obvious that climate is not the only variable affecting arable production. Account must also be taken of such factors as variations in the extent of land under cultivation and of the level of investment in seed, fertiliser, technology and labour, the possible effects of disease (which may or may not itself be influenced by climate), and even the intrusion of non-agricultural human activities such as devastating warfare. In any given case, the effects of climate will be masked by the operation of these other variables.

Though it is impossible in this brief space to discuss them, it is plain that the complexities of the relationship between climate and all other human activities are at least equal to those affecting arable production, and in many cases greater. In particular, it is evident that the more remote the relationship between climate and a given activity, the greater the number of complicating variables. In accounting for the occurrence of a phenomenon such as a rebellion or a grain riot, for example, many other factors apart from climate must invariably be taken into consideration (MacKay, this volume).

Simple correlations The complexity of the links between climate and human affairs means that to seek simple correlations between gross climatic variables and series of economic or other data is of limited value; indeed, it may be very misleading. This is particularly true of correlations in the sense of rough congruences between curves representing gross meteorological indicators and long time series of data on human activities only remotely (if at all) connected with climate. Thus Anderson (this volume: 338) notes that the intriguing congruence (albeit a crude one) between estimated 50-year mean annual temperatures for central England and the real wages of building workers for some periods since the mid-sixteenth century, is probably meaningless. The same is no doubt true of the inverse relationship, observed by Lamb (1977: 264) between central England temperatures and the mean age at first marriage in the parish of Colyton, Devon (England) in the seventeenth and eighteenth centuries.

Statistically tested correlations between gross climatic variables and data on economic activities more closely linked with atmospheric changes may be more useful. However, even this tool is so blunt that, though the existence of a statistically significant correlation may encourage further investigation, the absence of correlation does not necessarily indicate that a link with climate is absent. This point is vividly illustrated by de Vries (1980: 603–6). For example, he shows that a simple correlation analysis between winter temperature in Holland and butter prices at Leiden 1635–1839 reveals no significant relationship. He infers that the probable weaknesses of this test were that it took no account of the particular meteorological elements and periods of the year critical to dairy farming, and moreover assumed a linear relationship throughout the full range of values. A slightly more refined technique yielded better results: there did prove to be a statistically significant relationship between late frosts persisting into March and a rise in butter prices of about 10 per cent above the 15-year moving average. Even so the relationship appeared weak, because the price of butter was conditioned by many other variables apart from climate.

Models of causality De Vries' work emphasises the need for relatively sophisticated tests to identify and measure the impact of climate. In order to be useful these tests must be based on a detailed model of causality which specifies the nature of the hypothesised links between climate and human activity. Ideally they should test not only the climate link but all the variables affecting a given set of economic data (de Vries, 1980: 606–8).

In practice the construction and use of such models will be constrained by the availability of data: for example, de Vries (1980: 608) points out that, though it is theoretically possible to devise an econometric model of the variables (including climate) affecting rye prices at Utrecht in the seventeenth and eighteenth centuries, in reality this is not feasible because data on several of the key variables are lacking. However, in many cases the problem is not only that of inadequate data but also an imperfect understanding of the economic and social processes involved. The feasibility of constructing an adequate model depends on a number of factors.

1. The complexity of the task obviously varies according to the number of economic and social variables that are included in the investigation. The extreme case would be to attempt to gauge the effects of climatic variations on all aspects of the life of a given society. The construction of a model adequate for this task is not even remotely feasible in the foreseeable future. Less ambitious investigations, focusing on one or a small number of economic or social variables are more realistic. The potential value of such investigations, though limited, is nevertheless considerable given our present state of near-total ignorance concerning climatic impact in the past.

2. First- and second-order climatic effects (see Fig. 1.2) can be modelled more easily than the links between climate and more remotely-connected human activities. The more remote the activity, the greater the number of complicating variables.

3. The problems vary according to the geographical scope of the investigation. In general, it would appear easier to model the effects of climate for a small area than for a larger one (so long as climatic and other relevant data specific to the locality are available). The problems involved in studying a large area such as an entire country or even a continent are obvious. Weather and climate patterns may vary markedly

within national boundaries, the influence of these patterns will vary according to the varieties of regional landform and even local topography, and it will be necessary to take account of regional variations in economic and social structures.

Though limiting the area investigated inevitably limits the significance of the conclusions, in the present state of knowledge even findings based on circumscribed areas would be welcome. (Historians, aware of the complexities of studying in detail the processes of economic and social development and of geographical restrictions on the availability of data are currently very sympathetic to regional and local studies.) Anderson (this volume: 352) specifically recommends local investigations as a potentially fruitful approach to climatic impact studies, and it is significant that many of the contributors to this volume confine themselves to particular areas.

4. It is also plain that it is easier to investigate the effects of short-term (intra-annual, annual and inter-annual) climatic fluctuations. In the case of secular variations the number and complexity of the factors masking the climatic effect, including technological development and changes in the structure of the economy, are normally so great as to vitiate any attempt at rigorous measurement. This point is argued vigorously by de Vries (1980: 624–5) and Anderson (this volume).

5. However, point 4 may not hold for certain marginal societies (or groups); that is, those societies heavily dependent on agriculture or other economic activities conducted in conditions close to the climatic limits beyond which such activities are physically unviable. *A priori*, such marginal societies would appear to be particularly vulnerable to the effects of climatic variation. In such circumstances, the number of complicating variables may be effectively smaller, and the impact of climate itself correspondingly easier to gauge.

Marginal situations may prove valuable as 'laboratory' test cases. If the effects of long-term climatic changes appear absent or small even in such situations, it may *a fortiori* be concluded that secular climatic variations in more environmentally sheltered societies may be safely ignored. If, on the other hand, the effects of such climatic variations on marginal societies appears important, this will provide at least some basis for arguing the more general significance of climatic changes in human history and may encourage further consideration of the admittedly more complex cases of societies relatively well sheltered from climatic stress. These ideas are strongly urged by Parry (this volume).

Alternative strategies 1: semi-descriptive case studies The difficulties of constructing detailed causal models to identify and measure climatic impact, especially for larger areas, a wide range of economic and social variables, and extended periods, prompts the use of less precise methods which may nevertheless be capable of yielding worthwhile results. One approach is the use of semi-descriptive case studies, of which the most notable so far completed are by Post (1977) and Pfister (1975, and this volume). The method is to concentrate on particular periods of climate 'crisis' in which atmospheric variations appear to have been associated with marked changes in the economy and society of a given area. As far as possible, the links between climatic variations and the society in question are specified rigorously; for example, variations in crop yields are examined in relation to a detailed analysis of such meteorological variables as temperature and precipitation. However, at the point where rigorous modelling ceases to be feasible (on account of the large number of variables involved, the complexity of the links with climate, imperfections of the data, etc.), the case studies take on a more impressionistic character. Hence the term 'semi-descriptive'.

Such studies can be immensely stimulating and perceptive. However, their value is inevitably limited to the extent that rigorous analysis is abandoned. In some ways, too, they may be misleading. Concentration on particularly acute periods of crisis may give a false impression of the importance of climate in human affairs. Moreover, the descriptive element in the method can easily degenerate into the cataloguing of detail after detail of disruption and misery. The abundance of such detail, and the use in the description of such essentially rhetorical epithets as 'calamitous', 'threatening', 'disastrous' (all these examples are taken from Post (1977:6–24)) may in the mind of an unwary reader obscure the lack of a precise framework for gauging the importance of climatic impact.

Alternative strategies 2: Occam's razor Anderson (this volume) has suggested an approach reminiscent of Occam's razor (that is, for the purposes of explanation, things not known to exist should not, unless it is absolutely necessary, be postulated as existing) in order to gauge the possible long-term importance of climatic variations in circumstances in which rigorous testing and assessment are not possible. The method is to scrutinise long-term social and economic changes and to examine how far they are explicable in non-climatic terms. The aim is to isolate possible explanatory lacunae which appear to require the invocation of climatic change as a relevant variable. If none is found, it may be concluded that climatic change was of minimal importance. Clearly, the method is highly subjective. Nevertheless, it may be valuable as a means of eliminating cases in which climatic change was almost certainly of negligible importance and focusing attention on cases where the issue is more contentious.

Review of case studies

In the light of the foregoing discussion of methodology, we may review a number of specific case studies on climatic impact, including not only those which figure in this volume but also a number which have been published previously. The studies may be classified according to the time scale on which they consider climatic influence – short term (annual or intra-annual); medium term (inter-annual); and longer term (extending over periods of decades or centuries).

Short-term influences The study of short-term influences may itself be subdivided into: the examination of the possible impact of isolated climatic events, including such phenomena as single storms or periods of storminess, and individual months or seasons of aberrant weather; and time-series analysis of the impact of seasonal or annual climatic fluctuations.

The role of isolated climatic events may be considered very briefly. Clearly, the impact of such events is in normal circumstances likely to be small in relation to the totality of economic and other historical processes. However, there may conceivably be grounds for ascribing greater importance to a few isolated climatic events which happened to impinge on key historical situations. For example, it might be argued that the North Sea storms which helped to shatter the Spanish Armada in the summer and autumn of 1588 exerted an important influence on the course of English history; while Stolfi (1980) has recently drawn attention to the significance of the severity of the Russian winter of 1941–42 in halting the progress of Hitler's Russian campaign.

However, the number of climatic events which can plausibly be regarded as significant in this way appears to be small, and the precise degree of their significance a matter of debate. For example, many other factors besides the North Sea storms contributed to the failure of the Armada in 1588; and even if the Armada had landed it is by no means certain that the subsequent history of England, far less that of Europe, would have been significantly altered (Parker, 1976). Thus the argument that the storms had an important impact on human history is tenuous at best. In general, it would seem reasonable to regard the impact of isolated weather events on history as of fleeting and random importance only: it is significant that Stolfi's (1980) article is written with a view to demonstrating the historical significance not of *climate*, but of *chance*.

A much better case can be made for regarding as important the short-term impact of month-to-month, season-to-season, and year-to-year variations in the weather. Ashton (1959), Jones (1964), and others have urged recognition of the widely ramified influence of weather fluctuations on arable and pastoral husbandry, industrial activities and transport; and even historians sceptical of the significant impact of long-term climatic changes commonly admit the importance of short-term variations (e.g. Le Roy Ladurie, 1972: 119). However, it is striking that comparatively little research has been devoted to investigating in precise detail just how important such fluctuations were.

A recent attempt by de Vries (1980) to do something to fill this gap, by applying correlation and regression analysis to time series of data on weather, grain, butter and fuel prices, burial statistics, and transport data for seventeenth–nineteenth-century Holland, has thrown some doubt on the traditional assumption that weather fluctuations were self-evidently of major economic importance. In particular, he has challenged the widespread notion that 'the economic history of *ancien régime* Europe is essentially the summation of its harvests, which, in turn, were sensitive registers of its climate history' (in the sense of weather fluctuations) (de Vries, 1980: 601; cf. Hoskins, 1964, 1968). His scepticism rests on the fact that the links which his statistical analysis revealed between climate variations and social and economic data were relatively weak. He interprets these findings as an indication that, although the idea of weather-dominated economies may hold true for closed, technologically primitive, subsistence societies, in most areas of early modern Europe 'the level of economic integration was sufficient – including trade, markets, inventory formation, and even futures trading – to loosen greatly the asserted links between weather and harvests and between harvests and economic life more generally' (de Vries, 1980: 602).

Post (1980: 721) has voiced the obvious objection to this argument: the case of Holland may be untypical since the exceptionally sophisticated Dutch economy was unusually well insulated against such external shocks as climatic variation, not least on account of Holland's commanding position in the international grain trade. This objection, though inherently plausible, needs to be tested by rigorous statistical studies for less economically sophisticated areas of pre-industrial Europe. At present such studies are lamentably lacking, both in this volume and in the field as a whole. A major priority in future research should be to try to remedy this deficiency in our knowledge.

Something of an antidote to de Vries' scepticism is offered by a study made by Lee (1981) of the relationship between meteorological variables and short-run fluctuations in vital rates over the period 1659–1840 in England, a country which in the seventeenth century rivalled Holland's economic development and subsequently exceeded it.

Lee found no significant relationship between vital rates and rainfall data, unfortunately available to him only in the form of an inhomogeneous annual series from 1727 (Nicholas & Glasspoole, 1931). But quite striking effects of temperature variation were apparent. Lee's analysis indicated that mortality was increased by cold temperatures in the months from December to May ('winter'), and by hot temperatures in the June to November period ('summer'). The main effect of winter temperatures was contemporaneous, but in the case of summer temperatures the effect was delayed by one or two months (Lee speculates that low winter temperatures killed older elements in the population by means of rapidly lethal diseases, while hot summer weather tended to kill infants and small children through debilitating digestive tract diseases which took longer to cause death). A 1 deg C warming of winter temperatures would reduce annual mortality by about 2 per cent; a 1 deg C cooling of summer would reduce annual mortality by about 4 per cent. Over the period 1665–1834 temperature explained a smaller proportion of the variance in annual mortality than did prices, but temperature and prices were equally important from 1745 to 1834. With regard to fertility the effects of temperature variation were more muted, about a quarter to a third the size of those for mortality. Lee concluded that overall the effects of temperature variations on vital rates were quite striking (especially given the fact that the climate of England is relatively moderate), to the extent that he felt justified in considering the implications of his results for long-term changes in fertility and mortality (see below).

Medium-term influences Increasingly, climatologists are focusing attention on medium-term interannual climatic fluctuations. Flohn (this volume) notes that, from the layman's point of view, variations in the frequency and intensity of extreme seasons and individual weather events are the chief manifestation of comparatively small medium- to long-term changes in the average climate. It would appear that such extreme episodes have an observable tendency to cluster in adjacent years, even though such clustering is not necessarily statistically significant (that is, could occur purely by chance; Flohn, this volume; Mooley & Pant, this volume).

A priori, such clusters of extreme episodes, extending over periods of up to a decade, would appear likely to have a greater impact on economies than either short-term fluctuations or relatively weak long-term variations in climate. Indeed, Post (1980) isolates such inter-annual clusters of extreme seasons as probably the key phenomenon for measuring climatic impact. He argues that, whereas pre-industrial societies might to a large extent adjust to cope with annual fluctuations, after a *succession* of severe, bad or poor weather years the systems of adaptation and adjustment would be overchallenged, leading to higher death rates and some decline in economic activity.

A number of recent studies have focused in detail on the economic effects of anomalous weather conditions on inter-annual time scales. Pfister (1975, and this volume) has analysed conditions in Switzerland during the Little Ice Age period, especially in the years 1570–1600, 1768–71, and 1812–17. Bowden (1967: 630–3) has examined clusters of anomalous years in England in 1546–53, and 1618–25, periods similar to each other in that in each case a run of exceptionally good harvests was followed by a succession of crop failures. Unfortunately Bowden lacked complete data on climatic fluctuations, but his brief analysis is nonetheless suggestive: he concluded that a climatically induced sequence of bounty and dearth had the effect of impoverishing first one section of the community and then other sections, and thus seems to have provided the optimum conditions for the onset of trade depression. The relationship between climatic stress and trade cycles is more extensively explored by Post (1977) in his exceptionally rich and ambitious study of the effects of abnormal atmospheric conditions in the period 1816–18 (including the especially anomalous year 1816, 'the year without a summer'). Employing the semi-descriptive case study approach, he examines the ramified effects of adverse climatic conditions on agricultural production, industrial activities, demographic patterns, and such socio-political activities as movements of popular protest in the whole of the western world. His central case is that 'the elements can become an uncontrolled independent variable affecting the intensity, duration, and perhaps the occurrence, of trade cycles' (Post, 1977: 26).

The response of de Vries is sceptical, especially with regard to the supposed influence of climate stress on business cycles. Reviewing not only the contributions of Post and Pfister, but also other attempts to relate economic cycles to climatic variations, he concludes that the evidence is poor. If indeed there is a connection, it can only be within the context of a model of economic cycles which sees such cycles as a statistical artifact created by the cumulative effects of random shocks to the economy (de Vries, 1980: 621).

More generally, de Vries focuses on the inherent weaknesses, discussed above, of Pfister and Post's use of the semi-descriptive case study approach. That is, he argues that these authors may create a false impression by concentrating on highly unusual crises like those of 1816–18, the more so since they are unable to show what proportion of the misfortunes they describe are attributable to climate as opposed to such non-climatic factors as the dislocations and readjustments in trade and industry which succeeded the end of the Napoleonic wars in 1815. In this context it should be noted that certain studies indicate that even quite severe periods of meteorological stress may (in the absence of other disruptive factors) have only limited effects on the economic and social structure of a given area. An example is the analysis by Sutherland (this volume) of the effects of adverse (though admittedly not spectacularly severe) weather in the 1780s on a community in Upper Brittany. In the face of droughts in 1782 and 1785–86 and of a poor harvest in 1788, the social and economic structure of the area round the town of Vitré proved remarkably elastic. The worst-hit victims of economic stress were infants and small children, who experienced high rates of mortality; but because general fertility was high these children could be easily replaced, and their loss was from a social structural point of view relatively insignificant. Other symptons of a crisis in rural society were absent. Sutherland's conclusion is that 'many peasant communities were not as vulnerable to the weather as we generally think'.

Opinion on the medium-term influence of climatic stress is thus divided. Any attempt to come to an overall conclusion would be premature because not enough detailed studies have been carried out. As in the case of short-term climatic influences, the need for further research is obvious.

Longer-term influences The possibility that climatic change is an important independent variable affecting the course of human history over extended periods of decades or centuries is a seductive one, and there have been a number of attempts to prove the connection. Utterström (1955), for example, argued that climatic changes may well have been of 'decisive' importance in influencing population movements in medieval and early modern Europe. Braudel (1949) hinted that the economic and social difficulties that were experienced in the Mediterranean area at the end of the sixteenth century, and that heralded the slow decline in the economic importance of this region in the seventeenth were partly caused by climatic change. Subsequently (Braudel, 1973: 275) he urged the importance of climatic change for the whole of Europe: 'the "early" sixteenth century was everywhere favoured by the climate; the latter part everywhere suffered atmospheric disturbance'. More recently, Parker (1979: 17–28) has firmly asserted a relationship between the 'crisis' allegedly observable in the economy and society of seventeenth-century Europe and climatic 'deterioration' – the ultimate blame for which he appears to attach to sunspots (cf. Eddy, reprinted in Parker & Smith, 1978). (The issue of sunspot–climate links is patently peripheral to these arguments, but it should be noted that relationships on both short (decadal or less) and longer time scales (as suggested by Eddy and others before) have yet to be established on firm statistical grounds. Short time-scale effects have been reviewed with considerable scepticism by Pittock (1978), and longer time-scale links have been discussed and largely dismissed in the careful analyses of Stuiver (1980) and Williams *et al.* (1981).)

Suggestions of long-term climatic effects have provoked scepticism and even sharp criticism. Perhaps the most influential critical comment was made by Le Roy Ladurie (1972: 292–3), who directed attention to the apparently small magnitude of long-term climatic variations in Europe since about AD 1000 and questioned whether 'a difference in secular mean temperature ... [of about] 1 °C [could] have any influence on agriculture and other activities of human society', especially given the fact of human adaptation. In response to this question, however, Post (1973: 728) has argued that it is misleading to concentrate on mean temperature values as such: 'interdecennary or even *annual* thermometric values and precipitation levels are insufficient data for understanding the dynamics of ecological effects. The microaspects are essential: annual temperature means often conceal critically large seasonal variations, which frequently mask destructive monthly deviations.' This observation is supported by the evidence presented by Lamb (1977: 465, and this volume) that in Europe in the early onset stages of the Little Ice Age between about 1300 and 1450, and in the climax phase in the sixteenth and seventeenth centuries, the *variability* of the weather was particularly marked (additional support for this idea is offered by Pfister, this volume). However, as discussed earlier, the evidence for these changes in variability is still inconclusive; although it is self-evident, for example, that a depression of the mean temperature will produce a greater frequency of severe cold events.

It is possible to consider secular changes in climate in terms of changing levels of variability rather than of the mean. Such changes may or may not be statistically related to changes in the mean; we need only assume that they can occur. The question of assessing the impact of longer-term climatic shifts resolves itself then into the problem of measuring the importance of individual clusters of extreme weather events of the type discussed in the previous subsection, and devising some means of assessing the cumulative effects of a succession of such clusters. Given that the economic and social impact of even a single cluster is difficult to establish, and that on secular time scales the effects of climatic stress will be obscured by a multitude of other variables, the practical problems involved in these operations are clearly enormous.

In the light of these difficulties, it is inevitable that the majority of studies attempting to measure, or even to identify the existence of longer-term climatic effects should yield either negative or inconclusive results. This is particularly true of studies relating to areas of the world and periods of time for which basic climatic and other relevant information is sparse. Thus, the effort by Bowden *et al.* (this volume) to establish a relationship between climatic stress and two major population cycles (6000–2500 BP and 2500–750 BP) in the Tigris–Euphrates Valley founders on the rock of inadequate data – neither the climatic record nor the pattern of economic and social change can be established with sufficient precision. The masterly review by Shaw (this volume) of hypotheses previously advanced to explain the alleged decline of north Africa as a major grain-producing region after the Roman period reveals the ambiguity of much of the evidence that has been deployed. Neither literary evidence on past climate, faunal data, evidence about water resources and water supply systems, nor data on grain surpluses can be taken to prove either the fact or the economic significance of climatic change in the region. The record of historical botany and geology may provide better evidence, but the picture is still far from clear.

We have already noted that Europe is better documented than most other parts of the world; and its economic and social history has been much studied. In terms of the availability of data, therefore, this continent would seem to be a favourable area to study. It is to Europe and adjacent areas that Anderson (this volume) applies his version of Occam's razor, surveying a number of key episodes which *prima facie* might be expected to demonstrate the importance of longer-term climatic change: the origins of agriculture in the Near East; the expansion of Scandinavian settlement around the tenth century AD; and the European economic recessions of the fifteenth and seventeenth centuries. He finds that in none of these cases is it necessary to invoke climate as an explanatory variable. He does not argue that climatic variation had no effect whatsoever, but feels justified in concluding that, had there been no climatic change, the history of Europe would not have been different in any general sense.

It might be objected that Anderson's survey is too superficial, and heavily dependent on secondary sources whose authors may themselves have failed to consider the climate question in adequate detail or with appropriate rigour. Anderson's dismissal of climate as a possible determinant of population movements (by common consent a powerful engine of economic and social change) contrasts with the willingness of Lee (1981) to take the issue seriously. Impressed by the strength of the impact of temperature on short-run demographic fluctuations in England, Lee feels justified in considering the possible implications of his results for long-run trends. (He cautions, however, that medical evidence stresses the effects of *variability* of temperature as opposed to its level, and that it may therefore be quite inappropriate to draw conclusions about long-term trends from the evidence relating to short-run fluctuations. The possibility that secular temperature changes may be conceived in terms of changing levels of variability is something which he does not consider.) His suggestion, expressed with appropriate reserve, is that a 1 deg C warming of winters and cooling of summers could produce an absolute increase in the population growth rate of about 0.20 per cent per year – 'not a large effect, but . . . far from negligible', especially if replicated in more than one country.

Anderson's scepticism with regard to the effects of climatic change on the history of Europe as a whole and of individual countries within it may or may not be justified. In any event, even he is prepared to admit that 'local, and even regional histories, could well have been very different in the absence of changes in climate'. Of interest in this context is a paper by Hoffman (1973) on the economy of the Duchy of Wrocław (in modern Poland) in the fifteenth century. Hoffman uses a methodological technique similar to that employed by Anderson, invoking the variable of climatic change to fill explanatory lacunae not covered by endogenous economic factors. He argues that in the period AD 1425–80 the Duchy of Wrocław suffered an economic crisis which cannot be explained in Malthusian terms: exogenous factors have to be invoked. He identifies these exogenous factors as devastating wars and adverse climatic conditions. Both, he argues, were important; but imperfections in the data make it impossible to assess their relative significance.

Anderson is also prepared to admit that 'marginal' areas could have been significantly affected by longer-term climatic changes, an idea which is supported by certain of the contributions to this volume. McGovern examines the question of the extinction of the Norse colony in Greenland by about AD 1500 and concludes that climatic stress was indeed a major factor. Parry mounts a strong argument that in an upland zone of southeast Scotland long-term climatic changes over the period 1200–1700 were sufficient to render considerable areas of land submarginal for the purposes of oat cropping, and help to explain the permanent abandonment of this land. He recognises that the precise chronology and extent of land abandonment was conditioned by a host of short-term factors or triggers: some of them may have been climatic, such as brief periods of particularly adverse weather; others, perhaps including outbreaks of disease (such as the Black Death) and devastating wars, were of a non-climatic nature. Parry argues, however, that the effects of such short-term factors in themselves would have been ephemeral: land temporarily abandoned would have been returned to cultivation. Permanent abandonment is explicable only where long-term changes in the resource base, perhaps (though by no means always) induced by climatic change, have reduced the stage upon which the proximate [or short-term] factors played (Parry, this volume 331).

Parry's argument and analysis cannot, however, be taken as proof of the importance of climatic change as a causal factor in the abandonment of land. Parry (this volume 323–5) examines climatic change as only one of four possible explanations; the others being errors in farming judgement, changes in farming objectives, and changes in farming systems. Although not specifically stated by Parry, changes in farming objectives includes the effect of possible changes in the demand for land. Declining demand (whether caused by a reduction in population levels as a result of disease, war, emigration, or other factors, or by the availability of alternative and more eligible sources of income) could lead to the abandonment of farms; and in these circumstances it would be reasonable to expect relatively unprofitable land close to the agricultural margins to be abandoned in preference to any other. Irrespective of climatic change, such factors could conceivably account for much of the observed pattern of land abandonment. Parry himself emphasises the need to analyse and compare the evidence for different exploratory hypotheses before accepting any one in particular.

It is sobering to realise that even in a marginal situation such as Parry studies, the effects of long-term climatic change cannot be identified with absolute certainty. The overall message of the contributions to this volume is to re-emphasise the conclusion reached a decade ago by Le Roy Ladurie (1972: 119) that the impact of secular climatic changes may well have been negligible (except, perhaps, at a local or regional level and in certain marginal situations), and in any event is extremely difficult to detect. On the other hand, there are sufficient positive indications to make the problem seem worthy of further study.

Human adaptation to climatic stress

The relationship between climatic stress and economic and social life should not be conceived as a one-way process. Man is a highly adaptive animal, capable of devising and deploying a wide range of technologies and social strategies to cope with a wide variety of environmental conditions. In view of this fact, and given the comparatively small range of climatic variations in historic times, it may be assumed that past human societies have to a considerable extent had the potential to adapt successfully to changes in climate. In the future, the scope of such potential may be increased by the development of sophisticated climate modification technologies. Thus the inter-relationships between climate and human society may be conceived in terms of a two-way

Fig. 1.3. Conceptual model of the impact of climate on Man and society showing possible feedbacks via adaptive strategies. Uncontrolled effects refer to inadvertent or unplanned modifications of the biosphere and/or the climate as a result of climatic stress or otherwise.

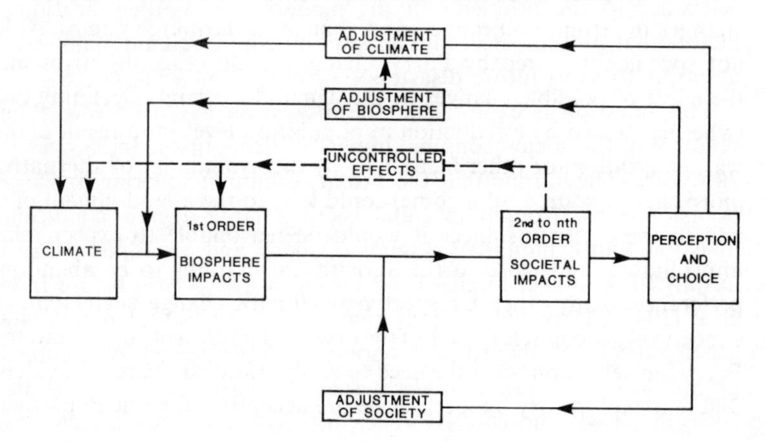

model involving elaborate feedback mechanisms, a simple version of which is presented in Fig. 1.3.

Approaches to climate–history studies that seek to assess the impact of climatic variations on human economic and social systems in the manner discussed in the preceding section frequently obscure the processes of human adaptation to climatic stress. Depending on whether the particular researcher is concerned to advocate or to reject the importance of climate in human affairs, the processes of adaptation may be either neglected as an embarrassing complication, or (as in the study by Anderson, this volume) emphasised in order to argue that worthwhile assessments of climatic impact are inordinately difficult.

An alternative approach is explicitly to focus attention on the processes of adaptation – or, on the other hand, of *failure* to adapt – that partly condition the impact of climatic stress in particular societies. This anti-deterministic approach, which has already been advocated by de Vries (1980: 625–30), is well represented among chapters in this volume. From such studies it is clear that the recognition that societies subjected to climatic stress must not be regarded as passive victims of external forces, but rather that such stress is in the nature of a challenge to which a variety of responses is possible, allows a more subtle appreciation of the role of climate in human affairs.

On this perspective, cases in which societies appear to have been seriously damaged by, or even totally succumbed to, climatic stress should not be taken to demonstrate the determining influence of climate. It is essential to consider ways in which these societies might have coped better, and to focus on the political, cultural, and socio-economic factors which inhibited them from doing so. Potentially such studies are of great relevance to the problems of the modern world. Identification of the factors that prevent successful adaptation could well aid planners in their attempts to avert future disasters.

This question is explored in the discussion by Mooley & Pant (this volume) of the socio-economic impact of droughts in India over the last 200 years. They argue that the extent of human suffering and mortality occasioned by drought in India is, to a large degree, conditioned by inflexible and exploitative systems of social and economic organisation, in part the result of foreign exploitation in the period before Independence (1947). They suggest that changes in agricultural practices and in other aspects of the rural economy, complemented by planned programmes to accumulate food reserves and cash funds to provide, when necessary, swift and effective relief in the drought-prone areas, could in the future significantly reduce the suffering occasioned by monsoon failure (the main determinant of drought in India).

Mooley and Pant's discussion is not specifically related to long-term climatic changes (on the contrary, they show statistically that the incidence of droughts in India in the period 1771–1977 appears to be random); and the climatic stresses that they study were not of a severity sufficient to destroy the entire society, though they were capable of inflicting immense suffering on substantial sections of it. By contrast, McGovern's chapter on Norse Greenland deals both with long-term climatic shifts and the total extinction of an entire society – albeit a very small one, located in a region where subsistence was arduous at best.

McGovern argues strongly that, even in this manifestly marginal case, it would be wrong to explain the collapse of the society solely in terms of a deteriorating climate, or even of climate in association with other external pressures. The Norse could have survived by shifting the economic balance of their society, de-emphasising stockraising in order to concentrate more on the exploitation of seals and other marine resources, and by adopting the use of skin boats, skin clothing, and other elements of indigenous Inuit culture and technology in order to facilitate the shift. But instead of pioneering these adaptive strategies, the political and religious élite in Greenland persisted in maintaining existing and increasingly inappropriate economic and cultural patterns. Their failure as managers of the community's scanty resources was the ultimate cause of the extinction of the Norse colonies.

The question of adaptation can be approached from other directions. In cases where marked climatic stress appears, in terms of gross economic or other indicators, to have had little impact, it may be valuable to investigate the processes of adaptation or of cost distribution whereby the climatic stress was minimised or absorbed. Again, the insights which might be derived from such studies could prove valuable in the context of present-day planning.

De Vries' (1980) contribution to the study of such adaptation focuses primarily on economic responses, specifically changes in agricultural practices and technology. He suggests (pp. 625–6) that certain changes in agricultural practices, such as variations in crop mixes, may sometimes be interpreted as adaptive strategies developed in response to climatic stress. He believes it possible, for example, that the tendency towards crop diversification identified in areas of midland England in the period 1550–1650 and hitherto explained in terms of shifts in demand towards cheaper grains, induced by a reduction in the purchasing power of much of the population (Skipp, 1978: 44–9), could represent farmers' attempts to reduce their vulnerability to a climate which had become more frequently threatening towards winter crops.

Sutherland (this volume) focuses not on changes in agricultural practices but more generally on the complex of social mechanisms that enabled the Vitré area of Brittany in the 1780s (and by implication other peasant societies) to cope with climatic stress. He found that in Vitré the poor were cushioned by secondary sources of income that were only partly dependent on the grain harvest; while employment for farmhands continued to be available throughout the period of meteorological stress. The richer peasantry, with substantial grain surpluses even in bad years, actually benefited from the high prices associated with poor harvests. In addition, poor relief systems and other charitable mechanisms, such as the willingness of landlords to postpone or waive demands for rent, helped to distribute the costs of climatic stress. Thus, a range of social and economic factors facilitated the absorption of the shocks administered by the climate and helped to maintain the social fabric more or less intact.

These approaches leave a number of questions unanswered. It is not clear by what means de Vries proposes to isolate adaptations stimulated by climatic stress from other elements of technological and economic change, except in certain specific cases. In the majority of cases the problems of detection of adaptive strategies would appear to be analogous to, but possibly even more complex than, those involved in identifying and measuring climatic impact. Clearly, there is a danger of arguing in a circle. Sutherland, while showing that the degree of climatic stress observable in the 1780s – which he admits was not spectacular – could be absorbed by the social system, is unable to consider the vital question which such an analysis provokes: what were the limits to the society's capacity to absorb stress? In other words, what degree of meteorological adversity would have been required to precipitate serious social and economic dislocation?

To generalise from these specific comments, it is plain that in order to advance the study of human adaptation to climatic variations to the point where the findings may be of real use to modern planners, it will be necessary to try to specify more clearly what degrees and types of climatic stress impose the greatest problems of successful response, and to seek to identify more rigorously the key features of more and less adaptive societies. To do this it will be necessary to attempt extensive cross-cultural comparisons. A preliminary step in this direction has been taken by Bowden *et al.* (this volume). Deploying material on agriculture in the US Great Plains in the period AD 1880–1979, on droughts in the Sahel region of Africa 1910–15 and 1968–74, and on the societies of the Tigris–Euphrates Valley over six millennia, they discuss the complementary hypotheses that over time societies adapt to cope with 'minor' climatic stresses (defined as events with a return period of the order of less than 100 years), but that thereby they do little to decrease, and may actually increase, their vulnerability to 'major' stresses of rarer frequency.

Because of inadequate data the results from the Tigris–Euphrates study proved inconclusive, but the evidence from the other areas was rather more suggestive. The part of the hypothesis relating to 'major' stresses could only be handled on a speculative level. However, it is argued strongly that both the Sahel and the US Great Plains have become less vulnerable to 'minor' climatic stress. In the case of the Great Plains, wheat yield evidence was analysed to discover whether the lessening of vulnerability could be explained by reference to developments in agricultural technology; the analysis failed to substantiate this possibility. Both in the Sahel and in the Great Plains the key factor in reducing vulnerability appeared to be the integration of each area into a wider economic system. The sufferings of the Sahel in 1968–74, though massive, were limited by external aid; on the US Great Plains, distress was reduced after about 1930 when the Union as a whole accepted the responsibility of providing relief for drought-afflicted areas, thus in effect sharing the costs of climatic stress. It proved possible to absorb even a 'major' climatic stress, the drought of the 1930s, in this way. Massive federal relief ensured that the potential catastrophe of these years took the form of a large ripple through the national economy, rather than a tidal wave of disaster located in the Great Plains region itself. However, Bowden *et al.* caution that in the future there may prove to be a limit to the effectiveness of such integrative mechanisms.

One point of interest that emerges from the study by Bowden *et al.* (though it is not emphasised in the theoretical apparatus of the chapter) is that newly settled regions (such as the Great Plains in the late nineteenth century) and areas where the economic system is subjected to rapid modification (as in the Sahel in the twentieth century) face particular problems of vulnerability to climatic stress. In such innovatory circumstances the nature of the climatic regime, or at least its likely hazards for the type of economic system which is being implemented, may be poorly understood, and mismanagement (leading sooner or later to disaster) is particularly likely to occur. The same point is emphasised in the chapter by Smith *et al.* (this volume) on climatic stress and Maine agriculture in the nineteenth century. A brief warming period in the 1820s raised unrealistic expectations that the region could support large-scale commercial agriculture, and the process of adjustment to the more normal cooler conditions which soon reasserted themselves was a painful one which took several decades to accomplish.

The question of human adaptation to climatic variation is a complex one, whose systematic study has only just begun. The value of this field of endeavour is obvious: apart from its intrinsic interest, it is of major relevance to the problems of modern world planning. Further research is urgently required. The studies so far completed, though limited in their scope and achievement, are of considerable interest and have added a major new dimension to the climatic impact debate.

Human perceptions of climate

The study of human adaptations to climatic variations inevitably raises questions about human perceptions of climatic phenomena. It is true that adaptations or other responses may occur in the absence of any clear perception of the climatic stimulus. It is sufficient that the *effect* of climatic change, such as variations in the performance of crops, should be noted in order for some adaptation or response to occur, whether in the form of a carefully thought out course of action or a more or less blind trial and error response. It may be assumed that most adaptations to long-term climatic changes (unless such changes manifested themselves in the form of increased variability on short time scales) must have occurred in the absence of any clear perception of the seculiar change, since without the aid of systematic instrumented records such changes cannot be observed.

However, shorter-term climatic variations are more readily open to human perception, and in the short run the accuracy of such perceptions and of future predictions based on them will partly condition the degree of success achieved by attempted adaptations. The case of Maine agriculture in the nineteenth century, discussed by Smith *et al.* (this volume: 456–9), the demonstrates how costly misconceptions could be.

But the theme of human perceptions of climate is of wider significance than this. So far our discussion has been based largely on an external view or 'observer's model' of human society, focusing on the socio-economic impact of climatic change and paying scant attention to the place which climate occupied in the mental world of contemporaries. The place of climate in the mind of Man has a history of its own. To date, this history has been little studied; but there are sufficient indications in the existing literature, including a number of essays in this volume, to suggest that the theme would amply repay further research.

One approach might be to examine the slow development of meteorological curiosity and to chart the process by which modern scientific meteorological methods were established. In Europe a key period in this process was in the seventeenth and early eighteenth centuries, which saw the beginnings and rapid development of instrumental meteorology. It seems clear that these achievements rested in part on a considerable acceleration of interest in meteorology in the sixteenth century, maifested in the practice of compiling non-instrumental meteorological series of which increasing numbers are coming to light all over Europe (Pfister, this volume: 215–18; Klemm, 1979; Landsberg, 1980: 633–4). The relationship between these series and (as far as is at present known) the less systematic and less numerous observational records of earlier periods needs to be explored; as does the intellectual context in which they were produced. At present, it would appear that astrological, astronomical and medical concerns were the main stimuli to early attempts to keep systematic weather records (Ingram *et al.*, this volume 188–9). Clearly, these questions relate to the wider theme of the history of the development of modern science, a subject which has aroused considerable interest among historians and scientists in recent years (e.g. Webster, 1974, 1975; Kutzbach, 1979).

But in searching for the origins of modern meteorology we must not neglect the study of past attitudes to climate which do not accord with modern scientific conceptions and which may, at first sight, appear irrational or superstitious. Such ideas are of great intrinsic interest and of value for an understanding of past concepts of the universe, the environment, and Man's place within them.

Even scientists with little direct interest in the mental world of past generations cannot afford to neglect such ideas, at least if they seek to use documentary evidence to reconstruct past climates. As Ingram *et al.* (this volume: 197) show, in order to exploit fruitfully the information on climatic variations embodied in such sources as chronicles, annals and diaries, it is essential to consider the historical milieux in which they were written and the structure of their authors' mental apparatus, for these factors dictated the selection of particular climatic phenomena for record and the form in which the observations were set down. The information included in chronicles and other sources may be gravely misinterpreted if these factors are ignored.

Yet, as MacKay (this volume: 368ff) argues – in a chapter which focuses explicitly on the question of human perceptions of climate – such issues can also be studied more positively, not as tiresome complications which hinder the derivation of 'objective' climatic data to suit the purposes of modern climatologists, but in order to elucidate the nature of past belief systems and the behavioural patterns to which they gave rise. MacKay focuses specifically on fifteenth-century Castile, revealing in the course of a complex and vivid argument how the population perceived climatic calamities in terms of punishment for sin, and, in response to climatic stress, took action to purge the community of corrupt elements in order to restore the order of the cosmos. A tendency to view atmospheric calamities as evidence of sin in the community was not confined to late medieval Spain but was a common phenomenon in pre-industrial Europe and elsewhere. Midelfort (1972: 30–66) has explored how, in Reformation Germany, divergent views on this issue were related to prosecutions for witchcraft. He found that Catholic theologians tended to ascribe such meteorological calamities as hailstorms to the actions of witches, who had to be extirpated if the storms were to be averted. Certain Protestant theologians, however, interpreted storms as direct judgements from God, intended to chastise, warn and test His people; the appropriate response was not to attach the blame to witches, but one of repentance and the offering of prayers to God to avert His wrath. (Whether this theological interest in hailstorms in the sixteenth century was in any way stimulated by an actual increase in the number or severity of such storms in that period is a matter for speculation; here the intention is simply to draw attention to a parallel to the sin-oriented conception of climate discussed by MacKay.)

To the modern observer such ideas may at first sight appear grotesque, even irrational. In fact they were, in the context of the belief systems prevalent at the time, perfectly rational. Moreover, we are in error if we suppose that non-scientific attitudes to climate have been wholly discarded in the world of today. Even among the modern scientific community, ideas about climate are inevitably influenced to some extent by current ideologies; indeed, ideological issues may be basic to the current intensification of debate about the role of climate in history. The last 20 years have seen a major shift in attitudes among sections of the scientific and non-scientific communities away from a confident belief in Man's ability to control the universe through increasingly elaborate technologies, towards a more anxious view which sees Man on the verge of ruin as a result of his rapacious and uncontrolled exploitation of the Earth's resources and thoughtless development of destructive technologies. The possibility of disaster brought on by climatic stress (whether of natural occurrence or artificially induced by such factors as the injection of carbon dioxide into the atmosphere) is regarded as one among a number of possible agents of Nemesis, others being the exhaustion of oil supplies and nuclear catastrophe. Some will regard such ideas about the destructive potential of climate as eminently realistic; others will regard them as far-fetched. Though the specific evidence which each side adduces may be criticised and evaluated, there is at present no way of deciding in favour of one or the other viewpoint – the differences between these two approaches to the existing data are fundamentally of a philosophical nature, and not susceptible to scientific testing.

It is not suggested that these particular philosophical issues (here presented in a very crude and abbreviated form) necessarily underpin any of the contributions to this volume or the climate–history debate in general. But philosophical stances of some sort, perhaps with vaguely defined moral overtones, undeniably condition differing interpretations of the often ambiguous evidence. In this sense, the kind of analysis applied by MacKay to attitudes to climate in fifteenth-century Castile is of some relevance to our own deliberations. In order to reach as clear a view as possible of the role of climate in human affairs, past, present and future, it is essential that we try to understand our own preconceptions and attitudes. It is hoped that, in presenting as wide a range as possible of approaches to the study of past climates, this volume will help to clarify the issues.

Acknowledgements

We thank our colleagues in the Climatic Research Unit for stimulating discussions on the subject of climate and history, and for specific comments on this chapter. Dr M. L. Parry and Mr V. F. G. Morgan provided constructive criticisms. We also thank Professor E. A. Wrigley, of the SSRC Cambridge Group for the History of Population and Social Structure, for making available to us, before publication, R. Lee's analysis of demographic trends. Much of the work in this chapter was financed by a grant from the Rockefeller Foundation.

References

Alexandre, P. (1976a). *Le climat au moyen âge en Belgique et dans les régions voisines (Rhénanie, Nord de la France)*. Centre Belge d'Histoire Rurale Publication No. 50.

Alexandre, P. (1976b). Le climat dans le sud de la Belgique et en Rhénanie de 1400 à 1600. Prélude au 'Petite Age Glaciaire' de l'époque moderne. *Annales du XLIV-ième Congrès*, Huy, 18–22 Aôut 1976. Fedaration des Cercles d'Achaeologie et d'Histoire de Belgique, ASBL.

Alexandre, P. (1977). Variations climatiques au Moyen Age. *Annales: Économies, Sociétés, Civilisations*, **32**, 183–97.

Ashton, T. S. (1959). *Economic fluctuations in England, 1700–1800*. Oxford: Clarendon Press.

Bell, W. T. & Ogilvie, A. E. J. (1978). Weather compilations as a source of data for the reconstruction of European climate during the medieval period. *Climatic Change*, **1**, 331–48.

Bernabo, J. C. (1977). *Sensing climatically and culturally induced environmental changes using palynological data*. Unpublished PhD thesis, Brown University, Rhode Island.

Bowden, P. (1967). Agricultural prices, farm profits, and rents., In *The Agrarian History of England and Wales, Vol. 4, 1500–1640*, ed. J. Thirsk, pp. 593–695. Cambridge: Cambridge University Press.

Braudel, F. (1949). *La Mediterranée et le Monde Mediterranéen a l'Époque de Philippe II*. Paris: Librairie Armand Colin.

Braudel, F. (1973). *The Mediterranean and the Mediterranean World in the Age of Philippe II*. 2 vols. London: Fontana/Collins.

Britton, C. E. (1937). A meteorological chronology to AD 1450. *Meteorological Office Geophysical Memoirs, No. 70*. London: HMSO.

Bryson, R. A. & Murray, T. J. (1977). *Climates of Hunger*, Wisconsin: University of Wisconsin Press.

Chappell, J. E. (1970). Climatic change reconsidered: another look at 'The Pulse of Asia'. *Geographical Review*, **60**, 347–73.

Chu Ko-chen, (1973). Preliminary study on the climatic fluctuations during the last 5,000 years in China. *Scientia Sinica*, **16**, 226–56.

Court, A. (1979). Comments on 'The association between mean temperature and interannual variability' (with reply by H. van Loon). *Monthly Weather Review*, **107**, 90.

Dansgaard, W., Johnsen, S. J., Reeh, N., Gundestrup, N., Clausen, H. B. & Hammer, C. U. (1975). Climatic changes, Norsemen and modern man. *Nature*, **255**, 24–8.

Denton, G. H. & Karlén, W. (1973). Holocene climatic variations – their pattern and possible cause. *Quaternary Research*, **3**, 155–205.

Denton, G. H. & Karlén, W. (1977). Holocene glacial and tree line variations in the White River Valley and Skolai Pass, Alaska and Yukon Territory. *Quaternary Research*, **7**, 63–111.

Douglas, K. S. & Lamb, H. H. (1979). Weather observations and a tentative meteorological analysis of the period May to July 1588. *Climatic Research Unit Research Publication (CRU RP) No. 6a*. Norwich: University of East Anglia.

Douglas, K. S., Lamb, H. H. & Loader, C. (1978). A meteorological study of July to October 1588: the Spanish Armada Storms. *Climatic Research Unit Research Publication (CRU RP) No. 6*. Norwich: University of East Anglia.

Duchon, C. E. & Koscielny, A. J. (1979). Comments on 'The association between mean temperature and interannual variability' (with reply by J. Williams & H. van Loon). *Monthly Weather Review*, **107**, 496–8.

Flohn, H. (1949). Klima und Witterungsablauf in Zürich im 16 Jahrhundert. *Vierteljahrsschrift der Naturforschungs-Gesellschaft in Zürich*, **94**, 28–41.

Fredskild, B. (1973). Studies in the vegetational history of Greenland: palaeobotanical investigations of some Holocene lake and bog deposits. *Meddelelser om Grønland*, **198** (4).

Fritts, H. C. (1976). *Tree Rings and Climate*. New York: Academic Press.

Gottschalk, M. K. E. (1971). *Stormvloeden en rivieroverstromingen in Nederland*. Part 1: *de periode vóór 1400*. Assen: Van Gorcum.

Gottschalk, M. K. E. (1975). *Stormvloeden en rivieroverstromingen in Nederland*. Part II: *de period 1400–1600*. Assen: Van Gorcum.

Gottschalk, M. K. E. (1977). *Stormvloeden en rivieroverstromingen in Nederland*. Part III: *de periode 1600–1700*. Assen; Van Gorcum.

Hellman, G. (1908). The dawn of meteorology. *Quarterly Journal of the Royal Meteorological Society*, **34**, 221–32.

Hennig, R. (1904). Katalog bemerkenswerter Witterungsereignisse von den ältesten Zeiten bis zum Jahre 1800. *Abhandlungen des Königlich Preussischen Meteorologischen Instituts. Bd. II, No. 4.* Berlin: A. Asher.

Hoffman, R. C. (1973). Warfare, weather, and rural economy: the Duchy of Wrocklaw in the mid-fifteenth century. *Viator*, **4**, 273–301.

Hoskins, W. G. (1964). Harvest fluctuations and English economic history, 1480–1619. *The Agricultural History Review*, **12**, 28–46.

Hoskins, W. G. (1968). Harvest fluctuations and English economic history, 1620–1759. *The Agricultural History Review*, **16**, 15–31.

Huntington, E. (1907). *The Pulse of Asia*. Boston, Mass: Houghton Mifflin.

Huntington, E. (1915). *Civilisation and Climate*. New Haven.

Ingram, M. J., Underhill, D. J. & Wigley, T. M. L. (1978). Historical climatology. *Nature*, **276**, 329–34.

Jones, E. L. (1964). *Seasons and Prices. The Role of the Weather in English Agricultural History*. London: George Allen & Unwin.

Jones, P. D. & Wigley, T. M. L. (1980). Northern hemisphere temperatures, 1881–1979. *Climate Monitor*, **9**, 43–7.

Karlén, W. (1976). Lacustrine sediments and tree-limit variations as indicators of Holocene climatic fluctuations in Lappland, Northern Sweden. *Geografiska Annaler*, **58 A**, 1–34.

Karlén, W. (1979). Glacier variations in the Svartisen area, northern Norway. *Geografiska Annaler*, **61 A**, 11–28.

Karlén, W. & Denton, G. H. (1976). Holocene glacial variations in Sarek National Park, northern Sweden. *Boreas*, **5**, 25–56.

Kington, J. A. (1981). Daily weather mapping from 1781. *Climatic Change* (in press).

Klemm, F. (1979). Die Entwicklung der meteorologischen Beobachtungen in Südwestdeutschland bis 1700. *Annalen der Meteorologie* (Neue Folge), No. 13. Offenbach am Main: Deutsche Wetterdienstes.

Kutzbach, G. (1979). *The thermal theory of cyclones. A history of meteorological thought in the nineteenth century*. Boston: American Meteorological Society Historical Monograph Series.

Labrijn, A. (1946). Het klimaat van Nederland gedurende de laatste twee en een halve eeuw [The climate of the Netherlands during the last two and half centuries]. *Meded en Verh. No. 49*, KNMI, De Bilt.

La Cour, P. (1876). *Tyge Brahes Meteorologiske Dagbog holdt paa Uranienborg for Aarene 1582–1597*. Copenhagen (Konglige Danske Videnskabernes Selskab: *Collectanea Danica*).

La Marche, V. C. (1973). Holocene climatic variations inferred from tree-line fluctuations in the White Mountains, California. *Quaternary Research*, **3**, 632–60.

La Marche, V. C. (1974). Palaeoclimatic inferences from long tree-ring records. *Science*, **183**, 1043–8.

Lamb, H. H. (1969). The new look of climatology. *Nature*, **223**, 1209–15.

Lamb, H. H. (1977). *Climate: Present, Past and Future*, Vol 2. London: Methuen.

Lamb, H. H. & Johnson, A. I. (1966). Secular variations of the atmospheric circulation since 1750. *Meteorological Office Geophysical Memoir, No. 110*. London: HMSO.

Landsberg, H. E. (1980). Past climates from unexploited written sources. *Journal of Interdisciplinary History*, **10**, 631–42.

Lee, R. (1981). In *The Population History of England 1541–1871, a Reconstruction*, (E. A. Wrigley & R. S. Schofield. London: Edward Arnold) (in press).

Le Roy Ladurie, E. (1972). *Times of Feast, Times of Famine: A history of climate since the year 1000*. London: George Allen & Unwin.

Le Roy Ladurie, E. & Baulant, M. (1980). Grape harvests from the fifteenth through the nineteenth centuries. *Journal of Interdisciplinary History*, **10**, 839–49.

McQuigg, J. D. (1975). *Economic impacts of weather variability*. Atmospheric Sciences Department, University of Missouri-Columbia.

Manley, G. (1974). Central England temperatures: monthly means 1659 to 1973. *Quarterly Journal of the Royal Meteorological Society*, **100**, 389–405.

Mercer, J. H. (1968). Variations of some Patagonian glaciers since the Late-Glacial. *American Journal of Science*, **266**, 91–109.

Mercer, J. H. (1972). The lower boundary of the Holocene. *Quaternary Research*, **2**, 15–24.

Midelfort, H. C. E. (1972). *Witch hunting in southwestern Germany 1562–1684: the social and intellectual foundations*. Stanford: Stanford University Press.

Mitchell, J. M., Dzerdzeevskii, B., Flohn, H., Hofmeyr, W. L., Lamb, H. H., Rao, K. N. & Wallén, C. C. (1966). Climatic Change. *World Meteorological Organisation Technical Note 79*. (WMO No. 195, T.P. 100). Geneva, Switzerland..

Moodie, D. W. & Catchpole, A. J. W. (1975). Environmental data from historical documents by content analysis: freeze-up and break up of estuaries on Hudson Bay 1714–1871. *Manitoba Geographical Studies*, **5**.

Moodie, D. W. & Catchpole, A. J. W. (1976). Valid climatological data from historical sources by content analysis. *Science*, **193**, 51–3.

Nicholas, F. J. & Glasspoole, J. (1931). General monthly rainfall over England and Wales, 1727 to 1931. *British Rainfall, 1931*, 299–306.

Nichols, H. (1975). Palynological and palaeoclimatic study of the Late Quaternary displacement of the Boreal Forest-Tundra ecotone in Keewatin and Mackenzie, NWT, Canada. *Occasional Paper No. 15*, Institute of Arctic and Alpine Research, Boulder, Colorado.

Parker, G. (1976). If the Armada had landed. *History*, **61**, 358–68.

Parker, G. (1979). *Europe in Crisis, 1598–1648* (Fontana History of Europe). London: Fontana.

Parker, G. & Smith, L. M. (eds.) (1978). *The General Crisis of the Seventeenth Century*, London: Routledge & Kegan Paul.

Parry, M. L. (1978). *Climatic Change, Agriculture and Settlement*. Folkestone: Dawson.

Pearson, R. (1978). *Climate and Evolution*. London: Academic Press.

Pfister, C. (1975). *Agrarkonjunktur und Witterungsverlauf im Westlichen Schweizer Mittelland*. Geographica Bernensia, G2. Bern.

Pfister, C. (1978). Climate and economy in eighteenth century Switzerland. *Journal of Interdisciplinary History*, **9**, 223–43.

Pittock, A. B. (1978). A critical look at long-term sun-weather relationships. *Reviews of Geophysics and Space Physics*, **16**, 400–420.

Pittock, A. B. (1980). Enigmatic variations (a review of *Sun, Weather, and Climate* by J. R. Herman & R. A. Goldberg). *Nature*, **283**, 605–6.

Post, J. D. (1973). Meteorological historiography. *Journal of Interdisciplinary History*, **3**, 721–32.

Post, J. D. (1977). *The Last Great Subsistence Crisis in the Western World*. Baltimore: Johns Hopkins.

Post, J. D. (1980). The impact of climate on political, social and economic change: a comment. *Journal of Interdisciplinary History*, **10**, 719–23.

Ratcliffe, R. A. S., Weller, J. & Collison, P. (1978). Variability in the frequency of unusual weather over approximately the last century. *Quarterly Journal of the Royal Meteorological Society*, **104**, 243–55.

Röthlisberger, F. (1976). Gletscher- und Klimaschwankungen in Raum Zermatt, Ferpècle und Arolla. *Die Alpen*, **52**, 59–152.

Sawyer, J. S. (1980). Climatic change and temperature extremes. *Weather*, **35**, 353–7.

Schweingruber, F. H., Fritts, H. C., Bräker, O. U., Drew, L. G. & Schär, E. (1978). The X-ray technique as applied to dendroclimatology. *Tree-Ring Bulletin*, **38**, 61–91.

Schweingruber, F. H., Bräker, O. U. & Schär, E. (1979). Dendroclimatic studies on conifers from central Europe and Great Britain. *Boreas*, **8**, 427–52.

Skipp, V. (1978). *Crisis and Development. An ecological case-study of the Forest of Arden, 1500–1674*. Cambridge: Cambridge University Press.

Slicher van Bath, B. H. (1977). Agriculture in the Vital Revolution. In *The Cambridge Economic History of Europe*, ed. E. E. Rich & C. H. Wilson, vol. 5, pp. 42–132, Cambridge: Cambridge University Press.

Starr, T. B. & Kostrow, P. I. (1978). The response of spring wheat yield to anomalous climate sequences in the United States. *Journal of Applied Meteorology*, **17**, 1101–15.

Stolfi, R. H. S. (1980). Chance in history: the Russian winter of 1941–42. *History*, **65**, 214–28.

Stuiver, M. (1980). Solar variability and climatic change during the current millennium. *Nature*, **286**, 868–71.

Stuiver, M. & Quay, P. D. (1980). Changes in atmospheric carbon-14 attributed to a variable sun, *Science*, **207**, 11–19.

Suess, H. E. (1980). Radiocarbon geophysics. *Endeavour*, *New Series*, **4**, 113–17.

Swain, A. M. (1978). Environmental changes during the past 2000 years in north central Wisconsin: analysis of pollen, charcoal and seeds from varved lake sediments. *Quaternary Research*, **10**, 55–68.

Thompson, L. M. (1975). Weather variability, climatic change and grain production. *Science*, **188**, 535–41.

Tilly, C., Tilly, L. & Tilly, R. (1975). *The Rebellious Century 1830–1930*. London: Dent.

Utterström, G. (1955). Climatic fluctuations and population problems in early modern history. *Scandanavian Economic History Review*, **3**, 3–47.

Van den Dool, H. M., Krijnen, H. J. & Schuurmans, C. J. E. (1978). Average winter temperatures at De Bilt (The Netherlands): 1634–1977. *Climatic Change*, **1**, 319–30.

Van Loon, H. & Williams, J. (1978). The association between mean temperature and interannual variability. *Monthly Weather Review*, **106**, 1012–17.

De Vries, J. (1977). Histoire du climat et économie: des faits nouveaux, une interprétation différente. *Annales: Economies, Sociétés, Civilisations*, **32**, 198–227.

De Vries, J. (1980). Measuring the impact of climate on history: the search for appropriate methodologies. *Journal of Interdisciplinary History*, **10**, 599–630.

Wales-Smith, B. G. (1971). Monthly and annual totals of rainfall representative of Kew, Surrey, for 1697 to 1970. *Meteorological Magazine*, **100**, 345–62.

Wales-Smith, B. G. (1973). An analysis of monthly rainfall totals representative of Kew, Surrey from 1697 to 1970. *Meteorological Magazine*, **102**, 151–71.

Wales-Smith, B. G. (1980). Revised monthly and annual totals of rainfall representative of Kew, Surrey for 1697 to 1870 and an update analysis for 1697 to 1976. *Meteorological Office Hydrological Memorandum No. 43.*

Wardle, P. (1973). Variations of the glaciers of Westland National Park and the Hooker Range, New Zealand. *New Zealand Journal of Botany*, **11**, 349–88.

Webster, C. ed. (1974). *The Intellectual Revolution of the Seventeenth Century.* Past and Present Series. London: Routledge & Kegan Paul.

Webster, C. (1975). *The Great Instauration. Science, Medicine and Reform, 1626–1660*. London: Duckworth.

Weikinn, C. (1958). *Quellentexte zur Witterungsgeschichte Europas von der Zeitwende bis zum Jahre 1850, Vol. 1.* Berlin: Akademie-Verlag.

Wigley, T. M. L. & Atkinson, T. C. (1977). Dry years in south-east England since 1698. *Nature*, **265**, 431–4.

Williams, L. D., Wigley, T. M. L. & Kelly, P. M. (1981). Climatic trends at high northern latitudes during the last 4000 years compared with ^{14}C fluctuations. In *Sun and Climate*, Centre National d'Etudes Spatiales, Toulouse, France (in press).

Wilson, C. H. & Parker, G., eds. (1977). *An Introduction to the Sources of European Economic History 1500–1800*, vol. 1: *Western Europe*. London: Weidenfield & Nicolson.

Wrigley, E. A. & Schofield, R. S. (1981). *The Population History of England 1541–1871, a Reconstruction*. London: Edward Arnold.

PART II

RECONSTRUCTION OF PAST CLIMATES

2

The use of stable-isotope data in climate reconstruction

J. GRAY

Abstract

Variations in stable-isotope ratios of many natural systems have been shown to reflect climatic conditions prevalent during deposition. In particular, datable stratified systems present potentially useful sources of material whose isotope ratios may be measured. If, as in the ideal case, an isotope ratio can be related via a temperature-dependent fractionation factor to a temperature of deposition then a direct linkage to prevailing temperature at the time of deposition is possible. Actual cases are usually more complex. However, useful climatic information has been obtained from a variety of stratified deposits including deep-ocean and lake sediments, polar ice sheets, speleothems, peat deposits and annual growth rings in trees. This chapter concentrates on those systems yielding climate information about the last 5000 years.

Such systems fall into three general groups: meteoric water variations (e.g. polar ice cores); $CaCO_3$ deposits (e.g. speleothems) and biological indicators (e.g. tree rings).

Polar ice sheets contain a record of all precipitation occurring in a given area over many thousands of years. The $^{18}O/^{16}O$ variations in the ice from dated cores have been related to climatic parameters in a number of cases although detailed interpretation of an isotope record in terms of a single climate variable, such as temperature, is in most cases not possible. This is due to the occurrence of a number of interdependent physical processes which play a role in determining isotope ratios.

The underlying principles of the $CaCO_3$ method have been known for many years and recently have been applied to speleothem deposits. Interpretation in terms of climate variables is complicated by lack of certainty of the composition of the ground-seepage water from which precipitation of the $CaCO_3$ occurred. Recent efforts in this field have concentrated on determining isotope ratios of included water to allow unequivocal determination of the temperature of deposition.

Most recently efforts have been made to determine the usefulness of biological systems such as tree rings. While being inherently more complex, biological systems present intriguing possibilities in that they possess more than one isotopic indicator which may be independent, thus potentially yielding absolute temperature information. The most recent results from tree rings and peat deposits are discussed.

Introduction

The chemical properties of an element are largely determined by its extranuclear electrons. In a neutral atom the number of such electrons is equal to the number of protons in the nucleus (the atomic number). A given element may have a varying number of neutrons in the nucleus giving rise to isotopes of that element whose properties are similar but not identical. Thus, whereas there are 92 naturally occurring elements, there are known to be some 270 naturally occurring isotopes of these elements. Studies of the slight differences in the chemical and physical behaviour of the isotopes of a given element have progressed largely in parallel with the developments of nuclear techniques, particularly those of mass spectrometry. The study of stable isotopes can be said to have begun with J. J. Thompson's experiments in 1910 when he demonstrated the existence of two isotopes of neon, namely ^{20}Ne and ^{22}Ne (a third isotope ^{21}Ne has since been discovered). It is now known that almost all the elements have more than one naturally occurring isotope. The element tin for instance has ten.

Physical and chemical properties of atoms depend to some extent on the nuclear mass of the atom undergoing the process. The greater the mass difference between two isotopes in relation to their absolute mass, the greater will be the isotope effect. For this reason stable-isotope studies have tended to concentrate on commonly occurring low atomic mass elements such as hydrogen ($^{2}H/H$ or D/H), carbon ($^{13}C/^{12}C$), nitrogen ($^{15}N/^{14}N$), oxygen ($^{18}O/^{16}O$) and sulphur ($^{34}S/^{32}S$).

Historically, oxygen isotope ratios have proven the most useful in relation to climate studies because, first, there exists a large reservoir of oxygen in the oceans; secondly, oxygen is very reactive and forms compounds with most other elements; and lastly, the natural abundance variations are relatively easy to measure. The natural abundances of the three oxygen isotopes are (^{16}O) 99.759 per cent, (^{17}O) 0.037 per cent and (^{18}O) 0.2039 per cent. The usual approach is to measure variations in the ratio $^{18}O/^{16}O$ by isotope mass spectrometry. In practice, rather than measure the absolute value of the ratio it is simpler to measure the $^{18}O/^{16}O$ isotope ratio relative to a standard and to quote the result as a 'del' value (per mil) defined by:

$$\delta^{18}O_{sample} = [(^{18}O/^{16}O)_{sample}/(^{18}O/^{16}O)_{standard} - 1]\,10^{3}\,‰.$$

The most commonly used standard is SMOW (standard mean ocean water). A positive $\delta^{18}O$ value means the sample is enriched in ^{18}O relative to sea water (i.e. isotopically 'heavier') and a negative $\delta^{18}O$ value means the sample is depleted in ^{18}O relative to sea water (isotopically 'lighter'). In a similar manner, other stable-isotope values are expressed as del values relative to a standard: carbon ($\delta^{13}C$) and hydrogen ($\delta^{2}H$, or, more commonly, δD) are the most significant climatologically, next to oxygen.

Natural isotope variations

In a chemical reaction, the various isotopes of the elements present in the system tend to distribute themselves to different extents in different compounds so as to minimise the free energy of the system. Thus in the equilibrium chemical reaction, $H_2O + CO_2 \leftrightarrows H_2CO_3$, the CO_2 will become enriched in ^{18}O relative to H_2O by about 41‰ at 25 °C. In general, the molecule containing the highest oxygen bond multiplicity will become enriched in the heavier isotope. This enrichment (denoted by a fractionation factor α) will usually decrease with increasing temperature since the difference in free energy between two isotopic species becomes less significant as the total energy of the system increases. This is illustrated in Table 2.1 where the fractionation factor α for the above reaction is shown for various equilibrium temperatures (Bottinga, 1968). When α is close to one the isotopic difference between species 'A' and species 'B' is given approximately by the relation

$$\delta^{18}O_A - \delta^{18}O_B \approx 10^3(\alpha_{AB} - 1).$$

Table 2.1 shows, for example, that at 25 °C

$$\delta^{18}O_{CO_2} - \delta^{18}O_{H_2O} \approx 40.7‰$$

so that oxygen in the CO_2 is about 41 ‰ heavier than that in the H_2O. The CO_2–H_2O system, therefore, is a potential indicator of the temperature of equilibration.

Physical processes are equally important in producing natural variations in isotope abundances. The two most important oxygen isotopic forms of water are $H_2^{16}O$ (99.8 per cent) and $H_2^{18}O$ (0.2 per cent). The natural abundances of these molecules vary from site to site because the processes of evaporation and condensation result in fractionation. When water evaporates, the vapour is depleted in ^{18}O because the vapour pressure of $H_2^{16}O$ is about 1 per cent greater than that of $H_2^{18}O$. The remaining water which is subject to evaporation becomes steadily enriched in ^{18}O. As the depleted vapour cools and condenses, the reverse process occurs and the $H_2^{18}O$ will tend to concentrate in the condensed phase. If condensation occurs immediately after evaporation the conden-

Table 2.1. *Variation of $\alpha(CO_2–H_2O)$ with temperature (Bottinga, 1968)*

α	T(°C)
1.0466	−2
1.0461	0
1.0455	7
1.0407	25
1.0401	29

sate will have essentially the same isotopic composition as the source water.

As condensation occurs, however, the vapour will become progressively more depleted in ^{18}O. Thus when water evaporates from oceans, the first condensate will tend to have the same $\delta^{18}O$ value as the ocean. As the vapour is carried to higher latitudes the $\delta^{18}O$ value will become more negative as will that of the precipitation. Thus by progressive cooling the air mass carrying the water vapour gives off precipitation whose ^{18}O composition decreases as the temperature of the air mass decreases. This simple model shows how the $\delta^{18}O$ of precipitation can be related to latitude and furthermore how it can be related to local temperature at the point of precipitation. This is illustrated in Fig. 2.1 (Hage, Gray & Linton, 1975). Despite the simplicity of the model the correlation between $\delta^{18}O$ of precipitation and air temperatures can be very high. Thus in many cases measurements of natural variations of isotope abundances can be related to temperature.

To determine past climate it is necessary to find natural systems which have recorded these variations in datable strata. Such a system must satisfy a number of criteria: (*a*) the isotope variations found within the system must be determined solely by climatic factors or in such a way that non-climatic influences can be filtered out; (*b*) the isotope record must be permanent, thus it must be proven that no further fractionation processes (such as equilibration or exchange) have occurred after the strata have been deposited, (*c*) the record preserved by the strata should be reasonably complete and continuous; (*d*) the strata must be datable so that a time scale can be attached to the record; (*e*) the time resolution of the stratified system must be commensurate with the climate information required.

Natural systems which to a greater or lesser extent satisfy the above criteria include: deep-ocean and lake sediments, polar ice sheets, speleothem deposits of $CaCO_3$, peat deposits, annual coral rings and annual growth rings of trees. Discussion of all these systems is not possible here so only those systems yielding climate information about the last 5000 years will be emphasised. The systems can be divided into three categories: (*a*) meteoric water variations, (*b*) $CaCO_3$ deposits and (*c*) biological indicators.

Meteoric water variations

The study of isotopic variations in precipitation as stored in polar ice sheets was pioneered independently by Dansgaard and Epstein (see for example Dansgaard, Johnsen, Clausen & Gundestrup, 1973; Hammer *et al.*, 1978; Dansgaard *et al.*, 1975; and Epstein, Sharp & Gow, 1965, 1970).

Dansgaard *et al.* (1973) have listed the observable effects caused by fractionation occurring during evaporation and condensation of water

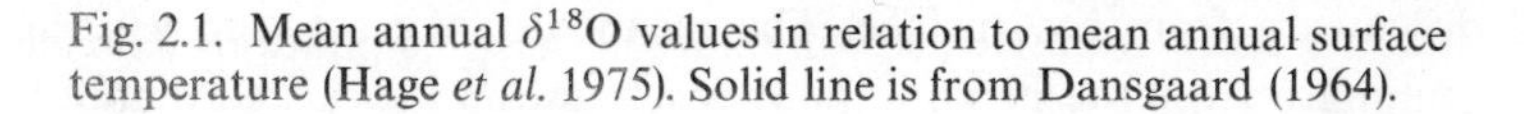

Fig. 2.1. Mean annual $\delta^{18}O$ values in relation to mean annual surface temperature (Hage *et al.* 1975). Solid line is from Dansgaard (1964).

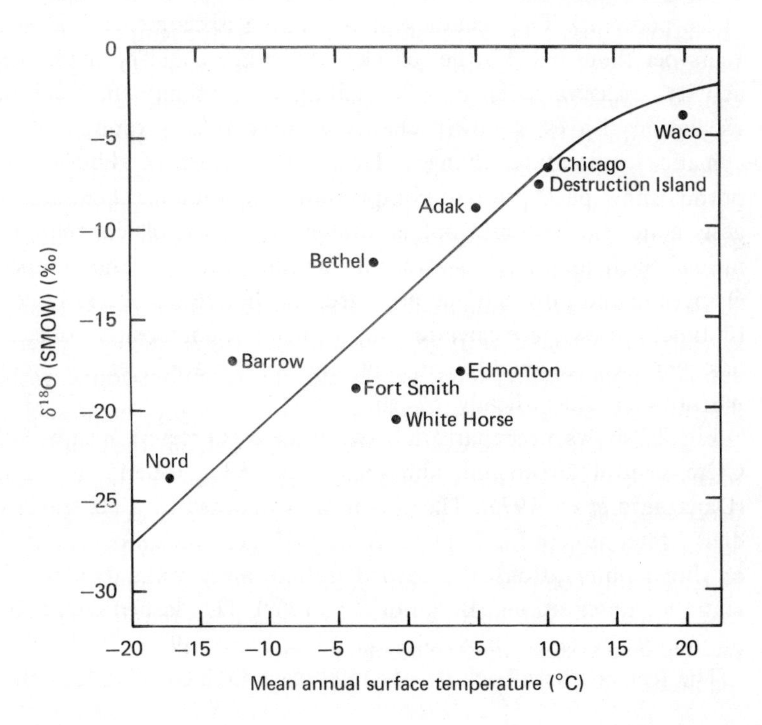

vapour as it is carried in an air mass (but see also Merlivat & Jouzel, 1979 and Siegenthaler & Oeschger, 1980).

(*a*) latitude effect: lower δs in precipitation at higher latitude;

(*b*) altitude effect: lower δs in precipitation at higher altitude;

(*c*) seasonal effect: lower δs in winter precipitation than in summer precipitation because in winter the conditions in the subtropical source area of the moisture are approximately the same as in summer, but during their movement towards higher latitudes the air masses cool faster and deeper in winter than in summer;

(*d*) climate effect: lower δs under cold climatic conditions than under warm conditions.

The dating of polar ice cores is a relatively easy matter in the case where a detailed isotopic analysis of the ice core reveals a seasonal δ-cycle and thereby the annual layering. In this way, a core from a region of high accumulation and no melting can be dated absolutely, simply by counting the annual layers downward from the top. Fig. 2.2 (W. Dansgaard, personal communication) shows a δ-sequence from Milcent, central Greenland which has been dated by counting summer maxima downward from the surface. The mean distance between a series of δ-minima reveals the mean annual precipitation (in this case about 54 cm of ice per year). The accuracy of the dating is claimed to be $\pm$ a few years per thousand. Further checks on the age could be made using β-activity measurements or ^{14}C dating. Smoothing the seasonal δ-oscillations leaves a slowly changing curve that is thought to reveal climatic temperature changes. However, because of diffusion in the porous snow pack prior to compression into solid ice, Dansgaard suggests conclusions should not be drawn on a basis of less than 10-year means. Such long-term δ-variations should reveal secular temperature changes at the snow surface provided: (*a*) the area is of nearly constant (in time) snow surface elevation, (*b*) the isotopic composition of sea water has remained reasonably constant, and (*c*) the atmospheric circulation has not changed radically.

Fig. 2.3 shows a comparison between the $\delta^{18}O$ record in snow fallen at Crête, central Greenland, and temperatures for Iceland and England (Dansgaard *et al.*, 1975). The curves are smoothed by a 60-year lowpass digital filter, except for England AD 800–1700. The full curves are based on direct observations, the dashed–dotted line is estimated from systematic ice observations (Bergthórsson, 1969). The dashed curves depend on indirect evidence (Bergthórsson, 1969; Lamb, 1966).

The Iceland curve back to AD 1590 has a high correlation with the δ-

curve ($r = 0.88$) as does the directly measured part of the England curve (back to AD 1680 ($r = 0.81$)). However, the extremely cold period AD 1550–1700 in England apparently has no parallel in central Greenland.

These examples illustrate the applicability of stable-isotope studies of

Fig. 2.2. Seasonal $\delta^{18}O$ variations in ice from Milcent, central Greenland, deposited AD 1210–40 dated by counting δ summer maxima downward from the surface. (Redrawn from unpublished results from W. Dansgaard)

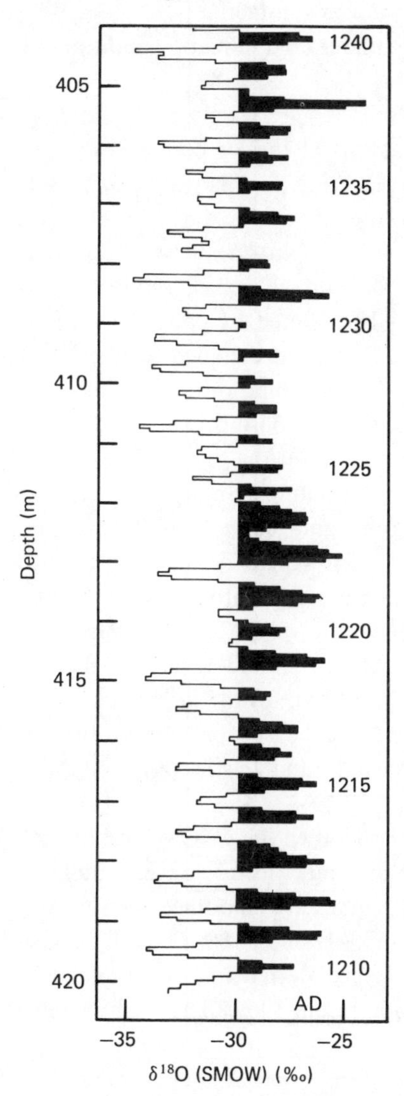

Fig. 2.3. Comparison between $\delta^{18}O$ curve for snow fallen at Crête, central Greenland, and temperature for Iceland and England. (Redrawn from Dansgaard *et al.*, 1975)

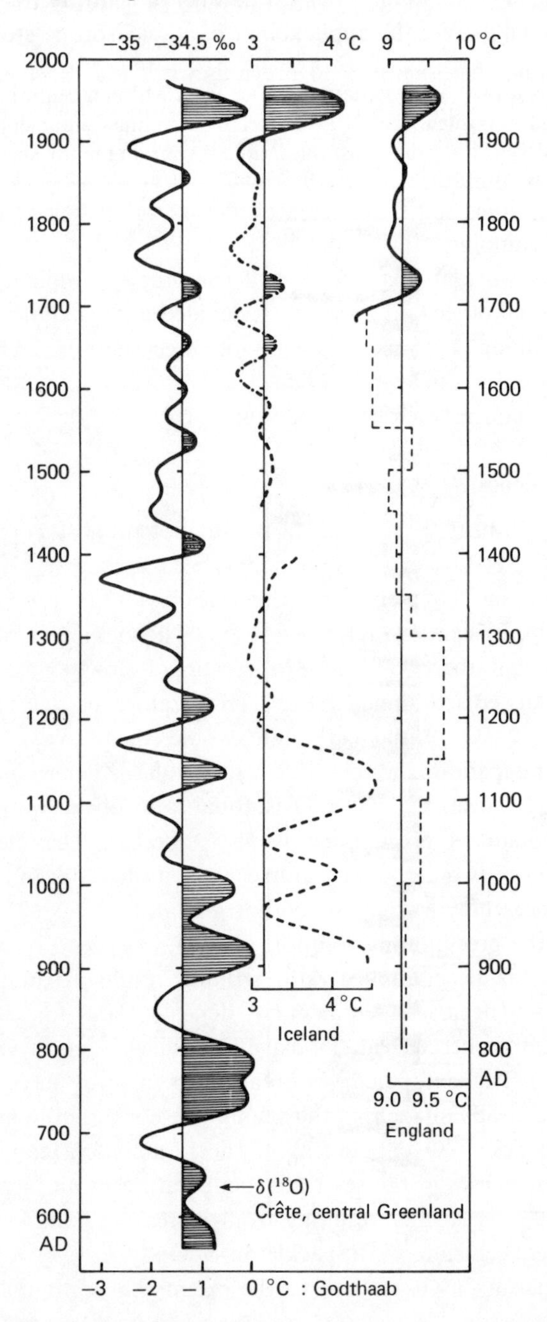

ice cores from Greenland to the determination of past climate. In Antarctica, however, it has recently been shown by Kato (1978) that the transportation of water vapour to the Antarctic ice sheet is one of the most important factors controlling the ^{18}O composition of Antarctic snow. It was found that, contrary to previous work (e.g. Epstein *et al.*, 1965) where the ^{18}O composition of precipitation was related only to its temperature of formation, the ^{18}O composition of the snow is largely controlled by the supply of ^{18}O-rich water vapour resulting from the approach of a cyclone and is related strongly to the distance from the open sea to the sampling station.

It is clear from work of this kind that, despite the enormous achievements and potential of ice core research, a detailed understanding of the processes controlling the ^{18}O content of precipitation and post-depositional processes is necessary for each site before the $\delta^{18}O$ data can be unequivocally converted to a temperature curve.

$CaCO_3$ deposits

The temperature dependence of the fractionation of oxygen isotopes in the CO_2–H_2O–$CaCO_3$ system has been used extensively for palaeotemperature measurements since the method was first proposed by Urey, Lowenstam, Epstein & McKinney (1951). The method is based on the observation that the $^{18}O/^{16}O$ isotope ratio of slowly precipitated calcium carbonate is determined by the temperature of the solution. Measurement of $\delta^{18}O$ of the carbonate would then give a precise temperature, if temperature were solely responsible in determining the $^{18}O/^{16}O$ ratio. However, the $^{18}O/^{16}O$ isotopic ratio of the water from which the carbonate is precipitated is also reflected. The measured difference in $\delta^{18}O$ of two carbonate samples is then the result of changes in both the temperature at which mineral precipitation occurred and the ^{18}O content of the precipitating solution.

This method has been applied with striking results to planktonic foraminifera from deep-sea sedimentary deposits (see for example Emiliani, 1966; Broecker & van Donk, 1970; Shackleton & Opdyke, 1973; Erez, 1979). The isotope curves obtained in this way have always suffered from the disadvantage that they cannot be converted to absolute temperatures because ^{18}O enrichment of the foraminifera can be the result of either decreases in the sea-water temperature or an increase in the ^{18}O content of sea water resulting from increasing continental ice volumes which concentrate ^{16}O (Shackleton & Opdyke, 1973).

A further criticism has been that, to the extent that foraminifera do indicate the palaeotemperatures at all, it is oceanic temperatures that are indicated whereas it is on the continents where most of the other climate indicators and climate records are found.

It has recently been shown, however, that the carbonate-water isotope thermometer may be applied to continental deposits. Two methods have emerged: Stuiver (1970) has shown that carbonates and molluscs from lake sediments in North America show increased ^{18}O content during the Hypsithermal interval (around 6000 BP). This is interpreted as suggesting higher mean annual temperatures for this interval than found at present. The interpretation is complicated by variations in ^{18}O content of the water which are in the opposite direction to and of the same magnitude as the changes in fractionation due to temperature.

The second method involves $\delta^{18}O$ measurements of $CaCO_3$ deposited as speleothems or cave travertines (see for example Hendy & Wilson, 1968; Hendy, 1971; Thompson, Schwarcz & Ford, 1974, 1975; Harmon, Thompson, Schwarcz & Ford, 1978). Speleothems are deposits of calcium carbonate (calcite or aragonite) formed in limestone caves. Such caves are found in widespread areas of the continents and on oceanic islands. Temperatures in the inner parts of such caves usually represent mean annual temperature at the ground surface above the cave.

Speleothems can form in oxygen isotopic equilibrium with the seepage water from which they precipitate provided the speleothem grows in an area having no direct contact with the external atmosphere and where there is little movement of air through it. This ensures the cave atmosphere is nearly saturated with water vapour and has a high CO_2 partial pressure. Under these conditions equilibrium precipitation of $CaCO_3$ occurs as a result of slow exsolution of CO_2 from the seepage water.

$$Ca^{2+} + 2HCO_3^- \leftrightarrows H_2O + CO_2 + CaCO_3\downarrow.$$

Hendy (1971) has suggested two criteria for recognising speleothems formed under these conditions:

(*a*) uniformity of $^{18}O/^{16}O$ ratios along a single growth layer and

(*b*) lack of correlation between $^{18}O/^{16}O$ and $^{13}C/^{12}C$ ratios along a single growth layer.

Under equilibrium conditions the $^{18}O/^{16}O$ ratio of $CaCO_3$ (p) and of the water (ω) are related to the temperture of deposition (T; deg K) by the expression (O'Neil, Clayton & Mayeda, 1969):

$$\ln[(^{18}O/^{16}O)p/(^{18}O/^{16}O)\omega] = AT^{-2} + B.$$

If the constants A and B are known, then measurement of $^{18}O/^{16}O$ yields temperature data with a theoretical precision of about 0.5 deg C.

However, except for recent deposits, the isotopic composition of the seepage water will not be known and assumptions about the effect of changing climate on the isotopic composition of meteoric precipitation must be made before a temperature scale can be attached to the isotope record. This complication arises because decreases in the mean global temperature (as well as local temperature at the site of deposition) have at least three distinct effects on the isotopic composition of the speleothem:

(*a*) an increase in the isotopic fractionation factor $\alpha(CaCO_3–H_2O)$ (O'Neil *et al.*, 1969);

(*b*) a decrease in the $^{18}O/^{16}O$ ratio of precipitation (Dansgaard, 1964); and

(*c*) an increase in the $^{18}O/^{16}O$ ratio of sea water owing to storage of isotopically light water on the continents as glacial ice (Shackleton & Opdyke, 1973).

Attempts have been made (Schwarcz, Harmon, Thompson & Ford, 1976; Harmon, Schwarcz & O'Neil, 1979) to infer the $^{18}O/^{16}O$ ratio of the seepage water by measurements of the D/H ratio of fluid inclusions extracted from the calcium carbonate. ($^{18}O/^{16}O$ ratios of fluid inclusions cannot be measured directly at this time and it is possible that over long periods of time ^{18}O exchange may occur between the included water and the carbonate.) Two implicit assumptions must be made. First, the extracted water is assumed to have the same isotopic composition as the water from which the $CaCO_3$ precipitated (i.e. the speleothem is a closed system); and, secondly, the $^{18}O/^{16}O$ ratio of the water can be inferred from its measured D/H ratio. This inference is usually based on the relation $\delta D = 8\delta^{18}O + 10$ (Craig, 1961), but it is well known that both the slope and intercept of this equation vary in different regions (Hage *et al.*, 1975). However, using this relation to infer $\delta^{18}O$ values of the fluid inclusions and hence the seepage water and by measuring $\delta^{18}O$ of the carbonate deposits, Schwarcz *et al.* (1976) have calculated a temperature of deposition similar to that actually measured in a particular cave.

The dating of speleothems is usually carried out using the $^{230}Th/^{234}U$ method (Thompson *et al.*, 1975; Nguyen & Lalou, 1969; Harmon, Thompson, Schwarcz & Ford, 1975) or the ^{14}C method. For deposits laid down 10 000 years to 300 000 years ago the $^{230}Th/^{234}U$ method is most useful. Deposits with ages less than 10 000 years can only be dated by this method where U concentrations are relatively high (10–100 ppm). Such deposits do exist, but the precision of the dates is of the order ± 200 years at best.

The use of the ^{14}C dating technique is best suited to deposits laid down in the last 35 000 years. However, the carbon deposited as carbonate on a stalagmite consists of a mixture of ancient carbon from the limestone carbonate, which contains essentially no ^{14}C, and modern carbon, respired by plant roots (Wilson, Hendy & Reynolds, 1979). Assuming saturation of $CaCO_3$ in the water percolating through the limestone into the cave, a little more than half the carbon in the resultant solution would be of recent biogenic origin and a little less than half derived from limestone (Hendy, 1971). This theoretical situation is not always the case due to exchange of CO_2 with cave air. Hendy (1971) has described how corrections can be made using $^{13}C/^{12}C$ ratios. In the case of a stalagmite which is still growing when collected, an even more accurate estimate can be made by determining the ^{14}C content of the outer layer of the stalagmite.

This approach has been used by Wilson *et al.* (1979) to obtain climate data from stalagmites growing in a cave in New Zealand. The results are shown in Fig. 2.4. The $\delta^{18}O$ measurements were made on samples taken at regular intervals down the axis of the stalagmite and a time base was obtained by ^{14}C dating. The resulting isotope–temperature curve has been compared with the curve given for central England by Lamb (1965). The isotope curve is scaled in terms of temperature by calibrating with modern meteorological records at a site near the cave. It is believed the estimate agrees well with the known temperature dependence of the fractionation factor of calcite suggesting little or no change in the seepage water composition took place and that $\delta^{18}O$ fluctuations in the stalagmite studied seem to be mainly due to temperature changes. The two curves appear to be broadly similar suggesting that such climate fluctuations as the Medieval Warm Period and the Little Ice Age are not just local European phenomena.

In summary, it appears that isotopic analysis of speleothems is a promising approach to determining past climates provided dating techniques can be applied to rapidly growing speleothems with sufficient accuracy to give good time resolution. The isotope chemistry is relatively simple, and it is possible that fluid inclusion work can be extended and improved to yield reliable information on the isotopic history of the seepage water.

Biological indicators

The use of isotope data obtained from biological systems to determine past climate is a recent development. This reflects the complex isotopic–geochemical problems involved. However, high rates of deposition and good time resolution suggest great potential for these methods. Furthermore, biological systems present intriguing possibilities in that they contain three potential isotope indicators which in any given system may be independent of one another (Libby, 1972). These are the isotope ratios $^{13}C/^{12}C$, $^{18}O/^{16}O$ and D/H, all of which may be measurable in a single sample. The potential exists, therefore, for measuring the isotope ratio of any two of the elemental constituents in a plant and establishing an absolute temperature relationship.

The systems studied to date are tree rings and to a lesser extent peat deposits. Tree-ring isotopes have great potential as climate indicators for a number of reasons: (*a*) they have a time resolution of at least one year; (*b*) they may be dated absolutely to one year; (*c*) a long continuous time period is available for investigation (the bristlecone pine chronology from the White Mountains in California, for example, has a length of

Fig. 2.4. Fifty-year running mean $\delta^{18}O$ profile through a stalagmite from New Zealand (*a*) is compared with Lamb's (1965) curve for central England (*b*) (Wilson *et al.*, 1979). PDB (Pee Dee Belemnite) is a fossil shell used as a standard.

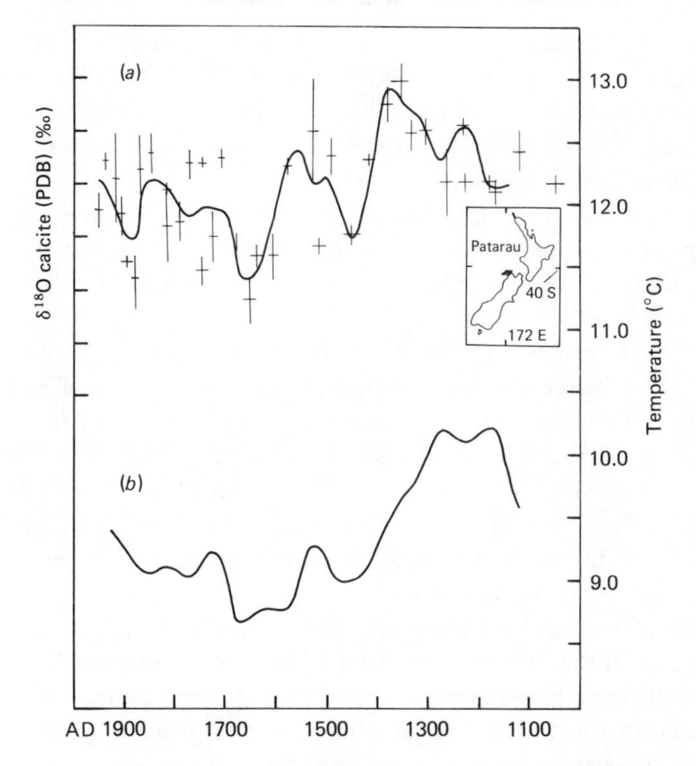

7000 years (Ferguson, 1968)); and (*d*) samples are widely distributed over almost all continental areas.

The factors which might be expected to relate isotope ratios of plant material to climatic parameters are complex and include: (*a*) temperature-dependent fractionation factors for the metabolic reactions occurring during photosynthesis and subsequent conversion of sugars which may affect ^{13}C, ^{18}O and D concentrations; (*b*) variations in the relationship between ground-meteoric water and local temperature, which will affect D and ^{18}O concentrations; and (*c*) kinetic isotope effects associated with evapotranspiration of leaf water which will affect D and ^{18}O concentrations. It is possible therefore that climatic factors such as mean annual temperature, local temperature during growth period, relative humidity, wind speed, amount of precipitation and amount of solar radiation may play an interdependent role in determining isotope ratios in plant material. Fortunately the results obtained to date suggest that isotope ratios of plant materials may be interpreted in somewhat simpler terms than these.

Peat studies – whole plant isotope ratios

In the early 1970s Schiegl reported measurements of D/H ratios of organic matter (Schiegl & Vogel, 1970; Schiegl, 1972, 1974) including peat deposits and tree rings. These results (and others subsequently reported by Libby *et al.* (1976)) were obtained using whole plant material and suffer from two complications.

First, whole plant material in general consists of more than one chemical compound. Wood for instance consists of roughly 40–50 per cent cellulose and hemicelluloses, the rest being mainly lignin with small amounts of other material. It has been shown by Gray & Thompson (1977) that the lignin fraction of whole wood is some 10‰ lighter in ^{18}O than the cellulose fraction and that the relative amounts of lignin and cellulose in wood are not climate dependent. Thus, while a strong correlation between $\delta^{18}O$ of cellulose and mean annual temperature has been found (Gray & Thompson, 1976), no significant correlation exists between the lignin ^{18}O isotopic composition and temperature. Analysis of whole wood was shown to result in much poorer correlations with temperature than with cellulose.

Secondly, it has been shown (Epstein, Yapp & Hall, 1976; Epstein & Yapp, 1977) that about 30 per cent of the hydrogen atoms in cellulose are exchangeable and tend to reflect the isotopic composition of the water with which they last came in contact, thus masking any climate signal retained by the hydrogen atoms bound to carbon.

Nevertheless, Schiegl was able to establish that the natural deuterium content of organically bound hydrogen shows a systematic variation with the D/H ratio of precipitation. Since this is often related to local mean annual temperature (Dansgaard, 1964) he established that plant material may record climate variations. The method was subsequently applied to peat deposits in the Netherlands (Schiegl, 1972).

The chemical transformation of plant debris to mature peat is effected by microbial agents, chiefly bacteria and fungi. The change is accompanied by an increase in carbon content of the organic substance. This introduces a complication into the interpretation of isotope ratios of whole peat which must be compensated for (Schiegl, 1972). The results showed that by comparing isotope results (corrected for chemical changes) with pollen analysis of the peat samples a relation between δD of the whole peat material and temperature was evident.

Peat cellulose studies

Recently, analyses of α-cellulose extracted from peat samples have been carried out which overcome difficulties due to varying chemical composition (Chatwin, 1979). Measurements of $\delta^{18}O$ were made on a 3.5 m frozen peat core taken near Fort Simpson, Northwest Territories, Canada. Prior ^{18}O analysis of cellulose from living plants, representing all the major peat-forming plants at the site showed them to have the same $^{18}O/^{16}O$ ratio so that ^{18}O analysis was carried out on cellulose extracted from the bulk peat. The preliminary $\delta^{18}O$ measurements together with ^{14}C ages are shown in Fig. 2.5. Peat accumulation began 10 380 years ago and terminated approximately 400 years ago. The $\delta^{18}O$ curve is tentatively extrapolated to present values from living plants.

A hiatus, possibly due to a period of freezing and uplift or to drying of the peat land is found at a depth of 1.6 m. This level is marked by a 2 cm layer of woody nanolignin peat. It is of interest that the change of peat type from sphagnum to a forest peat occurs immediately above the layer corresponding to the minimum in the $\delta^{18}O$ curve (presumably corresponding to the minimum in temperature) which occurred between 4000 and 3000 BP. This suggests a brief period of permafrost invasion and colonisation by black spruce. A maximum variation in $\delta^{18}O$ of 3.25‰ is found for the 10 000-year period. The results suggest a relatively cool period from 10 000 to 7000 BP with a climatic warming following and culminating at about 5500 BP, the temperatures being apparently less than present-day temperatures at all times.

In an attempt to place a temperature scale on the isotope results, samples of sphagnum moss from various sites across Canada have been analysed for ^{18}O content. These results, together with mean annual temperatures for the collection sites, are shown in Fig. 2.6. The data are reasonably defined by the equation: $\delta^{18}O = 0.52t + 20.2$ (correlation coefficient 0.98). These results will be discussed further in the section dealing with tree-ring work.

Tree rings

Early isotope work on tree rings (Schiegl, 1974; Libby *et al.*, 1976; Farmer & Baxter, 1974; Freyer & Weisberg, 1975; Pearman, Francey & Fraser, 1976) was reported on whole wood. As discussed

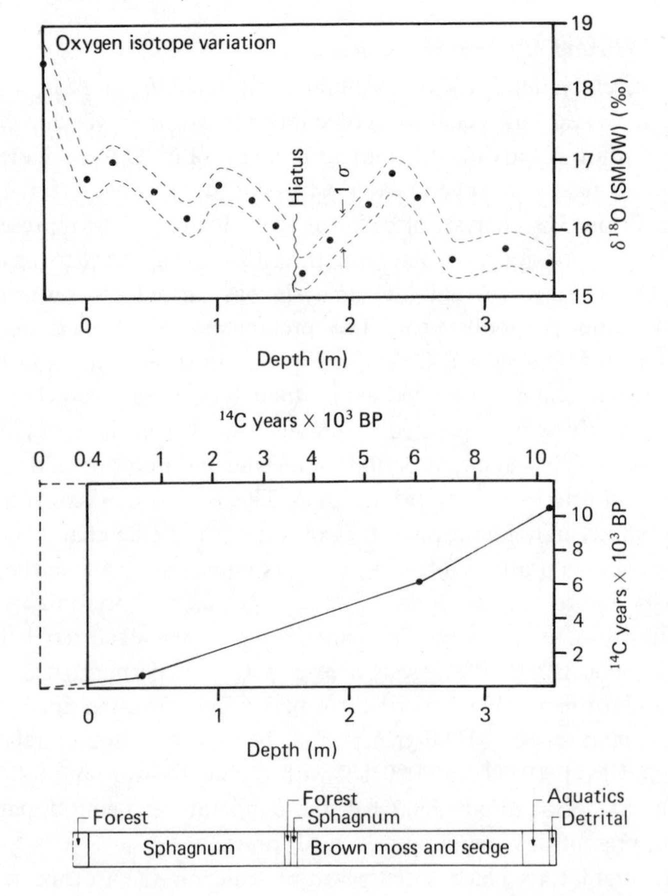

Fig. 2.5. $\delta^{18}O$ measurements on a peat core from Northwest Territories, Canada, together with the ^{14}C age profile. (Redrawn from Chatwin, 1979)

above, this is unsatisfactory, and most of the recent measurements have been made on α-cellulose.

${}^{13}C/{}^{12}C$ studies on tree-ring α-cellulose

${}^{13}C/{}^{12}C$ ratios have been reported for α-cellulose and lignin extracted from a few rings of *Pinus radiata* from New Zealand (Wilson & Grinsted, 1977). By correlating with seasonal temperatures, a temperature coefficient for both of about 0.2‰/deg is obtained. The results (Fig. 2.7) shows that lignin is generally 3‰ lower than cellulose and, since the lignin to cellulose ratio varies in a tree, the use of whole wood can introduce additional scatter. It is postulated that air temperature causes the observed variation of $\delta^{13}C$ and observed seasonal changes in atmospheric $\delta^{13}C$ (Keeling, 1960) are ruled out since the tree-ring variation is in the opposite direction and an order of magnitude larger.

D/H ratio of tree-ring α-cellulose

The use of D/H ratios of α-cellulose from biological material is complicated by the presence of hydroxyl (–C—OH) hydrogen atoms which are exchangeable with sap water in addition to non-exchangeable –C—H hydrogen atoms. Epstein & Yapp (1976) have described a method

Fig. 2.6. Variations of $\delta^{18}O$ measurements on cellulose extracted from modern sphagnum moss samples with mean annual temperature at the collection site. (Redrawn from Chatwin, 1979)

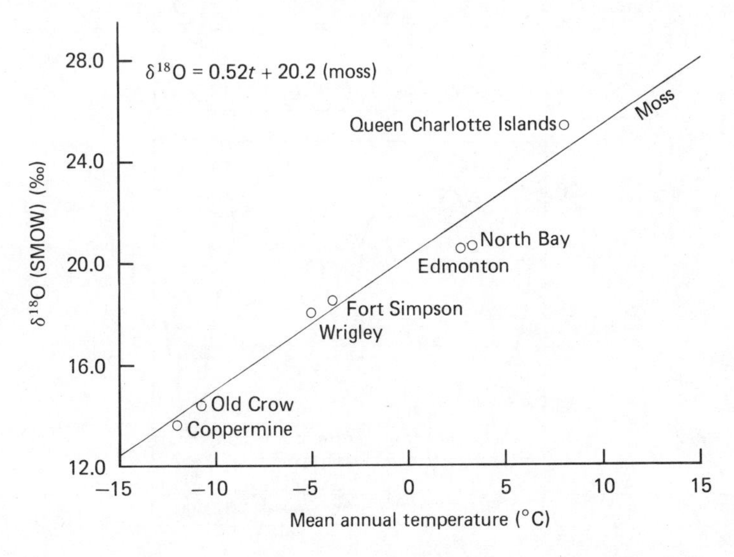

of removing the exchangeable hydrogen atoms by nitrating the hydroxyl groups and measuring the δD variations in the remaining –C—H hydrogens. Since the H atoms in –C—OH groups are exchangeable with sap water (and with reagents in the laboratory) they carry no useful climate information and indeed may tend to obscure the climate information carried by 'ring' hydrogens written as –C—H. Thus they are eliminated by nitration (or any other method which substitutes a non-hydrogen containing group) so that the hydrogen atoms originally in the C—OH group are *lost* and are not measured. What is measured is the δD of the cellulose nitrate resulting from this reaction. A positive correlation between the D/H ratios of the cellulose nitrate and D/H ratios of meteoric water supplying the tree was obtained. This suggests that a positive correlation with mean annual temperature will be found in many regions. Fig. 2.8 (Epstein & Yapp 1976) shows the δD record

Fig. 2.7. Variations in $\delta^{13}C_{PDB}$ in cellulose (*a*) and lignin (*b*) across tree rings from *Pinus radiata*. (Wilson & Grinsted, 1977)

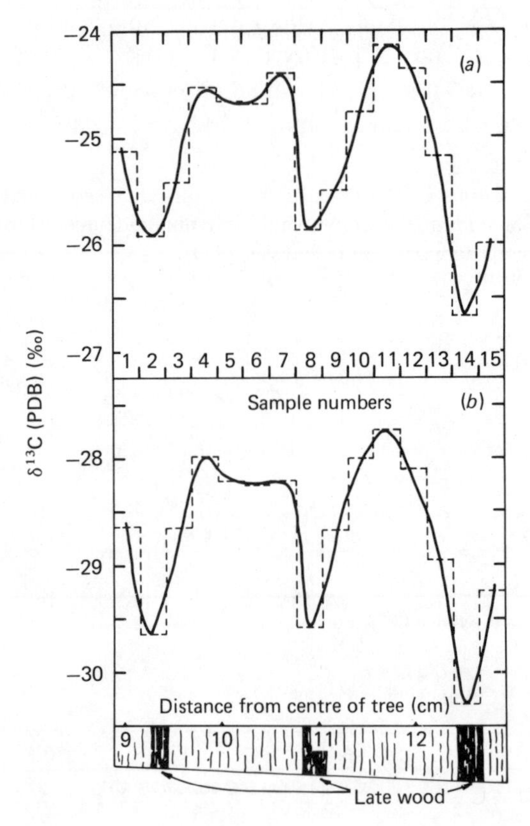

(10-year groups of rings) together with the 40-year running mean δD values in a 1000-year record obtained from two bristlecone pines whose growth periods overlapped by about 100 years, from the White Mountains in California. This record is compared with Lamb's (1966) winter temperature curve for central England and with absolute ring-width variations in a bristlecone pine from the White Mountains in California (La Marche, 1974).

Epstein & Yapp (1976) have also obtained δD records from a Scots

Fig. 2.8. (*a*) The δD record of isotopically non-exchangeable hydrogen of cellulose (cellulose nitrate) in two bristlecone pines which grew in the White Mountains of California. (*b*) The 40-year running average of the δD record. (*c*) Winter temperatures of central England (December, January, February) from a variety of indicators (Lamb, 1966). (*d*) Absolute ring width variations in a bristlecone pine from White Mountains, California (La Marche, 1974). (Epstein & Yapp, 1976)

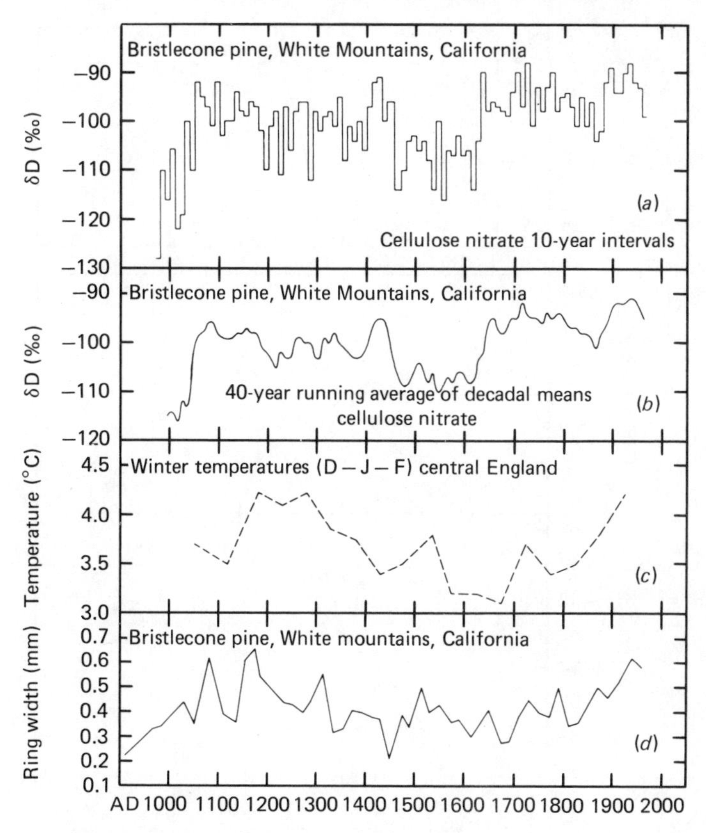

pine located near Loch Affric, Scotland (Fig. 2.9). The average δD value of the Scots pine is about $-60‰$ as compared to $-95‰$ in the bristlecone pine for the corresponding time interval. The difference is reasonably consistent with the difference of the δD values of meteoric waters in their respective localities, $-42‰$ for Loch Affric and -80 to $-100‰$ for the waters in the bristlecone pine area. A comparison of the 40-year running average of the δD data of the two pines produces a linear relationship which suggests that long-term isotopic climatic trends as recorded by the two species of pine are similar, while the short-term fluctuations are not related. Epstein concludes that widely different types of climatic environments will produce δD trends which might reflect even

Fig. 2.9. (*a*) The δD record of cellulose nitrate from 5-year segments of a Scots pine growing in Scotland. (*b*) The 40-year running average of decadal means for winter temperatures (October through March) of Edinburgh, Scotland. (*c*) The 40-year running average of decadal means of δD of the Scots pine. (Epstein & Yapp, 1976)

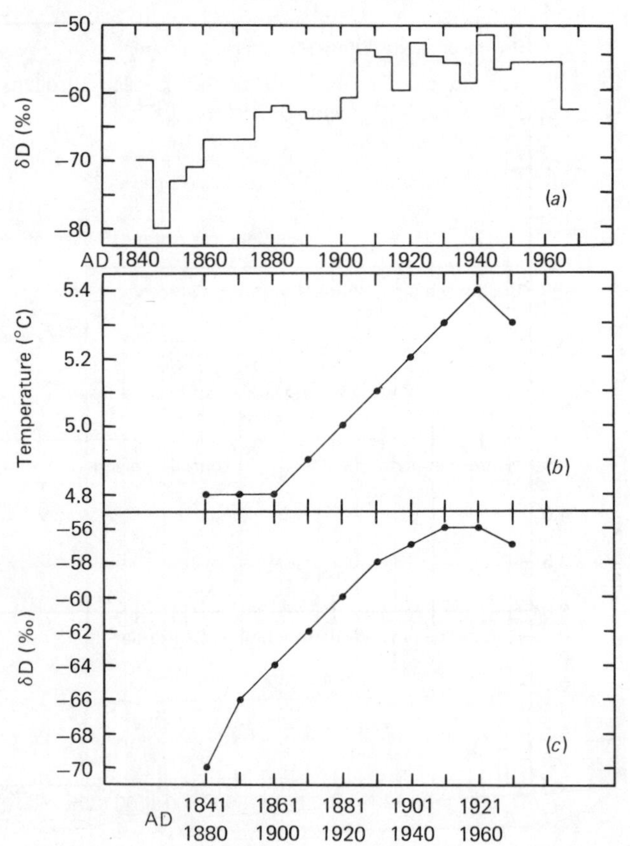

hemispheric long-term climatic changes, but superposed upon these long-term trends there may exist short-term δD variations which could reflect temperature changes over a much more restricted area.

A second method of dealing with the exchangeable hydrogen atoms in cellulose has been reported by Wilson & Grinsted (1975). Their approach is to use an isotope-dilution technique involving equilibration of the –C—OH hydrogen atoms with water of known hydrogen isotopic composition. This converts the –C—OH hydrogens to a known quantity so that the –C—H can be inferred by the difference between δD for total hydrogen and δD for the labelled water. Unfortunately, the results obtained by this method (Fig. 2.10) contradict those obtained by the nitration method in that a negative correlation between δD and temperature is obtained. This result is, furthermore, opposite to that expected on thermodynamic grounds. It is possible that the explanation may lie in the experimental procedures themselves. A. T. Wilson has pointed out (personal communication) that there are three types of hydrogen in wood α-cellulose: non-exchangeable –C—H, amorphous hydroxyl and crystal-

Fig. 2.10. Variation in isotopic ratio of the –C—H hydrogens in cellulose across tree rings. (Wilson & Grinsted, 1975)

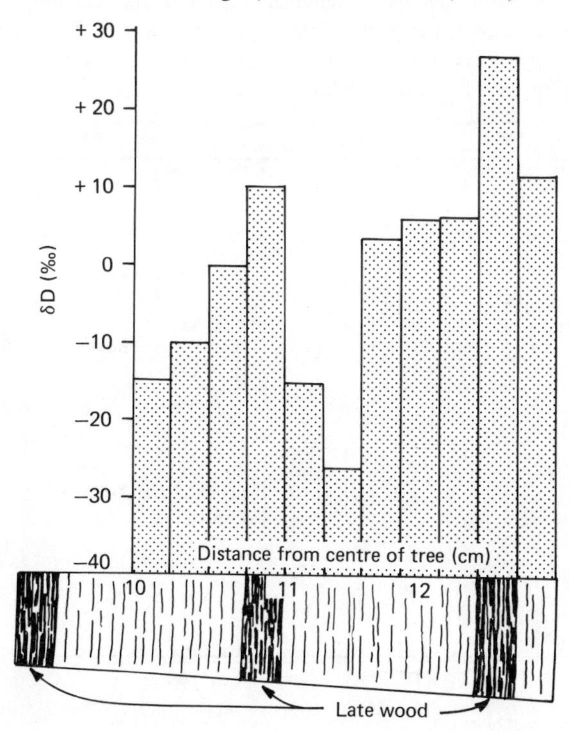

line hydroxyl atoms. Only the amorphous hydroxyl H atoms are exchangeable with sap water. Thus, while the nitration technique measures only –C—H atoms, the equilibration technique, after corrections for the isotopic composition of the equilibration water, measures all the hydrogen atoms in the cellulose. Thus, if the crystalline –OH hydrogen atoms have a very large negative temperature coefficient outweighing the positive temperature coefficient of the –C—H hydrogen atoms, the net result would be as found by Wilson. It appears that both techniques should be used on the same samples to test this hypothesis.

$^{18}O/^{16}O$ *Studies on tree-ring α-cellulose*

Oxygen isotope ratio measurements on α-cellulose extracted from trees have been reported by Gray & Thompson (1976, 1977) and Epstein, Thompson & Yapp (1977).

Gray & Thompson (1976) have reported $\delta^{18}O$ values of cellulose from successive 5-year groups of rings in a spruce tree which grew in Edmonton, Alberta (Fig. 2.11). A linear relationship between $\delta^{18}O$ and

Fig. 2.11. A comparison between the mean annual temperature (MAT) curve obtained from meteorological data and that obtained from $\delta^{18}O$ measurements of cellulose. The broken part of the isotope curve is calculated for some out-of-sequence rings that were not analysed. (Gray & Thompson, 1976)

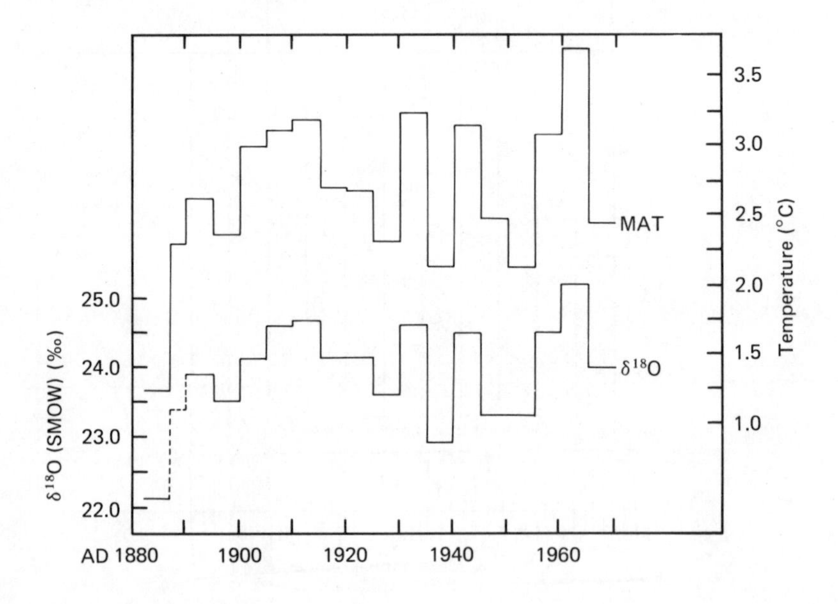

mean annual temperature was obtained (Fig. 2.12). This has since been demonstrated for other trees from the same site and from a number of other sites in North America.

The overall mean $\delta^{18}O$ and annual temperature data for a number of such sites are plotted in Fig. 2.13. The least squares equation defining these data is $\delta^{18}O = 0.54t + 22.9‰$ (SMOW) (correlation coefficient is 0.98). These data are compared with the sphagnum moss data mentioned above. The slopes of the two lines are the same within experimental error while the intercept of the peat $\delta^{18}O$ line is approximately 2.7‰ lower than that for the tree rings.

The linear relations between $\delta^{18}O$ of the tree-ring cellulose, the sphagnum moss cellulose and temperature suggest that it is the isotopic composition of the meteoric (soil) water which plays a dominant role in determining the ^{18}O content of the cellulose. The slope of the overall $\delta^{18}O$ v. mean annual temperature curve (0.5‰/deg C) is not the same as the equivalent slope at any individual site. Thus at Edmonton the slope is 1.2‰/deg C and Fort Vermillion 1.0‰/deg C. The reason for this is not clear. It appears that, in addition to the overall relation with meteoric water, local environmental factors or even microclimatic effects may play a role in determining the relation between $\delta^{18}O$ and tempera-

Fig. 2.12. Variations of $\delta^{18}O$ with September to August mean temperature for 5-year periods. The equation was obtained by a least-squares fit to the data points. (Gray & Thompson, 1976)

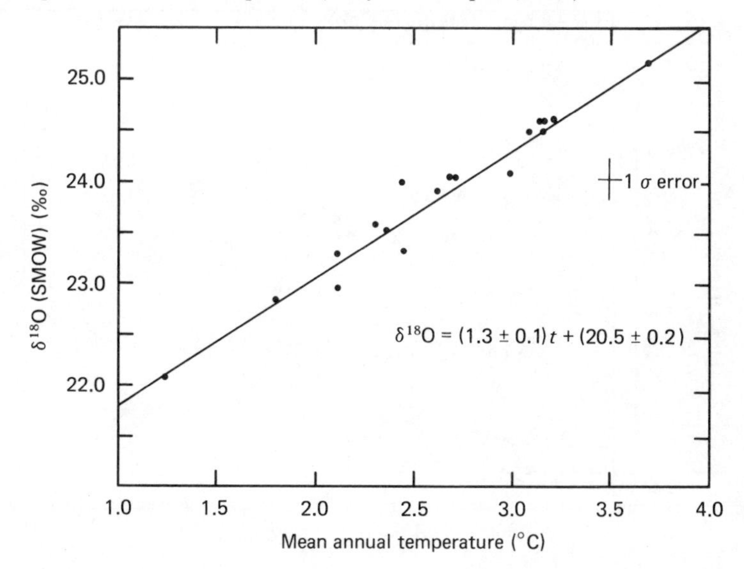

ture. Nevertheless, the empirical relations obtained here suggest that in many of the regions studied, temperature information is stored in the $\delta^{18}O$ of the cellulose.

This conclusion has been questioned by Epstein *et al.* (1977) who contend that kinetic isotope effects associated with evapotranspiration of water at the leaf surface play a dominant role in determining the $\delta^{18}O$ of the cellulose.

Evapotranspiration is known to be controlled mainly by water availability, relative humidity and wind speed and is only indirectly related to temperature (Craig, Gordon & Horibe, 1963). The linearity of the $\delta^{18}O$ v. mean annual temperature plots and particularly the similarity of the slopes of the tree-ring and sphagnum moss data (Fig. 2.13) suggest that enrichment by evapotranspiration must be relatively constant. At a single site this may simply reflect the constant average relative humidity over long periods of time. The apparently constant ^{18}O enrichment at widely separated sites is more difficult to explain. The sphagnum moss samples were collected in peat bogs where humidity is generally high and evapotranspiration will be minimised. It is possible

Fig. 2.13. Variation of cellulose $\delta^{18}O$ values with mean annual temperatures at various sites: ● tree rings; ○ sphagnum moss (peat producing plants); △ aquatic plants. (J. Gray and P. Thompson, unpublished results)

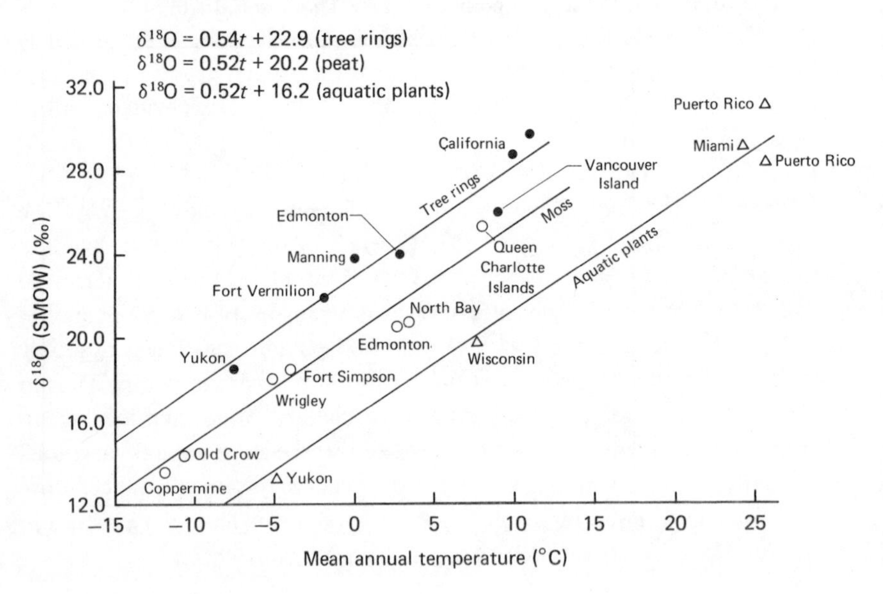

that the 2.7‰ displacement of the tree rings' $\delta^{18}O$ line reflects evapotranspiration effects. If this is the case it again points to a surprisingly constant enrichment from site to site. It also makes it unlikely that $\delta^{18}O$ values of cellulose will yield information on variation in relative humidity as has been suggested (Epstein *et al.*, 1977; DeNiro & Epstein, 1979). Aquatic plants have been shown to reflect the isotopic composition of the water in which they grow (Epstein *et al.*, 1977). Aquatic $\delta^{18}O$ data compiled from the literature (Epstein *et al.*, 1977) and from unpublished results obtained by P. Thompson while at the California Institute of Technology are plotted against estimates of mean annual temperature for the growth sites in Fig. 2.13 where they are compared with the tree-ring and sphagnum moss lines. Although poorly defined ($r = 0.98$ for five datum points) the slope of the aquatic data is essentially the same as that for the other two data sets. The intercept is about 4‰ below the sphagnum moss line and 6.7‰ below the tree-ring line. Aquatic plants cannot undergo evapotranspiration so the displacements of both the sphagnum moss and tree-ring line may be explained in terms of enrichment by evapotranspiration. This is possible only if the amount of enrichment is roughly constant over a wide range of mean annual temperatures. If this is the case the relation between $\delta^{18}O$ of cellulose in plant material and mean annual temperature demonstrated by Gray & Thompson (1976) is more firmly established.

Several models have been proposed to account for the oxygen isotopic composition of α-cellulose from trees and other plants. These models are based on the assumption that the oxygen of α-cellulose is derived solely from CO_2 (Gray & Thompson, 1976) or from oxygen atoms of CO_2 and H_2O in the ratio 2:1 (Epstein *et al.*, 1977). Recent evidence from laboratory experiments (DeNiro & Epstein, 1979) has shown the second model to be incorrect and that the α-cellulose oxygen is derived from CO_2 which has undergone complete exchange with the oxygen of the water in the plant. Thus the $^{18}O/^{16}O$ value of the water in the plant is the primary influence on the $^{18}O/^{16}O$ ratio of cellulose in terrestial plants. Further, since the effects of evapotranspiration are apparently rather constant, then the $^{18}O/^{16}O$ ratio of the meteoric water is probably the primary influence on the $^{18}O/^{16}O$ ratio of cellulose in plants. Thus in a region where a relation between ground-meteoric water and local mean annual temperature is well established, and where large annual variations in relative humidity do not occur, the $\delta^{18}O$ curve obtained from cellulose in tree rings can be interpreted in terms of mean annual temperature (see also Burk & Stuvier, 1981).

Conclusion

This review has concentrated on isotopic methods which yield climate information for relatively recent times (present to 5000 BP). It is necessarily incomplete therefore for two reasons. First, the interest in this area of research is such that new work is being reported at an unprecedented rate and, secondly, extensive studies are carried out on systems which yield climate information on a much longer time scale than is appropriate in this review. The best that can be achieved then is to outline the underlying principles and to select some of the more interesting applications for discussion.

It is clear that stable isotope work represents perhaps the best method available at this time for determining past climate. This is evidenced both by the results obtained and by the accelerating efforts being made around the world to find new natural systems whose isotope records may be interpreted in terms of climatic variables.

Nature has left these records and it is the scientist's challenge to interpret and understand them. Despite the difficulties the potential rewards are so large that it is clear that the effort must and will continue.

References

Bergthórsson, P. (1969). An estimate of drift ice and temperature in Iceland in 1000 years. *Jökull (Reykjavik)*, **19**, 94–101.

Bottinga, Y. (1968). Calculation of fractionation factors for carbon and oxygen isotopic exchange in the system calcite – carbon dioxide – water. *Journal of Physical Chemistry*, **72**, 800–8.

Broecker, W. S. & van Donk, J. (1970). Insolation changes, ice volumes, and the ^{18}O record in deep-sea cores. *Reviews of Geophysics and Space Physics*, **8**, 169–98.

Burk, R. L. & Stuiver, M. (1981). Oxygen isotope ratios in trees reflect mean annual temperature and humidity. *Science*, **211**, 1417–19.

Chatwin, S. C. (1979). Permafrost aggradation and degradation in Sub-Arctic peatland. PhD thesis, Department of Geology, University of Alberta, Edmonton, Canada.

Craig, H. (1961). Isotopic variations in meteoric waters. *Science*, **133**, 1702–3.

Craig, H., Gordon, L. I. & Horibe, Y. (1963). Isotopic exchange effects in the evaporation of water. *Journal of Geophysical Research*, **68**, 6079–87.

Dansgaard, W. (1964). Stable isotopes in precipitation. *Tellus*, **16**, 436–68.

Dansgaard, W., Johnsen, S. J., Clausen, H. B. & Gundestrup, M. (1973). Stable isotope glaciology. *Meddelelser om Grønland*, **197**, No. 2, 1–53.

Dansgaard, W., Johnsen, S. J., Reeh, N., Gundestrup, N., Clausen, H. B. & Hammer, C. U. (1975). Climatic changes, Norsemen and modern man. *Nature (London)*, **255**, 24–8.

DeNiro, M. J. & Epstein, S. (1979). Relationship between the oxygen isotope ratios of terrestrial plant cellulose, carbon dioxide and water. *Science*, **204**, 51–3.

Emiliani, C. (1966). Paleotemperature analysis of Caribbean cores P. 6304–8 and P. 6304–9 and a generalized temperature curve for the past 425000 years. *Journal of Geology*, **74**, 109–26.

Epstein, S., Sharp, R. P. & Gow, A. J. (1965). Six-year record of oxygen and hydrogen isotope variations in South Pole firn. *Journal of Geophysical Research*, **70**, 1809–14.

Epstein, S., Sharp, R. P. & Gow, A. J. (1970). Antarctic ice sheet: stable isotope analyses of Byrd Station cores and interhemispheric climatic implications. *Science*, **168**, 1570–2.

Epstein, S., Thompson, P. & Yapp. C. J. (1977). Oxygen and hydrogen isotopic ratios in plant cellulose. *Science*, **198**, 1209–15.

Epstein, S. & Yapp, C. J. (1976). Climatic implications of the D/H ratio of hydrogen in C–H groups in tree cellulose. *Earth and Planetary Science Letters*, **30**, 252–61.

Epstein, S. & Yapp, C. (1977). Isotope tree thermometers. *Nature*, **266**, 477–8.

Epstein, S., Yapp, C. J. & Hall, J. H. (1976). The determination of the D/H ratio of non-exchangeable hydrogen in cellulose extracted from aquatic and land plants. *Earth and Planetary Science Letters*, **30**, 241–51.

Erez, J. (1979). Modification of the oxygen-isotope record in deep-sea cores by pleistocene dissolution cycles. *Nature*, **281**, 535–8.

Farmer, J. G. & Baxter, M. S. (1974). Atmospheric carbon dioxide levels as indicated by the stable isotope record in wood. *Nature*, **247**, 273–5.

Ferguson, C. W. (1968). Bristlecone pine. Science and esthetics. *Science*, **159**, 839–46.

Freyer, H. D. & Weisberg, L. (1975). Anthropogenic carbon-13 decrease in atmospheric CO_2 as recorded in modern wood. In *Isotope Ratios as Pollutant Source and Behaviour Indicators*, IAEA Vienna, pp. 49–62. New York: Unipub.

Gray, J. & Thompson, P. (1976). Climatic information from $^{18}O/^{16}O$ ratios of cellulose in tree rings. *Nature*, **262**, 481–2.

Gray, J. & Thompson, P. (1977). Climatic information from $^{18}O/^{16}O$ analysis of cellulose, lignin and whole wood from tree rings. *Nature*, **270**, 708–9.

Hage, K. D., Gray, J. & Linton, J. C. (1975). Isotopes in precipitation in Northwestern North America. *Monthly Weather Review*, **103**, 958–66.

Hammer, C. U., Clausen, H. B., Dansgaard, W., Gundestrup, N., Johnsen, S. J. & Reeh, N. (1978). Dating of Greenland ice cores by flow models, isotopes, volcanic debris and continental dust. *Journal of Glaciology*, **20**, 3–26.

Harmon, R. S., Schwarcz, H. P. & O'Neil, J. R. (1979). D/H ratios in speleothem fluid inclusions: A guide to variations in the isotopic composition of meteoric precipitation? *Earth and Planetary Science Letters*, **42**, 254–66.

Harmon, R., Thompson, P. T., Schwarcz, H. P. & Ford, D. C. (1975). Uranium series dating of speleothems: a review. *National Speleological Society Bulletin*, **37**, 21–33.

Harmon, R., Thompson, P., Schwarcz, H. P. & Ford, D. C. (1978). Late pleistocene paleoclimates of North America as inferred from stable isotope studies of speleothems. *Quaternary Research*, **9**, 54–70.

Hendy, C., (1971). The isotopic geochemistry of speleothems – 1. The calculation of the effects of different modes of formation on the isotopic composition of speleothems and their applicability as paleoclimate indicators. *Geochimica et Cosmochimica Acta*, **35**, 801–24.

Hendy, C. H. & Wilson, A. T. (1968). Paleoclimatic data from speleothems. *Nature*, **219**, 48–51.

Kato, K. (1978). Factors controlling oxygen isotopic composition of fallen snow in Antarctica. *Nature*, **272**, 46–8.

Keeling, C. D. (1960). The concentration and isotopic abundance of carbon dioxide in the atmosphere. *Tellus*, **12**, 200–3.

La Marche, Jr, V. C. (1974). Paleoclimatic inferences from long tree-ring records. *Science*, **183**, 1043–8.

Lamb, H. H. (1965). The early medieval warm epoch and its sequel. *Palaeogeography, Palaeoclimatology and Palaeoecology*, **1**, 13–37.

Lamb, H. H. (1966). *The Changing Climate*, Ch. 7. London: Methuen.

Libby, L. M. (1972). Multiple thermometry in paleoclimate and historic climate. *Journal of Geophysical Research*, **77**, 4310–17.

Libby, L. M., Pandolfi, L. J., Payton, P. H., Marshall III, J., Becker, B. & Giertz-Sienbenlist, V. (1976). Isotopic tree thermometers. *Nature*, **261**, 284–8.

Merlivat, L. Jouzel, J. (1979). Globel Climatic Interpretation of Deuterium-Oxygen 18 Relationship for Precipitation. *Journal of Geophysical Research*, **84,** 5029–33.

Nguyen, H. V. & Lalou, C. (1969). Comportement géochimique des isotopes des familles de l'uranium et du thorium dans les concrétionnements de grottes: application à la datation des stalagmites. *Comptes Rendus Paris*, **269**, 560–3.

O'Neil, J. R., Clayton, R. N. & Mayeda, T. K. (1969). Oxygen isotope fractionation in divalent metal carbonates. *Journal of Chemical Physics*, **51,** 5547–58.

Pearman, G. I., Francey, R. J. & Fraser, P. J. B. (1976). Climatic implications of stable carbon isotopes in tree rings. *Nature*, **260**, 771–3.

Schiegl, W. E. (1972). Deuterium content of peat as a paleoclimatic recorder. *Science*, **175**, 512–13.

Schiegl, W. E. (1974). Climatic significance of deuterium abundance in growth rings of Picea. *Nature*, **251**, 582–4.

Schiegl, W. E. & Vogel, J. C. (1970). Deuterium content of organic matter. *Earth and Planetary Science Letters*, **7**, 307–13.

Schwarcz, H. P., Harmon, R. S., Thompson, P. & Ford, D. C. (1976). Stable isotope studies of fluid inclusions in speleothems and their paleoclimatic significance. *Geochimica et Cosmochimica Acta*, **40**, 657–65.

Shackleton, N. J. & Opdyke, N. D. (1973). Oxygen isotope and paleomagnetic stratigraphy of equatorial Pacific core V28-238: oxygen isotope temperatures and ice volumes on a 10^5 and 10^6 year scale. *Quaternary Research*, **3**, 39–55.

Siegenthaler, U. & Oeschger, H. (1980). Correlation of ^{18}O in precipitation with temperature and altitude. *Nature*, **285**, 314–17.

Stuiver, M. (1970). Oxygen and carbon isotope ratios of fresh-water carbonates as climatic indicators. *Journal of Geophysical Research*, **75**, 5247–57.

Thompson, P., Ford, D. C. & Schwarcz, H. P. (1975). $^{234}U/^{238}U$ ratios in limestone cave seepage waters and speleothem from West Virginia. *Geochimica et Cosmochimica Acta*, **39**, 661–9.

Thompson, P., Schwarcz, H. P. & Ford, D. C. (1974). Continental pleistocene climatic variations from speleothem age and isotopic data. *Science*, **184**, 893–5.
Urey, H. C., Lowenstam, H. A., Epstein, S. & McKinney, C. R. (1951). Measurement of paleotemperatures and temperatures of the upper cretaceous of England, Denmark, and the Southern United States. *Geological Society of America Bulletin*, **62**, 399–425.
Wilson, A. T. & Grinsted, M. J. (1975). Paleotemperatures from tree rings and the D/H ratio of cellulose as a biochemical thermometer. *Nature*, **257**, 387–8.
Wilson, A. T. & Grinsted, M. J. (1977). $^{13}C/^{12}C$ in cellulose and lignin as palaeothermometers. *Nature*, **265**, 133–5.
Wilson, A. T., Hendy, C. H. & Reynolds, C. P. (1979). Short-term climate change and New Zealand temperatures during the last millenium. *Nature*, **279**, 315–17.

3

Glaciological evidence of Holocene climatic change

STEPHEN C. PORTER

Abstract

Because glaciers typically respond to changes in their climatic environment by growing or shrinking in size, they can be used as sensitive palaeoclimatic indicators. Evidence gained from studies of ice cores and glacial deposits permits the construction of continuous or discontinuous time series from which both the direction and magnitude of climatic change can be inferred. However, glacier surges unrelated to climate, lags in the dynamic response of a glacier terminus to a climatic change, and chronological uncertainties that result from problems in dating glacial deposits make assessment of the climatic significance of glaciological data difficult. In few areas do historical observations extend back beyond the nineteenth century, so the history of glacier activity is commonly reconstructed from morainal evidence and dated by radiocarbon, dendrochronology, or lichenometry. Radiocarbon dates frequently are ambiguous due to variations in the atmospheric production rate of ^{14}C, and uncertainties in the lag time between moraine stabilisation and establishment of tree seedlings may render tree-ring dating of limited value; however, fast-growing lichens often provide a basis for assigning close minimum limiting ages to glacial deposits that are less than about a thousand years old.

Evidence for repeated fluctuations of glaciers during the Little Ice Age of the last four to five centuries is widespread in glaciated alpine regions. Most glaciers achieved maxima in the seventeenth, eighteenth, or nineteenth centuries and began a period of marked recession during the second half of the nineteenth century. Apparent non-synchrony of second-order advances among geographic areas may reflect different climatic histories, differences in response lags of temperate and subpolar glaciers, or uncertainties in dating. Earlier intervals of glacier expansion similar in magnitude to the Little Ice Age occurred about 1100–1200, 2800–3000, and 5000–5300 years ago in a number of areas, but as yet no convincing well-dated global pattern has been demonstrated. During glacial maxima the snowline typically was lowered about 100–200 m, representing a depression equivalent to about 15 per cent of that at the maximum of the last glaciation.

An inadequate data base severely restricts any attempt to make a comprehensive global synthesis of Holocene glacier fluctuations. With additional carefully collected information from selected sites along three major latitudinal and longitudinal transects, important insights can be gained regarding the intra- and interhemispherical synchrony and relative magnitude of glacier advances, possible periodicities of glacier variations and their causes, the extent of ice recession between advances, and the magnitude and climatic significance of snowline fluctuations.

Glaciers and climate

Because of their sensitivity to climate changes of various magnitudes and different time scales, glaciers constitute an important source of palaeoclimatic data. Their widespread geographic distribution makes them suitable both for establishing local climate-proxy time series and for assessing the nature of global climate fluctuations. Nearly continuous monitoring of the terminal positions of many glaciers during recent decades (e.g., Müller, 1977) makes it possible to compare climate and glacier variations on a worldwide basis, but for intervals prior to the present century such data are scarce and the past history of glacier fluctuations must be gleaned from historical documents, field studies of glacial deposits, or from laboratory investigations of ice cores. Although important advances in understanding glacier history have been achieved through such studies, detailed and reliable data are still so sparse that it is difficult to assemble a comprehensive overview of glacier fluctuations spanning past millennia.

Despite the obvious palaeoclimatic insights to be gained from glaciological and glacial–geologic investigations, certain inherent limitations exist in using glaciers and their deposits for reconstructing past climatic changes. These limitations involve problems related mainly to glacier mass balance and flow dynamics, and to the dating of glacial deposits. Unless such problems are recognised and properly evaluated, climatic reconstructions based on glacial data can be misleading or of limited value.

The relationship of glacier fluctuations to climatic variations can be visualised in a simplified manner by means of a process-response model (Fig. 3.1) that relates climate to glacier mass balance, glacier flow, and glacial deposition. The response history of a given glacier reflects local meteorological conditions of the glacier basin that control various mass and energy fluxes at the glacier surface and which lead ultimately, at the end of a balance year, to an annual net mass balance for the glacier. The dynamic response of the glacier, especially its terminus, to changes in mass balance is complex, may include various feedbacks, and typically involves a lag which, depending on such factors as the size, geometry, and thermal characteristics of the glacier, may range from several years to decades or even centuries. Consequently, determination of the age of a glacial deposit, even with considerable accuracy, may not mean that the date of a related climatic change can also be specified. Furthermore, certain glaciers exhibit an inherent instability, apparently unrelated to climate, that results in periodic surges (Meier & Post, 1969). Such surge events can produce moraines that are essentially indistinguishable from moraines built during a climate-induced advance.

Dating problems increase with increasing age of deposits, and for this reason the greatest opportunity for obtaining well-dated records of glacier fluctuations centres on the last several centuries, an interval encompassing the later part of the Little Ice Age. With the exception of ice cores from glaciers, which may preserve recognisable annual layers that permit precise dating, most glacial records more than several centuries old can seldom be dated to within less than half a century with any confidence. For this reason it is generally necessary to view the history of glacier fluctuations on several separate time scales. Despite these limitations, glaciers can offer important information about both the direction and magnitude of climatic changes, and often with a greater precision than is possible for certain other palaeoclimatic proxy data.

Fig. 3.1. Chain of processes relating glacial–geologic features to climate in a glacierised basin.

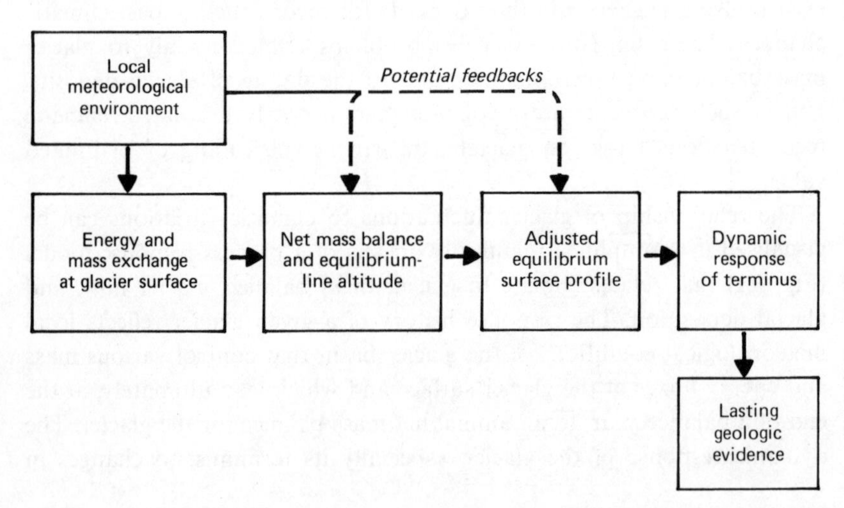

Sources of data

Palaeoclimatic information from glaciers and glacial deposits is obtained using three principal approaches. Ice cores taken from existing glaciers offer the opportunity of directly analysing firn and ice samples to construct continuous time series (Fig. 3.2) depicting variations in specific climate or climate-related parameters such as rates of accumulation and air temperature. Landforms and sediments produced by glaciers can be used to construct continuous and discontinuous records of glacier fluctuations and also make it possible to reconstruct the area, volume, and long profile of former glaciers that reflect climatic regimes different from that of the present. Such information can also be used to calculate equilibrium-line altitudes (snowlines) of former glaciers from which estimates of the magnitude and character of climate change can be inferred.

Fig. 3.2. Examples of continuous and discontinuous time series derived from glacial data; (*a*) oxygen-isotope record from Devon Island ice cap (Paterson *et al.*, 1977; Fig. 4); (*b*) lichen-dated moraine chronology from Swedish Lapland (Karlén & Denton, 1976: Fig. 25).

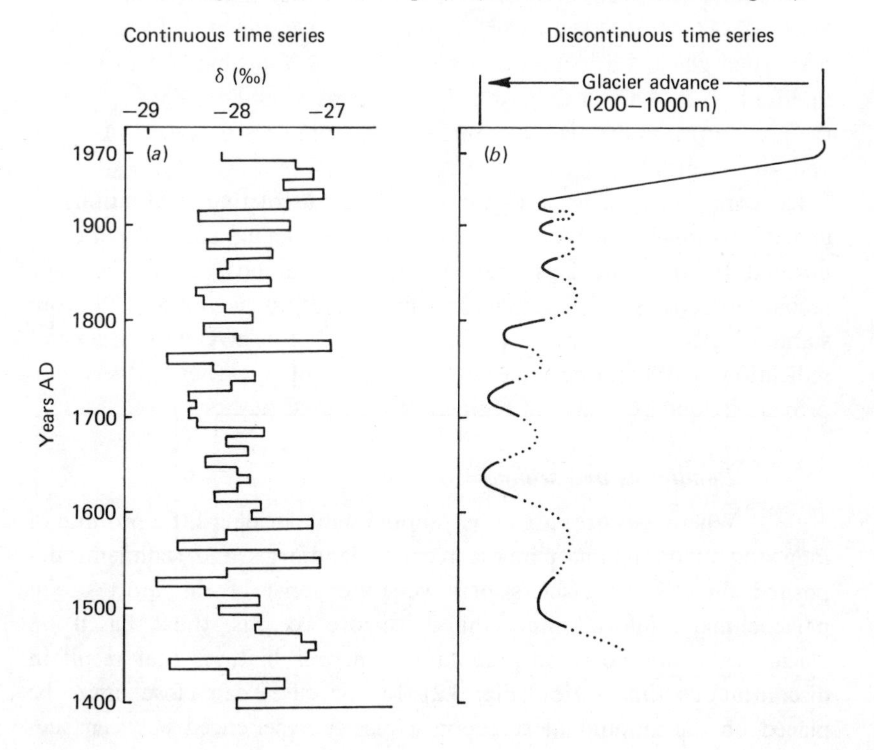

Ice cores

The acquisition and analysis of ice cores from glaciers provides an unparalleled opportunity for evaluating the history of changing climate within the drainage basin of a single glacier or segment of an ice sheet. Studies of cores obtained from the Antarctic and Greenland ice sheets (Johnsen, Dansgaard, Clausen & Langway, 1972; Hammer *et al.*, 1978), from ice caps in arctic Canada (Paterson *et al.*, 1977), and from alpine glaciers in middle latitudes (Oeschger *et al.*, 1977) have generated palaeoclimatic time series that are resolvable at the scale of individual years. Under the most favourable conditions, as in the case of the Camp Century core from Greenland, the record of annual fluctuations extends back some 5000 to 8000 years. From the resulting data, information on mean annual precipitation, snow accumulation rates, mean air temperature (based on oxygen-isotope measurements), atmospheric chemistry, particulate influx, prevailing wind trajectories (from incorporated volcanic-ash layers), and ice volume (based on altitudes of the former glacier surface) can all be inferred. Because most ice cores have come from polar and subpolar glaciers, data are limited mainly to the polar regions and to a few high-altitude sites at lower latitudes. Temperate glaciers are less suitable for some types of ice-core studies and the records from such glaciers are comparatively short, often no more than several hundred years. For these reasons, palaeoclimatic ice-core data spanning the last several millennia will come mainly from sites poleward of 50° N and 40° S latitude.

Ice core studies offer the promise of reconstructing fluctuations of climatic parameters within narrow limits and on an annual scale, uncomplicated by the time lags that are inherent in most glacial–geologic reconstructions. Such direct climatic information is not obtainable from standard glacial–geologic data which, at best, can give only a general indication of the former mass-balance state of a glacier, rather than provide unique estimates of temperature or precipitation.

Landforms and sediments

Where ice cores are either unobtainable or offer little promise of long and unambiguous climatic records, landforms and sediments deposited directly by glaciers may generate some of the most useful palaeoclimatic information. Unlike ice-core records, those based on glacial sediments commonly contain numerous hiatuses that result in discontinuous time series (Fig. 3.2). In few cases can close limits be placed on the amount of recession a glacier experienced between successive advances, so a continuous curve of glacier fluctuations based on the geologic record generally cannot be assembled. Nevertheless, ice-marginal features related to culminations of advances can sometimes be dated closely and can be used to infer the chronology of climatic changes.

End moraines, because they mark the limits of glacier advances, have frequently been used as a basis for evaluating glacier fluctuations. Although moraines are often obvious and easily mapped features, care must be exercised in their identification, for other landforms of non-glacial origin may readily be confused with them. Because rockfall deposits, protalus ramparts, rock glaciers, and other landforms related to mass-wasting processes can superficially resemble moraines, evidence for differentiating among such features must be sought in each case to avoid misinterpretation. Lateral moraines in mountainous regions frequently are high, massive ridges that are built up by successive ice advances over long periods of time. Their composite character is shown by the presence of buried soils that mark former land surfaces (Röthlisberger, 1976). Possibly some large lateral moraines of alpine glaciers are cored by drift deposited during both Holocene and late-glacial times. Surface sediment on the distal flanks of such moraines may vary considerably in age depending on the degree to which new debris covered older deposits during each successive advance.

Outwash terraces that reflect aggradation and degradation during episodes of glacier advance and recession also can record a discontinuous chronology of glacier fluctuations, and may be especially useful where end moraines are lacking or poorly preserved.

Finally, sediments that accumulate in lakes adjacent to or downstream from a glacier may record fluctuations of the glacier margin through variations in particle size, thickness, and content of organic matter (e.g., Karlén, 1976). Such sediments may retain a continuous record of glacier fluctuations that can be compared directly with variations in pollen influx and relative pollen percentages which reflect local vegetation changes. In basins where rhythmic sedimentary layering is present and is demonstrably annual in character, precise dating may permit evaluation of times of environmental changes, changes in rates of processes, and differences in response times of physical and biological systems.

Snowline fluctuations

The equilibrium-line altitude (ELA) of a glacier is the altitude of the boundary between the zone of accumulation and the zone of ablation. On temperate glaciers, the ELA closely approximates the altitude of the snowline or firn limit. It is therefore a climate-sensitive parameter that can be used to assess changes in climate through time. If the ELA of some former interval of glacier advance can be reconstructed and compared with the present ELA, a measure of the relative magnitude of climate change can be obtained.

Of various approaches to snowline reconstruction, two that are simple and straightforward generally give reliable results. In one method, the equilibrium line is assumed to lie at the median altitude of a glacier. The computation involves using the altitude of the downvalley limit of an end moraine and the altitude of the highest portion of the existing glacier (or, where a glacier is no longer present, the inferred headward limit of the former glacier) to compute a median altitude. The method is most suitable for small, geometrically simple glaciers having a normal area/altitude distribution. Under the best circumstances, ELAs can be reconstructed with an accuracy of about ± 50 m. The major difficulty involves estimating the headward limit of the glacier, often a subjective judgement involving considerable potential error. Nevertheless, reasonably consistent results can be obtained if a population of glaciers is used so that a local mean ELA is determined.

An alternative method generally giving more reliable results uses the glacier accumulation-area ratio (AAR). This parameter is the ratio of the accumulation area of a glacier to total area, both values obtainable from aerial photographs or topographic maps by planimetry. AARs of modern glaciers under steady-state conditions generally are between 0.5 and 0.8 (Meier & Post, 1962; Porter, 1975); an assumed mean AAR of 2/3, or 0.65, is often employed. Ice-limit data are used to reconstruct the areal configuration of a former glacier, and from altitude information on the glacier margin, a contour map is produced (Fig. 3.3). The ELA, based on the assumed AAR, is then calculated using area/altitude data obtained by planimetry. Depending on the size and complexity of the glacier, and on the contour interval employed, an accuracy of $\leqq 50$ m often can be achieved.

To compare the difference between present and former ELAs in an area, the past and present regional ELA gradient (regional slope of the ELA surface) must be calculated. If this is not done the degree of difference in ELAs may be overestimated, for the ELA gradient sometimes exceeds 10 m/km (Porter, 1975: Fig. 9, 1977: Fig. 2).

Although differences in ELA calculated by either method can give an estimate of the magnitude of climatic change, it is misleading to translate such differences into temperature values by applying present and (or) inferred former lapse rates. With adequate meteorological data, the relationships between present and former snowlines and various climatic variables can be evaluated by regression analysis (Porter, 1977). However, because of the complex relationship between glacier mass balance and meteorology (Fig. 3.1), a unique set of climatic values generally cannot be derived from estimates of ELA depression.

Principal dating methods and problems

The ultimate value of palaeoclimatic time series derived from glaciological information rests on the degree to which they can be accurately dated. There is little problem in the case of ice cores or varved lake sediments that permit resolution of individual years, but most glacial sediments and landforms are not easily or closely datable. The

Fig. 3.3. Use of glacial–geologic data to reconstruct the areal extent and topographic configuration of a former glacier. Estimated position of equilibrium line separating accumulation area (Sc) from ablation area (Sa) is based on assumed steady-state accumulation-area ratio (AAR) of 0.65. Position of contours at glacier margin is controlled by ice-limit data; shape of contours on glacier surface, bowed upglacier or downglacier from equilibrium line, is based on analogy with modern glaciers.

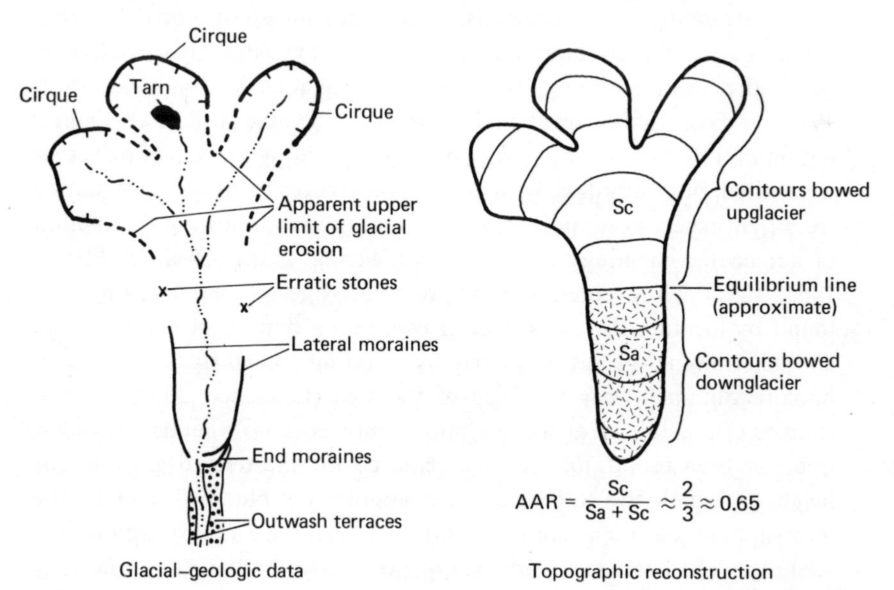

approaches that have proved most widely applicable for deposits spanning the last 6000 years involve use of historical data, botanical methods, radiocarbon dating, and tephrochronology.

Documentary data

Documentary records offer some of the most reliable information from which glacier variations during the later part of the Little Ice Age can be reconstructed. In the Alps, for example, documentary accounts, paintings, sketches, and lithographs offer evidence of the size of some glaciers as far back as the early seventeenth century. When carefully assessed and supplemented by recent photographs and records of annual monitoring of ice-front positions, such information can provide a reasonably detailed picture of the frontal variations of specific glaciers (e.g., Zumbühl, 1975). However, in many key areas, including most of the North and South American cordillera, the high mountains of central Asia, and the polar latitudes, documentary records seldom extend back much beyond the beginning of the present century, and, for the vast majority of glaciers, they are restricted to the recent era of aerial photography and satellite imagery.

Botanical methods

In many forested regions, such as northwestern North America, it may be possible to date times of glacier advance directly or indirectly by dendrochronology (e.g., Heusser, 1956; Sigafoos & Hendricks, 1961, 1972). In favourable circumstances, one may be able to date the culmination of a glacier advance either by (*a*) determining the age of trees that were tilted, but not killed, by a glacier (Lawrence, 1950), or (*b*) by dating trees that experienced an increased rate of growth following destruction of adjacent competing trees by an advancing glacier (Bray & Struik, 1963: 1251). More commonly, approximate ages of end moraines are found by locating the oldest tree growing on a deposit of unknown age and determining the age of the tree by counting annual rings in a section or core obtained near the base of the tree (Lawrence, 1950). Several sources of possible error include uncertainty as to (*a*) whether the oldest tree has been found, (*b*) the time required for the tree to grow to the height at which the section or core sample was obtained, and (*c*) the interval between stabilisation of a substrate and the establishment of tree seedlings. The latter factor is a major potential source of error and may substantially limit the precision of the dating technique in some areas. For instance, Sigafoos & Hendricks (1969) found that under conditions of abundant seed supply, a fine-grained sediment surface, and a moist maritime climate, germination of seedlings on low-altitude valley floors at Mt Rainier, Washington, occurred within about 5 years of site stabilisation, whereas at high-altitude sites on recent moraines the interval was 23 years or more. Trees at 1525–1675 m altitude on one moraine started growing 10–15 years later than trees at a site 365 m lower on the same moraine. Because of the general uncertainty attached to this lag factor, tree-ring counts often cannot provide minimum ages for a deposit with accuracy better than the nearest quarter-century. Accordingly, care must be taken to regard all moraine dates based on such tree-ring counts as strictly minimum ages that are subject to substantial error.

Lichenometry is an alternative dating method that is applicable under a wide range of conditions (Webber & Andrews, 1973). With adequate control, growth curves for different lichen species can be developed for use in dating moraines or other features of unknown age (e.g. Denton & Karlén, 1973b; Porter, 1981). In most cases the lag time between stabilisation of a moraine and initial establishment of lichens probably is very short. At middle-latitude sites under conditions of rapid growth ($\geq$40 mm/century), lichen thalli large enough to be seen with the unaided eye commonly appear within about a decade after the ice has receded. In such areas it may be possible to date a young surface ($\leq$300 years) to within 10 years or less (Porter 1981). However, because lichen growth tends to diminish with time, possibly at an exponentially decelerating rate, dating error also increases with increasing age of deposits. Where growth rates are low, as in the High Arctic, the potential temporal range of the method may equal or exceed 5000 years; but where growth rates are rapid, as in some maritime middle-latitude mountain ranges, the effective span of lichenometry may be 1000 years or less. Species having different growth rates can often be used together, with a rapid-growing species permitting close dating of the younger deposits and another, slower-growing species providing less-precise dates for older deposits.

Because several different methodologies have been used in lichenometric dating of glacial deposits, the resulting time series are of varying reliability. Detailed comparison of such time series is of little value, for one often cannot attach potential error limits to derived dates. If a uniform field method can be agreed upon and widely adopted, lichen dating should provide a basis for greatly expanding the number of glacial time series that will ultimately permit both interregional and inter-hemispherical correlations of recent glacier variations.

Radiocarbon dating

The radiocarbon method constitutes the most common basis for dating glacial–geologic time series. Where conditions permit preservation of organic matter in association with glacial deposits, limiting ages of sediments and landforms can be obtained (Fig. 3.4(*b*), (*c*), (*d*)). In certain circumstances, the time of sediment deposition can be specified from dates of contemporaneous samples (Fig. 3.4(*a*)). Radiocarbon dates often provide the only available temporal control and form the basis for many interregional correlations and comparisons with palaeoclimatic time series established from other types of proxy data. In areas where botanical dating methods are not applicable, radiocarbon may offer the only prospect for dating prehistoric deposits.

Despite the advantages of radiocarbon dating, unavoidable problems restrict the usefulness of the method for close dating of sediments deposited during the last 6000 years. Sample contamination by modern organic matter, usually not a serious problem for young sediments, can be avoided by careful field sampling and laboratory pre-treatment of samples. Of greater concern is calibration of the radiocarbon time scale, for it is now well known that changes in the atmospheric ^{14}C reservoir have affected the ^{14}C ages of dated samples. Radiocarbon dates can be converted to calendar dates by using tables based on analyses of rings from ling-lived trees (e.g. Ralph, Michael & Han, 1973). However, the several conversion tables available do not always agree. Furthermore, there are numerous intervals where ambiguous ages can result. Stuiver (1978), for instance, has shown that during the last 450 years (an interval encompassing most of the Little Ice Age) it generally is not possible to obtain a unique and unambiguous calendar age from a radiocarbon date, especially if the standard error of the radiocarbon date is considered. As an example, a radiocarbon age of 220 ± 50 years could, in fact, represent a calendar age anywhere in the ranges of 150–210, 280–320, or 410–420 years (Fig. 3.5). An actual calendar age of 220 years, however, is unlikely! Because of the extreme variability of atmospheric ^{14}C during the Little Ice Age, there is little opportunity of obtaining unique calendar ages from radiocarbon dates within this important time interval.[1] As a result, the validity of Little Ice Age glacier chronologies based only on radiocarbon dating should be questioned, as should correlations employing such chronologies. Although detailed calibration curves like that shown in Fig. 3.5

are not yet available for all of the last 6000 years, there may be a number of intervals within that time span where dates will be ambiguous. While this may be a serious problem only if dating precision better than the nearest century is required, analysis of possible short-term climate-related glacier variations having a duration of a century or less may not be feasible on the basis of radiocarbon dating alone.

Fig. 3.4. Examples illustrating stratigraphic relationship of ^{14}C samples to glacial deposits and derivation of (*a*) contemporary, (*b*) bracketing, and (*c*), (*d*) limiting dates.

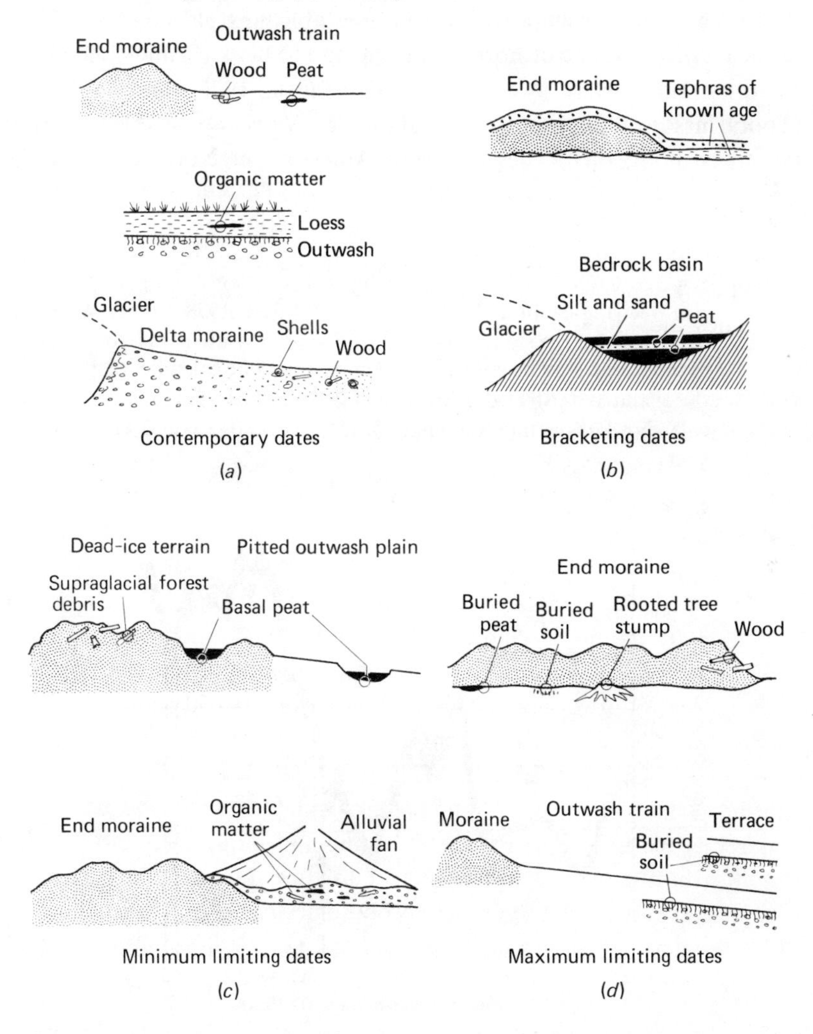

Tephrochronology

Tephra layers produced by explosive volcanic eruptions are often widely distributed downwind from source vents and may offer a convenient basis for dating and correlating glacial deposits. When the age of a tephra layer is known either from documentary information, tree-ring counts, or bracketing radiocarbon dates, the layer can be used to assess the age of landforms or sediments with which the layer is associated (Figs. 3.4(*b*), 3.6). Volcanic regions characterised by numerous eruptions closely spaced in time are best-suited for tephrochronologic dating, for closer bracketing ages may then be possible. Tephrochronology has proved useful in establishing glacial chronologies in the Cascade Range of northwestern United States (Crandell & Miller, 1974), in the eastern Aleutian Islands (Black, 1976), and in Iceland (Thorarinsson, 1964, 1966). It also should be widely applicable for dating moraines along the Andes of South America, on North Island (New

Fig. 3.5. Curve showing relationship between conventional ^{14}C years and tree-ring-calibrated calendar years (Stuiver, 1978: Fig. 1). Width of curve is twice the standard deviation in the measurements. Any radiocarbon year may be equivalent to more than one calendar year; in the example illustrated, a ^{14}C age of 220 ± 50 years is equivalent to all calendar dates within the ranges of 150–210, 280–320 and 410–420 years.

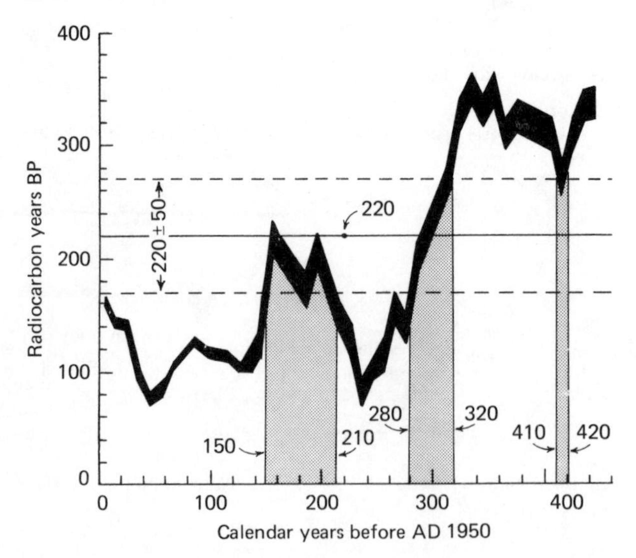

Zealand), and in several parts of the North American cordillera that have not yet been studied in detail.

Comparison of methods

No single method is ideally suited for dating deposits embracing all of the middle and late Holocene (Fig. 3.7). Although radiocarbon dating can be employed over this entire period, the serious ambiguity of dates within at least part of this time interval limits its usefulness for close dating. The botanical methods are most suitable for dating late-Holocene deposits, especially those of the Little Ice Age, whereas documentary records seldom extend back more than a century of two. Tephra layers generally do not permit close dating and their age, in turn, is commonly controlled by bracketing (and potentially ambiguous) radiocarbon dates. Accordingly, it is often necessary to use several dating methods in combination within a field area.

In establishing a chronology of glacier fluctuations based on geologic evidence, an attempt is made to date as closely as possible the times of moraine construction. But if the ultimate intent is to evaluate past climate, what climatic event is thereby dated? Clearly it is not the climatic change that led to the advance, for the lag in response of a glacier terminus to that climatic event may be significant. For instance, in the case of small middle-latitude temperate glaciers of the Austrian

Fig. 3.6. Use of tephra layers for dating Holocene moraines. Illustrated is an example from the Cascade Range, Washington, where several widespread units of known age in and near Mt Rainier National Park permit bracketing ages to be assigned. Indicated ages are approximate (Mullineaux, 1974).

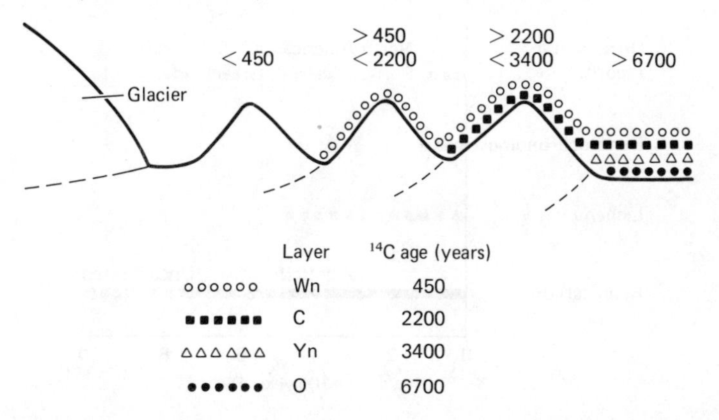

Alps that have high activity indices, this lag is estimated to be approximately 7 years (Posamentier, 1977); for larger glaciers in the Mt Blanc massif the lag is 10 years or more (Reynaud, 1977; S. C. Porter and G. Orombelli, unpublished data). The age of a moraine may, instead, be closer to the time marking the change to a negative mass-balance regime that led to glacier recession (Fig. 3.8). Because vegetation generally becomes established as ice recedes and the debris stabilises, the oldest lichens and trees on a moraine not only provide minimum ages for the moraine, but also for the beginning of a climatic trend that favoured glacier retreat (Fig. 3.8). Radiocarbon dates of basal organic matter from lakes or bogs on or behind moraines also give limiting ages for the beginning of ice recession, but such dates may not provide *close* minimum ages (Porter & Carson, 1971).

Maximum limiting ages for moraines generally are obtainable only through radiocarbon dating of organic matter that was killed and overridden or incorporated in glacial sediments (Fig. 3.4); together with minimum limiting ages obtained either from lichens, tree rings, or radiocarbon analyses, they can be used to infer the approximate age of the culmination of an advance. In exceptional cases, living trees that were affected by an advancing glacier can be used to date the time the glacier achieved its maximum expansion. Seldom, however, is it possible to

Fig. 3.7. Effective ranges of methods commonly used to date Holocene glacial deposits. For any given area, the tree-ring and lichen-dating methods depend, respectively, on the age span of different tree species and on the growth rates and longevity of different lichen species. For the last *c* 7500 years, radiocarbon ages can be calibrated by using tree-ring series (e.g. Ralph *et al.*, 1973).

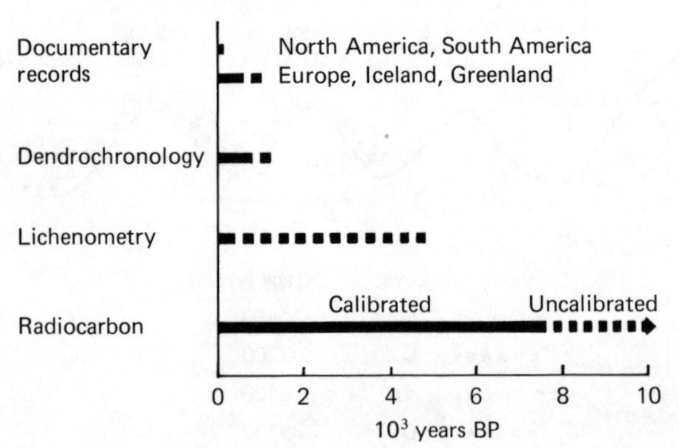

obtain a close date for the beginning of an advance and thereby infer the age of the climatic change that led to a positive mass-balance regime for the glacier.

When one considers the large potential uncertainty in any ^{14}C date that includes both the standard error and past variations in the atmospheric radiocarbon reservoir, there is seldom much hope of bracketing the age of a moraine more closely than to within a few hundred years by

Fig. 3.8. Schematic curves depicting a hypothetical climatic fluctuation that leads to a cycle of advance and recession for a typical alpine glacier. Beginning of advance or of retreat of the glacier terminus lags behind the change in climate responsible for a change in glacier mass balance, the length of the lag (in this example about 7 years) depending on the response characteristics of the glacier. Establishment of pioneer lichens and trees used to assess the minimum age of a moraine occurs sometime after stabilisation of the deposit.

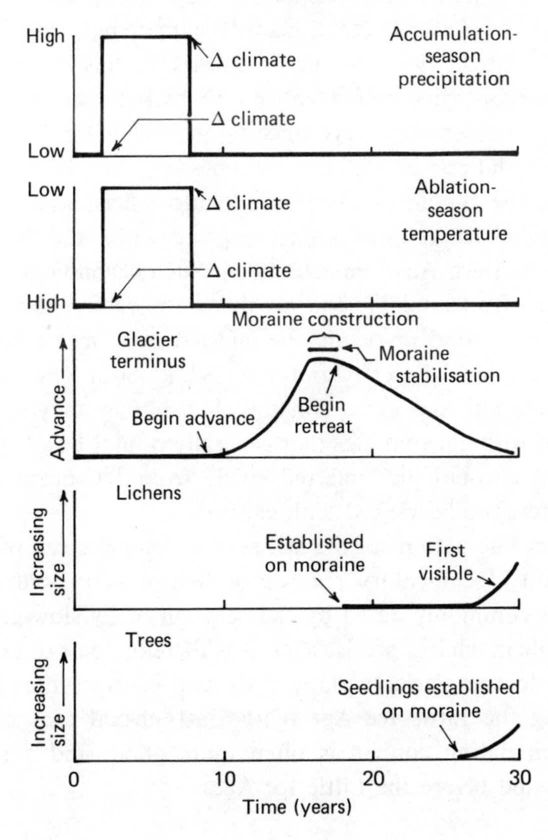

^{14}C dating. Somewhat closer minimum-age control for young moraines ($\leq$500 years old) may be achieved by tree-ring dating, but for areas where a well-calibrated lichen growth curve can be developed on the basis of documentary data, lichenometry may offer the best opportunity for close minimum dating of such moraines.

Glacier variations during the past 6000 years

Although Holocene moraines have been studied by numerous investigators over a period of many years, comparatively few well-dated chronologies have emerged that permit reliable interregional comparisons. Well-dated chronologies are here regarded as those in which moraines built during the last two to three centuries are confidently dated to within the nearest 10 years or less, or, for earlier periods, to within at least the nearest 25 to 50 years. Only if glacial deposits are dated to within such limits of precision can one realistically hope to compare the record of glacier fluctuations with climate records and with other types of well-dated palaeoclimatic time series. Because many of the available glacial chronologies are controlled only by radiocarbon or tree-ring dates having unacceptably large uncertainties, climatic reconstructions based on correlations of such chronologies may not be reliable.

In some areas, the timing of glacier fluctuations has been inferred indirectly from radiocarbon-dated pollen sequences (e.g. the 'Rotmoos fluctuation' of the eastern Alps, Patzelt, 1974). Such chronologies suffer from two sources of uncertainty, one involving the radiocarbon dates themselves, and the second involving the inferred relationship between vegetation changes and glacier variations. While both glaciers and vegetation may respond to a given climatic change, they may do so in different ways and with different lags that generally cannot be adequately evaluated. Glacial chronologies inferred solely from ^{14}C-dated pollen records should therefore be viewed with caution.

Although glacier fluctuations of the last several centuries can often be closely dated using documentary records or lichenometry, older fluctuations are most commonly dated by radiocarbon or by slow-growing lichens that provide much less precise ages. It is therefore best to examine the history of Holocene glacier variations on two separate time scales, one encompassing the Little Ice Age of the last several centuries, an interval for which dating control is often quite good, and a second embracing the period before the Little Ice Age.

Little Ice Age

In most areas of the world where recent moraines have been studied, evidence points to one or more minor ice advances within the last several centuries, a phase of glacial activity referred to by Francois Matthes (1939: 520) as the Little Ice Age. In the Alps, the Little Ice Age is widely regarded as spanning the period from about AD 1590–1850 (Le Roy Ladurie, 1971). However, although this interval clearly includes the culminating advances of the Little Ice Age, the climatic deterioration that led to these advances may have begun much earlier (e.g., Dansgaard *et al.*, 1975: 26). Evidence from the Alps (Röthlisberger, 1976; S. C. Porter and G. Orombelli, unpublished data), Scandinavia (Karlén, 1979) and from the Cascade Range of western United States (Miller, 1969; Sigafoos & Hendricks, 1972) indicates that significant expansions of glaciers occurred as early as the thirteenth or fourteenth centuries. Isotopic measurements of the Camp Century ice core from the Greenland Ice Sheet suggest that the comparatively cold Little Ice Age climate began at least as early as the fifteenth century (Dansgaard *et al.*, 1971), whereas at the site of the Crête core, it began in the late twelfth century (Dansgaard *et al.*, 1975). At about that time sea ice began to expand around Greenland, and its southern limit moved southward, possibly forcing abandonment of traditional sailing routes between Iceland and the Norse colonies in west Greenland (see McGovern, this volume). The Little Ice Age may therefore have begun at least two to three centuries earlier in the North Atlantic region than is generally inferred. Although the termination of this climatic episode is often placed at about AD 1850, major halts or readvances of glaciers that occurred both in the northern and southern hemispheres at the end of the nineteenth century and in the first third of the twentieth century point to continual variations in climate during the waning phase of this glacial episode.

Representative time series for the main part of the Little Ice Age in the North Atlantic region and western Europe illustrate the high frequency of glacier variations that characterised this interval (Fig. 3.9). Although a broad similarity of pattern is evident, when viewed in detail an apparent lack of synchrony is discernible. In part this may reflect measurement of different parameters (intervals of advance, culminations of advance, isotopic values) or different response lags, but it may also reflect poor dating precision where documentary information is lacking. Where moraines are dated by slow-growing lichens, for instance, a measurement error of 1 to 2 mm could result in a dating error of as much as one or two decades. The apparent lack of synchrony between the Alps and Swedish Lapland in the period before about AD 1800 may largely reflect such dating uncertainties. Alternatively, it could be due to differences in response rate of glaciers existing under two different environmental

Fig. 3.9. Selected isotopic and glacial–geologic time series for the past three and a half centuries from the North Atlantic region and western Europe. Isotopic variations in the upper part of the Camp Century, Greenland, ice core are from Dansgaard *et al.* (1971); intervals of glacier advance based on documentary data from Iceland are from Thorarinsson (1943); lichenometric and historically dated moraine chronology from Swedish Lapland was reported by Karlén & Denton (1976); lichenometric and historically dated moraine chronology from the Alps is based on a summary by Le Roy Ladurie (1971) supplemented by unpublished data of S. C. Porter and G. Orombelli from the Italian side of the Mt Blanc massif; mean ablation-season temperature and mean accumulation-season precipitation for the Great St Bernard Pass are derived from records summarised by Janin (1970).

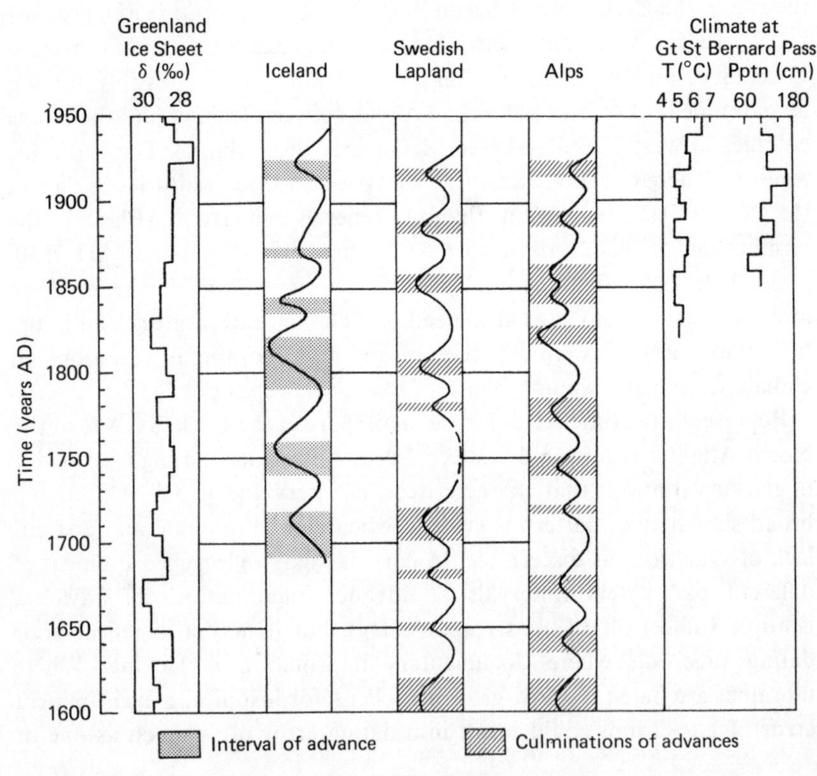

regimes, to a response to different climate variables (e.g. precipitation v. temperature), or to non-synchronous climatic fluctuations in different regions. Even where a reasonable degree of synchrony can be demonstrated between different areas, the times of greatest advance may not coincide. For example, according to Thorarinsson (1943) the maximum advances of most glaciers in Iceland occurred in AD 1750–60, whereas on the Italian flank of the Mt Blanc massif most glaciers achieved their Holocene maximum sometime between about AD 1815 and 1825 (S. C. Porter and G. Orombelli, unpublished data). On Mt Rainier in the Cascade Range, several large valley glaciers reached their Holocene maximum about AD 1820–30, but smaller cirque glaciers in the same area are bordered by moraines as much as several millennia old. In such cases, it is often apparent that small advancing glaciers were unable to surmount and override older moraine ridges and therefore built nested moraines that decrease in age as one approaches an existing glacier. Large valley glaciers, however, may override older end moraines and either obliterate them or bury them with younger drift.

Most glaciers apparently reached their Little Ice Age maxima during the seventeenth, eighteenth, or nineteenth centuries, and then experienced an interval of dramatic recession beginning toward the middle or end of the nineteenth century (e.g., Porter & Denton, 1967). Apparently most glaciers were in an advanced state from at least the end of the sixteenth century (and possibly earlier) until the end of the nineteenth century. The second-order recessions and readvances within this broad time interval produced a complex moraine record in many areas that must largely reflect short-term climatic variation within the Little Ice Age. However, because few such records have been carefully studied and accurately dated, it is still uncertain whether these short-term second-order glacier advances were synchronous events either globally or regionally.

A direct comparison between fluctuations of glaciers and of climate can be made where adequate records exist. A continuous climatic record extending back to AD 1818 has been kept at Great St Bernard Pass on the Swiss–Italian border (Fig. 3.9). If the climate record is compared with the record of glacier fluctuations from this same part of the Alps, it is seen that intervals of glacier advance followed decades that were characterised by high accumulation-season precipitation and low ablation-season temperature. A direct cause-and-effect relationship is suggested, for it is these two climatic factors that most affect the mass balance of temperate middle-latitude glaciers. The limited meteorological record precludes assessment of this relationship back beyond the middle of the last century, but because the three most recent episodes of ice advance followed intervals marked by cloudy, cool summers and wet winters, earlier episodes of advance may also have been related to such conditions.

Glacier fluctuations prior to the Little Ice Age

End moraines lying beyond those attributable to Little Ice Age advances and generally having denser vegetation cover, older trees, larger lichens, stronger soil-profile development, or more subdued character have been described from many parts of the world and indicate that glacial episodes similar in character and magnitude to the Little Ice Age occurred earlier during the Holocene. The age of such moraines is less easily determined, and rarely are close limiting dates available. In many cases, ages are inferred by comparison with sequences in other areas where some dating control does exist. Surveys of available glacial chronologies (Porter & Denton, 1967; Denton & Porter, 1970; Denton & Karlén, 1973a) have indicated that such deposits apparently cluster in several age groups centering about 1100–1200, 2800–3000, and 5000–5300 calendar years ago (Fig. 3.10). In some areas, specific stratigraphic names have been assigned the moraines to distinguish them from moraines of the Little Ice Age. Although a consistent pattern is reasonably clear within certain geographic regions such as northwestern North America (Fig. 3.10), a convincing global pattern has not yet been demonstrated. In some areas limiting dates suggest either that pre- Little Ice Age moraines fall between the principal age clusters, or that moraines dating to the dominant intervals are lacking (e.g. Grove, 1979). Evidence for an event of Little Ice Age magnitude about 5000 years ago exists in the southern hemisphere (Mercer, 1968; Wardle, 1973), but throughout much of the northern hemisphere evidence for such an event is either absent, unrecognised, or inadequately dated. Few moraines dating to the first half of the Holocene have been identified, suggesting either that significant widespread glacial advances did not occur then, or were less-extensive than those of later date. Most described and dated Holocene advances of mountain glaciers fall within the interval of the last 5300 years, a period which in North America is commonly referred to as Neoglaciation (Fig. 3.10; Porter & Denton, 1967).

The relative magnitude of different Holocene glacial events may vary from glacier to glacier within a given region, or from region to region. The record of an individual glacier or even several glaciers may not be representative of a region as a whole; only detailed studies of a large

Fig. 3.10. Holocene glacier fluctuations in western North America. Some specific areas where glacial deposits in several age ranges have been mapped are indicated in bold type. Formal stratigraphic names used to designate deposits of different age are given in parentheses. Areas in which reasonably close age control is available are starred; for other areas, ages are inferred either on the basis of relative-dating criteria or from dates that do not closely limit or bracket the ages of deposits. Estimated snowline depression is based on selected calculated values for reconstructed glaciers of the Little Ice Age, estimates for earlier advances are based on areal extent of glaciers relative to those of the Little Ice Age maximum.

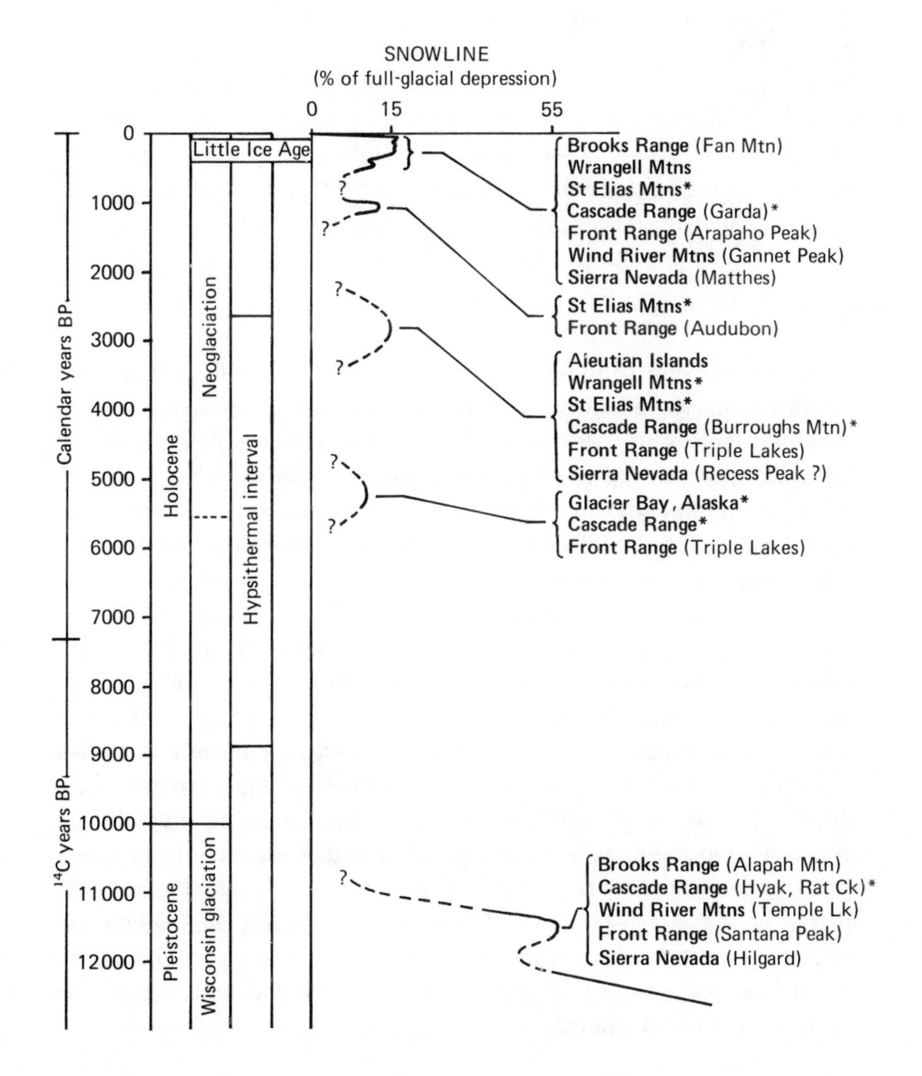

population of glaciers ($\geqq 20$) is likely to provide a reasonably complete picture of the relative magnitude of various glacial advances. In assembling such comparative data, it is important to exclude information from surging and calving glaciers, the behaviour of which may be anomalous and unrepresentative of the regional pattern. Advances that culminated about 5000 years ago were apparently the most extensive of the last 6000 years in New Zealand (Wardle, 1973) and in the southern Andes (Mercer, 1976), whereas in the Cascade Range (Miller, 1969) they were less extensive than those of the Little Ice Age. Similarly, the advance or advances that culminated close to 2800 years ago were the greatest for some glaciers in the Cascade Range (Crandell & Miller, 1974), in the Wrangell Mountains (Denton & Karlén, 1977), and in the Aleutians (Black, 1976). However, the greatest Holocene advance of numerous glaciers was during the Little Ice Age. In many areas moraines representing two or more intervals of significant Holocene advance are closely nested, implying that the different glacial episodes were of similar magnitude; the discernable differences probably resulted largely from local or regional climatic and geographic factors.

Snowline variations

A measure of the magnitude of Holocene climate changes that generated glacier advances can be obtained from measurements of snowline depression. Most published studies concern glaciers of the Little Ice Age, but because the major glacial episodes of the last 6000 years were apparently similar in magnitude, estimates for the maximum Little Ice Age snowline depression also apply broadly to culminations of earlier advances. Results from the Cascade Range (Scott, 1977; S. C. Porter, unpublished data), the Colombian Andes (Herd, 1974), the Southern Alps of New Zealand (Porter, 1975), and the Alps (Heuberger, 1968) suggest that snowlines were lowered as much as 100–200 m during the last several centuries, an amount representing 15 ± 5 per cent of the total snowline depression during the last glacial age. In terms of snowline depression, therefore, the Little Ice Age climate change brought these mountain areas about a sixth of the way toward full glacial conditions (Fig. 3.10). On the basis of the known glacial–geologic record, snowlines probably fluctuated through a maximum altitudinal range of about 250 m during the last 6000 years. However, considerably more data are needed before it will be possible to specify regional or global differences in snowline fluctuations that might provide important clues to worldwide patterns of climatic change.

Prospects for acquisition of data

Despite decades of research on Holocene glacier fluctuations, detailed chronologies have been obtained for very few areas. For some regions, Holocene chronologies are based on records from only one or a few glaciers, whereas for others, numerous glaciers have been studied. Because there has been no universally adopted approach to the study of Holocene glacial deposits, different field methods have been used, and data have been presented in different ways. Different dating methods have been employed, sometimes uncritically, and the resulting chronologies are often difficult to evaluate because of dating uncertainties. If the full potential of glacial–geologic studies for palaeoclimatic reconstruction is to be realised, a uniform approach to data acquisition and dating is desirable, and special care should be given to dating events as closely as possible using methods that will produce the smallest possible errors.

Although information of varying quality is available for many parts of the world, some important geographic gaps in knowledge remain. For example, there is still little information on late Holocene glacier variations from arctic areas, and the record from high mountains of central Asia and South America is still poorly known. For a proper understanding of the global pattern of Holocene glacier fluctuations, a broader geographic coverage is clearly desirable. The distribution of present glaciers dictates that the most promising areas for study lie along three intercontinental transects (Fig. 3.11). A cordilleran transect, running the length of the Americas and encompassing the Coastal and Rocky Mountain systems of North America and the Andean chain of South America, includes a vast number of glaciers that should be capable of providing a detailed picture of interhemispherical similarities and differences in glacier response to climate change during the last half of the Holocene. Although important data are emerging from southern Alaska, the Canadian and Colorado Rockies, the Cascade Range, and the southernmost Andes, other areas along this transect are either inadequately studied or poorly dated.

Two latitudinal transects crossing the Arctic and central Eurasia include glaciers having different sizes and response characteristics. Glacial records from these transects may turn out to be nonsynchronous, reflecting different climatic histories for different latitudes, and (or) a different response lag for subpolar and temperate glaciers. Good data are available for several locations along the transects, most notably in Scandinavia and in the Alps, but many other areas should have considerable potential for producing long and reasonably complete chronologies of glacier fluctuations.

With an increase in the number of well-dated records of local glacier populations, it should be possible to evaluate important questions involving (*a*) the timing and relative magnitude of Holocene glacier

Fig. 3.11. Latitudinal and longitudinal transects along which favourable sites may be located for the development of palaeoclimatic time series from glaciologic data. Data sets developed along these transects would permit intra- and interhemispheric comparisons to be made that would bear on such problems as possible synchrony of glacier response and the relative magnitude of climate-related glacier fluctuations during the Holocene. Solid circles are areas where well-dated and well-documented information is available for a large population of glaciers. AL-Aleutians, AR-Alaska Range, BR-Brooks Range, LM-Logan Mountains, IR-Icefield Ranges, CM-Coastal Mountains of Canada, CR-Canadian Rocky Mountains, CA-Cascade Range, RM-US Rocky Mountains, SN-Sierra Nevada, MV-Mexican Volcanoes, CO-Colombian Andes, EA-Equadorian Andes, PA-Peruvian Andes. NC-Northern Chilean/Argentine Andes, CC-Central Chilean/Argentine Andes, SC-Southern Chilean/Argentine Andes. BI-Baffin Island, WG-West Greenland, EG-East Greenland, IC-Iceland, SW-Northern Sweden, NZ-Novaya Zemlya, SZ-Severnaya Zemlya, WA-Western Alps, EA-Eastern Alps, CS-Caucasus, HK-Hindu Kush, HW-Western Himalaya, HE-Eastern Himalaya, PA-Pamir, AL-Altai, KO-Kodar Mountains, SK-Suntar Khayata, KA-Kamchatka.

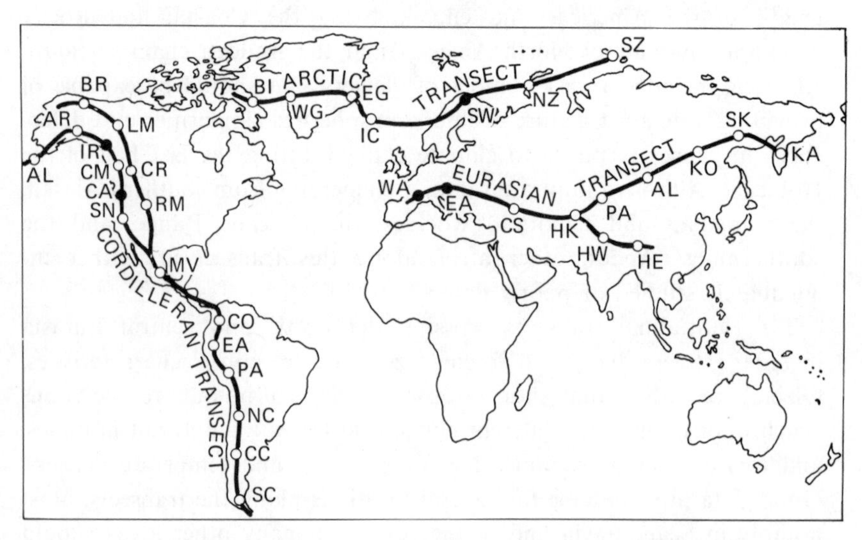

fluctuations within and between hemispheres, (*b*) possible periodicities of glacier advances and their causes, (*c*) the extent of ice recession between major intervals of advance, and (*d*) the magnitude and possible regional differences in the amount of snowline depression during successive advances. The answers to such questions can provide important insights into the fundamental problem of climatic change on short time scales.

Acknowledgements

I thank J. Roger Bray, Minze Stuiver, Richard B. Waitt, Jr, and T. M. L. Wigley for helpful comments on a draft of this chapter which they kindly reviewed.

Note

1. Wood specimens offer a possible exception, for unique ages might be obtained by closely dating several widely spaced rings or groups of rings which, in combination, could permit the calendar age of the sample to be determined.

References

Black, R. F. (1976). Geology of Umnak Island, eastern Aleutian Islands, as related to the Aleuts. *Arctic and Alpine Research*, **8**, 7–35.

Bray, J. R. & Struik, G. J. (1963). Forest growth and glacial chronology in eastern British Columbia, and their relation to recent climatic trends. *Canadian Journal of Botany*, **41**, 1245–71.

Crandell, D. R. & Miller, R. D. (1974). Quaternary stratigraphy and extent of glaciation in the Mount Rainer region, Washington. *US Geological Survey Professional Paper 847*, 59 pp.

Dansgaard, W., Johnsen, S. J., Clausen, H. B. & Langway, C. C., Jr (1971). Climatic record revealed by the Camp Century ice core. In *The Late Cenozoic Glacial Ages*, ed. K. K. Turekian, pp. 37–56. New Haven: Yale University Press.

Dansgaard, W., Johnsen, S. J., Reeh, N., Gundestrup, N., Clausen, H. B. & Hammer, C. U. (1975). Climatic changes, Norsemen and modern man. *Nature*, **255**, 24–8.

Denton, G. H. & Karlén, Wibjörn (1973a). Holocene climatic variations – their pattern and possible cause. *Quaternary Research*, **3**, 155–205.

Denton, G. H. & Karlén, Wibjörn (1973b). Lichenometry; its application to Holocene moraine studies in southern Alaska and Swedish Lapland. *Arctic and Alpine Research*, **5**, 347–72.

Denton, G. H. & Karlén, Wibjörn (1977). Holocene glacial and tree-line variations in White River valley and Skolai Pass, Alaska and Yukon Territory. *Quaternary Research*, **7**, 63–111

Denton, G. H. & Porter, S. C. (1970). Neoglaciation. *Scientific American*, **222**, 100–10.

Grove, J. M. (1979). The glacial history of the Holocene. *Progress in Physical Geography*, **3**, 1–54.

Hammer, C. U., Clausen, H. B., Dansgaard, W., Gundestrup, N., Johnsen, S. J. & Reeh, N. (1978). Dating of Greenland ice cores by flow models, isotopes, volcanic debris, and continental dust. *Journal of Glaciology*, **20**, 3–26.

Herd, D. G. (1974). Glacial and volcanic geology of the Ruiz-Tolima volcanic complex, Cordillera Central, Columbia. Unpublished PhD Dissertation, University of Washington, Seattle.

Heuberger, Helmut (1968). Die Alpengletscher im Spat- und Postglazial. *Eiszeitalter und Gegenwart*, **19**, 270–5.

Heusser, C. J. (1956). Postglacial environments in the Canadian Rocky Mountains. *Ecological Monographs*, **26**, 263–302.

Janin, Bernard (1970). *Le Col du Grand-Saint Bernard, climat et variations climatiques*. Aoste: Marguerettaz-Musumeci. 112 pp.

Johnsen, S. J., Dansgaard, W., Clausen, H. B. & Langway, C. C., Jr (1972). Oxygen isotope profiles through the Antarctic and Greenland Ice Sheets. *Nature*, **235**, 429–34.

Karlén, Wibjörn (1976). Lacustrine sediments and tree-limit variations as indicators of Holocene climatic fluctuations in Lappland, northern Sweden. *Geografiska Annaler*, **58**, 1–34.

Karlén, Wibjörn (1979). Glacier variations in the Svartisen area, Northern Norway. *Geografiska Annaler*, **61A**, 11–28.

Karlén, Wibjörn & Denton, G. H. (1976). Holocene glacial variations in Sarek National Park, northern Sweden. *Boreas*, **5**, 25–56.

Le Roy Ladurie, E. (1971). *Times of Feast, Times of Famine; a History of Climate Since the Year 1000*. New York: Doubleday. 426 pp.

Lawrence, D. B. (1950). Estimating dates of recent glacier advances and recession rates by studying tree growth layers. *American Geophysical Union Transactions*, **31**, 243–8.

Matthes, F. E. (chairman) (1939). Report of Committee on Glaciers. *American Geophysical Union Transactions*, **20**, 518–23.

Meier, M. F. & Post, A. S. (1962). Recent variations in mass net budgets of glaciers in western North America. *IUGG/IASH Committee on Snow and Ice, General Assembly, Obergurgl. IASH Publication 58*, pp. 63–77.

Meier, M. F. & Post, A. S. (1969). What are glacier surges? *Canadian Journal of Earth Sciences*, **6**, 807–17.

Mercer, J. H. (1968). Variations of some Patagonian glaciers since the Late-Glacial. *American Journal of Science*, **266**, 91–109.

Mercer, J. H. (1976). Glacial history of southernmost South America. *Quarternary Research*, **6**, 125–66.

Miller, C. D. (1969). Chronology of Neoglacial moraines in the Dome Peak area, North Cascade Range, Washington. *Arctic and Alpine Research*, **1**, 49–66.

Müller, Fritz (1977). *Fluctuations of glaciers, 1970–1975*. Paris: IAHS-UNESCO, **3**, 269 pp.

Mullineaux, D. R. (1974). Pumice and other pyroclastic deposits in Mount Rainier National Park, Washington, *US Geological Survey Bulletin*, **1326**. 83 pp.

Oeschger, H., Schotterer, U., Stauffer, B., Haeberli, W. & Röthlisberger, H. (1977). First results from alpine core drilling projects. *Zeitschrift für Gletscherkunde und Glazialgeologie*, **13**, 193–208.

Paterson, W. S. B., Koerner, R. M., Fisher, D., Johnsen, S. J., Clausen, H. B., Dansgaard, W., Bucher, P. & Oeschger, H. (1977). An oxygen-isotope climatic record from the Devon Island ice cap, arctic Canada. *Nature*, **266**, 508–11.

Patzelt, Gernot (1974). Holocene variations of glaciers in the Alps. *Colloques Internationaux du centre National de la Recherche Scientifique No. 219: Méthodes quantitatives d'étude des variations du climat au cours du Pléistocène*, 51–9.

Porter, S. C. (1975). Equilibrium-line altitudes of late Quaternary glaciers in the Southern Alps, New Zealand. *Quaternary Research*, **5**, 27–47.

Porter, S. C. (1977). Present and past glaciation threshold in the Cascade Range, Washington, USA: topographic and climatic controls, and paleoclimatic implications. *Journal of Glaciology*, **18**, 101–16.

Porter, S. C. (1981). Lichenometric studies in the Cascade Range of Washington: establishment of *Rhizocarbon geographicum* growth curves at Mount Rainer. *Arctic and Alpine Research* **13**, 11–23.

Porter, S. C. & Carson, R. J., III (1971). Problems of interpreting radiocarbon dates from dead-ice terrain, with an example from the Puget Lowland of Washington. *Quaternary Research*, **1**, 410–14.

Porter, S. C. & Denton, G. H. (1967). Chronology of Neoglaciation in the North American Cordillera. *American Journal of Science*, **265**, 117–210.

Posamentier, H. W. (1977). A new climatic model for glacier behaviour of the Austrian Alps. *Journal of Glaciology*, **18**, 57–65.

Ralph, E. K., Michael, H. N. & Han, M. C. (1973). Radiocarbon dates and reality. *MASCA Newsletter*, **9**, 1–19.

Reynaud, Louis (1977). Glacier fluctuations in the Mont Blanc area (French Alps). *Zeitschrift für Gletscherkunde und Glazialgeologie*, **13**, 155–66.

Röthlisberger, Friedrick (1976). Gletscher- und Klimaschwangungen im Raum Zermatt, Ferpecle und Arolla. *Die Alpen*, 3/4 Quartal 1976, 59–152.

Scott, W. E. (1977). Quaternary glaciation and volcanism, Metolius River area, Oregon. *Geological Society of America Bulletin*, **88**, 113–25.

Sigafoos, R. L. & Hendricks, E. L. (1961). Botanical evidence of the modern history of Nisqually Glacier, Washington. *US Geological Survey Professional Paper 387-A*. 20 pp.

Sigafoos, R. L. & Hendricks, E. L. (1969). The time interval between stabilization of alpine glacial deposits and establishment of tree seedlings. *US Geological Survey Professional Paper 650-B*, B89–B93.

Sigafoos, R. L. & Hendricks, E. L. (1972). Recent activity of glaciers of Mount Rainier, Washington, *US Geological Survey Professional Paper 387-B*. 24 pp.

Stuiver, Minze (1978). Radiocarbon timescale tested against magnetic and other dating methods. *Nature*, **273**, 271–4.

Thorarinsson, Sigurdur (1943). Oscillations of the Iceland glaciers in the last 250 years. *Geografiska Annaler*, **25**, 1–54.

Thorarinsson, Sigurdur (1964). On the age of the terminal moraines of

Brúarjökull and Hálsajökull, a tephrochronological study. *Jökull*, **14**, 67–75.
Thorarinsson, Sigurdur (1966). The age of the maximum postglacial advance of Hagafellsjökul eystri, a tephrochronological study. *Jökull*, **16**, 207–10.
Wardle, P. (1973). Variations of the glaciers of Westland National Park and the Hooker Range, New Zealand, *New Zealand Journal of Botany*, **11**, 349–88.
Webber, P. J. & Andrews, J. T. (1973). Lichenometry, a commentary. *Arctic and Alpine Research*, **5**, 295–302.
Zumbühl, H. J. (1975). Die Schwankungen des Unteren Grindelwaldgletschers in den Historischen Bild- und Schriftquellen des 12. bis 19. Jahrhunderts. *Zeitschrift für Gletscherkunde und Glazialgeologie*, **11**, 12–50.

4

The use of pollen analysis in the reconstruction of past climates: a review

H. J. B. BIRKS

Abstract

Pollen analytical data obtained from lake and bog sediments provide direct information relevant to the reconstruction of the past flora and vegetation of the area under study. As both modern floras and modern vegetation are related in a broad way to modern climate, pollen analysis can provide indirect information relevant to the reconstruction of past climates over particular time spans (10^2–10^5 years) with a sample resolution of 10–1000 years.

Various approaches are available in reconstructing past climates from palynological data. The floristic or 'indicator-species' approach is reviewed, in which the modern distributions of particular species are compared with contemporary climatic patterns. Limitations of this approach are considered in the light of recent experimental studies on the climatic tolerance of selected species of contrasting distributions. Studies involving the distribution and performance of a few selected species in relation to two or three ecologically important climatic variables are also discussed. The vegetational or 'multivariate' approach is described in which mathematical transfer functions are used to calibrate modern pollen assemblages in terms of modern climate. These functions are then used to transform fossil pollen data into quantitative estimates of past climate. Various refinements to this approach are considered, and the advantages, disadvantages, and assumptions of the approach are discussed. The assumption that plant distributions are in equilibrium with climate both today and in the past is examined, and is shown to be invalidated by several recent studies on the migration of trees in northwest Europe and in parts of eastern North America during the Holocene.

Introduction

The ultimate aim of many, if not all Quaternary palaeoecological studies is the reconstruction of past environments, including past climates, for particular points in time and space. Of the various types of fossils preserved in Quaternary terrestrial deposits, few are more abundant than pollen grains and spores of vascular plants, with the result that Quaternary terrestrial palaeoecology is dominated by the technique of pollen analysis.

The reason why pollen analysis provides a means for reconstructing past climates is that as a pollen assemblage at a particular time and space is a function of the regional flora and vegetation, and as the regional flora and vegetation are largely controlled by regional climate, there is thus some relationship, admittedly a complex and indirect one, between pollen and climate.

In the context of Quaternary palaeoclimatology, the terrestrial fossil-pollen record represents a potentially important source of climatic information both in terms of length of record (10^2–10^5 years; see Fig. 4.1) and sample resolution (10–1000 years). Instrumental and historical records of past climate and tree-ring data have the resolution of tens of years but a record of hundreds of years only (Fig. 4.1). The marine plankton record can provide a long record for 10^4–10^6 years but with a resolution of about 1000–5000 years per sample.

In some instances overlap between these various sources of quantitative climatic information can occur. For example, pollen records from annually laminated lake sediments have a sample resolution similar to tree rings (see Swain, 1973, 1978; Saarnisto, Huttunen & Tolonen, 1977; Cuynar 1978; Tolonen, 1978); tree-ring records from bristlecone pines

Fig. 4.1. The length of record and sample resolution of climatic information derivable from the marine plankton record, the pollen record, the tree-ring record, and historical and instrumental records. (Modified from an unpublished diagram by T. Webb)

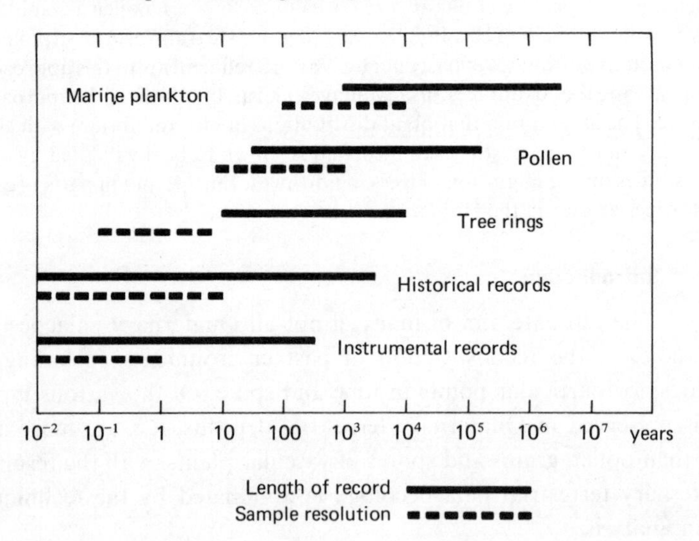

have a time span approaching the length of the pollen record (La Marche, 1974); and high sedimentation-rate marine cores can, in some instances, provide a sample resolution similar to that of the pollen record (Heusser, 1978). Such overlap is important, as it provides a means of comparing climatic estimates based on one aspect of an ecosystem with estimates derived from another aspect of an ecosystem.

The aim of this chapter is to review the various approaches to the reconstruction of past climates from Quaternary pollen analytical data. Such reconstructions can be based on either a few, selected species, the so-called floristic or 'indicator-species' approach, or on the fossil assemblage as a whole, the so-called vegetational assemblage or 'multivariate' approach. In either approach, information is required about the present-day climatic tolerances and requirements of the taxa found as fossils in the pollen record. The basic assumption of all these approaches is that of methodological uniformitarianism (see Rymer, 1978; Birks & Birks, 1980), namely that modern-day observations and relationships can be used as a model for past conditions and, more specifically, that the relationships between plants and climate have not changed with time, at least for the time period of the late Quaternary.

Indicator-species approach

The occurrence of fossil pollen or spores of a particular species whose modern ecological tolerances, including climatic ones, are known, provides a basis for reconstructing past environments. On the assumption of methodological uniformitarianism, the past environment is suggested to have been within the environmental range occupied by the species today (see Iversen (1964) for a discussion of the concept of indicator species). This floristic approach clearly requires knowledge of what environmental factors influence the distribution today of the species concerned. Providing such information is basically a problem in autecology, the ecology of individual species.

The commonest approach to providing such information is to compare the present-day distribution of the species concerned with the distribution of selected climatic variables (e.g. Dahl, 1964; Conolly & Dahl, 1970; Kolstrup & Wijmstra, 1977; Kolstrup, 1979, 1980). If the geographical trend of some climatic variables coincides closely with the distribution pattern of the species in question, a cause–effect relationship between the two is often assumed. For example, Conolly & Dahl (1970) related the distribution of *Betula nana* (dwarf birch) in Britain (Fig. 4.2) to the 22 °C maximum summer-temperature isotherm for the highest point in the areas where *B. nana* grows today. Fossil records indicate that it occurred in southeast England during the Devensian late-glacial. Today the maximum summer-temperature isotherm there is 30 °C, so Conolly and Dahl propose that there was a depression of 8 deg C in maximum summer temperatures in the Devensian late-glacial.

This type of approach is widely used in Quaternary palaeoecology. It is, however, unwise to assume that spatial correlations between species

Fig. 4.2. Map showing the distribution of *Betula nana* in the British Isles in relation to the 22 °C maximum summer temperature summit isotherm and the location of Devensian late-glacial fossil records. (From Conolly & Dahl, 1970)

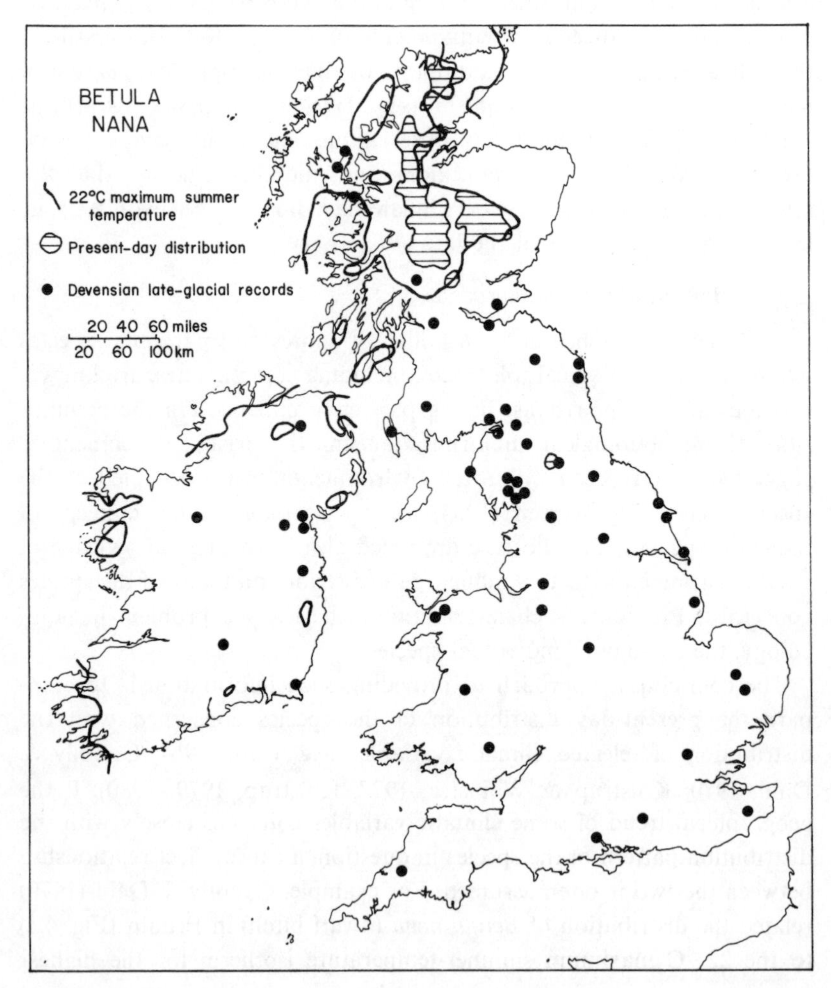

distribution and selected climatic variables, however precise, imply any simple causal connection for several reasons (see Ratcliffe, 1968; Pigott, 1975).

1. Many climatic variables are themselves correlated with each other. For example, solar radiation, day temperatures, and evaporation are often positively correlated, but negatively correlated with rainfall. Plants respond physiologically to all these variables and moreover to interactions between them.

2. The actual climate that a plant experiences differs from that recorded in standard meteorological stations. Moreover, the parts of a single plant can be simultaneously exposed to very different climatic conditions. For example, at noon on a warm sunny day, the leaves on the south side of a bush of *Corylus avellana* (hazel) may, according to Pigott (1975) be at 25–30 °C, those in the shade at 15–20 °C, and the roots at 10 °C. This range in temperature is as great as the differences in July mean temperatures between the Mediterranean and the North Cape. The ecologically limiting temperatures may thus be very different from that recorded by a meteorologist. Moreover, climatic extremes (which are not necessarily closely related to means), may be more important biologically as limiting factors than climatic averages for many species (see Dahl, 1951; Ratcliffe, 1968).

3. Even after micro-climatic measurements have been made at sites where the species grows and does not grow, any relationship between climate and distribution is hypothetical. Testing of this hypothesis can only be obtained by carefully designed and co-ordinated field and laboratory experiments (see Pigott (1970) for elegant examples of such co-ordinated studies). Such experiments are, however, rare.

4. The relationship of a species to climate may change within the range of the species due to interactions between climatic variables or to physiological ecotypes within the same species (Proctor, 1973). For example, arctic and alpine populations of *Oxyria digyna* (mountain sorrel) have different physiological tolerances (Mooney & Billings, 1961). Arctic populations have a higher rate of photosynthesis at lower temperatures, whereas alpine populations withstand higher temperatures and require more light for maximal photosynthesis (see Billings & Mooney (1968), Billings, Godfrey, Chabot & Bourque. (1971), Kershaw (1975), and Smith & Hadley (1974) for other examples of physiological ecotypes in several arctic, alpine, and boreal species).

5. The relationship of a species to climate may be indirect and may result from competition with other species whose growth is directly

influenced by climate. For example, Woodward & Pigott (1975) and Woodward (1975) have shown that the growth of the arctic-alpine *Sedum rosea* is largely insensitive to temperature, whereas the growth of the lowland *S. telephium* is greatly reduced by low temperatures. This reduced growth of *S. telephium* at low temperatures results in its inability to compete with upland species such as *S. rosea*, whereas at higher temperatures *S. telephium* grows more vigorously and is thus at a competitive advantage over *S. rosea*. Woodward (1979) has shown similar responses in growth to differences in temperature in the lowland grasses *Phleum bertolonii* and *Dactylis glomerata* compared with the upland species *P. alpinum*. These experiments indicate not only the insensitivity of *S. rosea* and *P. alpinum* to climate (cf. Dahl, 1951; Conolly & Dahl, 1970) but they suggest that their present distribution is limited primarily by competition with lowland species (see Pigott, 1978). As competition with other species has undoubtedly been different in the past, no palaeoclimatic inferences should be made on the basis of fossil occurrences of species such as *S. rosea* and *P. alpinum*.

6. It is presumed that during a period of climatic stability, species distributions may reach an equilibrium with climate. In the case of plants we are very ignorant as to whether such an equilibrium ever exists and if it does how long it takes to be attained. It may take tens, hundreds, or even thousands of years, depending on the species concerned, its reproductive strategy, its generation time, and competition with other plants. The occurrence of short-term fluctuations of climate raises the possibility of lag effects, and reduces the validity of apparent species–climate correlations as indicative of real limiting conditions. The nature of the equilibrium is also unknown (see Ratcliffe, 1968).

7. Climate may have its greatest influence not on the adult plant whose distribution is known, but on some critical phase in the life cycle, such as germination, pollination, or seed production. For example, *Tilia cordata* (lime) reaches its northern limit in Britain in the southern Lake District. In these localities it rarely produces fertile fruit even though it flowers and produces pollen every year. This failure appears to be due to lack of fertilisation, and laboratory experiments show that pollen of *T. cordata* requires temperatures above 16–18 °C for the development of the pollen tube and its growth down the long style of a *Tilia* flower (Paice, 1974; Pigott, 1975). These are about the maximum temperatures reached by *Tilia* flowers in the Lake District and then only for very short periods in the day, unless the weather is exceptionally warm. Thus pollen germination and fertilisation are critically dependent on the weather. Fertilisation rarely occurs, and fertile fruit is very rarely produced.

Individual lime trees may, however, live for many centuries. Because of their ability to regenerate from suckers, they are able to persist more or less indefinitely in a climate which inhibits fruit production. In the southern Lake District a plant–climate equilibrium does not exist, and may not exist for many hundreds or even thousands of years, as the normal life cycle of the plant (germination, growth, flowering, seed production) cannot be completed in the present-day climate. Fruit production frequently occurs further south in the Midlands of England, where mean summer-temperatures are about 2 deg C higher than in the Lake District.

It is very likely that the present-day distribution of *T. cordata* in northern England (Fig. 4.3) is a 'fossil' one from the Holocene (post-glacial) climatic optimum (Fig. 4.3) when temperatures were 2–3 deg C higher than they are today (Godwin, 1975). Such temperatures would have resulted in the production of fertile fruit and hence the colonisation of sites previously lacking *T. cordata*. It has, however, been able to persist in these northern localities by vegetative reproduction even though fertile fruit production occurs very infrequently. This failure to regenerate from

Fig. 4.3. Maps showing the present-day native distribution of *Tilia cordata* in the British Isles (from Perring & Walters, 1962) and the percentages of *Tilia* pollen (expressed as percentages of total tree pollen, including *Corylus*) at 5000 BP in the British Isles (from Birks, Deacon & Peglar, 1975) with incorporation of data from Paice (1974) and H. J. B. Birks (unpublished). The likely northern limit of *Tilia cordata* in England at 5000 BP is shown as a dashed line.

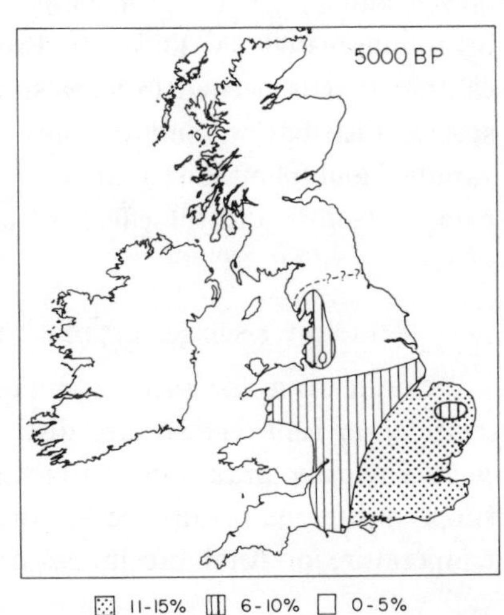

seed limits its ability to establish new colonies and restricts *Tilia* to situations where, for some reason, trees have been able to persist despite widespread forest destruction. Such situations include cliffs, steep ravines, screes, rocky slopes, and limestone pavements (Pigott & Huntley, 1978). This ecological and physiological work on *Tilia* is strongly suggestive of a 3000–5000-year lag in the attainment of an equilibrium between climate and distribution of *T. cordata* (see also Pennington, 1979). Any reconstruction of past climate based on *T. cordata* should consider these problems.

8. Many factors other than climate influence the distribution of plants, particularly edaphic, topographical, and historical factors. The last factor is particularly important, as it includes recent destruction of habitats. The absence of a species in an area can often be related to non-climatic factors (see Ratcliffe, 1968).

The physiological results obtained by Dilks & Proctor (1975) on the relationship of net assimilation rate to temperature in mosses of contrasting distribution emphasise many of the problems in the 'indicator-species' approach. They considered three species of *Andreaea*: *A. rothii*, which is widespread in the north and west of Britain; *A. alpina*, which is a montane species generally found above 500 m altitude from North Wales northwards; and *A. nivalis*, which is restricted to late snow-beds above 1100 m in the Scottish Highlands. Considerations of their present distributions would suggest that *A. rothii* would be more tolerant of high temperatures than *A. alpina* or *A. nivalis*, and that *A. nivalis* would be intolerant of high temperatures. Experimental data indicate the reverse, with the optimum temperature for net assimilation being 30 °C for *A. nivalis*, and 20–25 °C for *A. rothii* and *A. alpina*. These experiments, and others discussed by Dilks & Proctor (1975) emphasise the extreme caution that is needed in suggesting cause–effect relationships between species distribution and climate *solely* on the basis of one climatic variable and plant distributional data without consideration of other climatic factors and the effects of intraspecific competition.

Indicator-species approach using two or more climatic variables

In many instances it is more profitable for a palaeoecologist to consider present-day species distributions in relation to two or more physiologically important variables (see Hintikka, 1963), rather than simply using one readily measurable climatic variable such as mean July temperature or total precipitation. This 'bivariate' approach using indicator species was pioneered by Iversen (1944) in his classic work on *Viscum album* (mistletoe), *Hedera helix* (ivy), and *Ilex aquifolium* (holly). On the basis of detailed ecological observations over several years, Iversen related the distribution and performance of these shrubs to the mean temperature of the coldest month and to the mean temperature of the warmest month (Fig. 4.4). He delimited the 'thermal-limits' of these species within which the species flowered and produced seed. Of the three shrubs *Ilex* is intolerant of cold winters (its cambium is killed by frost) but it is tolerant of cool summers; *Hedera* is intolerant of winters with mean temperatures colder than −1.5 °C but it requires warmer summers than *Ilex*; *Viscum* is tolerant of cold winters but it requires warmer summers than either *Ilex* or *Hedera*.

The three shrubs are ideal indicator species because their pollen is identifiable to species level, their pollen is not blown great distances so problems of long-distance transport do not arise, and the plants are rapidly dispersed by birds eating their berries and hence the distributions may be in equilibrium with the climate.

Fig. 4.4. The thermal-limit curves for *Ilex aquifolium*, *Hedera helix*, and *Viscum album* in relation to the mean temperatures of the warmest and coldest months. Samples 1, 2, and 3 represent samples with pollen of *Ilex*, *Hedera*, and *Viscum*, *Hedera* and *Viscum*, and *Ilex* and *Hedera*, respectively. See text for further details . (From Iversen, 1944)

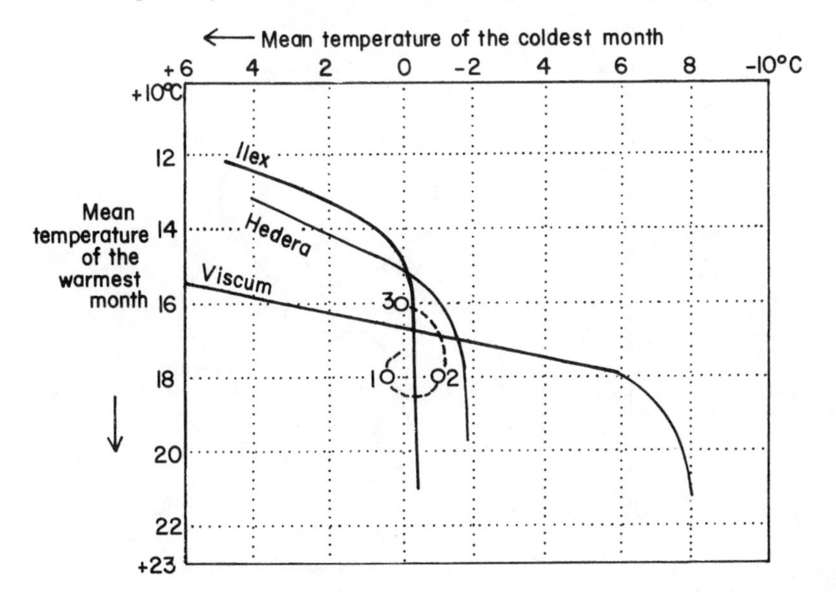

In reconstructing past climates from the pollen record, suppose pollen spectra 6000 years ago contained pollen of *Viscum*, *Hedera*, and *Ilex*. It can be suggested that the climate at that time fell within the thermal limits of all three species (point 1 on Fig. 4.4). If pollen spectra of 4000 BP contained pollen of *Viscum* and *Hedera*, the climate at that time may be estimated from the thermal limits of *Viscum* and *Hedera* (point 2 on Fig. 4.4). Similarly if pollen spectra of 1000 BP lacked pollen of *Viscum* but contained *Ilex* and *Hedera* pollen the climate can be inferred as that of point 3 on Fig. 4.4. Iversen used this approach to suggest that in the mid-Holocene of Denmark summers were 2–3 deg C warmer and winters 1–2 deg C warmer than today.

A similar approach was used by Churchill (1968) for relating the distribution of *Eucalyptus* species in southwest Australia to climate. Instead of using temperatures of the warmest and coldest months, Churchill related the distribution of different species to the mean rainfall of the wettest and driest months, as measures of the extremes of soil

Fig. 4.5. Grichuk's technique for estimating the January and July temperatures for a fossil pollen spectrum that contained pollen of *Ulmus*, *Alnus*, *Picea*, and *Myrica*. The shaded area indicates the temperatures today where all four species grow. (From Grichuk, 1969)

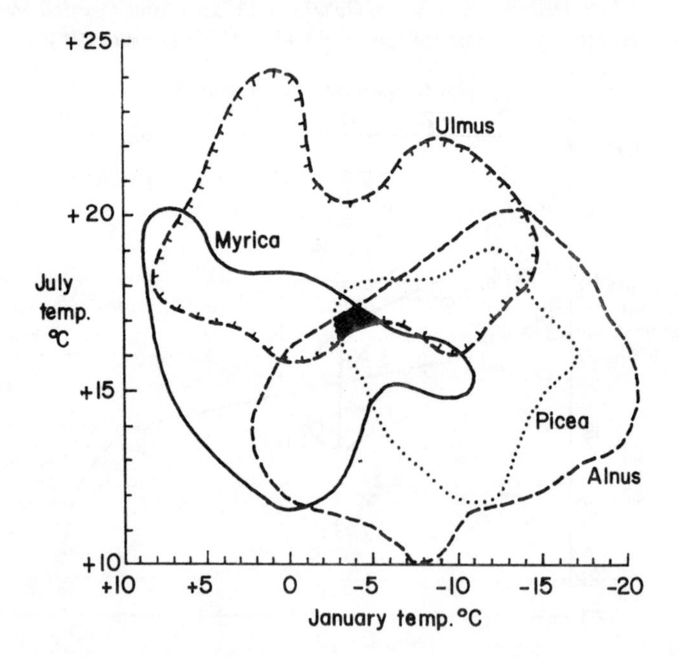

moisture. He showed that *E. diversicolor* required higher soil moisture than either *E. marginata* or *E. calophylla.* By studying the ratio of *E. diversicolor* pollen to *E. calophylla* pollen in peat deposits, Churchill derived a time series of changes in moisture from 6000 BP to the present day. Moist conditions prevailed from 6000 to 5000 BP. Thereafter it became increasingly dry until about 2500 BP, when conditions became wetter again until 1500 BP. After this there was a marked dry period until 500 BP, when conditions became wetter up to the present day.

A related approach but involving four climatic variables and several 'indicator species' has been developed by Grichuk (1969) to reconstruct mid-Holocene (5500 BP) climates for the northern hemisphere from pollen analytical data. The climatic variables considered were January and July mean temperatures, total annual precipitation, and the number of frost-free days. For each variable, he plotted the tolerance limits of the taxa found in pollen diagrams (see Fig. 4.5 for a pollen sample containing *Ulmus*, *Alnus*, *Picea*, and *Myrica*). When the total distribution of the four taxa is considered in relation to January and July temperatures, the area where all their thermal limits overlap is considered to represent the past climate at the time of the pollen sample.

Grichuk presents estimates of the four climatic variables based on 20 pollen sites in the northern hemisphere and deviations of these variables from present-day climate. The deviations in temperature with latitude at 5500 BP are shown in Fig. 4.6 along the 90° W longitude meridian for North America and the 35° E meridian for Europe. January temperatures were up to 4 deg C higher than today at 65–55° N in Europe and America, whereas south of 40° N mean January temperatures were lower (−3 to −7 deg C) than at present, especially in the southeast United States and in the Middle East. The estimates for deviations of the July

Fig. 4.6. Deviations from present-day values of January and July mean temperatures at 5500 BP for sites at different latitudes in North America and Europe along the meridians 90 °W and 35 °E, respectively. (From Grichuk, 1969)

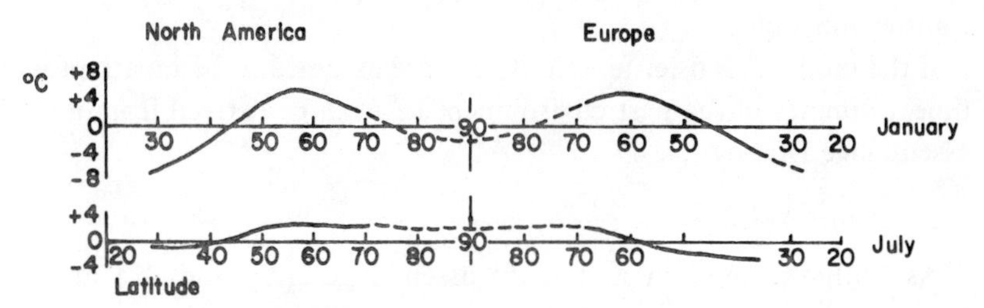

temperatures are slightly different. July temperatures were 1–2 deg C cooler in southern latitudes and 1–2 deg C warmer between 50° N and 70° N.

The major limitation of all these 'indicator-species' approaches is that they assume that the distribution of particular plants is controlled by one or a few climatic variables only. In the few cases where climatic controlling factors have been studied in detail (e.g. Forman, 1964, Pigott, 1970) many factors appear to be operative. Moreover, only a few taxa are considered, and little attention is given to the numerical frequencies of the different fossil pollen and spore types in fossil spectra. An alternative approach is thus to consider quantitatively the whole fossil assemblage in relation to a large number of environmental variables. Such a 'multivariate' or vegetational assemblage approach is discussed in the next section.

Multivariate approach

This approach considers the fossil assemblage as a whole and the numerical proportions of the different fossils in that assemblage. The approach has been used by pollen analysts in an intuitive, non-quantitative way for many decades. For example, pollen assemblages have been interpreted as reflecting tundra, park-tundra, birch forest, or pine forest. Inferences about the past climate have been based on the present-day climate in which these broad vegetation types are found today. In recent years, attempts have been made to quantify this approach by deriving transfer functions that directly relate assemblages of organisms to the environment. Such transfer functions can be derived mathematically for the present-day by relating modern assemblages of pollen over a broad geographical area to modern environmental variables using the equation

$$X_m T_m = E_m$$

where X_m and E_m are data matrices for modern biological and environmental variables, respectively, and T_m is the matrix of modern transfer functions.

If the modern transfer functions (T_m) are assumed to be invariant with time, estimates of the past environment ($\hat{E}_f$) can be derived from a fossil assemblage (X_f) by

$$\hat{E}_f = X_f T_m$$

As both the modern and fossil assemblages (X_m and X_f) and the modern environmental variables (E_m) are defined by many variables, multivariate methods for data analysis are essential to derive modern transfer functions (T_m). Several methods were developed independently to relate modern pollen data to regional climate by Cole (1969), Webb (1971), and Webb & Bryson (1972), to relate modern foraminiferal assemblages in the oceans to oceanographic data by Imbrie & Kipp (1971), and to relate tree-ring thicknesses to climatic data by Fritts, Blasing, Hayden & Kutzbach (1971). Recently Webb & Clark (1977) and Sachs, Webb & Clark (1977) have reviewed these and other methods. They show that they are all variants of the basic multiple linear regression model.

Traditionally pollen analysts have reconstructed past climate in two stages. Firstly the past vegetation (V_f) is reconstructed from the fossil pollen assemblage (P_f) using some modern representation factor (R_m) that corrects for the different pollen productivities of different taxa, i.e.

$$\hat{V}_f = P_f R_m$$

Secondly the past vegetation is used as a basis for reconstructing the past climate (C_f), by

$$\hat{C}_f = \hat{V}_f D_m$$

where D_m is some function that expresses the present-day relationship between vegetation and climate.

These equations can be combined into

$$P_f(R_m D_m) = \hat{C}_f = P_f T_m$$

where T_m are the modern transfer functions between modern pollen assemblages and modern climate. Rather than deriving T_m directly by comparing modern pollen with modern climate, pollen analysts have inferred past climate qualitatively or quantitatively using these two steps. Transfer functions are invariably established non-analytically as mental, qualitative judgements to relate pollen to vegetation and vegetation to climate. In some instances the relationship between pollen and vegetation has been quantified (e.g. Andersen, 1970; Bradshaw, 1978), but the relationship between vegetation and climate has been rarely quantified. Because of the difficulty of defining D_m, Cole (1969), Webb (1971), and Webb & Bryson (1972) followed the alternative approach, and related modern pollen assemblages to modern climate directly by means of transfer functions T_m, i.e.

$$P_m T_m = C_m$$

For T_m to be applicable to the reconstruction of past climate from fossil pollen data, various assumptions have to be made (see Webb & Bryson, 1972: 74–5): (*a*) climate is the ultimate cause of changes in the pollen record, (*b*) vegetation of the past, as reflected in the observed pollen assemblages has responded in a constant way to climate and has thus been in equilibrium with the climate, (*c*) linear relationships implicit in the mathematical techniques used to estimate T_m represent adequately the relationships between climatic variables and pollen. Webb & Bryson (1972) suggest that for the scale, both temporal and spatial, of their study these assumptions are valid. I shall return to this point later.

Webb & Bryson (1972) used a set of 34 modern pollen samples from central North America defined by eight variables. They related these samples to a series of modern climatic variables by means of canonical correlation analysis (a standard multivariate technique for relating two sets of multivariate observations) to derive modern transfer functions. These functions were then applied to fossil pollen sequences from Minnesota and Wisconsin to derive time series of quantitative estimates of past climate (Fig. 4.7).

For the numerical analysis, they excluded several pollen types for a variety of reasons. For example Cyperaceae pollen was excluded because

Fig. 4.7. Estimated values for 0–15 000 BP of the precipitation, precipitation minus potential evaporation, and the July mean temperature for three sites in Minnesota and Wisconsin derived by transfer functions of fossil pollen data. (From Webb & Bryson, 1972)

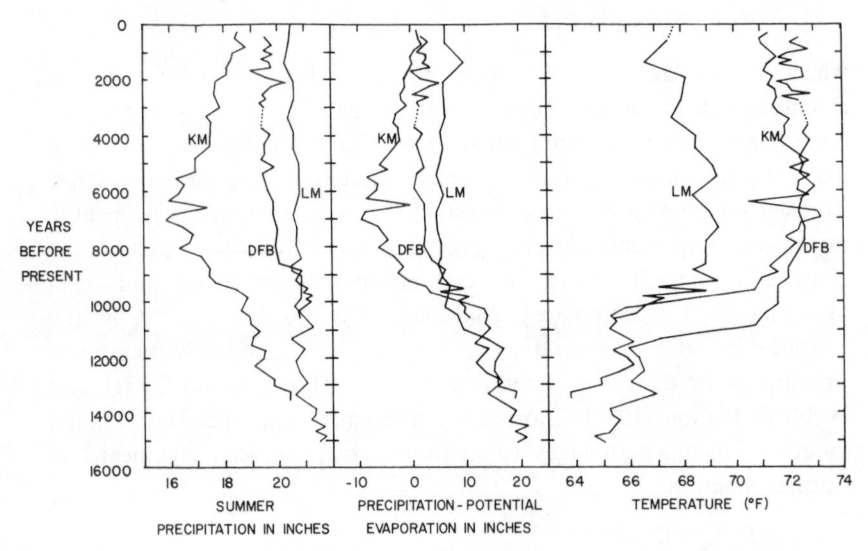

it was of local origin; *Ambrosia*-type pollen was excluded because present-day values are a function of human disturbance; *Populus*, Compositae, Rosaceae, and Leguminosae pollen were excluded because of consistently low (<2 per cent) values; *Larix*, *Fraxinus*, *Ostrya/Carpinus*, *Ulmus*, and *Corylus* pollen were excluded because their fossil values are two to five times greater than any modern values (i.e. modern values are not an exact analogue for past assemblages); Gramineae, *Alnus*, *Artemisia*, *Abies*, *Tilia*, and *Salix* pollen were excluded to 'increase the ratio of observations to variables and hence the statistical significance' of the results. The remaining eight pollen types were *Pinus*, *Picea*, *Betula*, *Quercus*, *Acer*, *Carya*, *Juglans*, and Chenopodiaceae. They were expressed as percentages of their sum plus the sum of *Abies*, *Salix*, and *Artemisia*, none of which were included in the analysis. The rationale for this procedure is far from clear. It should be noted that the pollen assemblage used by Webb & Bryson (1972) is a highly selective and thus a potentially biassed sample of the total pollen assemblage deposited in the Midwest of the United States during the last 15 000 years.

A wide variety of modern climatic variables was used in the derivation of the transfer functions, including duration of air masses, July mean temperature, rainfall during the growing season, length of growing season, moisture stress, snowfall, etc. The modern pollen and climatic data were subjected to canonical correlation analysis, and the canonical regressions obtained for the first two correlations were used as transfer functions. The accuracy of these was tested by applying them to an independent set of modern pollen data, and comparing the predicted climate with the actual climate. In general there was a close correspondence, with July temperature estimates within 0.6 deg C and precipitation estimates within 2 cm.

The estimates for past climate for the three fossil sequences considered by Webb & Bryson (Fig. 4.7) show that a sharp rise in temperature occurred within 200–300 years at the opening of the Holocene at about 10 000 BP. Prior to 10 000 BP precipitation decreased gradually. At about 9000 BP precipitation decreased rapidly at Kirchner Marsh, with a resulting greater moisture-stress, corresponding with the eastward expansion of the prairie at this time. After about 5000 BP precipitation increased gradually again at Kirchner Marsh. Temperature appears to have varied little throughout the Holocene, but with a slow fall after 6000 BP. This pattern corresponds broadly to Grichuk's (1969) reconstruction for these latitudes.

The results obtained by Webb & Bryson (1972) are generally consistent with climatic reconstructions inferred qualitatively by other workers. The quantitative approach of Webb and Bryson has several important advantages: (*a*) the results are repeatable and the estimates are based on the quantitative information for all the taxa considered, (*b*) the numerical methods force the investigators to be explicit about the methods and their assumptions, and (*c*) unsuspected features are revealed that justify further, more detailed work. The approach is potentially a valuable one and the methods can clearly be refined by using more extensive and more detailed modern pollen data (Davis & Webb, 1975; Webb & McAndrews, 1976), more detailed climatic information, and more versatile and less restrictive numerical techniques (Bryson & Kutzbach, 1974; Harris, Darwin & Newman, 1976; Gnanadesikan, 1977; Howe & Webb, 1977). Ultimately palaeoclimatic maps derived from maps of pollen data (e.g. Birks *et al.* 1975; Birks & Saarnisto, 1975; Bernabo & Webb, 1977) can be produced for the late Holocene at least, by the use of transfer functions. Such synoptic palaeoclimatic data can then be incorporated with palaeoclimatic data derived from independent sources, such as tree-line fluctuations based on buried tree-stumps (Karlén, 1976), glacial advances and retreats (Denton & Karlén, 1973), inorganic sedimentation in lake sediments (Karlén, 1976), faunal evidence (Kerney, 1968; Osborne, 1976; Casteel, Adam & Sims, 1977), peat-stratigraphical studies (Aaby, 1976), and fire history (Swain, 1973, 1978; Cwynar, 1978).

Discussion

The basic assumptions of the multivariate assemblage approach developed by Webb & Bryson (1972) are that climate is the ultimate cause of changes in the pollen record, that the relationship between vegetation (and hence pollen) and climate has not changed with time, and that linear relationships adequately represent the relationships between climatic variables and pollen assemblages. For these assumptions to be valid, modern vegetation must be in equilibrium with climate and, moreover, past vegetation must have been in equilibrium with past climate. Otherwise climate need not have been the only cause of pollen stratigraphical changes and, conversely, changes in the pollen-stratigraphical record do not necessarily reflect climatic changes. The discussion presented here concentrates on the question of vegetation–climate equilibria in the past.

It is likely that climate was *not* the only cause of many of the major pollen-stratigraphical changes during the Holocene in either northwest

Europe (Iversen, 1960) or eastern North America between 65° W and 90° W longitude (Davis, 1976, 1978). The well-documented changes in pollen stratigraphy over the last 10 000 years probably reflect primarily the migration of species into new areas from glacial refugia stimulated initially by large-scale regional climatic change at the onset of the Holocene. The northward extension of plant ranges *could* be interpreted simply as reflecting climatic amelioration at higher and higher latitudes as glacial climates gave way to interglacial climates. This simple interpretation cannot, however, apply to the known Holocene history of the majority of trees, either in eastern North America (Davis, 1978) or in northwest Europe (Iversen, 1960).

In eastern North America each species migrated in different directions and at different rates from its refugia (Fig. 4.8 and Davis, 1976) until it reached its climatic limits. There would clearly be a lag in attaining such equilibrium boundaries (Jacobson, 1979). The observed migration patterns are too complex and diverse to support any simple climatic interpretation. Moreover, the differences in migration rates of various species may be related, not only to the migration potential of the species, but also to the complex climatic patterns that might allow one species to migrate quickly while impeding the spread of other species. Davis (1978) presents preliminary evidence to suggest that the rate of westward migration of beech in Michigan in the last 3000 years may have varied in response to climatic change. In the case of some species, however, the climatic limits may not yet have been reached (Jacobson, 1979), even after 10 000 years. Such non-equilibrium distributions either today or in the past cannot provide any precise information about climate. They must be carefully distinguished from equilibrium situations where species have migrated into an area and have subsequently retreated. Such retreats, in the absence of evidence for human interference, can provide important and ecologically unambiguous evidence for environmental change. Examples in eastern North America include the retreat of *Pinus strobus* (white pine) in eastern Canada 5000 years ago (Terasmae & Anderson, 1970), the westward retreat of the prairie–forest boundary in Minnesota since 7000 BP (McAndrews, 1966; Wright, 1968), and the altitudinal shifts in forest vegetation in the White Mountains of New Hampshire (Davis, 1978; Davis, Spear & Shane, 1980). Examples in northwest Europe include the mid-Holocene southward retreat of the northern limit of *Corylus avellana* (hazel) in Scandinavia (Andersson, 1902), the retreat of the northern limit of *Pinus sylvestris* (pine) in northern Scandinavia since 5000 BP (Hyvarinen, 1975), and the altitu-

Fig. 4.8. Isochrones (in thousands of years BP) of the time of arrival and expansion of selected trees in eastern North America during the Holocene. Dashed lines and shaded areas show the present range. (From Davis, 1976)

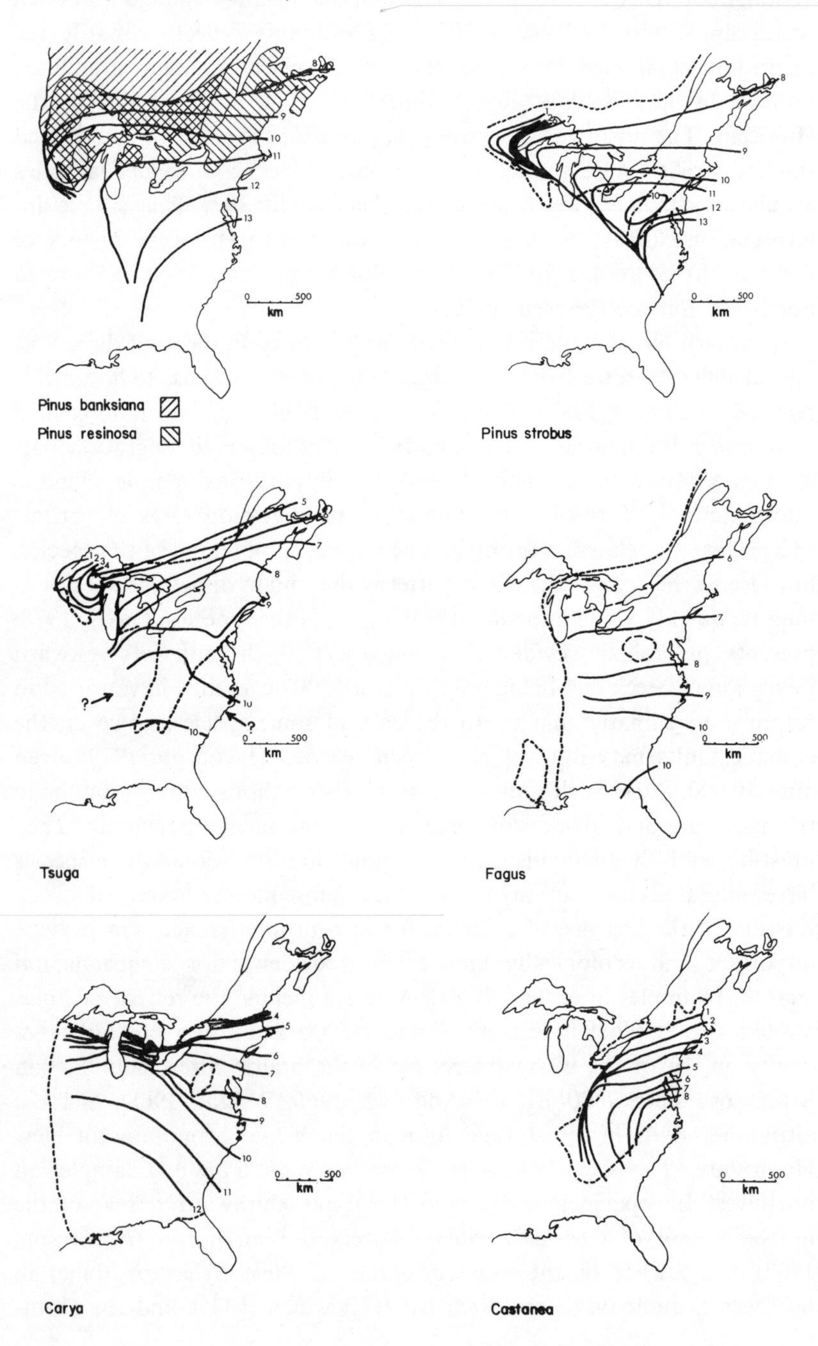

dinal descent of the tree line in the Scandinavian mountains since 5000 BP (Karlén, 1976).

Isopollen maps for European trees recently prepared by B. Huntley & H. J. B. Birks (unpublished data) for the last 13 000 years similarly show a wide range of rates and directions of migration, with species migrating at different rates from refugia in southern, southwestern, southeastern, eastern, and even western Europe. The migration routes of *Picea abies* (European spruce) are shown by its isopollen maps from 10 000 BP to the present day (Fig. 4.9). It is still migrating today in Scandinavia at its northern, western, and southern limits (Berglund, 1966; Moe, 1970). Its present-day distribution is presumably not in equilibrium with climate, and the simplest explanation is that spruce has not yet reached its climatic limits in Scandinavia.

The major pollen stratigraphical changes during the Holocene of northwestern Europe and much of eastern North America may thus simply reflect the arrival of migrating species rather than climatic change (Iversen, 1960; Davis, 1976, 1978). The arrival of a species that is limited by its migration rate cannot be interpreted as reflecting climatic change, nor is the absence of its pollen in older levels evidence for unfavourable conditions. In circumstances where there are delays in species migration due to slow rates of spread and physical barriers to migration, the competitive interactions between species would almost certainly have been different in the past, resulting in vegetational assemblages that have no modern counterparts. Moreover, the ecological amplitudes of some species may have been different in the absence of species that immigrated later.

The problem of fossil pollen assemblages and climatic patterns that have no modern analogues is a critical one in the multivariate assemblage approach, as it largely invalidates the approach. The high percentages of *Ulmus*, *Larix*, *Fraxinus*, and *Ostrya/Carpinus* pollen in the Wisconsinan late-glacial were ignored by Webb & Bryson (1972) because the values were higher than in any modern surface spectra from eastern North America and thus reflected a vegetational type that has no modern analogue. The high percentages of these pollen types in the late-glacial may indicate the former abundance and/or wider ecological amplitude of these trees at a time when *Betula*, *Pinus*, and *Quercus* (all later immigrants) were absent. Alternatively the pollen assemblage may reflect a vegetation type that developed in response to a unique combination of climatic and other environmental conditions at the time of the retreat of the Laurentide ice sheet in the Great Lakes region of North

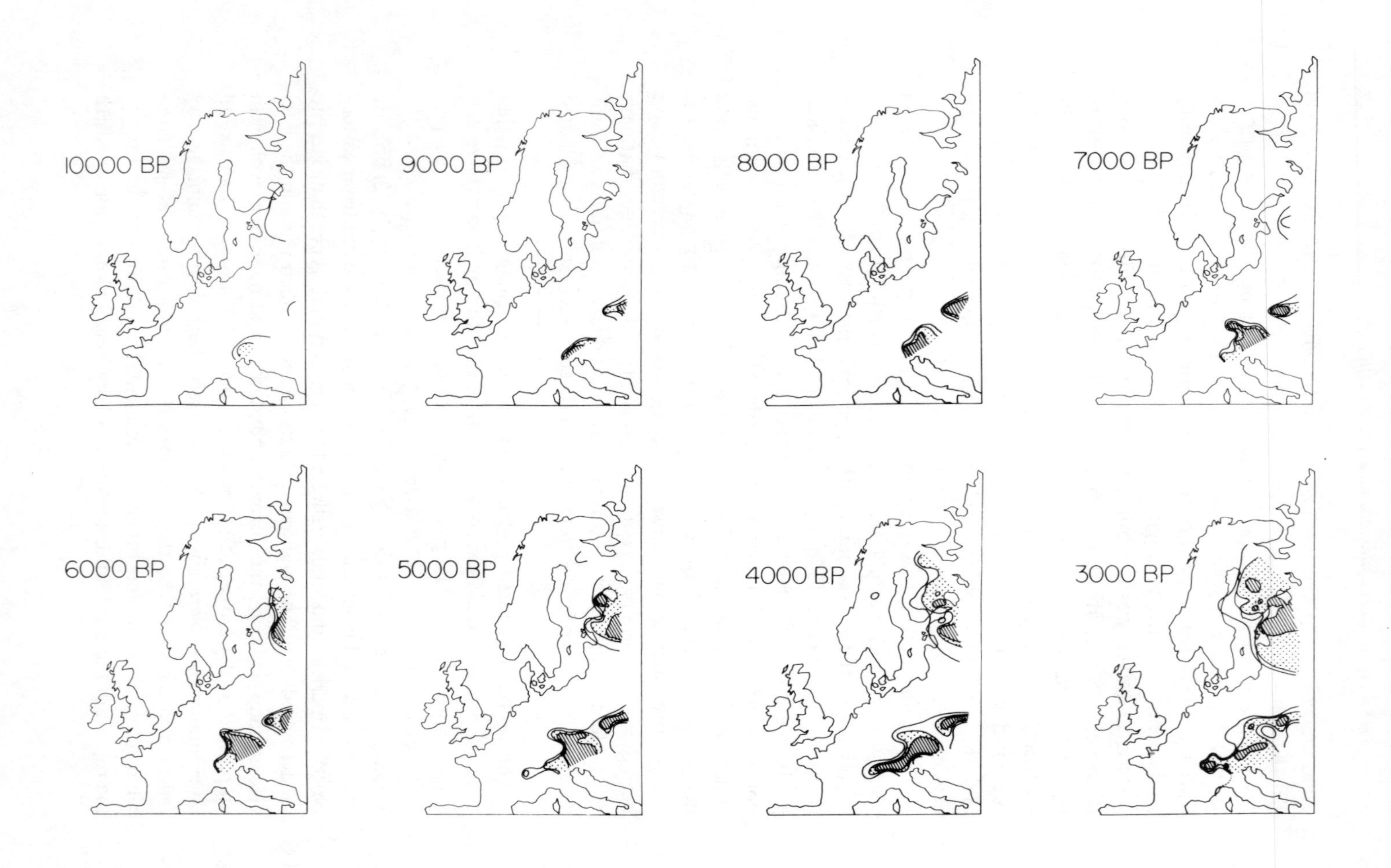
10000 BP
9000 BP
8000 BP
7000 BP
6000 BP
5000 BP
4000 BP
3000 BP

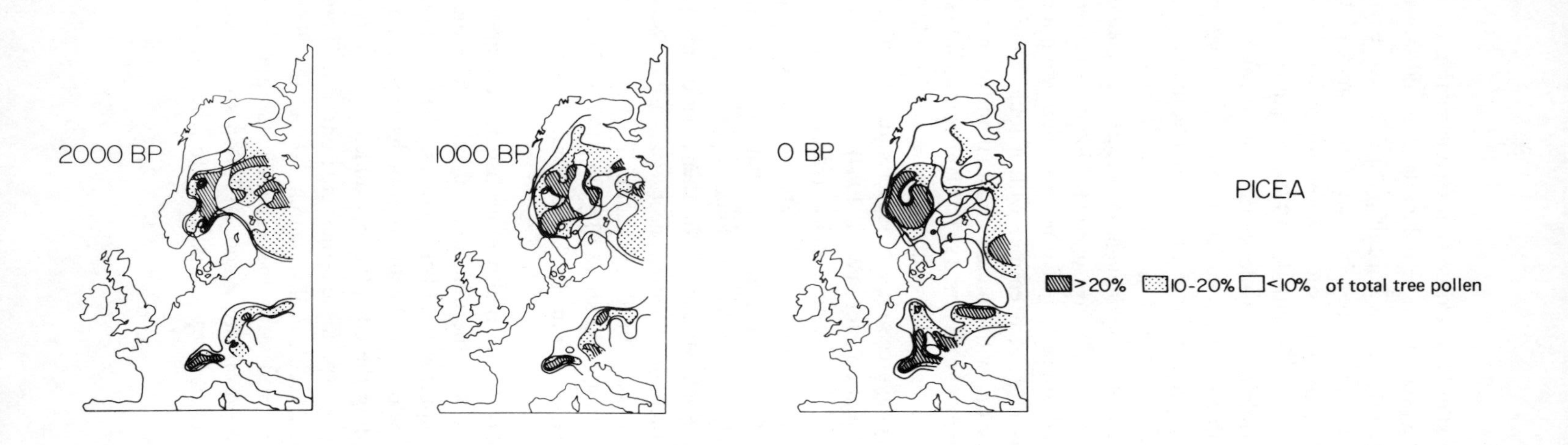

Fig. 4.9. Isopollen maps for 10 000–0 BP showing the percentages of *Picea* pollen (expressed as percentages of total tree pollen) for central and western Europe. (B. Huntley & H. J. B. Birks, unpublished data)

America. In either case we should be cautious in reconstructing the past climate by analogy with the present. These pollen data should, however, be utilised rather than being discarded if a full picture of past environments (including climate) is ever to be obtained. Other lines of independent evidence are clearly required in such circumstances, such as from stable-isotope studies (Eicher & Siegenthaler, 1976), faunal assemblages (Coope, 1975), and geomorphological evidence (Sissons, 1979).

The problem of no-analogues becomes even more acute when interglacial pollen assemblages are considered. For example, in the Gortian interglacial in Ireland species such as *Picea abies*, *Taxus baccata*, *Abies alba*, *Rhododendron ponticum*, *Ilex aquifolium*, *Bruckenthalia spiculifolia*, and *Hymenophyllum* spp. were growing together. Nowhere today can such an assemblage be found, suggesting that modern plant communities should *not* be regarded as integrated entities with a long, continuous history that are in equilibrium with climate and that move *in toto* in response to environmental change (Watts, 1973). Plant communities and the vegetation which they comprise clearly do not have any long history, but are simply temporary aggregations of species developed under particular historical and climatic factors (West, 1964).

It is, of course, possible to derive statistically significant correlations between modern pollen assemblages and a variety of climatic variables, just as it is possible to find high spatial correlations between the distribution of single species and selected climatic parameters in the indicator-species approach. Such correlations on their own do not prove in *either* approach that there is a cause–effect relationship between vegetation and climate, as is assumed by Churchill (1968), Grichuk (1969), Conolly & Dahl (1970), Webb & Bryson (1972), Bryson & Kutzbach (1974), Kay (1979), Kolstrup (1979, 1980) and Andrews, Mode & Davis (1980).

Despite the attractiveness of the multivariate assemblage approach with its quantitative results, and the difficulties of both the indicator-species and the assemblage approaches, I believe that in the long run, our most reliable reconstructions of past climate from pollen analytical data will come from understanding the complex and difficult, but very real, ecological interactions between species and their environment. There is a need for carefully designed ecological research, along the lines of Iversen's (1944) studies coupled with the experimental approach of Pigott (1970, 1975) and Woodward & Pigott (1975) on species–climate interactions, particularly of species of interest to the Quaternary palaeoecologist.

There is also a need for other ecological factors to be considered before any attempts at quantitative climatic reconstructions can be made. The problems of competition and the varying ecological amplitudes of species in different competitive regimes have already been discussed. The importance of herbivorous animals, both vertebrates and insects and of pathogens in controlling the abundance and distribution of plants has largely been ignored by palaeoecologists. Notable exceptions are Davis's (1981) work on the dramatic decline of *Tsuga* pollen in eastern North America 4800 years ago and Anderson's (1974) study of the well-documented decline of *Castanea dentata* in New England between 1904 and 1940, as reflected in the pollen-stratigraphical record. Both catastrophic declines in pollen frequencies could easily have been misinterpreted as reflecting climatic change if the palaeoecology of the events had not been so intensively and critically studied.

There is clearly a need for active co-operation and interchange of ideas and approaches between climatologists and ecologists, for as Faegri (1950) said 'Palaeoclimatology suffers from the disadvantage that those who can judge the evidence (biologists, geologists, etc.) cannot judge the conclusions, and those who can judge the conclusions (meteorologists) cannot judge the evidence.'

Acknowledgements

I wish to acknowledge the many stimulating and valuable discussions about palaeoclimatic reconstructions that I have had with Dr Hilary H. Birks, Professor Margaret B. Davis, Dr Brian Huntley, Mr Henry F. Lamb, Professor W. A. Watts, and Dr Thompson Webb III. Several of the ideas presented here have developed in connection with research on 'Pollen, vegetation, and climatic maps for Europe 0–12 000 years ago' supported by the Natural Environment Research Council. I am grateful to Dr Hilary H. Birks and Dr Brian Huntley for their critical reading of the manuscript.

References

Aaby, B. (1976). Cyclic climatic variation in climate over the past 5500 yr reflected in raised bogs. *Nature*, **263**, 281–4.

Andersen, S. Th. (1970). The relative pollen productivity and pollen representation of north European trees, and correction factors for tree pollen spectra. *Danmarks Geologiske Undersøgelse*, II series, **96**. 99 pp.

Anderson, T. W. (1974). The chestnut pollen decline as a time horizon in lake sediments in eastern North America. *Canadian Journal of Earth Sciences*, **11**, 678–85.

Andersson, G. (1902). Hasseln i Sverige fordom och nu. *Svenska Geologiske Undersøgelse*, Series Ca, No. 3. 168 pp.

Andrews, J. T., Mode, W. N. & Davis, P. T. (1980). Holocene climate based on pollen transfer functions, eastern Canadian Arctic. *Arctic and Alpine Research*, **12**, 41–64.

Berglund, B. E. (1966). Late-Quaternary vegetation in eastern Blekinge, south-east Sweden. II Post-glacial time. *Opera Botanica*, **12**(2). 190 pp.

Bernabo, J. C. & Webb, T. (1977). Changing patterns in the Holocene pollen record of northeastern North America: a mapped summary. *Quaternary Research*, **8**, 64–96.

Billings, W. D., Godfrey, P. J., Chabot, B. F. & Bourque, D. P. (1971). Metabolic acclimation to temperature in arctic and alpine ecotypes of *Oxyria digyna*. *Arctic and Alpine Research*, **3**, 277–89.

Billings, W. D. & Mooney, H. A. (1968). The ecology of arctic and alpine plants. *Biological Reviews*, **43**, 481–529.

Birks, H. J. B. & Birks, H. H. (1980). *Quaternary Palaeoecology*. London: Edward Arnold.

Birks, H. J. B., Deacon, J. & Peglar, S. M. (1975). Pollen maps for the British Isles 5000 years ago. *Proceedings of the Royal Society of London* B, **189**, 87–105.

Birks, H. J. B. & Saarnisto, M. (1975). Isopollen maps and principal components analysis of Finnish pollen data for 4000, 6000, and 8000 years ago. *Boreas*, **4**, 77–96.

Bradshaw, R. H. W. (1978). Modern pollen representation factors and recent woodland history in S.E. England. University of Cambridge, unpublished PhD thesis. 170 pp.

Bryson, R. A. & Kutzbach, J. E. (1974). On the analysis of pollen-climate canonical transfer functions. *Quaternary Research*, **4**, 162–74.

Casteel, R. W., Adam, D. P. & Sims, J. D. (1977). Late-Pleistocene and Holocene remains of *Hysterocarpus traski* (Tule Perch) from Clear Lake, California, and inferred Holocene temperature fluctuations. *Quaternary Research*, **7**, 133–43.

Churchill, D. M. (1968). The distribution and prehistory of *Eucalyptus diversicolor* F. Muell., *E. marginata* Donn ex Sm., and *E. calophylla* R. Br. in relation to rainfall. *Australian Journal of Botany*, **16**, 125–51.

Cole, H. (1969). Objective reconstruction of the paleoclimatic record through application of eigenvectors of present-day pollen spectra and climate to the late-Quaternary pollen stratigraphy. University of Wisconsin, unpublished PhD thesis.

Conolly, A. P. & Dahl, E. (1970). Maximum summer temperature in relation to the modern and Quaternary distributions of certain arctic-montane species in the British Isles. In *Studies in the vegetational history of the British Isles*, ed. D. Walker & R. G. West. Cambridge University Press.

Coope, G. R. (1975). Climatic fluctuations in northwest Europe since the Last Interglacial, indicated by fossil assemblages of Coleoptera. In *Ice Ages: Ancient and Modern*, ed. A. E. Wright & F. Moseley. Geological Journal Special Issue **6**, 153–68.

Cwynar, L. (1978). Recent history of fire and vegetation from laminated sediment of Greenleaf Lake, Algonquin Park, Ontario *Canadian Journal of Botany*, **56**, 10–21.

Dahl, E. (1951). On the relation between summer temperature and the distribution of alpine vascular plants in the lowlands of Scandinavia. *Oikos*, **3**, 22–52.

Dahl, E. (1964). Present-day distribution of plants and past climate. In *The Reconstruction of Past Environments*, ed. J. J. Hester & J. Schoenwetter. Fort Burgwin.

Davis, M. B. (1976). Pleistocene biogeography of temperate deciduous forests. *Geoscience and Man*, **13**, 13–26.

Davis, M. B. (1978). Climatic interpretation of pollen in Quaternary sediments. In *Biology and Quaternary Environments*, ed. D. Walker & J. C. Guppy. Australian Academy of Sciences, Canberra.

Davis, M. B. (1981). Outbreaks of forest pathogens in Quaternary history. *Proceedings of the IV International Conference on Palynology, Lucknow, India 1976–77* (in press).

Davis, M. B., Spear, R. W. & Shane, L. C. K. (1980). Holocene climate of New England. *Quaternary Research*, **14**, 240–50.

Davis, R. B. & Webb, T. (1975). The contemporary distribution of pollen in eastern North America: a comparison with the vegetation. *Quaternary Research*, **5**, 395–434.

Denton, G. H. & Karlén, W. (1973). Holocene climatic variations – their pattern and possible cause. *Quaternary Research*, **3**, 155–205.

Dilks, T. J. K. & Proctor, M. C. F. (1975). Comparative experiments on temperature responses of bryophytes: assimilation, respiration and freezing damage. *Journal of Bryology*, **8**, 317–36.

Eicher, U. & Siegenthaler, U. (1976). Palynological and oxygen isotope investigations on late-glacial sediment cores from Swiss lakes. *Boreas*, **5**, 109–17.

Faegri, K. (1950). On the value of palaeoclimatological evidence. *Centenary Proceedings of the Royal Meteorological Society* 1950, 188–95.

Forman, R. T. T. (1964). Growth under controlled conditions to explain the hierarchical distributions of a moss, *Tetraphis pellucida*. *Ecological Monographs*, **34**, 1–25.

Fritts, H. C., Blasing, T. J., Hayden, B. P. & Kutzbach, J. E. (1971). Multivariate techniques for specifying tree-growth and climate relationships and for reconstructing anomalies in paleoclimate. *Journal of Applied Meteorology*, **10**, 845–64.

Gnanadesikan, R. (1977). *Methods for statistical data analysis of multivariate observations*. London: Wiley.

Godwin, H. (1975). *The History of the British Flora*, 2nd edn. Cambridge University Press.

Grichuk, V. P. (1969). Opyt rekonstruktsü nekotorykh elementov klimata severnogo polusharüa v Atlanticheskü period golotsena. In *Golotsen VIII Kongressu INQUA, Paris*, ed. M. I. Neishtadt. Izd-vo Nauka.

Harris, W. F., Darwin, J. H. & Newman, M. J. (1976). Species clustering and New Zealand Quaternary climate. *Palaeogeography, Palaeoclimatology, Palaeoecology*, **19**, 231–48.

Heusser, L. (1978). Pollen in Santa Barbara Basin, California: A 12 000 year record. *Geological Society of America Bulletin*, **89**, 673–8.

Hintikka, V. (1963). Über das Grossklima einiger Pflanzenareale in zwei Klimakoordinatensystemen Dargestellt. *Annales Botanici Societatis Zoologicae Botanicae Fennicae 'Vanamo'*, **34**(5). 64 pp.

Howe, S. & Webb, T. (1977). Testing the statistical assumptions of paleoclimatic transfer functions. *Fifth Conference on Probability and Statistics, American Meteorological Society, Boston*, pp. 152–7.

Hyvarinen, H. (1975). Absolute and relative pollen diagrams from northernmost Fennoscandia. *Fennia*, **142**. 23 pp.

Imbrie, J. & Kipp, N. G. (1971). A new micropaleontological method for quantitative paleoclimatology: application to a late Pleistocene Caribbean core. In *The Late Cenozoic Glacial Ages*, ed. K. K. Turekian. Yale University Press.

Iversen, J. (1944). *Viscum*, *Hedera* and *Ilex* as climate indicators. A contribution to the study of the post-glacial temperature climate. *Geologiska Föreningens i Stockholm Förhandlingar*, **66**, 463–83.

Iversen, J. (1960). Problems of the early post-glacial forest development in Denmark. *Danmarks Geologiske Undersøgelse*, IV series, **4**(3). 32 pp.

Iversen, J. (1964). Plant indicators of climate, soil, and other factors during the Quaternary. *Report of the sixth International Congress on Quaternary, Warsaw 1961*, vol. 2, *Palaeobotanical Section*, pp. 421–8.

Jacobson, G. (1979). The palaeoecology of white pine (*Pinus strobus*) in Minnesota. *Journal of Ecology*, **67**, 697–726.

Karlén, W. (1976). Lacustrine sediments and tree-limit variations as indicators of Holocene climatic fluctuations in Lappland, northern Sweden. *Geografiska Annaler*, **58A**, 1–34.

Kay, P. A. (1979). Multivariate statistical estimates of Holocene vegetation and climatic change, forest-tundra transition zone, NWT, Canada. *Quaternary Research*, **11**, 125–40.

Kerney, M. P. (1968). Britain's fauna of land mollusca and its relation to the post-glacial thermal optimum. *Symposium of the Zoological Society of London*, **22**, 273–91.

Kershaw, K. A. (1975). Studies on lichen-dominated systems. XIV. The comparative ecology of *Alectoria nitidula* and *Cladina alpestris*. *Canadian Journal of Botany*, **53**, 2608–13.

Kolstrup, E. (1979). Herbs as July temperature indicators for parts of the Pleniglacial and late-glacial in the Netherlands. *Geologie en Mijnbouw*, **58**, 377–80.

Kolstrup, E. (1980). Climate and stratigraphy in northwestern Europe between 30 000 BP and 13 000 BP with special reference to the Netherlands. *Mededelingen Rijks Geologische Dienst*, **32**, 182–253.

Kolstrup, E. & Wijmstra, T. A. (1977). A palynological investigation of the Moershoofd, Hengelo, and Denekamp interstadials in The Netherlands. *Geologie en Mijnbouw*, **56**, 85–102.

La Marche, V. C. (1974). Paleoclimatic inferences from long tree-ring records. *Science*, **183**, 1043–8.

McAndrews, J. H. (1966). Postglacial history of prairie, savanna, and forest in northwestern Minnesota. *Memoirs of the Torrey Botanical Club*, **22**(2). 72 pp.

Moe, D. (1970). The post-glacial immigration of *Picea abies* into Fennoscandia. *Botaniska Notiser*, **123**, 61–6.

Mooney, H. A. & Billings, W. D. (1961). Comparative physiological ecology of arctic and alpine populations of *Oxyria digyna*. *Ecological Monographs*, **31**, 1–29.

Osborne, P. J. (1976). Evidence from the insects of climatic variation during the Flandrian period: a preliminary note. *World Archaeology*, **8**, 150–8.

Paice, J. P. (1974). The ecological history of Grizedale Forest, Cumbria, with particular reference to *Tilia cordata* Mill, University of Lancaster, unpublished MSc thesis, 277 pp.

Pennington, W. (1979). The origin of pollen in lake sediments, an enclosed lake compared with one receiving inflow streams. *New Phytologist*, **83**, 189–213.

Perring, F. H. & Walters, S. M. (1962). *Atlas of the British Flora*. London: T. Nelson.

Pigott, C. D. (1970). The response of plants to climate and climatic change. In *The Flora of a Changing Britain*, ed. F. H. Perring. Faringdon: Classey.

Pigott, C. D. (1975). Experimental studies on the influence of climate on the geographical distribution of plants. *Weather*, **30**, 82–90.

Pigott, C. D. (1978). Climate and vegetation. In *Upper Teesdale. The area and its natural history*, ed. A. R. Clapham. Glasgow: Collins.

Pigott, C. D. & Huntley, J. P. (1978). Factors controlling the distribution of *Tilia cordata* at the northern limits of its geographical range. I. Distribution in north-west England. *New Phytologist*, **81**, 429–41.

Proctor, M. C. F. (1973). Summing up: an ecologist's viewpoint. In *Quaternary Plant Ecology*, ed. H. J. B. Birks & R. G. West. Oxford: Blackwell.

Ratcliffe, D. A. (1968). An ecological account of Atlantic bryophytes in the British Isles. *New Phytologist*, **67**, 365–439.

Rymer, L. (1978). The use of uniformitarianism and analogy in palaeoecology, particularly pollen analysis. In *Biology and Quaternary Environments*, ed. D. Walker & J. C. Guppy. Australian Academy of Sciences, Canberra.

Saarnisto, M., Huttunen, P. & Tolonen, K. (1977). Annual lamination of sediments in Lake Lovojärvi, southern Finland, during the past 600 years. *Annals Botanica Fennici*, **14**, 35–45.

Sachs, H. M., Webb, T. & Clark, D. R. (1977). Paleoecological transfer functions. *Annual Review of Earth and Planetary Science*, **5**, 159–78.

Sissons, J. B. (1979). Palaeoclimatic inferences from former glaciers in Scotland and the Lake District. *Nature*, **278**, 518–21.

Smith, E. M. & Hadley, E. B. (1974). Photosynthetic and respiratory acclimation to temperature in *Ledum groenlandicum* populations. *Arctic and Alpine Research*, **6**, 13–27.

Swain, A. M. (1973). A history of fire and vegetation in northeastern Minnesota as recorded in lake sediments. *Quaternary Research*, **3**, 383–96.

Swain, A. M. (1978). Environmental changes during the last 2000 years in north-central Wisconsin: analysis of pollen, charcoal, and seeds from varved lake sediments. *Quaternary Research*, **10**, 55–68.

Terasmae, J. & Anderson, T. W. (1970). Hypsithermal range extension of white pine (*Pinus strobus* L.) in Quebec, Canada. *Canadian Journal of Earth Sciences*, **7**, 406–13.

Tolonen, M. (1978). Palaeoecological studies on a small lake, S. Finland, with special emphasis on the history of land use. *Annales Botanicae Fennici*, **15**, 177–208, 209–22, 223–40.

Watts, W. A. (1973). Rates of change and stability in vegetation in the perspective of long periods of time. In *Quaternary Plant Ecology*, ed. H. J. B. Birks & R. G. West. Oxford: Blackwell.

Webb, T. (1971). The late- and postglacial sequence of climatic events in Wisconsin and east-central Minnesota: quantitative estimates derived from fossil pollen spectra by multivariate statistical analysis. University of Wisconsin, unpublished PhD thesis.

Webb, T. & Bryson, R. A. (1972). Late- and postglacial climatic change in the Northern Midwest, U.S.A.: Quantitative estimates derived from fossil pollen spectra by multivariate statistical analysis. *Quaternary Research*, **2**, 70–115.

Webb, T. & Clark, D. R. (1977). Calibrating micropaleontological data in climate terms: a critical review. *Annals of the New York Academy of Sciences*, **288**, 93–118.

Webb, T. & McAndrews, J. H. (1976). Corresponding patterns of contemporary pollen and vegetation in central North America. *Geological Society of America Memoir*, **145**, 267–99.

West, R. G. (1964). Inter-relations of ecology and Quaternary palaeobotany. *Journal of Ecology*, **52** (Supplement), 47–57.

Woodward, F. I. (1975). The climatic control of the altitudinal distribution of *Sedum rosea* (L.) Scop. and *S. telephium* L. II. The analysis of plant growth in controlled experiments. *New Phytologist*, **74**, 335–48.

Woodward, F. I. (1979). The differential temperature responses of the growth of certain plant species from different altitudes. I. Growth analysis of *Phleum alpinum* L., *P. bertolonii* D.C., *Sesleria albicans* Kit., and *Dactylis glomerata* L. *New Phytologist*, **82**, 385–95.

Woodward, F. I. & Pigott, C. D. (1975). The climatic control of the altitudinal distribution of *Sedum rosea* (L.) Scop. and *S. telephium* L. I. Field observations. *New Phytologist*, **74**, 323–34.

Wright, H. E. (1968). History of the Prairie Peninsula. In *The Quaternary of Illinois*, ed. R. E. Bergstrom. University of Illinois College of Agriculture Special Publication 14. 179 pp.

5

Reconstructing seasonal to century time scale variations in climate from tree-ring evidence

HAROLD C. FRITTS, G. ROBERT LOFGREN
AND GEOFFREY A. GORDON

Abstract

Ring-width variations from conifers in arid North America provide information on short time-scale variations in palaeoclimate. Dendroclimatic data are especially useful because: (*a*) they can be dated to the exact year, (*b*) they can resolve seasonal fluctuations in climate, (*c*) the climatic signal can be separated from non-climatic noise, and (*d*) the data are available from a variety of climatic regimes throughout the world. Dendroclimatic reconstructions are obtained by calibrating 65 ring-width chronologies from western North America with spatial variations in seasonal temperature, precipitation, or sea level pressure for North America and the North Pacific. Spatial variations in the tree-ring record are transformed into statistical estimates of variations in the seasonal meteorological variables from 1600 to 1962. The synoptic-scale climatic anomalies that are detected may last from one season to several centuries. Verification of the reconstructions is accomplished using independent meteorological data, historical information, or proxy data. The well-dated, spatially continuous seasonal maps of climate derived from tree rings allow the association of isolated historical facts with meaningful synoptic-scale climatic patterns. The dendroclimatic reconstructions along with historical and other proxy data can serve as a reliable record bridging the gap between the modern meteorological record and the long-term record of traditional palaeoclimatology.

Introduction

Proxy records of climatic fluctuations are of different lengths and can resolve variations at different time scales. For example, ocean-bottom cores provide millennia-long records with a time resolution of from hundreds to thousands of years. Pollen sequences are usually shorter with a time resolution of 50 to more than 100 years. Varved sediments or ice cores are variable in length with a time resolution of about one year. Tree-ring chronologies are often shorter, but their time resolution is to a specific season and year. Historical data vary in length but can provide the highest resolution of all, sometimes to the exact moment of occurrence.

Until recently, short time-scale palaeoclimatic variations have received little attention. With the possible exception of Europe or Asia, instrumental and historical records are too short to evaluate the important variations at time scales of decades or longer. Data prior to 1900 are sparse and the longest records are often geographically or temporally discontinuous. Our knowledge of decadal to century time scale variations in climate is based largely upon this limited information. Tree-ring studies offer an exciting opportunity for lengthening and extending the spatial coverage of high-resolution information which will be especially useful in studying decade to century-long climatic variations.

Tree rings as high-resolution proxies of climatic variations

Much climatic information recorded by trees growing in stressful temperate forest environments can be extracted from the size, structure, and chemical (including isotopic) composition of the annual growth layers called tree rings. A variety of methods has been developed for dating the tree rings and for extracting the climatic information in the form of estimates of specific climatic factors such as precipitation, temperature, sea level pressure, and indicators of drought (La Marche, 1974; Stockton & Meko, 1975; Fritts, 1976; Fritts, Lofgren & Gordon, 1979). The precisely dated and continuous climatic information from tree rings is an important addition to other available data on seasonal to centenary climatic variations.

The discipline which uses tree rings to study climate is referred to as *dendroclimatology*. Dendroclimatology has been described in detail by Fritts (1976), La Marche (1978) and Hughes *et al.* (1981), so the details need not be discussed here. However, it is important to point out that rigorous methods of tree-ring dating are used to identify the specific year in which each ring was formed. In addition, the downward trends in growth associated with increasing tree age are removed by a standardising procedure called *indexing*. This transforms each ring-width sequence into indices which form a nearly stationary time series. The time series are averaged by year among ten or more trees within a site to obtain a *ring-width chronology*. When chronologies are developed exclusively for dating purposes (i.e., not necessarily for climate reconstruction), a simpler standardising procedure is sometimes used, and the chronologies are not stationary. These chronologies cannot readily be used to reconstruct climatic variations. More than a thousand standardised chronologies are now available from western North American conifers and collecting, research, and analysis programs have begun in most temperate regions of the world (La Marche, 1978).

Reconstructing spatial variations in climate

When a large number of tree-ring chronologies are available for a region and have been shown to represent the effects of a specific climate variable, the spatial anomalies in tree growth may be mapped and used to deduce climate anomalies (Fritts, 1965, 1976). However, trees rarely show a simple response to climate, so that ring-width data cannot, in general, be interpreted in terms of a single climate variable. The relationship between a ring-width chronology and climate at a single site can be statistically estimated by means of a *response function* (see Fritts, Blasing, Hayden & Kutzbach, 1971; Fritts, 1974, 1976). Response functions demonstrate that the partial correlation between ring width and temperature or precipitation is dependent on the season of the year, the species, and the site in which the trees grow. Because of the complexity of single-site response functions, Fritts and co-workers have developed a method for climate reconstruction based on direct analysis of spatial patterns of tree growth and of climate. The two are related by what is called a *transfer function* (Fritts *et al.*, 1971). The response function is designed to portray statistically the cause-and-effect relationships (although it does not prove such a relationship exists), while the transfer function reverses the process allowing one to deduce, statistically, climatic variations from the correlated variations in ring width.

The statistical technique of principal component analysis, or eigenvector analysis, is used to extract and simplify the spatial variations in both tree-growth and climatic data. When the climate for more than one station or variable is reconstructed at one time, a multivariate canonical regression scheme is used to relate the most important patterns of anomalous growth to the most important patterns of anomalous climate and to calculate a transfer function which converts the growth patterns into estimates of climate (Blasing, 1978; Fritts *et al.*, 1979). Transfer functions using different numbers and kinds of predictors and predictands constitute different models. The procedure of deriving a transfer function for a particular model is called *calibration.*

Biological factors

The tree-ring data consist of 65 ring-width chronologies obtained from Douglas fir (*Pseudotsuga menziesii*), bigcone spruce (*P. macrocarpa*), ponderosa pine (*Pinus ponderosa*), pinyon pine (*P. edulis*), limber pine (*P. flexilis*), bristlecone pine (*P. longaeva*), Jeffrey pine (*P. jeffreyi*), and white fir (*Abies concolor*) from sites throughout semi-arid western North America (Fig. 5.1) (Fritts & Shatz, 1975).

A simplification of the biological model used in this study is summarised in Fig. 5.2 (Fritts, 1976). The letter t in the figure denotes the climate year in which the tree rings grew, starting with September of the prior year and ending with the August concurrent with the cessation of growth. Climatic conditions during climate year t limit one or more plant processes involved in the formation of the annual growth ring for several years, and so are recorded in several values of the tree-ring chronology. The multivariate regression scheme used to calibrate such a model (Fig. 5.3) must use growth values in more than one year to reconstruct a particular year's climate.

Fig. 5.1. Site locations of the 65 replicated ring-width chronologies from western North America.

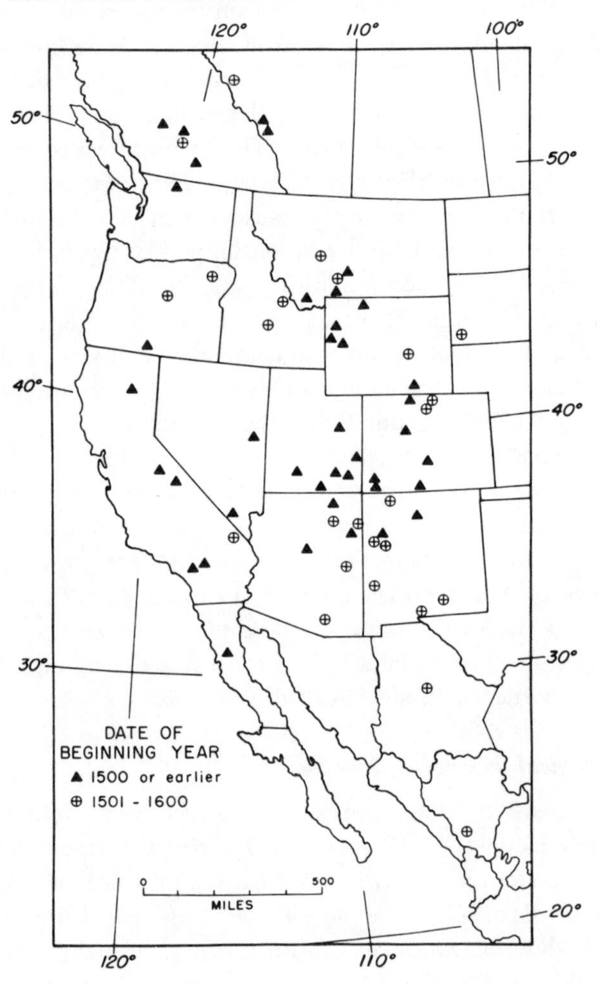

Climate data grids

The grids of climate data chosen for calibration were as follows:
(*a*) sea level pressure at 10° longitude and latitude grid points from 80° W to 100° E and from 20° N to 70° N, except every 20° of longitude at 60° and 70° latitude north (a total of 96 grid points; see Fig. 5.4);
(*b*) temperature for 77 stations in the United States and southwestern Canada;
(*c*) precipitation for 96 stations in the United States and southwestern Canada.

Monthly data were averaged or totalled by season according to the following seasonal definitions: winter: December–February; spring: March–June; summer: July–August; and autumn: September–November.

Calibration of models

The period common to both the tree-ring indices and the climate data is used for the calibration. The complete set of tree-ring chrono-

Fig. 5.2. Model of the effect of climate on growth in a given year, t. The influence of climate is greatest in year t and less in years $t+1$ up to year $t+k$, after which there is no substantial influence of year t climate. The growth of year t is also related to growth in the prior year, $t-1$, through autocorrelated phenomena.

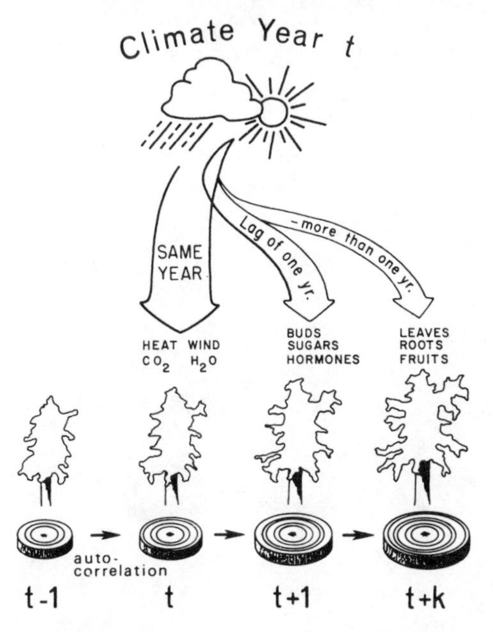

logies extended from 1601 to 1963. The pressure data began in 1899 and the complete set of temperature and precipitation data began in 1901. Eigenvectors were extracted for the entire record which was arbitrarily defined for the climatic data to end in 1970. However, the calibration period started in either 1899 or 1901 and ended in 1961, 1962, or 1963 depending upon the particular predictand–predictor combination used. Transfer functions are obtained by least squares canonical regression using ring-width chronologies as the statistical predictors. Separate calibrations are made using each climate variable and each season (Fritts *et al.*, 1971; Fritts, 1976; Blasing, 1978). Because climate is a spatially continuous phenomenon which can be correlated over a large area, climate data surrounding, as well as coincident with, the immediate region of the tree sites were calibrated with the ring-width chronologies. Tree-ring predictors including different lags (Fig. 5.2) were calibrated and the best calibrations were selected.

One statistic used to evaluate a calibration is the percentage of climate variance accounted for by the tree rings: that is, the percentage by which the

Fig. 5.3. Model of the multivariate statistical transfer function of the biological relationships described in Fig. 5.2. The growth indices, *x*, are multiplied by coefficients, *b*, and the sums of the products used as estimates of year *t* climate.

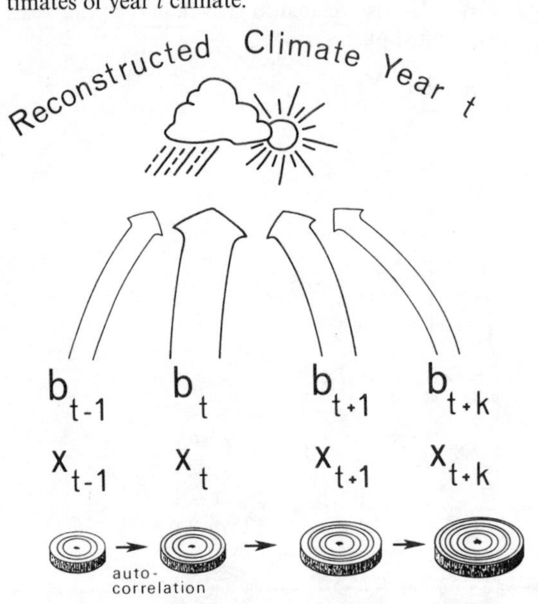

original variance is reduced by the model during the calibration period. The models used in this chapter were selected largely on the basis of this statistic. Fig. 5.4 is a map of the percentage variance calibrated for three winter models. Note how the relatively small tree-ring network (Fig. 5.1) accounts for quite large amounts of climate variance in regions quite far removed from the three sites (also see Kutzbach & Guetter, 1980).

Verification with independent data

Even if a large percentage of climate variance was accounted for over a calibration period, this does not guarantee that such success will carry over to a different period. To establish the reliability of a calibration, some form of independent verification is necessary. Verification can be approached in three ways: (*a*) statistical verification with independent meteorological data, (*b*) verification with independent proxy data, and (*c*) verification with the historical record. Each approach to verification has its strengths and weaknesses, and all three should be

Fig. 5.4. Percentage variance calibrated by the transfer function for the three climatic grids: sea level pressure for the calibration period 1899–1961; temperature and precipitation, 1901–1961.

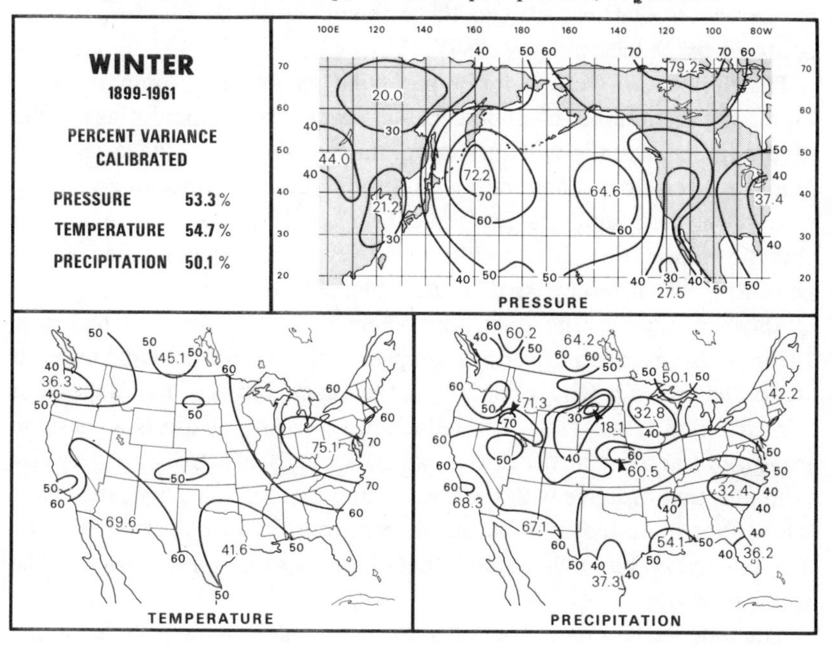

used whenever possible. In the following section we will give examples of each of these types of verification.

Statistical verification with independent meteorological data is the most objective and precise method of verification, and for that reason it is of greatest scientific value. The procedure requires observational measurements of the calibrated climate variable for intervals of time not included in the calibration. These measurements are compared directly with the reconstructions obtained by applying the calibrated transfer function to tree-ring data not used in the calibration, and measuring the similarity objectively with standard statistics. These statistics can then be tested to estimate the probability that the association between the actual and estimated data could have occurred by chance.

The statistics used in our verification procedure are: (*a*) the correlation coefficient, (*b*) counts of agreement in sign (Beyer, 1968), and (*c*) the product means (Fritts, 1976). The first two statistics are applied to both the original estimates and observed values and to their first differences (i.e., the difference between the current and the preceding value). This results in five verification tests for each variable. These five statistics are tested at the 5 per cent significance level. When a reconstruction passes a test at a particular station, there is a likelihood of only 5 per cent that the measured association between the estimates and the observed values occurred by chance.

The verification results for a particular model are summarised by expressing the number of stations passing each test as a percentage of the total number of stations (considering only stations which have seven or more years of independent data). As a rough guide, a percentage greater than 5 per cent supports the contention that the model does contain climatic information. The results of all tests may be combined and a percentage of all tests passed is also obtained.

Another statistic, the reduction of error, RE (Lorenz, 1956; Fritts, 1976), is also used in our verification procedure. RE is calculated for each station having independent data. The theoretical limits of RE range from negative infinity to $+1.0$. The reduction of error statistic is a sensitive measure of agreement because it can be strongly affected by a very few estimates that disagree with the observations, and it has diagnostic value. RE cannot be tested in a statistical sense, but because of its sensitivity, any positive value will indicate that there is some information in the reconstruction.

The indepéndent data from all stations and all years are pooled and a contingency table of equally probable classes is constructed. The agree-

Table 5.1. *Verification summary for a spring temperature model*

Test	Percentage passed during: independent period	calibration period
Correlation coefficient	30	100
Correlation on first differences	49	100
Sign agreement	9	99
Sign agreement on first differences	21	97
Product means	19	99
Total significant	26	99
Reduction of error statistics >0	38	100
Chi-square for pooled data	17	95

ment between actual and reconstructed values in the table is tested using the chi-square statistic and a 5 per cent significance level.

To demonstrate the results of the verification procedure, a spring temperature model calibrating 45 per cent of the variance was verified. The spring season had 53 stations which have seven or more years of independent data with an average length of record of 24.8 years. Verification results are summarised in Table 5.1. This particular model passed 26 per cent of the tests. Fewer correlation and sign agreement tests passed for the reconstructed data than for their first differences. This indicates that the year-to-year variation reconstructed by this model is usually in the same direction as that of the observational data, but some trends in the reconstructions and observational data differ.

Fig. 5.5 is a plot of reconstructed and observed temperature for Santa Fe, New Mexico, a station which passed all five tests and had a 44-year independent record. Since the regression estimates account for only 45 per cent of the variance, the magnitude of year-to-year variability in the reconstructed data is smaller than in the original data. In this particular example, the long-term trends apparent in the observational data are well reconstructed.

Analysis of reconstructed climatic variations

A basic premise of climatology is that a knowledge of past climate is a key to understanding future climate. We believe that there is a genuine need to involve historians in the study of past climate and, in particular, in dendroclimatology. In the following pages we will describe different ways in which it has been possible to extract information on past climatic variations from our reconstructions which could be of value to historians. Where possible, we have selected examples which also involve historical references to past climatic variations.

Variations in specific seasons

Fig. 5.6 shows maps for the winter of 1889–90 including reconstructions of pressure, temperature, and precipitation along with available observational data. The reconstructed pressure anomalies show a weakened Aleutian low with below-normal pressures over the west coast and near-normal pressures for much of the United States. Such a pattern would be associated with fewer storms in the region of the Aleutian low and with an enhanced flow of moist air into the southwest and Rocky Mountains.

Above-normal temperatures are correctly reconstructed from the central Rocky Mountains eastward, while below-normal temperatures are reconstructed in the northern Rockies. However, the below-normal temperatures along the west coast have not been accurately reconstructed. The reconstructions of precipitation show good agreement with observed conditions except over the southwest. Again, the regression estimates underestimate the size of the anomalies (Fig. 5.5).

Fig. 5.5. Reconstructed and observed (real) spring temperatures (°C) for Santa Fe, New Mexico, for the independent and dependent periods. Verification statistics for these two periods are: r: correlation coefficient; r_d: correlation of the first differences; RE: reduction of error.

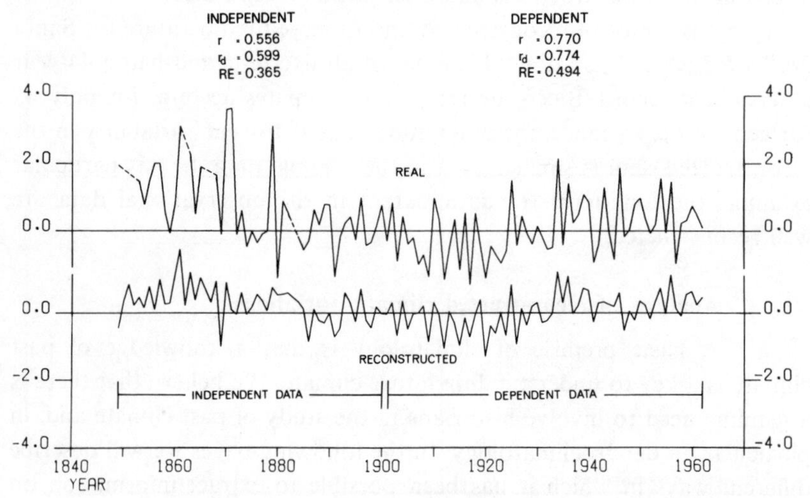

For this winter Ludlum (1970) reported record December rains for Los Angeles; Lake Tahoe rose 12 inches; a levee break flooded Yolo County, California; and the Los Angeles River flooded and changed its course. The Pacific Railroad reported 228 inches of snow at Cisco, California, with snow slides and derailments tying up the summit tracks for the last half of January.

Fig. 5.7 and 5.8 show the reconstructions for spring and summer of

Fig. 5.6. Reconstructed sea level pressure, temperature, and precipitation anomalies along with available meteorological observations of temperature and precipitation for winter 1889–90 expressed as departures from the mean periods 1899–1970 for pressure and 1901–70 for temperature and precipitation.

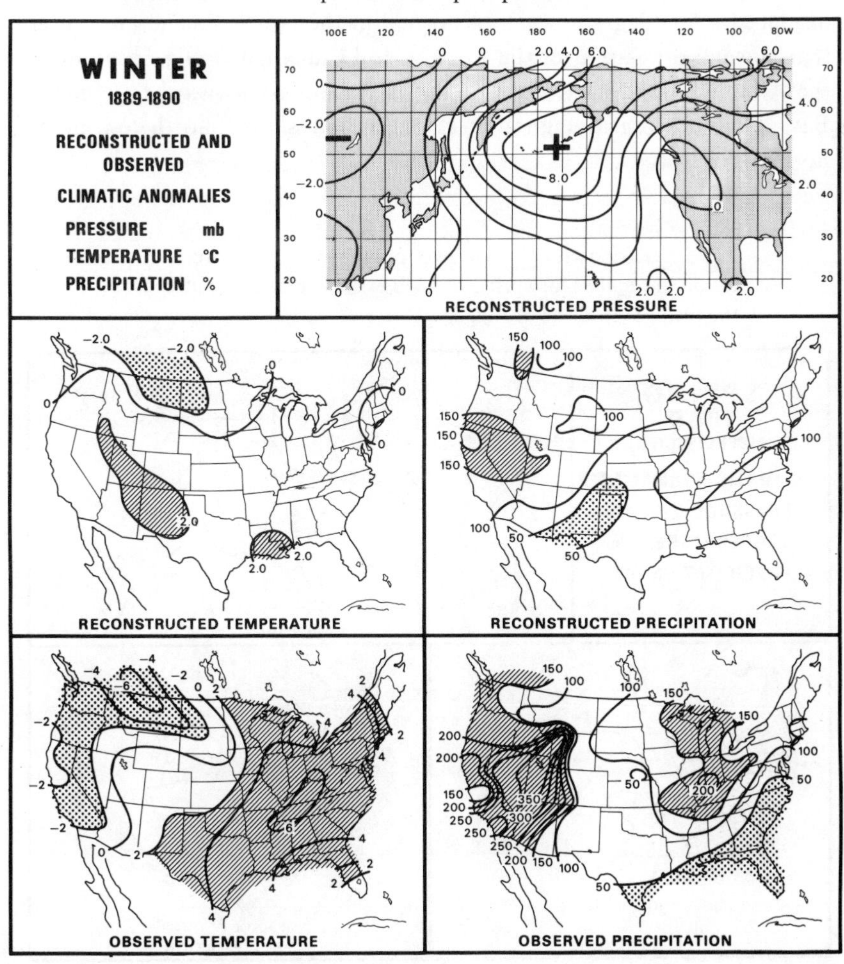

1849 for which there are fort records and diary accounts of migrants travelling to Oregon and to the gold rush in California (Lawson, 1974). The military fort precipitation records show that the spring of 1849 was relatively wet in western Missouri and Arkansas. The temperature records suggest that spring warming occurred earlier along the Santa Fe Trail than along the Oregon Trail, but travellers on neither trail experienced extremely high late-spring or summer temperatures. The diary accounts are more subjective but suggest that the migrants were buffeted by frequent frontal storms during late spring and by severe thunderstorms in summer.

Our reconstructed spring surface pressure patterns for 1849 (Fig. 5.7) show a weak anomaly pattern with a more intense subtropical high displaced a few degrees north of normal in the North Pacific and lower than normal pressure over the Alaskan and Canadian Arctic. In summer, the Arctic low anomaly pattern (Fig. 5.8) extends eastward to Hudson Bay. The subtropical high was stronger and shifted northwest of its normal position.

Fig. 5.7. Reconstructed sea level pressure, temperature, and precipitation anomalies for spring of 1849 expressed as departures from the mean periods as in Fig. 5.6.

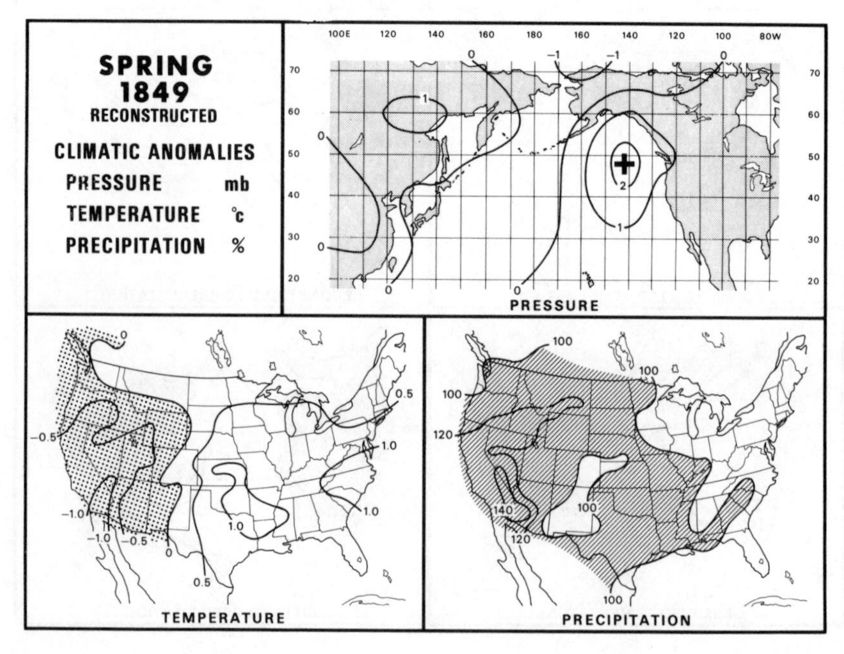

The summer pressure reconstruction shows an anomalous pressure gradient over Alaska and western Canada which would have created a flow of cold Arctic air into that region. Cooler temperatures are confirmed by the narrow rings of temperature-sensitive trees in the subarctic (Fritts, 1976). The pressure anomalies would also enhance a northerly flow of cold air into the Far West and a southerly flow of warm moist air into the Prairies.

The reconstructed precipitation anomaly pattern over the United States for spring shows the West to have been wet and the East to have been dry. For summer the reconstructed precipitation for the central prairie states was above normal. This coincides with the region covered by the early fort data which indicate a wet spring and summer. The temperature anomalies for spring and summer are similar. A cooler West than East is evident, but the anomalies are weak.

This example suggests how spatially discontinuous historical data on climatic variation for a restricted area can be interpreted in the context of spatially continuous dendroclimatic reconstructions.

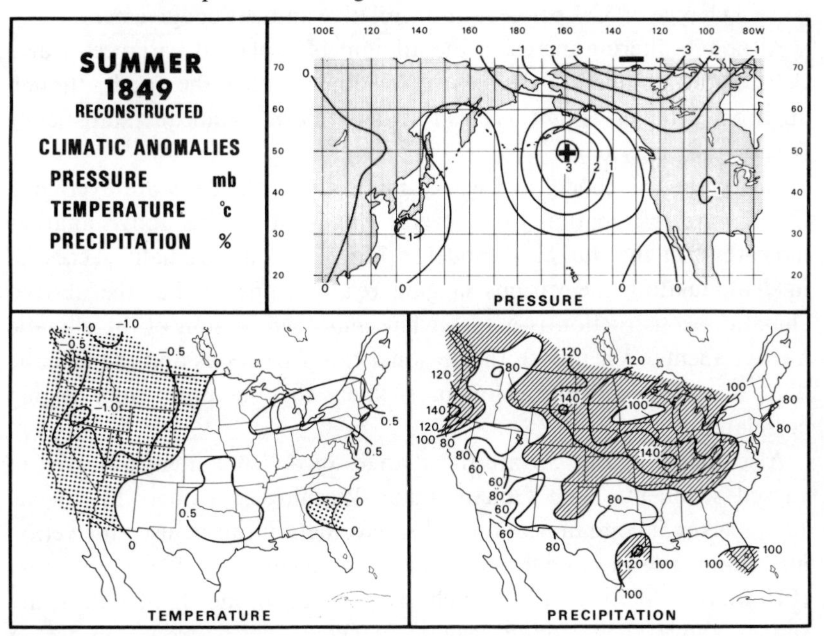

Fig. 5.8. Reconstructed sea level pressure, temperature, and precipitation anomalies for summer of 1849 expressed as departures from the mean periods as in Fig. 5.6.

Regionally averaged variations

When the reconstructions are averaged over space or time for the calibration period, the percentage of climate variance accounted for usually increases. This is seen when the average percentage variance calibrated at individual stations is compared with the square of the correlation coefficient between the estimated and observed data after averaging.

As an example, the reconstructions for nine individual stations near the Columbia River Basin were averaged to obtain Plot (*a*) in Fig. 5.9. The average percentage variance calibrated at the individual stations (i.e., the average for all stations used in calibration) was 41.3 per cent. The correlation between the averaged reconstruction and the averaged observational data for the nine Columbia Basin sites (dots on right-hand side of Fig. 5.9(*a*)) was 0.735, indicating a calibrated variance of 54.0 per cent. (Note that these are *not* independent data since they span the time period used for calibration. This exercise does not constitute a verification procedure but rather is a guide for model development.) When the averaged reconstruction and the averaged observational data were smoothed using a low-pass filter (Fig. 5.9(*b*)), the resulting correlation was 0.860, giving a calibrated variance of 74.0 per cent. In this example the percentage variance calibrated after both averaging and filtering amounted to a change of 32.7 per cent in the relative amount of agreement.

Although filtering increases the amount of explained variance, it also reduces the number of degrees of freedom so that the result, though apparently better is not *necessarily* any more significant statistically (although it may be).

The biological model in Fig. 5.2 implies that the trees are averaging the climate information over several ring widths. The above analysis indicates that the statistical model in Fig. 5.3 is only partially successful in demodulating the various lagging relationships so that the derived climatic reconstruction is a somewhat smoothed version of the climatic measurements. When both the seasonal reconstructions and the climatic data are treated with a low-pass filter, they become more highly correlated.

Also shown in Fig. 5.9 are the changes of pollen content observed in varved sediments from Gillette Lake, Washington (Albert M. Swain, 1978, personal communication). This information can be used for verifying the long-term climatic variations. The changes in *Pinus ponderosa* (PP) and *P. contorta* (LP), which grow on open and dry habitats, are shown above those for *P. monticola* (WP), which grows on wetter habitats. The longer term decreases or increases in these two contrasting pollen types are generally consistent with the filtered reconstructions of winter precipitation, with drier habitat pollen (PP, LP) increasing during intervals of lower than average precipitation and wetter habitat pollen (WP) increasing during intervals of higher than average precipitation. There are two intervals, at the beginning of the eighteenth century and in the twentieth century, when the percentage of *P. monticola* pollen did not

Fig. 5.9. Average yearly reconstructions of winter precipitation for nine meteorological stations near the Columbia Basin, expressed as departures from the 1901–70 observational record. Plot (*a*) is unfiltered, and Plot (*b*) has been treated with a 13-weight, low-pass filter passing 50 per cent of the variance at a period of eight years. The observational record from 1901 to 1970 is plotted as dots on the right. Vertical marks at the bottom of the plots designate those departures which equal or exceed the 95 per cent confidence interval. Year designations are for January of each winter season. Intervals of change in the percentage of pollen content of *Pinus ponderosa* (PP), *P. contorta* (LP), and *P. monticola* (WP) observed in the varved sediments from Gillette Lake, Washington, are indicated below the filtered time series.

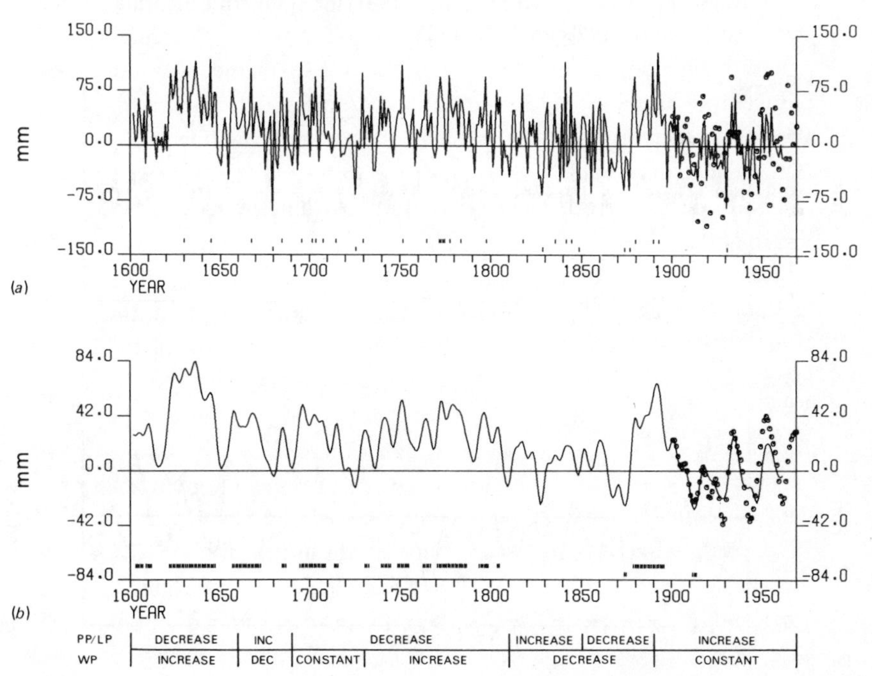

change. These intervals were reconstructed to have had near-average precipitation when there would have been less likelihood for a change in vegetation and pollen content. While the resolution of the pollen data is less than that of the tree-ring data, the information in both data sets is consistent. This adds credence to the low-frequency climate variations reconstructed for that particular region.

As a second example of this type of analysis, the average reconstructed winter precipitation for nine stations in the valleys of California is shown in Fig. 5.10. As in Fig. 5.9, the regionally averaged observational data are plotted to the right. Not all periods of drought are coincident in

Fig. 5.10. Average yearly reconstructions of winter precipitation for nine meteorological stations in the California Valleys expressed as departures from the 1901–70 observational record. Plot (*a*) is unfiltered and Plot (*b*) has been treated with a 13-weight, low-pass filter passing 50 per cent of the variance at a period of eight years. The observational record from 1901 to 1970 is plotted as dots on the right. Vertical marks at the bottom of the plots designate those departures which equal or exceed the 95 per cent confidence interval. Year designations are for January of each winter season. Dashed lines indicate filtered annual rainfall indices derived by Lynch (1931) for Southern California plotted at 5-year intervals for 1775–1925.

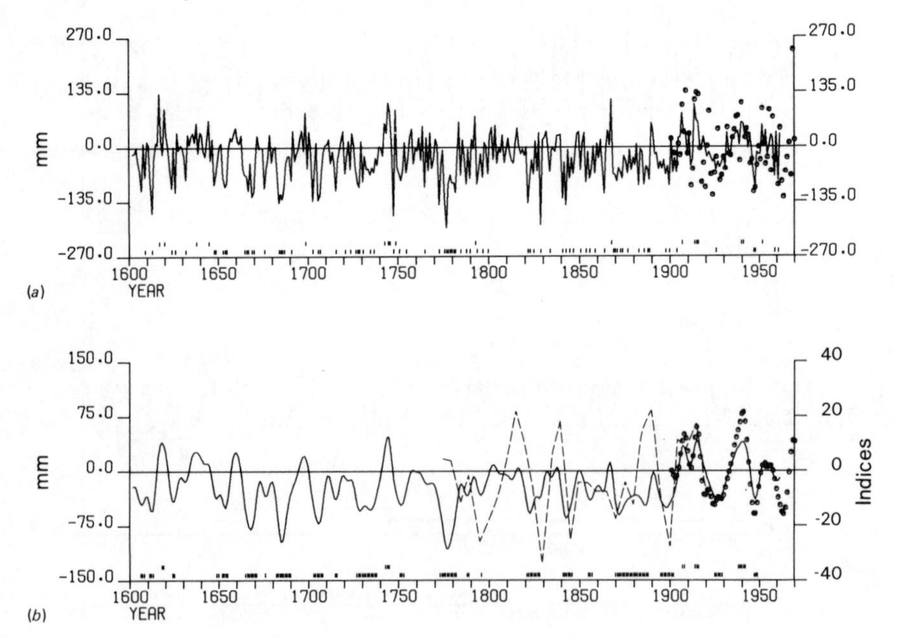

California and the Columbia River Basin. There are substantial differences in the reconstructed winter precipitation between these two regions. For example, a long-term downward trend in precipitation from 1602 through the 1880s is apparent only for the Columbia Basin.

Lynch (1931) developed indices of annual precipitation for Southern California using mission crop records from 1770 to 1832. From 1833 to 1850 his data included only historical references to weather conditions, floods, droughts, and crops. After that date actual precipitation measurements were incorporated in his indices. Filtered values of Lynch's indices are plotted as dashed lines joining the points at every five years in Fig. 5.10(*b*). We calculated the verification statistics between Lynch's indices and our winter reconstructions for California. Using unfiltered data, all five tests were significant at the 5 per cent level for 1770–1900, but only one test was significant for 1770–1830. Comparisons of the results for filtered and unfiltered sets indicated that the 1770–1830 indices, which were based on mission crop records, agreed with our reconstructed short time-scale variations but disagreed with the long time-scale values. The indices for 1831–50 agree with our long time-scale variations better than with the short time-scale values. After 1850 there was general agreement at all time scales. The correlation coefficients for the unfiltered and filtered sets for this time period were 0.486 and 0.485, which were significant at the 5 per cent level.

Because there are no obvious differences in the tree-ring sets for those time periods, it is possible that the long time-scale changes in the mission crop records are due to factors other than climate. However, Lynch's data for 1831–50 fitted the long time scale reconstructed changes more accurately than the short time-scale changes. This situation was improved after 1850 with the availability of precipitation measurements. The three large peaks in Lynch's data in the nineteenth century are not reproduced by the reconstructions of this model, indicating that we are underestimating precipitation during wet periods and might try calibration with the logarithm of precipitation.

Temporally averaged variations

In order to use the apparent increased reliability of the reconstructions averaged over time, maps have been made for decades, half-centuries, and the interval 1602–1900. Fig. 5.11 is one example of the averaged reconstructions for 1861–70, a decade for which we have some historical data. The tree-growth map on the left of the figure expresses the average growth as departures from the 1601–1963 mean. The climatic

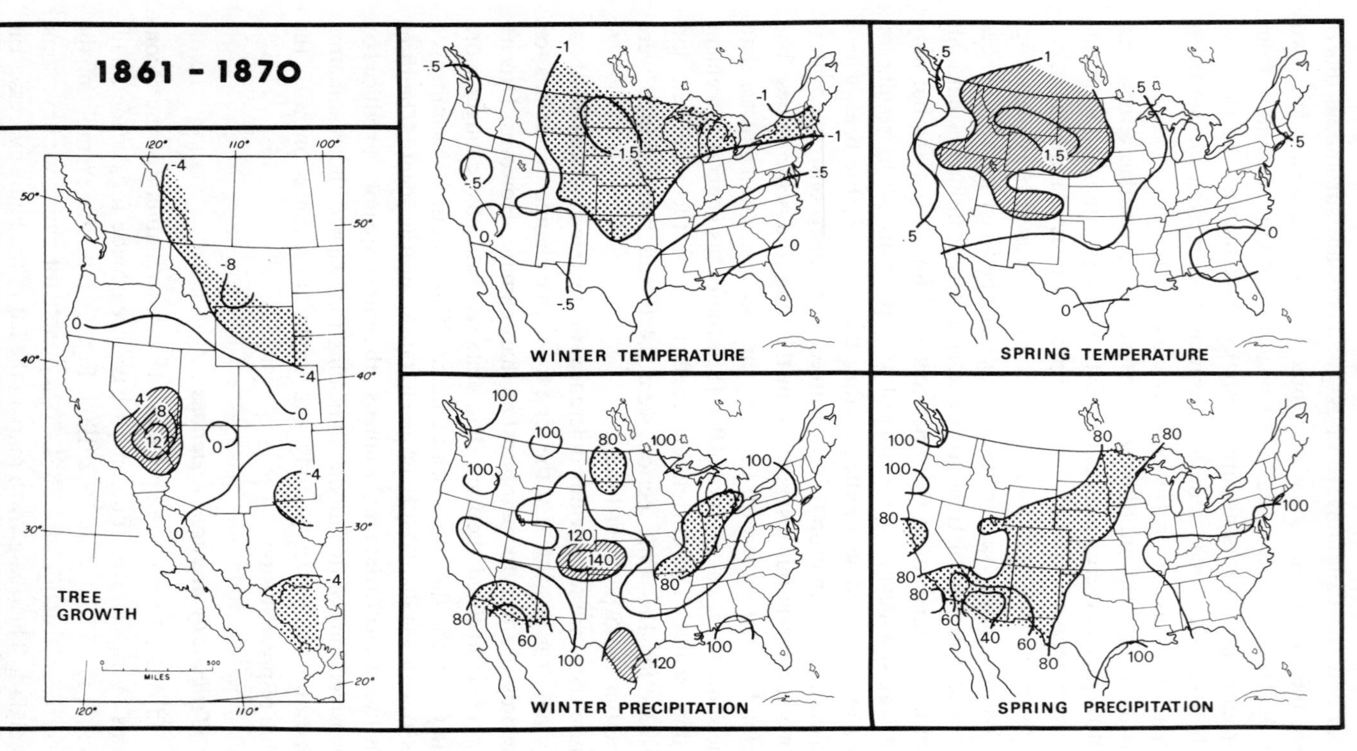

Fig. 5.11. Mean anomaly tree growth and mean reconstructed anomalies in temperature and precipitation for the winters and springs of 1861–70. Tree growth is expressed as normalised values multiplied by 10 using the 1601–1963 means and standard deviations. The climatic reconstructions are expressed as departures in deg C or as percentages from the mean period 1901–70.

data for winter and spring are mapped as anomalies from the 1901–70 normal figures.

Below-normal winter temperatures were reconstructed throughout most of the United States and southwestern Canada. Wet conditions were centred in Colorado and Kansas westward to Northern California, and dry conditions in the Great Lakes, the Mississippi and Ohio River valleys, and the extreme Southwest. Spring was reconstructed to have been warmer and drier than the twentieth century, especially in the West.

An examination of a few available historical references yielded no account of unusual temperature for the entire decade. However, several accounts of heavy precipitation were found. The Great Salt Lake was reported to have risen 10 feet during this decade (Antevs, 1938). There is one reference to a heavy snowstorm in San Francisco during the winter of 1869 (Ludlum, 1970) and to heavy rains in Utah in December, 1867, causing very high river discharge (Antevs, 1938). During the winter of 1862–63 there is a reference to a drought in Los Angeles, and both the California Historical Society (1924) and Lynch (1931) show lower than average precipitation indices. While there are too few data for verification, these references are included to illustrate the potential for historical comparison.

Reconstructions can also be averaged for years coincident with historically dated climatic events. These averaged reconstructions can then be examined to determine whether the resulting anomaly pattern bears a reasonable relationship to the referenced climate event. What follows is an example of this approach, conducted in collaboration with Dr P. M. Kelly (1977, personal communication). Full details will be published elsewhere.

Arakawa (1955) has compiled data concerning the severity of winters in Japan back to the seventh century AD. We first tested for differences in the winter circulation pattern between the years of mild and severe conditions in the twentieth century using actual pressure anomaly data by subtracting the average pressure anomaly for mild years from the average anomaly for severe years. The difference maps showed the mild winters to be associated with higher than normal pressure east of Japan which would bring southerly flow to the islands.

The difference maps for reconstructions were calculated in the same way, first for the twentieth century and second for the independent period 1602–1898. The reconstructed pattern in the twentieth century was nearly identical to that of the actual data. The pattern for the independent period was similar; however, the magnitudes of the anomalies were reduced. These results indicate that certain historical references can be interpreted in the context of related climate conditions.

Anomalies from the twentieth-century normals

Bryson & Hare (1974), Bryson (1974), and Lamb (1977) have warned that the interval from 1930 to 1960 has been unusual for large portions of the northern hemisphere. The 361-year-long tree-ring reconstructions provide us with a means of measuring how unusual the twentieth century is and documenting when, where, and in what season climatic statistics have varied from twentieth-century values.

In order to compare the statistics of the past to those of the present, we chose 1901–70 to be our normal period and averaged the station reconstructions for all four seasons and within eleven separate regions. The averaged reconstructions for the 1901–70 period can be compared to the averaged reconstructions in other periods to show the relative changes. However, it is easier to relate these comparisons to the variance measured by the instrumental record after adjusting each reconstruction by adding the difference between the means and multiplying by the ratio of the standard deviation of the instrumental data to the standard deviation of the reconstructions for the common period. This ensured that the mean and standard deviation of reconstructed and actual data for the calibration period were equal. It is done for comparative purposes to bypass the problem of reduced variability mentioned in relation to Fig. 5.5. The adjusted values are no longer considered regression estimates. We then calculated the means for 1961–70 and expressed them as departures from, or percentages of, the 1602–1900 means.

The annual temperature departures (Fig. 5.12) show that the twentieth century was warmer than the prior three centuries for only six of the eleven regions. Region 3, the Intermountain Basins, showed a large negative departure. Region 9, the Great Lakes, had the greatest departure of 0.93 deg C. These averages were largely the result of anomalies in winter temperatures. Those in other seasons often varied in the opposite direction to those in winter and the annual averages.

Precipitation in the twentieth century (Fig. 5.12) totalled over all four seasons was below those reconstructed for 1602–1900 for six regions. It was 19 per cent higher than the reconstructions for 1602–1900 for Region 2, California.

These data indicate that there have been sizeable differences between the means of the past and those of the present. The cooler world-average temperatures of the nineteenth century reported by Willett (1950) and extended to earlier centuries by others appear in our reconstruction to have been largely a winter phenomenon restricted to the central and eastern United States. The four-century trend in annual temperature was in the opposite direction for the western United States. Our reconstructions also suggest that the twentieth-century climate has been wetter than in the past for the southwestern portions of the country. If we were to use actuarial figures based upon the seventeenth through nineteenth centuries to project the future, California would appear most likely to have a sizeable average moisture deficit compared to the amounts measured in the twentieth century.

Conclusions

Spatial variations in climate can be reconstructed from ring-width data when there is sufficient spatial density of replicated, well-dated, and appropriately standardised tree-ring chronologies. Specific site responses to particular climatic variables and seasons are related to the spatial arrays of a particular climate variable in a canonical regression analysis. Each variable and each season require a separate

Fig. 5.12. Estimated mean change in annual temperature (top) and precipitation (bottom) between the twentieth century and the three previous centuries for eleven regions in the United States and western Canada. Variance scaled upward to correct for the variance loss in the regression estimates.

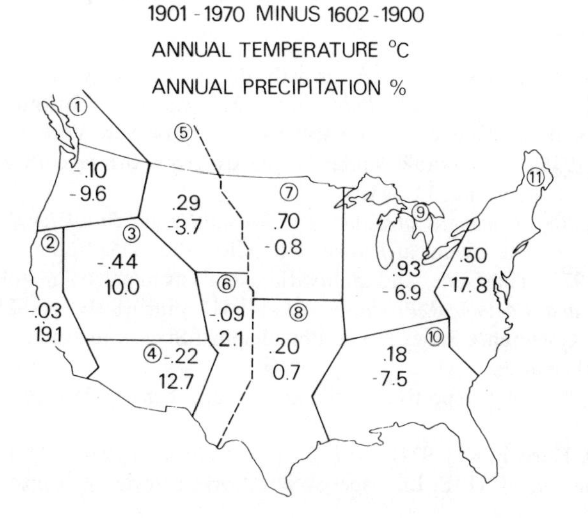

calibration. Reconstructions are verified with available independent instrumental observations, associated proxy records, or historical data.

The greatest amount of information contained in the dendroclimatic reconstructions is found in synoptic-scale climate variations which can last for time periods of one year to several decades. Individual station reconstructions showing seasonal and annual variability are also available.

Our reconstructions show that not all of the United States was warmer in the first 70 years of the twentieth century than in 1602–1900. Average conditions were colder, especially in winter, throughout the western United States. Annual precipitation was higher for the southwestern United States, especially in California. The East was drier.

We caution, however, that these reconstructions should not be viewed as final. They are our first approximations derived from trees restricted to western North America. In the future, we hope to sample sites from a more extensive spatial grid and to improve our techniques of modelling.

We believe the future is bright for closer co-operation between historians and dendroclimatologists. Dendroclimatic reconstructions can help resolve what may appear to be contradictory historical data, which, in fact, result from synoptic-scale climatic anomalies or from differences among the seasons. The collection and analysis of dendroclimatic, historical, and other proxy data should aid in the search for a fundamental understanding of important physical causes behind past climatic variations. Such an understanding is essential to climate modelling and to anticipating future climatic change.

References

Antevs, E. (1938). Rainfall and tree growth in the Great Basin, *American Geographical Society Special Publication* 21, Carnegie Institution of Washington and the American Geographical Society of New York. 97 pp.

Arakawa, H. (1955). Remarkable winters in Japan from the seventh century, *Geofisica Pura E Applicata*. **30**, 144–6.

Beyer, W. H. (1968). *Handbook of Tables for Probability and Statistics*, 2nd edn. Cleveland, Ohio: The Chemical Rubber Co. (CRC Press). 642 pp.

Blasing, T. J. (1978). Time series and multivariate analysis in paleoclimatology. In *Time Series and Ecological Processes*, ed. H. H. Shugart, Jr, pp. 213–228. SIAM-SIMS Conference Series No. 5. Philadelphia: Society for Industrial and Applied Mathematics.

Bryson, R. A. (1974). A perspective on climatic change. *Science*, **184** (4138), 753–60.

Bryson, R. A. & Hare, F. K. (1974). *World Survey of Climatology*, vol. 11, *Climates of North America*, ed. H. E. Landsberg. New York: Elsevier. 420 pp.

California Historical Society (1924). The memoirs of Lemuel Clarke McKeeby. *California Historical Society Quarterly*, **3**, 169.

Fritts, H. C. (1965). Tree-ring evidence for climatic changes in western North America. *Monthly Weather Review*, **93**(7), 421–43.

Fritts, H. C. (1974). Relationships of ring widths in arid-site conifers to variations in monthly temperature and precipitation. *Ecological Monographs*, **44**(4), 411–40.

Fritts, H. C. (1976). *Tree Rings and Climate.* London: Academic Press. 567 pp.

Fritts, H. C. & Shatz, D. J. (1975). Selecting and characterising tree-ring chronologies for dendroclimatic analysis. *Tree-Ring Bulletin*, **35**, 31–40.

Fritts, H. C., Lofgren, G. R. & Gordon, G. A. (1979). Variations in climate since 1602 as reconstructed from tree rings. *Quaternary Research*, **12**(1), 18–46.

Fritts, H. C., Blasing, T. J., Hayden, B. P. & Kutzbach, J. E. (1971). Multivariate techniques for specifying tree-growth and climate relationships and for reconstructing anomalies in paleoclimate. *Journal of Applied Meteorology*, **10**(5), 845–64.

Hughes, M., Kelly, P. M., Pilcher, J. & La Marche, V. C. Jr, eds. (1981). *Climate from Tree Rings.* Cambridge University Press.

Kutzbach, J. E. & Guetter, P. J. (1980). On the design of paleoenvironmental data networks for estimating large-scale patterns of climate. *Quaternary Research*, **14**(2), 169–82.

La Marche, V. C., Jr (1974). Paleoclimatic inferences from long tree-ring records. *Science*, **183**, 1043–8.

La Marche, V. C., Jr (1978). Tree-ring evidence of past climatic variability. *Nature*, **276**, 334–8.

Lamb, H. H. (1977). Understanding climatic change and its relevance to the world food problem Sixth G. E. Blackman Lecture, 17 Nov. 1976, University of Oxford. *Climatic Research Unit Research Publication No 5.* Norwich: University of East Anglia. 23 pp.

Lawson, M. P. (1974). The climate of the Great American Desert, reconstruction of the climate of Western Interior United States, 1800–1850. *University of Nebraska Studies: New Series No. 46.* Lincoln, Nebraska: University of Nebraska Press. 135 pp.

Lorenz, E. N. (1956). Empirical orthogonal functions and statistical weather prediction. MIT *Statistical Forecasting Project Scientific Report 1*, Contract No. AF 19(604)–1566.

Ludlum, D. M. (1970). A century of American weather: decade 1881–1890. *Weatherwise*, **23**(3), 131–5.

Lynch, H. B. (1931). Rainfall and stream run-off in Southern California since 1769, Metropolitan Water District of Southern California, Los Angeles, California. 31 pp.

Stockton, C. W. & Meko, D. M. (1975). A long-term history of drought occurrence in Western United States as inferred from tree-rings. *Weatherwise*, **28**(6), 244–9.

Willett, H. C. (1950). Temperature trends of the past century. In *Centenary Proceedings*, pp. 195–206. London: Royal Meteorological Society.

6

Archaeological evidence for climatic change during the last 5000 years

ROBERT McGHEE

Abstract

Climatic change is frequently used as an explanation for events in the archaeological record. Events such as abandonment of a region previously occupied, occupation of a region presently uninhabitable, sudden changes in population size, and the introduction of new agricultural crops or techniques, may be related to local environmental changes and ultimately to changes in climate. Yet during the last 5000 years, when climatic variations have been relatively minor and when most human populations were relatively sophisticated both socially and technologically, most such events are as likely to have been the result of social, political or cultural factors. Interpretations of several such events are briefly discussed. It is argued that useful archaeological evidence for climatic change is most likely to come from areas climatically marginal to human occupation, and areas occupied by peoples with relatively simple technologies and economies. Finally, it is suggested that archaeology may usefully contribute to the study of climate by providing other disciplines with information on environmental changes which probably resulted from human activity in various regions over the last 5000 years.

Introduction

Over the last few years there has been a general shift in the types of question which interest archaeologists. Now that archaeological work has succeeded in building at least vague chronological sequences of culture history for most regions of the Holocene world, the emphasis of interpretation has moved away from questions as to 'when' and 'how' certain events happened in the past, and towards attempts to explain 'why' these events occurred. One of the factors most frequently invoked to explain past cultural events is environmental change, which is often linked either explicitly or by inference to change in climatic conditions.

Problems of archaeological interpretation

Despite the fact that archaeologists commonly use environmental or climatic change as an explanatory device, archaeology has no theoretical basis or standardised set of techniques, such as those used by most other palaeoenvironmental disciplines, explicitly designed to detect or investigate such occurrences. Archaeological interpretations of past climatic change generally take the following form. The archaeologist first notes an event in the archaeological record, such as population decline or abandonment of a region, introduction of a new subsistence pattern, or widespread civil unrest, which requires explanation. He then notes that evidence produced by some other palaeoenvironmental discipline suggests that the region concerned underwent an environmental change at approximately the same time, and concludes that the two events are related: that social or cultural change was caused by environmental change, and ultimately by climatic change.

I would suggest that most such interpretations should be treated with some degree of suspicion, on at least two sets of grounds. First, there is the problem of the degree of precision of both the archaeological and palaeoenvironmental records. History and ethnography suggest that human responses to major environmental changes generally take place on a time scale which can be measured in months, or at most in years. Although a small proportion of archaeological events can be accurately dated through the analysis of written documents, the general precision of dating is to within several decades at best, and usually only to within a century or more.[1] Similarly, most palaeoenvironmental techniques appear to be capable only of detecting trends which occurred over a period of decades or centuries, or which cannot be dated with greater precision. Undoubtedly there have been many past environmental and climatic changes that have caused changes in human adaptations or ranges of occupation, but we are usually capable of detecting only vague chronological correlations between such events, and not cause–effect relationships.

A second reason for doubting facile correlations between past human and environmental events lies in the complexity of human cultural response, especially in the relatively sophisticated societies which occupied much of the Earth during the last 5000 years. Climatic changes over this period appear to have been minor, relative to those that occurred during and shortly after the last glacial period, and most human groups had developed societies, knowledge and technology which allowed them to cope with the changes in a variety of ways. Faced with a deteriorating environment, most societies must have had the choice of several options: either to attempt to maintain their life-style under reduced circumstances, trade with, or make war on, neighbouring groups in order to maintain a supply of necessary resources, adopt new techniques of food production, abandon their traditional area and move to an area which was not subject to environmental change, or any combination of these choices. Furthermore, new cultural patterns which would leave archaeological traces identical to those produced by the changes mentioned above must have occurred frequently without the stimulus of climatic or environmental change, but merely under the influence of social, demographic, economic or political forces. Consequently, while many of the events which we see in the archaeological record may have been climatically induced, we can rarely be certain that climatic change was involved.

Given the difficulties in archaeologically assessing the role of environmental or climatic change as an originator of cultural change in past human activities, can we hope to turn the question around and to use archaeological evidence to infer the nature of climatic changes in the past? It would seem that if it is so difficult to assign a climatological explanation to a past cultural event, it can be expected to be much more difficult actually to use that cultural event, as known through archaeology, as a proxy climatic indicator. Several such attempts are discussed in the following section, and the reader may form his own opinion as to the usefulness of archaeological evidence in inferring past climatic change.

Archaeological evidence for climatic change

This section presents several attempts which have been made at documenting past climates from archaeological evidence alone. The discussion is limited to cases in which climatically induced environmental change has been proposed as a reasonable explanation for archaeological events. Archaeological events for which a climatic cause has been less convincingly suggested, and those for which a relationship between archaeological and climatic events has been proposed purely on the basis of apparent temporal coincidence, are omitted.

The majority of the cases to be discussed involve desert, dry-land or arctic regions (see Fig. 6.1). Although it can be argued that climatic change has had as great an effect on human cultures in more temperate regions of the world, I suspect that archaeological evidence from such regions is less capable of detecting human reactions to climatic events. The spread of agriculture into these regions, changes in the relative importance of various agricultural crops, or changes in trade patterns may be seen as the result of shifts in cultural preferences or political and

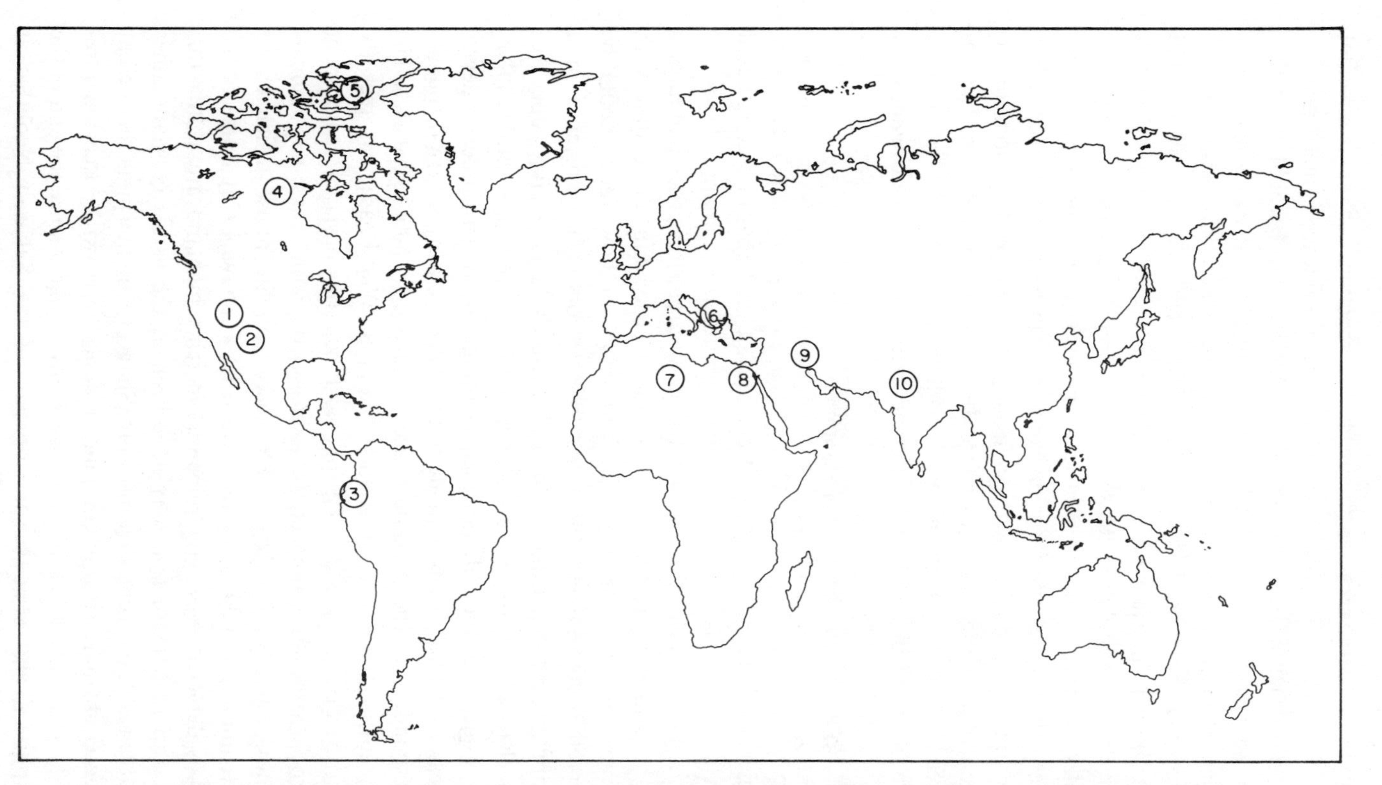

Fig. 6.1. Locations of archeological events discussed in the text: 1, Great Basin; 2, Paneblo area; 3, west coast South America; 4, Barren Grounds; 5, High Arctic Canada; 6, Greece; 7, Sahara; 8, Egypt; 9, Mesopotamia; 10, Indus Valley.

economic relationships, as plausibly as reactions to environmental change. In almost all cases, the archaeological evidence is inadequate to allow a distinction between these two types of event. I would argue that useful archaeological evidence for changes in past climates is most likely to come from environmentally marginal areas where small-scale climatic changes may have produced relatively drastic environmental effects, and particularly from such areas which were occupied by peoples with a relatively low population density and unsophisticated technology. Such peoples, with comparatively little technological and social 'insulation' between themselves and their environment, can be expected to have had fewer options in dealing with environmental change. Reactions to environmental change may therefore be more easily detected and interpreted from the archaeological evidence, as compared with those which occurred in more temperate and culturally advanced regions.

Desert and dry-land civilisations

The Saharan region One of the more prominent attempts in this field is that dealing with past climates of the Saharan region, based on evidence from rock paintings showing animals which are today found only in moister regions of Africa. Such use of prehistoric rock paintings is fraught with danger, since they are notoriously difficult to date with accuracy, and we can never be certain as to how they should be interpreted. As Raikes (1967) has pointed out, 'The existence in a London suburban sitting room of Landseer's 'Stag at Bay' does not mean that red deer roamed London in the 19th century.' Several scholars have argued that the Saharan rock paintings, along with other evidence, suggest a relatively moist climate in the Sahara during a period roughly corresponding to the Atlantic phase in Europe (*c.* 6000–3000 BC). The evidence for the existence of this 'Neolithic Wet Phase' has recently been questioned by Shaw (1976), partly on the basis of redating of the rock paintings. Many of the paintings depicting savannah animals are now dated to the Upper Pleistocene pluvial phase which ended approximately 11 000 years ago (Mori, 1974; Shaw, 1976). Some of the animals depicted may have remained in relict highland areas after this time, and their extinction may have been at the hands of Man rather than the result of decreasing moisture. Shaw (1976) interprets the present evidence as indicating a continuous desiccation, with occasional relatively moist interludes, of the Saharan environment since the end of the last pluvial, and sees the development of the Saharan Neolithic at approximately 6000 BC not as a response to moister conditions, but as an adaptation to a drying climate and the decline of animals which had been previously hunted. Nevertheless, archaeological evidence for the existence of a pastoral Neolithic economy over large regions of the Sahara appears to indicate that the area was moister than at present, and the decline of the Saharan Neolithic during the third millennium BC is probably related to the establishment of present environmental and climatic conditions.

Egypt Egypt provides relatively convincing archaeological evidence indicating desiccation, especially towards the end of the third millennium BC. The evidence from inscriptions is well summarised by Bell (1971), who considers the first clear sign of famine to be a relief from late Dynasty V in the twenty-third century BC, showing a group of emaciated people. Later, a group of tomb inscriptions referring to widespread disaster was carved within a few decades of 2150 BC. Most of these inscriptions speak of civil strife and movements of people as well as of famine, and it has been argued (Vandier, 1936) that the famines were caused by social and political collapse. Nevertheless, Bell asserts that there are sufficient texts which can be interpreted as referring to low water levels in the Nile, and perhaps to encroaching desert sands, to suggest that the civil breakdown was caused by failure of the Nile floods and ultimately by a decrease in rainfall over East Africa during this period. A second 'dark age' in the decades around 2000 BC may have had similar causes.

The evidence for famine during this period is supported by O'Connor's (1972) analysis of burials along a 35 km section of the Nile in Upper Egypt. He estimates that the rate of burial in this area was 1.2 per year during the twenty-fifth to twenty-third centuries BC, rising to 12.2 per year in the period between approximately 2160 and 2100 BC, then dropping to 3.5 per year during the twenty-first century and to 0.6 per year in the twentieth century BC. The burial rates for the twenty-second century BC are far higher than those for any other period studied, and O'Connor concludes that this is the only variation that may have been caused by environmental change.

The Middle East and India Bell (1971) suggests that the climatic change during the last centuries of the third millennium BC was not confined to Egypt, and relates its influence to the approximately contemporaneous end of the Early Helladic II period in the Aegean area; the Early Bronze II period of Anatolia; the downfall of the Sargonid empire in Mesopotamia; and the decline of the Harappan civilisation of the Indus Valley. Here the evidence is even less secure than that for Egypt. The collapse of the Sargonid empire about 2230 BC can more plausibly be seen as the result of political forces, since this first major attempt at building a Mesopotamian empire was fraught with invasion and internal revolt throughout its century of existence (Roux, 1964). In the eastern Mediterranean there is evidence of widespread destruction and perhaps of invasion, but no indication of the cause of these occurrences. The latest levels of the Harappan cities also show intentional destruction followed by decline or abandonment at a time which we now date, by adjusted radiocarbon dates, to approximately the twenty-second century BC (3700 BP). This event has been ascribed variously to invasions by Indo-European speaking peoples from the north (Wheeler, 1966; Allchin & Allchin, 1968); drastic tectonic movements which flooded the Indus Valley (Raikes, 1965); a major change in the course of the Indus River (Lambrick, 1967); or a significant decline in annual rainfall (Singh, 1971). There is obviously insufficient archaeological evidence presently available to suggest that the decline of Harappan civilisation was due to climatic change. Such a view would be supported if climatologists could demonstrate that a reduction in summer rainfall over East Africa with a consequent lowering in the level of the Nile floods, as suggested by the Egyptian documents of the period, would be inevitably associated with a decline in summer monsoon rainfall in the Indus Valley region.

Greece Climatologists have made an attempt to attach a climatic cause to cultural change with regard to the 'dark age' of Mediterranean and Near Eastern civilisation which began about 1200 BC (Bryson, Lamb & Donley, 1974). Using the anomalous weather patterns of the winter of 1954–55, and suggesting that such a pattern could theoretically have occurred with greater frequency during some period in the past, their work supported and extended Carpenter's (1966) argument relating the sudden decline of Mycenaean civilisation to a climatic change which brought prolonged drought to much of Greece. Carpenter suggests that a continuation of these conditions was responsible for the 'dark age' in Greek civilisation, which occurred between roughly 1200 and 750 BC, and similar climatic conditions may have produced a second 'dark age' between AD 600 and 800. The archaeological evidence bearing on Mycenaean decline is similar to that marking the end of Harappan civilisation: apparently intentional destruction at many sites, followed by abandonment or rebuilding and reoccupation at a reduced level. Although Carpenter's argument for a climatic cause of these events is persuasive, it is based on little direct evidence. In view of the preparations for defence seen in the thirteenth century BC tablets recovered from Pylos, and thirteenth century BC expansions of the fortifications in other Mycenaean cities, most scholars continue to believe that the direct cause for Mycenaean decline was local war, or raids and invasions by outsiders (Desborough, 1972; Betancourt, 1976). Whether these events were in turn related to environmental changes of the sort proposed by Carpenter cannot be reliably deduced from the present evidence. The signs of widespread disturbances throughout the eastern Mediterranean and in Anatolia at this time, however, are suggestive of a cause more basic than local warfare or invasion, and climatic change may have been a factor which triggered the events seen in the archaeological record.

The Andean region The rise and decline of civilisations in eastern Asia and the New World present little evidence for the influence of climatically induced environmental change. This may be due in part to our relatively incomplete archaeological knowledge of these civilisations, and perhaps in part to the fact that most of these developments occurred in areas that were not as climatically marginal as Egypt or the Indus Valley. The exception would seem to be the civilisations of the desert west coast of South America, and here there is a suggestion that climatic changes may have shaped the history of the area. Paulsen (1976), from a study of the use of wells on the coast of Ecuador, shows that these wells were not in use during the periods AD 600–1000 and after AD 1400, and infers from this a drier climate and a resulting change in local economies from agriculture to coastal fishing and gathering. She relates these drought periods, when the diverse communities of coastal Ecuador and Peru underwent decline, to the establishment of the two great highland states of the area, the Huari empire beginning about AD 600 and the Inca empire which was established in the fifteenth century. The link between a drying climate and the establishment of highland empires must have been extremely complex, however, and with the present evidence we cannot be certain of the nature of this link and of what other factors may have been involved.

The North American deserts Another region of the New World from which we might expect to extract information on past human adaptations to climatic change is the desert area of the southwestern United States. In the centuries prior to European settlement, much of the area known as the Great Basin was occupied by hunting and gathering peoples with a low population density and relatively unsophisticated technology, and archaeology indicates that this way of life extends for several millennia into the past. In the northeastern Great Basin, occupation shifted from the margins of the Great Salt Lake to upland areas about 2000 BC (3500 BP) (Madsen & Berry, 1975). A similar movement from riverine to desert environments is seen at the same time in Owens Valley on the southwestern edge of the Great Basin (Bettinger, 1977). These movements may have been reactions to an increase in moisture and the resulting increased productivity of desert food plants, and to a rise in the level of the Great Salt Lake which reduced the productivity of salt marsh resources. The next major event in the Great Basin appears to have taken place about AD 500–600 (1500 BP). In Owens Valley, settlement at this time began to concentrate in riverine environments, suggesting the onset of drier conditions (Bettinger, 1977). In the eastern Great Basin, this period saw the introduction of agriculture with the appearance of the Fremont culture (Madsen & Berry, 1975); this event may plausibly be seen as simply a spread of cultural knowledge or technology, but perhaps it may relate to adjustments to environmental desiccation. Farther to the east, on the southern Great Plains, Dillehay (1974) notes the absence of bison (*Bison bison*) hunting sites between AD 500 and 1200, and suggests that this may have been due to drier conditions which resulted in changes in the range of bison populations.

On the southern periphery of the Great Basin, the region occupied by Pueblo agriculturalists, a series of archaeological events which may be related to climatic change occurred in the centuries following AD 1000. Between AD 1000 and 1300, irrigation agriculture gradually replaced dry farming methods, and the twelfth and thirteenth centuries AD saw the abandonment of many of the larger settlements. A study of maize grown during this period shows a significant decrease in the size of cobs during the thirteenth century, and these events have been related to a decrease in winter rainfall in the area (Sears, 1958; Mackey & Holbrook, 1978), although factors such as warfare and invasion have also been suggested to account for the abandonment of sites (Jett, 1964).

Arctic North America

The High Arctic All of the environmentally marginal regions which we have discussed so far are dry or desert areas, where the environment is under moisture-stress. Another type of climatic stress exists in High Arctic North America, perhaps the most climatically marginal area occupied by Holocene man. The North American High

Arctic includes that part of the Canadian Arctic Archipelago lying to the north of Parry Channel, and the Pearyland region to the north of the Greenland ice cap. This area lies almost entirely to the north of 75° N latitude, and consists of a series of islands which are at present snow-free for only a few summer months, separated by narrow gulfs and channels which are frozen for 9–12 months per year. The climate is classified as Polar Desert, and most of the region receives less than 100 mm of precipitation annually. Temperatures above freezing occur for 2–3 months during the summer, and, very rarely, exceed 10 °C for short periods of time. During the winter night, which lasts several months, average temperatures approach −40 °C.

The High Arctic was unoccupied at the time of nineteenth-century European exploration, and it has been generally assumed that the region was simply too barren to support human life, even that of the Eskimos who had settled most other regions of Arctic North America. However, recent archaeological work has shown that the area has been occupied sporadically for at least the past 4500 years. The history of these occupations may be useful in documenting climatic change, if we assume that the various episodes of occupation and abandonment were in large part reactions to changing environmental conditions. We may also assume, with a fair degree of certainty, that the environmental variable which would have been most important to these people was the nature and extent of summer sea ice. This variable directly conditions the density and distribution of sea mammal populations, the primary food resource of most Arctic island peoples.

The extent of High Arctic summer sea ice appears to be heavily influenced by summer weather conditions. Although climatic studies in the region have barely begun, and a direct relationship between climatic patterns and distribution of summer sea ice has not been formally demonstrated, Alt (1978) has shown that summer weather patterns dominated by anticyclonic circulation over the Arctic islands correlates with a negative mass balance of the Devon Island ice cap. This in turn has been correlated with relatively large amounts of open water in the channels between the Arctic islands (Koerner, 1977). Conversely, a summer weather pattern dominated by a low pressure area over northeastern North America correlates with a positive mass balance of the Devon Island ice cap and heavy ice cover throughout the High Arctic. In another attempt at synoptic modelling, Bradley & England (1979) have suggested that when depression tracks in the North Atlantic are displaced northward, regeneration of these depressions along the Siberian coast brings relatively warm and moist summer conditions in the High Arctic; southward displacement of North Atlantic depression tracks results in relatively cool and dry summer conditions. It may be possible to infer the relative dominance of these two summer weather patterns during various periods of the past by examining the nature and extent of human occupation in the High Arctic.

The first known human occupation of this region occurred about 2600 BC (4000 BP), and was accomplished by Palaeoeskimos of the Independence I culture (Knuth, 1967; McGhee, 1976) whose roots lay far to the south and west in Alaska or Siberia. The expansion of these people across Arctic Canda during the third millennium BC was probably not a reaction to an improved environment, but to the development of the first efficient human adaptation to the Arctic coast and tundra. The bone refuse found in High Arctic Independence I sites indicates that these people were hunters of muskoxen (*Ovibos moschatus*), seals, and small animals and birds. The species represented are identical to those present in the area today, but the large number of muskoxen bones found in sites on Ellesmere Island and Pearyland suggests a larger muskoxen population than at present. The presence of bones of ringed seal (*Phoca hispida*), bearded seal (*Erignathus barbatus*) and walrus (*Odobenus rosmarus*) indicates the presence of open water during the summer months, and winter ice cover sufficiently extensive for the breeding of ringed seals. The seasonal extent of sea ice would appear to be similar to, or less than, that of the present time.

The Independence I occupation is well dated by a series of 12 radiocarbon dates ranging between 2600 and 2100 BC (4000–3600 BP). After 2100 BC there are no acceptable radiocarbon dates from the High Arctic for approximately a millennium, and we suspect that the area was either abandoned, or population declined to a point where the people became archaeologically invisible. The most plausible explanation for this event is environmental deterioration, probably involving an increase in the extent of summer sea ice and a consequent decline in sea mammal populations.

The next major occupation of the High Arctic is poorly dated, but appears to have occurred between roughly 1300 BC and 800 BC (3000–2600 BP). This Independence II occupation seems to have begun with renewed migrations from more southerly Arctic islands, and was based on a way of life very similar to that of the earlier Independence I people. From our present archaeological knowledge, this occupation does not appear to have been as heavy as that of the Independence I culture, and we might suggest that environmental conditions were not as advantageous to human occupation as they were a millennium earlier. After approximately 700 BC there is another gap in the archaeological record, with little evidence of occupation until after about AD 500 (1500 BP). This Palaeoeskimo Dorset culture occupation was less extensive than the earlier ones, despite the fact that these people had a more advanced technology and were efficient hunters of larger sea mammals such as walrus. The same range of animals is represented by the bones found on Dorset sites, and again we may infer climatic and environmental conditions rather similar to those of the present day.

About 1000 years ago High Arctic prehistory took a new turn with the sudden appearance of a new population, the ancestors of the present Eskimos of Arctic Canada and Greenland. These Thule culture people were natives of Alaska, where over the centuries they had developed a complex maritime hunting technology, and techniques for open-water hunting of large sea mammals including the bowhead or right whale (*Balaena mysticetus*). The rapid expansion of this population across Arctic Canada about AD 1000 (1000 BP) was probably conditioned by the expansion of an environmental zone in which this technology was useful (McGhee, 1970). Thule culture winter villages, each containing the bones of several large whales, are scattered throughout the Arctic islands in areas far beyond the present range of such whales. Evidence for the widespread use of the *umiak*, a large skin boat used for whaling and travelling, indicates the presence of open water in areas such as the northern coast of Greenland which today would not be navigable because of severe ice conditions.

By the thirteenth century these people were in contact with the Greenlandic Norse, another population which may have been encouraged to navigate the Arctic seas by improved ice conditions. The subsequent movement of Thule Eskimos southward along the west coast of Greenland, bringing them into conflict with the Norse colonies, may have been due to a deteriorating climate as has been suggested by several scholars. On the other hand, this may simply have been the final movement in the population expansion which had brought their ancestors from Alaska. In any case, the Thule people appear to have abandoned the High Arctic by approximately AD 1600, and there is no indication of subsequent occupation of the region. In more southerly Arctic regions, Thule people at this time began to readapt their economy; open-water whaling was abandoned in most areas, and replaced by interior hunting during the summer and winter hunting of ringed seals from snow-house villages on the sea ice. Such an adaptation would be consistent with a major expansion in the extent of summer sea ice, reducing or eliminating the populations of large sea mammals in most of the Central and High Arctic (McGhee, 1970). The same deteriorating environmental and climatic conditions may have been a factor in the extinction of the Greenland Norse colonies (but cf. McGovern, this volume).

In summary, the history of human occupation of the High Arctic is consistent with the hypothesis that summer sea ice in the region was relatively less extensive prior to 2100 BC, perhaps during the periods 1200–700 BC and AD 500–1000, and certainly from AD 1000 to AD 1600. During the remainder of the last 5000 years, sea ice distribution may have been as extensive or more extensive than at present. This may in turn reflect changes in the dominance of the two summer weather patterns described by Alt (1978).

The Barren Grounds Moving 1000 km to the south of the High Arctic, we find another marginal region which may be useful in the study of climatic history. The Barren Grounds region to the west of Hudson Bay is today an extensive tundra area bounded on the south and west by the boreal forest. This region was occupied at least as early as 5000 BC by Indian hunters who lived primarily on the herds of caribou (*Rangifer tarandus*) which migrate annually from the forest to summer calving grounds on the tundra (Gordon, 1975, 1976). This Indian occupation seems to have continued until the late third millennium BC, when the area was abandoned; the most recent radiocarbon date on this occupation is 2200 BC (3700 BP). The next major occupation of the Barren Grounds was by Palaeoeskimos, southern relatives of the Independence I people, who occupied caribou hunting camps on the tundra and within the present northern limit of the boreal forest between approximately 1500 BC and 1000 BC (3200–2800 BP). The Palaeoeskimos then abandoned the region, which was reoccupied by Indian caribou hunters who continued to live in the area from 800 BC (2600 BP) until the nineteenth century AD (Wilmeth, 1978).

We may suggest that the fluctuations in human occupation of this region may be most plausibly related to changes in the size or migration patterns of the caribou herds which were the basic food resource of the area. The weather conditions to which caribou are most susceptible probably include cold and wet summers which increase calf mortality on the tundra calving grounds (Gordon, 1976). An increase in the dominance of Alt's (1978) cyclonic summer weather pattern at around 2200 BC, as was previously suggested with relation to the end of the Independence I occupation of the High Arctic, might have increased the frequency of such conditions in the Barren Grounds region. The subsequent occupation of the region by Palaeoeskimos, whose way of life in all other areas appears to have been adapted to life on the tundra rather than in the boreal forest, suggests that tundra was more extensive between 1500 and 1000 BC, and we might infer a continued dominance of cold and wet summer weather patterns. The return of Indian populations to the Barren Grounds by at least 800 BC suggests that by this time the environment and climate of the region were approaching modern conditions.

Summary

As can be seen from the above discussion, the climatic events which can be postulated purely on the basis of archaeological evidence are few, and the nature of these events can be suggested only in very general terms. Nevertheless, there seem to be several periods during which widespread changes in the archaeological record may be related to climatic change. These may be summarised as follows.

2200–2000 BC (3700–3500 BP)

Old World dry lands Egyptian accounts tell of a period when the Nile was consistently low, probably as a result of decreased summer rainfall over East Africa. The decline of Harappan civilisation may have been caused in part by a related failure of the summer monsoon rains in north-western India, although this relationship is speculative and not supported by any convincing evidence.

Arctic North America Palaeoeskimo occupation of the High Arctic declined or ceased at this time, possibly as a result of the onset of cooler summer conditions. These conditions may also have caused the abandonment of the Barren Grounds by Indian caribou hunters.

2000–1300 BC (3500–3000 BP)

Arctic North America The abandonment of much or all of the High Arctic throughout this period, and occupation of the Barren Grounds by Palaeoeskimos rather than by Indians, suggests that cooler summer conditions may have continued.

1300–800 BC (3000–2500 BP)

Old World dry lands The first major 'dark age' in Greece, and turmoil throughout the eastern Mediterranean, was perhaps caused in part by drier winter conditions, although again there is no direct and convincing evidence to support this hypothesis.

Arctic North America This period saw the return of Palaeoeskimo populations to the High Arctic, and the reoccupation of the Barren Grounds by Indian hunters by at least 800 BC, both events perhaps related to warmer summer conditions.

800 BC–AD 500 (2500–1500 BP)

Arctic North America Abandonment of the High Arctic again suggests the onset of cooler summers in this region.

AD 500–1000 (1500–1000 BP)

Old World dry lands By this time, archaeological evidence is superseded by historical evidence in most parts of the Old World, and the history of this region can be dealt with more competently by historians.

Arctic North America About AD 500–600 the High Arctic was once again occupied by Palaeoeskimos, and we may suggest that summer conditions were becoming warmer.

New World dry lands Adaptation to drier conditions may have occurred in the Great Basin, and a drier climate is also suggested for the west coast of South America, beginning about AD 600. The archaeological evidence for such change is, however, very weak.

AD 1000–1600 (1000–300 BP)

Arctic North America This period saw the spread of the whale-hunting Thule culture, indicating that summer conditions in the area must have been considerably warmer than at present. This warm period, and the subsequent 'Little Ice Age' during which the whaling culture disappeared from most of Arctic North America, is well attested in European historical accounts.

New World dry lands Continued drying, probably due to a decrease in winter precipitation, is again suggested in the American southwest during the first half of this period. Another dry period is suggested for the west coast of South America, beginning about AD 1400.

Conclusion

It is obviously impossible, on the basis of archaeological evidence alone, to suggest a worldwide picture of climatic change over the last 5000 years. What we have is scattered bits of information pertaining to critical seasonal climatic conditions in local regions. It must be left to the climatologist to decide whether or not these pieces of information are consistent with regional or worldwide climatic patterns, and with changes in these patterns which can be deduced from other palaeo-environmental evidence. If it can be convincingly demonstrated that the archaeological evidence is consistent with a general picture of climatic change, it may be worth our while to restudy the archaeological events described above as incidents illustrating human adaptability. By seeing how human beings of various cultures and in various economic and environmental situations have reacted to climatic changes, we may gain some insights into how our own cultural systems can be expected to react to similar changes in our climate and environments.

Finally, I feel that archaeology has one other useful contribution to make to the study of past environments and climates. Over at least the last 5000 years, Man has been an important factor in changing local environments in most regions of the world (Raikes, 1967; Bryson & Murray, 1977). In areas such as Egypt or Mesopotamia, which supported major agricultural civilisations, irrigation, land clearance, drainage and grazing have profoundly altered the hydrology and biota of large regions. Agricultural clearance and the introduction of exotic domesticates have altered the flora over much of the tropical and temperate world. Even in areas to which agriculture was only recently introduced, hunting practices such as the burning of forest or grassland have probably contributed to environmental change; for example, Lampert (1975) has suggested that the Australian landscape is in large part a result of repeated firings by Man over the last 30000 years. Palaeobotanists, hydrologists and soil scientists must be aware of this influence of Man on local environments, if they are to interpret correctly the causes of past environmental change. Archaeologists may be able to make useful suggestions regarding the probable nature of such human impact on local environments over the last several thousand years.

Note

1. The dating of archaeological events. Archaeological events are usually dated by one of two methods: (*a*) absolute dates are derived from historical records, or occasionally by other techniques such as dendrochronology; and (*b*) dates are obtained by the radiocarbon dating technique. In a study such as this, temporal comparisons must be made between events dated by both methods. In order to minimise the incompatibility between the two systems, all radiocarbon dates and sequences mentioned in this study have been calculated using the 5730 year half-life of radiocarbon, and calibrated using the MASCA calibration curve (Ralph, Michael & Han, 1973). Absolute dates and adjusted radiocarbon dates are reported as BC/AD; in order to facilitate comparison with reports using unadjusted radiocarbon dates, these dates are also reported in standard radiocarbon years as (BP). Because of the general nature of this study, all radiocarbon dates are rounded off to the nearest century; to report exact dates might give an illusion of a greater degree of precision than we can presently attain.

References

Allchin, B. & Allchin, R. (1968). *The Birth of Indian Civilization*. Harmondsworth: Pelican.

Alt, Bea Taylor (1978). Synoptic climate controls of mass-balance variations on Devon Island ice cap. *Arctic and Alpine Research*, **10**, 61–80.

Bell, Barbara (1971). The dark ages in ancient history, 1: the First Dark Age in Egypt. *American Journal of Archaeology*, **75**, 1–26.

Betancourt, Philip P. (1976). The end of the Greek Bronze Age. *Antiquity*, **50**, 40–7.

Bettinger, R. L. (1977). Aboriginal human ecology in Owens Valley: prehistoric change in the Great Basin. *American Antiquity*, **42**, 3–17.

Bradley, R. S. & England, J. (1979). Past glacial activity in the High Arctic. *Geografiske Annaler*, **61A** (3–4), 187–201.

Bryson, R. A., Lamb, H. H. & Donley, David L. (1974). Drought and the decline of Mycenae. *Antiquity*, **48**, 46–50.

Bryson, R. A. & Murray, T. J. (1977). *Climates of Hunger*. Madison: University of Wisconsin Press.

Carpenter, Rhys (1966). *Discontinuity in Greek Civilization*. Cambridge University Press.

Desborough, V. R.d'A. (1972). *The Greek Dark Ages*. New York: St Martin's Press.

Dillehay, Tom D. (1974). Late Quaternary bison population changes on the Southern Plains. *Plains Anthropologist*, **19**, 180–96.

Gordon, Bryan H. C. (1975). *Of men and herds in Barrenland prehistory*. Ottawa: Archaeological Survey of Canada, Mercury Series 28.

Gordon, Bryan H. C. (1976). *Migod: 8000 years of Barrenland Prehistory*. Ottawa: Archaeological Survey of Canada, Mercury Series 56.

Jett, Stephen C. (1964). Pueblo Indian migrations: an evaluation of the possible physical and cultural determinants. *American Antiquity*, **29**, 281–300.

Knuth, Eigil (1967). *Archaeology of the Musk-Ox-Way*. Paris: Contributions du Centre d'Etudes Arctiques et Finno-Scandinaves, 5, Ecole Pratique des Hautes Etudes.

Koerner, R. M. (1977). Devon Island Ice Cap: core stratigraphy and paleoclimate. *Science*, **196**, 4285, 15–18.

Lambrick, H. T. (1967). The Indus flood-plain and the 'Indus' civilization. *Geographical Journal*, **1**, 483–95.

Lampert, R. J. (1975). Trends in Australian prehistoric research. *Antiquity*, **49**, 197–206.

Mackey, James C. & Holbrook, Sally J. (1978). Environmental reconstruction and the abandonment of the Largo-Gallina area, New Mexico. *Journal of Field Archaeology*, **5**, 29–49.

Madsen, David B. & Berry, Michael S. (1975). A reassessment of north-eastern Great Basin prehistory. *American Antiquity*, **40**, 391–405.

McGhee, Robert (1970). Speculations on climatic change and Thule culture development. *Folk*, **11–12**, 173–84.

McGhee, Robert (1976). Paleoeskimo occupations of Central and High Arctic Canada. In *Eastern Arctic Prehistory: Paleoeskimo Problems*, ed. Moreau S. Maxwell, pp. 15–39. Washington: Memoirs of the Society for American Archaeology, 31.

Mori, Fabrizio (1974). The earliest Saharan rock-engravings. *Antiquity*, **48**, 87–92.

O'Connor, David (1972). A regional population in Egypt to circa. 600 BC. In *Population Growth: Anthropological Implications*, ed. Brian Spooner, pp. 78–100. Cambridge, Mass: MIT Press.

Paulsen, Allison C. (1976). Environment and empire: climatic factors in prehistoric Andean culture change. *World Archaeology*, **8**, 121–32.

Raikes, Robert (1965). The Mohenjo-Daro floods. *Antiquity*, **39**, 196–203.

Raikes, Robert (1967). *Water, Weather and Prehistory*. London: John Baker.

Ralph, E. K., Michael, H. N. & Han, M. C. (1973). Radiocarbon dates and reality. *MASCA Newletter* **9**, 1–19.

Roux, Georges (1964). *Ancient Iraq*. London: George Allen & Unwin.

Sears, Paul B. (1958). Environment and culture in retrospect. In *Climate and Man in the Southwest*, ed. Terah L. Smiley. Tucson: University of Arizona Bulletin 28 (4).

Shaw, Brent D. (1976). Climate, environment and prehistory in the Sahara. *World Archaeology*, **8**, 133–49.

Singh, Gurdip (1971). The Indus Valley culture. *Archaeology and Physical Anthropology in Oceania*, **6**, 177–88.

Vandier, Jacques (1936). *La Famine dans l'Egypte Ancienne*. Cairo: L'Institut Français d'Archéologie Orientale.

Wheeler, Sir Mortimer (1966). *Civilizations of the Indus Valley and beyond*. London: Thames & Hudson.

Wilmeth, Roscoe (1978). *Canadian Archaeological Radiocarbon Dates* (revised version). Ottawa: Archaeological Survey of Canada, Mercury Series 77.

7

The use of documentary sources for the study of past climates

M. J. INGRAM, D. J. UNDERHILL
AND G. FARMER

Abstract

Documentary evidence is an important source of detailed information on past climates, particularly for the period from about AD 1100 to the beginnings of the era of instrumental meteorology. This chapter is concerned with the study and climatic interpretation of this evidence. It comprises four sections. The first provides a survey of the available sources, offering some insights into the historical milieux which produced them and indicating in very general terms the kinds of information they contain. There are, however, many pitfalls involved in using historical records, and in the second section we discuss the most basic of them, the problems involved in assessing the reliability of sources as records of events. The third section deals with the more difficult problems of source interpretation and analysis, some attention being given to the use of content analysis techniques. Finally, we discuss the scientific analysis of the materials to produce climatologically valuable data. In turn we consider the analysis of early instrumental observations; the construction of meteorological series from the relatively few types of documentary data which provide continuous series of more or less homogeneous information; and the analysis of fragmentary and non-homogeneous series using indexation procedures and spatial mapping techniques (the latter making possible the analysis of qualitative material in terms of atmospheric circulation patterns).

Introduction

The possibility that the world is now undergoing a period of significant climatic shift has greatly stimulated scientific interest in the study of the climate and weather of the past. But the evidential problems are great. Reliable instrumental observations of meteorological phenomena cover only a minute fraction of the earth's climatic record, and to penetrate further into the past it is necessary to use a variety of materials, all of which involve major interpretative difficulties (Fig. 7.1). To date, scientists trying to reconstruct past climates have concentrated most attention on types of evidence which can be handled by the normal field and laboratory techniques of direct observation, measurement, and quantitative manipulation. Hence there has been a massive application of methodological expertise to the study of the 'natural record' of climatic change embodied in such phenomena as tree rings, fossil floral and faunal remains, glaciers, ice sheets, and lake varves (the enormous advances made in techniques for the study of such field data are reviewed in Chapters 2–5 of this volume). However, it has long been recognised that *documentary* evidence is potentially another major source of information on past climates, especially with regard to the details of short-term shifts and fluctuations and particularly for the millennium or so immediately preceding the era of modern instrumental meteorology. As early as 1926, Brooks (himself building on extensive earlier research) included much documentary data in his pioneering synthesis *Climate through the Ages*; while in succeeding decades, Brooks' lead was enthusiastically followed by Flohn (e.g. 1950), Manley (e.g. 1952), Schove (e.g. 1949), and (above all) Lamb (e.g. 1961), who in the face of the widespread indifference of the scientific community urged the interest and value of documentary climatology. It is a measure of the achievement of these pioneers that, with the recent explosion of interest in research into past climates, more and more attention is being paid to the exploitation of documentary material.

Fig. 7.1. Sources of information for the reconstruction of past climates. The separate categories often overlap, and some of the more important interactions are shown.

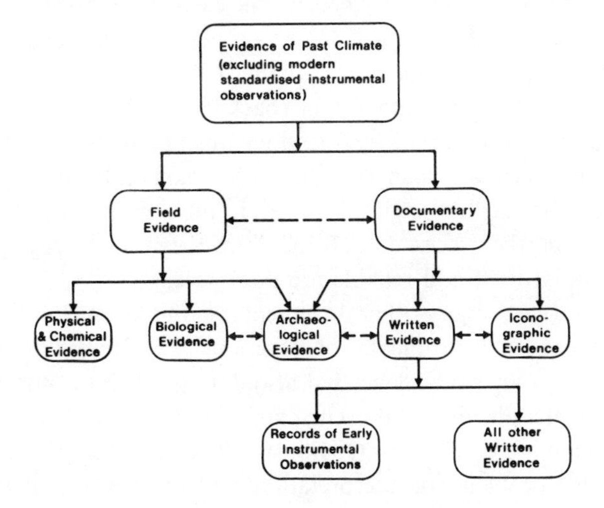

But the use of materials foreign to the established methodologies of empirical science has not proved easy. Scientists have had to learn painful lessons in the use of historical records, while historians themselves (with the notable exception of Le Roy Ladurie (e.g. 1972) and a handful of others) have been slow to volunteer their expertise. However, in the very recent past there has been an encouraging growth in interdisciplinary co-operation, a joint effort which promises to produce major rewards. In this chapter, we attempt to review past and current work in the light of the latest available information on what constitutes sound methodology. We begin with an extensive survey of the sources, including as far as possible some insights into the historical circumstances which produced them; continue with sections on the techniques of source criticism and of historical interpretation; and conclude with a discussion of the scientific analysis of the materials.

Survey of documentary sources

Records of instrumental observations are the most obvious type of documentary evidence for the study of past climates, and of course form the most satisfactory of all the available kinds of data. Unfortunately the geographical and chronological scope of such materials is very limited. It is true that wind vanes and some form of rain gauge were apparently known in antiquity, while fragments of wind direction records actually survive from the time of Ptolemy (Hellmann, 1916). But it was not until the mid-seventeenth century AD, following the development of the barometer and the liquid-in-glass thermometer, that the age of instrumental meteorology effectively began. The earliest records (*c.* 1650–*c.* 1750) relate to parts of Europe and to certain places in the eastern USA (von Rudloff, 1967; Landsberg, Yu & Huang, 1968). In most other parts of the world, instrumental data extend back no further than the late eighteenth century, and in many locations (including much of Asia, Africa and the Americas) span less than 100 years (for summary, see Lamb, 1977). The earliest series, moreover, pose many problems of interpretation, and the precise dates from which they may be regarded as acceptably standardised and reliable are debatable.

There are, however, many other documentary sources of climatic data apart from records of instrumental observations. They may include not only direct meteorological evidence but also data on such phenomena as glacier movements, phenological events, and other more or less indirect indicators of climatic conditions. Obviously all such materials are likely to pose complex problems of interpretation, and these will be discussed in later sections. At this stage the aim is simply to indicate the great range of potentially relevant sources, and to mention (with only a minimum of critical comment) some of the more notable attempts to use them.

Of these available documentary sources, the overwhelming majority are written. However, some *pictorial documents* have been used as evidence of past climates. For example, Butzer (1958) has exploited rock drawings, tomb reliefs and paintings, and pictorial decorations on pottery for evidence of shifts in the distribution of the larger fauna of ancient Egypt and the Sahara region, and hence of changing wind and rainfall patterns. Le Roy Ladurie (1972) and others (Messerli, Messerli, Pfister & Zumbühl, 1978) have assembled impressive collections of paintings, prints and photographs which reveal in considerable detail the movement of Alpine glaciers since the sixteenth century. Lamb (1967) has sought supporting evidence for the onset of a period of colder climate in western Europe in the late sixteenth century in the winterscapes of Pieter Bruegel the elder, and has suggested that the depiction of cloud cover by Dutch and English landscape painters 1550–1940 provides an interesting record of the prevailing character of summers over that period. This work has been developed further with reference to English art by Thornes (1978).

Written sources for the study of past climates are legion. Their precise nature varies throughout the world and according to the period in which they were written, but virtually all the materials known to exist can be accommodated in the following categories.

Myths, legends and imaginative literature

The vast majority of written sources of potential value to climatologists demonstrably relate to real events, to phenomena which actually happened at some time in the past. However, there are various classes of written source whose relationship to reality is more questionable or more complex. These include such materials as myths and legends, often having the superficial appearance of historicity and possibly embodying some historical truth or half-truth; and literary productions such as poems, plays, stories and even philosophical writings and religious works which – it can be argued – incidentally embody useful information about the material world as their writers saw it. Examples of the use of such materials are provided by Bergeron, Fries, Moberg & Ström (1956), who have speculated that certain Norse legends had their origin in a prehistoric experience of climatic deterioration; and by Chu Ko-Chen (1973), whose analysis of climatic change in China 1100 BC to AD 1400 is based partly on poetic references to the incidence of such flora as plum, bamboo and orange trees.

Ancient inscriptions

This is a hybrid category in which a miscellany of very ancient records may be conveniently grouped. Examples from ancient Egypt include the fragments of a large stone stele, carved during the Fifth Dynasty (*c*. twenty-fifth century BC), which recorded Nile flood levels for every year back to about 3050 BC. This source, as well as numerous tomb inscriptions and papyri, has been exploited by Bell (1970, 1971) in an attempt to argue that the First Dark Age of Egypt was brought on by a prolonged and intense drought. For China, the earliest records bearing on past climate take the form of carved oracle bones dating from before 1000 BC. These oracle bones include prayers for rain and snow and descriptions of crops and agricultural methods; they are also said to embody a ten-day weather record for 20–29 March in the sixth year of Wen Ting of the Ying Dynasty (1217 BC) (Chu Ping-Lai, 1968; Chu Ko-Chen, 1973).

Annals, chronicles and histories

The habit of recording notable events in the form of annals, chronicles and histories of greater or lesser sophistication has been widespread from early times. Natural events – whether because they threatened food supplies or physical security, because they were regarded as of religious or occult significance, or simply because they were out of the ordinary – have almost always been of interest to the compilers of such documents, which hence form an important source of evidence for historical climatologists. China boasts probably the longest and richest continuous tradition of compiling annals and histories, some Chinese materials dating from about 1000 BC (Chu Ko-Chen, 1973). Europe is also highly favoured in this respect. Some materials exist for southern Europe from about 500 BC, while in northern Europe the historical tradition is roughly coterminous with the Christian era, becoming particularly rich after about AD 1100.

But not all chronicles are equally informative about past climate. In the case of western European materials, for example, it is generally found that the more sophisticated productions, especially works with pretensions to form interwoven ‘histories’ rather than mere year by year accounts of events, tend to give relatively little attention to weather. More jejune local annals, by contrast, frequently include comparatively detailed accounts of weather and related phenomena. Among the richest of all types of chronicle from a meteorological point of view are urban annals, of which numerous examples survive from late medieval and early modern Europe (*c.* 1350–1700): towns were critically dependent on food supplies and hence on the vagaries of climate. But, even in chronicles containing a relative abundance of meteorological detail, the weather was rarely the *main* focus of interest; characteristically the climatic information available from these narrative sources is very fragmentary, and often a high proportion of notices are reports of isolated meteorological events such as notable thunderstorms.

Records of public administration and government

This category embraces a huge span of records from imperial, national and local archives. Governments in the past inevitably took an interest in weather and related natural events, as a direct result of their concern with tax yields, the public disorder which might result from weather-induced harvest failures or other catastrophes, the maintenance of communications and of public buildings and works, the effects of good and bad weather on military and naval campaigns, and similar problems. Thus, direct descriptions of weather conditions may be found in such sources as the reports of diplomats like the Venetian ambassadors resident in England in the sixteenth and seventeenth centuries (Baker, 1932); the reports of local officials to their superiors, like the 'sheriffs' of eighteenth-century Iceland whose official letters to their Danish overlords included descriptions of the weather of inclement seasons and their economic and social effects (Astrid Ogilvie, personal communication); and official surveys and gazetteers, including the rich series which survive for many cities and regions of China from the fifteenth century AD onwards (Chu Ko-Chen, 1973; Wang Shao-wu & Zhao Zong-ci, this volume). Of course the weather reports to be found in such sources are normally selective and often very fragmentary, reflecting the specific, short-term interests of hard-pressed administrators; but they are nevertheless of potential value. Records which contain more indirect evidence of climate and weather can also be found in the archives of central and local government. These include tithe lists, harvest price series, and (of immense importance) local registers of wine harvest dates in France, Switzerland and certain other parts of Europe from the sixteenth century onwards (Le Roy Ladurie & Baulant, 1980; Pfister, this volume).

Private estate records

Sources such as farm accounts, estate surveys, and the reports of bailiffs and estate managers to their lords or employers often include evidence bearing on past climate. Especially noteworthy is the material included in the thirteenth-, fourteenth- and fifteenth-century account rolls of the estates of the Bishopric of Winchester, the Abbey of Westminster and certain other major English ecclesiastical landholders. The data include direct references to weather, floods, parched or sodden ground, etc., noted on the rolls to explain unusually heavy expenditure or poor returns, but the documents also embody information about more indirect indicators of climatic conditions such as seeding rates, ploughing dates and corn yields (Titow, 1960, 1970; Brandon, 1971; Stern, 1978).

Personal papers

Personal writings are of great importance as sources of climatic information. Letters, especially those sent to correspondents at a considerable distance and intended to serve as newsletters, often include much weather data. A good example is the fine series of letters written by John Chamberlain of London, which spans the period from the 1590s almost to his death in 1627 (McClure, 1939). Even more useful are diaries, journals and commonplace books, which in Europe begin to survive in significant numbers from the end of the middle ages and become plentiful in the period from the late sixteenth through the eighteenth centuries (for a discussion of English diaries, see Macfarlane, 1970). Of course not all writings of this type are of climatic interest; some are wholly personal, others devoted to political or other matters. But fortunately a great many early diaries have a decidedly outward-looking orientation, some of the oldest being in fact very similar to local annals, and often give some attention to meteorological matters. Typical examples from the fifteenth, sixteenth and seventeenth centuries include the so-called journal of a bourgeois of Paris, 1405–1449 (Shirley, 1968); the diary of Luca Landucci of Florence, 1450–1516 (Badia, 1927); and the 'dietario' of the Catalan Juan Porcar, 1589–1629 (Castañeda y Alcover, 1934). Even among these more 'public' diaries, however, the personal characteristics of the writer inevitably determined just how much attention he gave to matters of climatological interest. Writers of eminence, living close to the centre of political or religious affairs, frequently relegated meteorological matters to a place of minor importance. More humble writers such as burghers, parish priests, and yeomen or peasants often gave more attention to weather; though characteristically their notices of heavy rains, droughts, great frosts and floods have to rub shoulders with reports of major political happenings, religious celebrations, executions, spectacles, fires, monstrous births and other events, and the diarists often forgot meteorology when other interests were pressing. The most informative diaries from a climatological point of view (apart from the specialised meteorological journals discussed later) were usually written by people deeply involved in activities to which the weather was vital: farmers, vine growers, millers and the like. Such interested observers frequently described the performance of crops and the movements of prices, and made monthly, weekly, or even daily notes on weather patterns.

Early journalism

In certain respects the tradition of local chronicles and annals was carried on into the early modern world not only by diarists (as suggested above) but also by printed broadsheets, pamphlets, and early newspapers; and indeed this tradition is still a recognisable component of modern provincial journalism. In any event the products of early journalistic writing cannot be ignored by the historical climatologist. For example, many remarkable episodes in the climatic record of late sixteenth- and seventeenth-century England, notably the snowy winters of 1579 and 1615, the great frosts of 1608 and 1684, the disastrous floods of 1580 and 1607 and the stormy winter of 1613, are illuminated by detailed pamphlets (Shaaber, 1929); while frequent weather reports for England and further abroad are to be found in such journals as *The Gentleman's Magazine* (monthly from 1731). Eighteenth-century Scottish newspaper accounts of cold and snowy winters and other notable climatic episodes have been extensively exploited by Pearson (e.g. 1973).

Scientific and proto-scientific writings (apart from instrumental series)

The key feature of writings in this category is their approach towards regular, comprehensive observations of meteorological and parameteorological phenomena, usually with the purpose – in fact often unrealised – of analysing the data to discover patterns, meanings and relationships. In other words, their approach was more or less 'scientific', though this term should be used with care. Nowadays historians stress that progress towards what we recognise as modern scientific method took a very winding path, involving many excursions into what have today been written off as magical and superstitious blind alleys; it is to take account of these that the term 'proto-scientific' is useful.

A variety of motives stimulated the keeping of relatively systematic meteorological observations. One was the pure spirit of inquiry which seems to have animated amateur scholars, many of them priests or country gentlemen, in increasing numbers in western Europe from the sixteenth to the nineteenth centuries. An example was Christopher Sanderson of County Durham in England. His diary (manuscript located in Gateshead Central Library, England) reveals an active and percipient mind in the process of development towards scientific method. From random notes on remarkable droughts, floods and the like from the 1650s, his weather observations became more and more detailed until in the period 1682–89 he produced a daily meteorological journal. A second motive towards meteorological inquiry was the search for a more efficient agriculture, which in Europe gathered momentum in the seventeenth and eighteenth centuries; it stimulated numerous series of more or less detailed observations, some of which were printed in textbooks on agricultural improvement (e.g. Lisle, 1757). A third major spur to the recording of meteorological phenomena was astrology. This fundamentally ancient but constantly elaborated system of ideas, now fallen into scientific disrepute, for centuries exercised the greatest minds and was midwife both to astronomy and meteorology. The earliest known European journal of the weather, kept by William Merle at Oxford and in Lincolnshire from 1337 to 1344, may well have owed its existence to astrological interests (Symons, 1891). Astrology was equally a force in the sixteenth and seventeenth centuries. The meteorological contribution of such astrologers as John Goad and John Gadbury is well known, the latter being responsible for a very important continuous series of non-instrumental observations 1668–1700 (Manley, 1974). Even as the prestige of astrology decayed in the eighteenth century it could still serve to stimulate meteorological records: the interesting series of observations published in *The Ladies Diary* from 1717 to 1745 was intended to show 'how different the Weather has proved from what ... [the astrologers] have pretended to foretel ...'. Yet another important stimulus to meteorological inquiry was medicine. From the time of Hippocrates' *De aëre, aquis et locis* (fifth century BC) through the eighteenth century AD, it was an important axiom of European medical thinking that climate and weather were of major significance in affecting the human constitution and the incidence of disease. From the late middle ages onwards medical treatises frequently included at least crude and disjointed observations of meteorological phenomena, while an increasing methodological rigour in the seventeenth and eighteenth centuries eventually produced such detailed series as those of Huxham (1739) and Wintringham (1727). Indeed, medical men such as Jurin were very prominent among the early exponents of instrumental meteorology.

Maritime, commercial and exploration records

In referring to instrumental series the discussion has almost come full circle, but one final category of documentary evidence must be isolated: maritime, commercial and exploration records. To be sure, these include much material which could well be considered in the 'scientific' category, as well as a great deal of a less systematic nature. But they are best considered under a separate heading to highlight the fact that, unlike the sources hitherto mentioned, the information they contain bears largely (though not exclusively) on the climate of ocean areas and is hence of especial interest to the historical climatologist.

Such maritime, commercial and exploration records, which relate mainly to the period from the sixteenth century onwards, exist in considerable quantities in North American and European record repositories. They include the archives of major trading and exploration organisations like the Hudson's Bay Company and the English and Dutch East India Companies; the records of the English Board of Admiralty and corresponding naval offices in other countries; the published and unpublished reports of explorers; and numerous other records of fishing and commercial enterprises. The individual sources are of many types and embody a rich variety of different kinds of information. For example, ships' logs, naval officers' diaries and commercial day journals supply direct evidence, sometimes highly systematic, on wind directions and other atmospheric and oceanic conditions (including information about sea ice); while records dealing with the movements and catches of fishing and whaling fleets provide evidence about the past distribution of marine species from which data on sea temperatures and ocean currents may be inferred. To date these sources remain relatively poorly exploited, but a beginning has been made. Oliver & Kington (1970) have used ships' logs to supplement land-based observations in constructing daily weather maps of the North Atlantic–western European sector of the northern hemisphere; Fiona Macdonald (personal communication) has employed records of the English East India Company to investigate the incidence of monsoons in the seventeenth century; while a number of researchers have addressed themselves to the records of the Hudson's Bay Company (see review in Catchpole & Moodie, 1978).

This concludes our survey of sources, but a few general comments are called for before we proceed to a discussion of verification and analysis. How far back into the past can the climatologist hope to penetrate with the aid of the kinds of documentary material which have been described? Obviously the answer varies with location. However, it is not yet possible to give a comprehensive guide to the dates from which materials survive for different areas because much essential exploratory work to discover possible sources has not yet been carried out. It is important to emphasise this point and to stress that source research is likely to pay major dividends. In recent years, scholars have begun to turn up a vast amount of material, relating to India, Africa, the Middle East and elsewhere, whose existence was previously hardly suspected. Even in relatively well-researched parts of the world it is plain that there is still much to be discovered: Pfister (this volume), for example, has greatly added to the material for Switzerland. Many other parts of Europe are more or less untapped treasure houses for the historical climatologist: the rich archives of Italy, for example, cry out for serious research, while some indication of the great mass of information potentially available for Spain has come from the researches of MacKay (this volume) on late medieval Castile, and from the ongoing work of Wendy Bell (personal communication) on ecclesiastical archives in sixteenth- and seventeenth-century Murcia.

However, enough is already known of the documentary sources for climatic reconstruction to give a rough indication of the main patterns of distribution. The information summarised in Table 7.1 shows that in certain areas the scope of the historical climatologist is limited to only a few decades preceding the establishment of standardised instrumental observations. But for large areas of the world there are materials extending over centuries, and in a few favoured locations over thousands of years.

But even in areas for which early materials survive, the distribution of sources is by no means uniform over time. For example, the European evidence for the period before about AD 1000 is very scanty. It is appreciably richer, but still very imperfect, for the period AD 1000–1500. Thereafter many more materials are available, but it is only in the late seventeenth and the eighteenth centuries that sources become really abundant to the extent that many parallel series of observations are often

Table 7.1. *Earliest dates from which written evidence of climatic phenomena survives in selected areas*

Area	Earliest written evidence (approximate dates)
Egypt	3000 BC
China	2500 BC
Southern Europe	500 BC
Northern Europe	0
Japan	AD 500
Iceland	AD 1000
Northern America	AD 1500
South America	AD 1550
Australia	AD 1800

available for a comparatively small area. Moreover, there are great differences in quality between different types of material, with nearly all the richest and least problematic sources relating to the more recent periods.

Data verification

Before approaching problems of interpretation it is necessary to discuss a more basic difficulty. Documentary sources of information about past climate are not equally reliable, and indeed much material which purports to record historical events is gravely misleading. Historians have developed a standardised methodology for evaluating sources and rejecting unreliable information. It is plainly essential that climatologists using documentary materials should similarly recognise the need for such source criticism and be aware of the available critical techniques. But, as Gottschalk (1971, 1975, 1977), Alexandre (1974, 1976) and Bell & Ogilvie (1978) have emphasised, at least until very recently many scientists have been ignorant on one or both of these counts. It is true that one of the hallmarks of some of the latest work in documentary climatology has been strict attention to the problems of data verification. Nevertheless, the principles of historical source criticism are still sufficiently unfamiliar to many scientists to justify a brief survey of the subject here.

For the purposes of discussion it is convenient to distinguish between *source compilations and syntheses* and *sources* proper.

Source compilations and syntheses

There exist many eighteenth-, nineteenth- and twentieth-century compilations and syntheses of historical climatic information (see lists in Lamb, 1977 and Bell & Ogilvie, 1978). Though some of these works include elements of interpretation and analysis, their key feature for our purposes is the presentation of large quantities of raw information about past meteorological and related phenomena culled from previous writers. At first sight these compendia seem to provide a conveniently ready-made data bank, and it is therefore not surprising that they have been much used by scientists seeking to reconstruct past climates. In reality, many of these compilations and syntheses are highly defective as data sources, and work founded on them is likely to be inaccurate. Their weaknesses, which as far as the medieval period is concerned have been analysed in detail by Bell & Ogilvie (1978), may be summarised as follows. First, their transmission of the original data is often unsatisfactory. As will be seen later, rigorous analysis of documentary sources often demands close attention to the words and forms of expression originally used. But compilations often consist of abridged, translated, paraphrased or summarised material, with each departure from the original involving some loss or distortion of information. Second, many of the works under discussion include errors introduced by the author or compiler, the most common mistake being misdating. (Both the first and second of these defects are often difficult to remedy, because compilations frequently fail to give adequate references to the sources on which they are based.) Third and most fundamentally, the majority of compilations do not distinguish adequately (if at all) between reliable and unreliable sources. It is true that some very recent works, notably Gottschalk (1971, 1975, 1977) and Alexandre (1976, 1977), are models of critical awareness. As a whole, however, the available compilations and syntheses comprise a mishmash of valuable and worthless data, and to use them uncritically for purposes of climatic reconstruction is unacceptable. This point is illustrated by Table 7.2, which indicates the alarming amount of unreliable material contained in sections of two widely used compilations. (In constructing this table we have subjected the compilations of Hennig (1904) and Weikinn (1958) for the period AD 0–1000 to a critical historical analysis, using the standard methods of source criticism outlined below, and counted the number of items which survive this criticism. Such items are described in the table as 'verified'.)

Table 7.2. *Historical reliability of two European weather compilations for the period AD 0–1000*

Century (AD)	Hennig				Weikinn				Total no of seasons for which reliable descriptions are available
	Total no. of seasons described	Seasons verified No.	%	Additional information[a] by season	Total no. of seasons described	Seasons verified No.	%	Additional information[a] by season	
0–99	20	10	50	2	8	7	88	5	12
100–199	15	1	7	2	2	1	50	2	3
200–299	9	0	0	1	1	1	100	—	1
300–399	16	4	25	3	7	6	86	1	7
400–499	22	3	14	—	2	2	100	1	3
500–599	44	11	25	5	14	10	71	6	16
600–699	23	3	13	1	3	0	0	4	4
700–799	31	4	13	6	9	5	56	5	10
800–899	47	27	57	23	35	32	91	18	50
900–999	45	13	29	6	8	6	75	13	19
Total	272	76	28	49	89	70	79	55	125

[a] The 'additional information' refers to reliable material which we have extracted from other primary sources.

Sources

When scientists try to penetrate beyond compilations and modern syntheses to the materials on which they are based, or if they seek new data, they are faced with the problem of assessing the reliability of sources. How is this problem to be resolved?

The first requisite is an accurate version of the text. Sources are sometimes available only in their original form or old copies (often manuscript, but from the end of the fifteenth century onwards sources may include printed materials), but may also exist in the form of later printed editions produced to make a source more easily available to a wider circle of scholars. But it is important to recognise that many editions, especially those produced before the mid-nineteenth century but also some more recent ones, include serious errors, distortions and omissions caused by the misreading of the original text, the use of imperfect copies, the unwarranted conflation of texts, and major editorial interventions such as the rearrangement or abridgement of the original. For example, the edition of the diary of Ralph Josselin (1616–83) edited by Hockliffe (1908) omits much valuable weather information which the editor regarded as 'of no interest whatsoever'; fortunately the full text is now available in an excellent edition by Macfarlane (1976).

In using edited texts, therefore, it is essential to check the scholarly credentials of the editor and his work and to ascertain that in other ways the edition is adequate for the purpose in hand. This may be done by reference to standard bibliographies, such as Graves (1975) for medieval England, and by inspection of the edition itself – a good edition will contain a lengthy introduction in which the writer clearly explains how he has established his text and what rules of transcription, etc., he has followed. In the absence of a satisfactory edition, it will of course be necessary to go back to the original manuscript or a photographic copy thereof. Use of the original is anyway to be recommended where practicable, especially if a particular text is subjected to very detailed analysis.

But it cannot be assumed without close examination that even original manuscripts or impeccably edited texts contain information which is both reliable and valuable. In other words, the authority of texts varies greatly. The problems of assessment can be complex, though this is not invariably the case. No one would deny that the credentials of many texts, especially those which are relatively recent, exist in their original form, and involve no great problems as to nature and authorship, are comparatively easy to establish. This is true, for example, of many sixteenth-century and later diaries, which often survive in the author's own handwriting and include notes about the writer and the circumstances in which he kept his record. At the other extreme, many narrative sources such as chronicles pose major problems of elucidation. Annalists and chroniclers frequently tried to present not only information about contemporary happenings but also a chronology of major events from times remote from their own. Hence they often copied earlier manuscripts, mostly without acknowledgement, frequently misunderstanding and distorting the earlier material. Copying errors inevitably occurred: numbers and names were often misread, and corrupt passages developed as scribes struggled to make sense of their predecessors' mistakes. Legends, rumours and downright fabrications were on occasion included to swell the story. Some chronicle series, indeed, were subject to constant addition and revision over centuries, resulting in interrelated groups of documents each of which is a complex tissue of original matter and borrowings, reliable and unreliable information (for further discussion and illustration, see Bell & Ogilvie, 1978).

Various criteria may be used to assess reliability and value. In their entirety the tests are numerous and the methods of applying them complex. A number of handbooks are available (e.g. Bernheim, 1889; Langlois & Seignobos, 1898; Samaran, 1961). But the subject is best studied in action, by reference to the critical essays which introduce well-edited texts (e.g. Plummer & Earle, 1952). Here only the basic principles are indicated.

The most important critical tests are those based on the principles of contemporaneity, propinquity and faithful transmission. Recorded statements cannot be regarded as reliable and valuable unless it can be shown either that the writer lived close in time and space to the events he purports to describe, and recorded his observations immediately or within a short space of time after these events had taken place; or that he had access to first-hand oral or written reports and can be presumed to have accurately transmitted the information derived from them.

Many materials emerge badly from such tests. It is frequently found that sources are in whole or in part either slavishly derivative and therefore of no independent value; inaccurately and corruptly derivative; wholly or partly spurious; or of literary, philosophical, or religious significance but of no value as records of events. Numerous examples could be given; but in this brief survey a single illustration, an instance of fabrication, must suffice. The *Historia Croylandensis*, a chronicle of

Crowland Abbey (Lincolnshire) for the period *c.* AD 626–1091 with a continuation to 1117, was once thought to be a valuable historical work. In the nineteenth century, however, it was shown to be an elaborate late-medieval forgery, compiled to further a lawsuit in which the abbey was engaged (Tout, 1934).

Much of the documentary data which climatologists have hitherto used have been drawn, either directly or more often by way of compilations, from such unreliable sources, and are therefore subject to error. The most important errors are inaccurate or uncertain dating of particular events; spurious multiplication of events through failure to recognise that accounts recorded under different dates in various sources in fact relate to a single event; acceptance of accounts which are distortions or amplifications of original observations (for example, purely local events may appear through the distortions of later writers as major catastrophes covering a large area); and inclusion of events for which there is no reliable evidence whatever (such 'non-events', moreover, tend to be artificially multiplied through being assigned to different dates in subsequent, derivative accounts).

These errors are not randomly distributed and are therefore not self-correcting. On the contrary, the problem of unreliable data is so great that even 50-year series are liable to be inaccurate if unverified material is used (for illustration of these points see Ingram, Underhill & Wigley, 1978). It is essential to build up a data bank of historical information of unquestionable reliability, especially since the problems of meaningful interpretation of much documentary data are in any case very great. It is to these problems of interpretation that we now turn.

Historical analysis

The preceding section was concerned with assessing the reliability of evidence in the sense of establishing that recorded events did actually occur at a certain time in the past. But invariably material verified in this way must be subjected to further historical analysis to determine as far as possible its *meaning* in terms of current scientific knowledge.

Modern meteorological observations, which embrace a range of variables that scientific research has identified as of key significance, which are monitored by accurate instruments, and which are expressed with the precision either of numbers or of a carefully standardised technical vocabulary, provide the nearest approach to an objective and comprehensive record of climatic phenomena. Historical climatic data from documentary sources, on the other hand, to a greater or lesser extent embody distortions and omissions reflecting the personal and cultural biases of past observers. The categories used to conceptualise environmental phenomena vary from society to society, between groups of different cultural background within a society, and to some extent from individual to individual; moreover, they change over time. Again, in selecting phenomena for record, observers are consciously or unconsciously influenced by a variety of criteria for judging the importance and interest of events, criteria which in part reflect the intellectual and social milieu in which the observer is located and in part his own individual preferences and prejudices. A further problem arises through the use of language and writing. Prevailing linguistic conventions impose severe limitations on the expression of ideas; this fact emerges vividly from reading the work of early scientists as they struggled to formulate a technical vocabulary adequate to express the concepts which they were developing. Additional problems may arise when an observer is not recording in his native tongue; while conventions in the use of written language may dictate the suppression or distortion of ideas – for example, a writer's desire for regular rhyme or metre or for an 'elevated' style may result in the cloaking of his basic ideas in the language of elegant obscurity.

Plainly, in the final analysis these factors limit the quality and quantity of the information which can be derived from documentary evidence of past climate. But while they remain unidentified or imperfectly understood, they also pose problems of *interpretation*, in effect obscuring what scientifically useful information the materials *do* contain. This obscurity must, as far as possible, be removed by subjecting the sources to rigorous analysis to establish the meanings of words and categories and to determine the nature of biases. This may be achieved by reference to specialised historical knowledge of the intellectual and social milieux which gave rise to the texts, and by internal analysis of individual sources and comparison between sources relating to the same or analogous phenomena. It is impossible to provide here a comprehensive discussion of all the issues which may arise, but the analyst must invariably consider the following basic factors.

(*a*) Why the information was recorded, and how the purposes of the writer biased his perception and selection of phenomena. For example, examination suggests that the content of Ralph Josselin's record of mid-seventeenth century weather was influenced by a number of factors: by his role as a small farmer, which led him to give greater emphasis to certain types of weather, and to certain months or seasons, than to others; by his interest in weather and other natural phenomena as symbols of God's pleasure or displeasure, a factor which evidently led to exaggeration in certain cases; and by such beliefs as the notion that the phases of the moon affected rainfall, an idea which further distorted his perception and selection of phenomena (Macfarlane, 1970, 1976).

(*b*) How phenomena were categorised. Even categories which at first sight appear relatively precise may require close scrutiny. Moodie & Catchpole (1975), for example, in analysing texts which describe the freezing and breaking up of ice in the estuaries of Hudson Bay 1714–1871, noted that even modern standards and norms for isolating these events are not strictly consistent, while extensive analysis was necessary to interpret the historical observations. Even more difficult problems arise in the analysis of such relatively imprecise categories as 'drought', 'flood', 'sea ice', etc.

(*c*) The significance of terms of degree. Statements like 'the greatest flood which hath occurred in the memory of any man living' occur far too often to be taken at face value. This and other impressionistic words and expressions denoting degree require close scrutiny to establish their level of precision or imprecision.

(*d*) The dating and duration of phenomena. A particular problem arises here in that many different forms of calendar reckoning have been employed in the past (cf. the discussion in Lamb, 1977), and plainly it is vital to establish which system was employed by the writer of a particular source. It is well known that the current Gregorian calendar, introduced in the countries of Europe from the sixteenth century onwards, ran increasingly in advance of the Julian system which it superseded, and adjustment has to be made to bring Julian dates in line with modern reckoning. Account must also be taken of the fact that practices of reckoning when the year began were very diverse and liable to sudden change; in medieval and early modern Europe the commonest dates for New Year's Day were 25 December, 1 January, 1 March and Easter Sunday. But the beginning of the year was also liable to be expressed in terms of local events such as (in towns) the date of mayoral elections; while calendars based on reigns and dynasties were also common, especially in areas outside Europe.

Again, words and phrases denoting duration must be critically evaluated. It is clear, for example, that when longer-term meteorological events were described, dates were frequently rounded off by reference to the nearest important saint's day or other calendrical landmark, with the result that apparently precise dates are in reality significant only as approximations. And the terms winter/spring/summer/autumn (fall) – and their equivalents in other languages – must be interpreted carefully. In sixteenth- and seventeenth-century England, for example, 'winter' was generally used to denote the long period from late October through to February or March; while in medieval documents the term *autumnus* was normally used to denote the harvest season.

(*e*) The location and geographical extent of observations. Past observers were not necessarily aware that weather conditions vary markedly from region to region, and they often did not carefully distinguish between purely local phenomena, 'foreign' events outside their direct experience, and conditions which they were able to observe themselves but which also occurred in other areas. Careful analysis of the contents of individual sources, including non-climatic information which may be useful in revealing just how much the writer knew about the outside world, may be necessary to minimise uncertainty about these points.

There remains a further factor to be considered in relation to historical analysis. As far as possible, rules of interpretation must be codified and systematised to ensure that individual analysts, faced with the same body of materials, will derive essentially the same information from them. The reliable or 'objective' interpretation of information in this way is the basic aim of the method of *content analysis*, which was developed for use in the social sciences by Berelson (1952) and others and whose relevance to documentary climatology has been stressed by Moodie & Catchpole (1976). The method involves the use of strict frequency counting techniques, the compilation of comprehensive lexicons identifying and defining the terms used to describe phenomena, the assignment of values to terms of degree, the weighting of certain items of information relative to others on the basis of multivariate analyses, and similar procedures designed to systematise the analysis of materials and eliminate subjective interpretations on the part of individual analysts. Material organised in this way is coded numerically and subjected to consistency and validity checks, the whole procedure being facilitated by computer processing. Variations on this approach have to date been used not only by Moodie and Catchpole but also by Pfister (this volume) and others. So far, however, individual research teams have not mutually standardised their methods. Given the wide range of documentary materials available, each with their own strengths, weaknesses and idiosyncratic features, a watertight overall system may be difficult to achieve. However, it is certain that a constructive dialogue among content analysis users would be valuable to encourage consistency of methoc

Scientific analysis

A number of different approaches are required to analyse in scientific terms the various forms of climatic information obtainable from documentary sources. Firstly we may consider the specialised techniques involved in the interpretation and standardisation of early instrumental observations. Basically the method is to work back from a key series of good modern data to the more questionable early observations. By comparison and adjustment, using appropriate statistical techniques, it is hence possible to compile a continuous homogenised record from a miscellany of overlapping fragmentary series of observations. In carrying out the analysis it is necessary to consider the following factors.

(*a*) The characteristics of early meteorological instruments. Early instruments varied greatly in design and quality, accuracy and sensitivity; and they frequently aged in use to produce an increasingly degenerate record. Such weaknesses must be assessed and allowance made when interpreting the record. This task may require both documentary research to discover contemporary descriptions of or comments on instruments, and statistical analysis of the suspect record to isolate anomalous values.

(*b*) Methods used in the exposure of early instruments. Knowledge of exposure methods is clearly of vital importance. In the case of rainfall records, for example, major problems of interpretation are posed by the fact that up to about 1760 raingauges were normally sited on a roof or wall (Craddock, 1976). Towneley, who kept the earliest series of rainfall records known to survive, used a funnel on the roof of Towneley Hall from which a pipe descended nine yards vertically before turning in through a window and hence into a collecting vessel. It is generally accepted that, thereby, Towneley collected too little rain (Folland & Wales-Smith, 1977).

(*c*) The time of day that observations were taken. Considerable irregularities were common in the first century or so of instrumental observations.

(*d*) The nature of the scales used by early observers. Numerous different measuring systems were used, often posing problems of conversion to modern units. Conversion data for a large number of early measurements have, however, been compiled by Lamb & Johnson (1966).

(*e*) The location of the original observations. Plainly, account must be taken of the general characteristics of the location of the series analysed, including height above sea level and distance from whatever observation station provides the basis for the key series. Moreover the researcher must glean as much information as possible, both from contemporary literary evidence and from direct observation today, about local peculiarities of site (frost hollows, south facing slopes, urban environment, etc.) which may affect the interpretation of the record.

From the many fine pieces of scientific research which embody these methods, we may single out Manley's construction of a continuous series of monthly mean temperature values for central England from 1659 (Manley, 1953, 1974). The key series used in this study was derived from the average of data recorded in Oxford and Lancashire over the period 1815–1974. For the years 1771–1815 the record was constructed by averaging monthly anomalies for a number of inland stations with records long enough to be bridged into the key series for 1815–40. The record for the period 1723–71 was based on Edinburgh, Greenwich and Lancashire means supplemented with data from Lyndon, Exeter, Plymouth and elsewhere. An unfortunate gap in the data from 1707 to 1722 was filled essentially by estimates based on values for Utrecht; while before 1707 the series depends on Derham's Upminster record 1699–1706 and on numerous short series of seventeenth-century observations from scattered locales reinforced by certain non-instrumental materials. The final series is regarded as reasonably accurate from about 1720 onwards, the earlier data being presented as somewhat less reliable.

Such instrumental series provide the ideal type of climatological information. The value of material from other documentary sources must be judged on how far it approximates to the essential characteristics of instrumental series. Two basic factors are involved. The first is how strongly the information is correlated with the standard meteorological variables of temperature, precipitation and pressure. At first sight this might seem to imply that any information which happens to be in some sense meteorological (i.e. directly descriptive of atmospheric phenomena) is, all other things being equal, preferable to parameteorological data (evidence about sea and river ice, the wetness or dryness of the ground, lake and river levels, etc.) and to even more indirect indicators of climate. In fact this is not so, and the value of a particular type of information as a proxy for one of the standard meteorological variables needs to be decided on the merits of the individual case. On the other hand, it does perhaps need stressing that many phenomena which *prima facie* appear to embody a useful climatic message in fact pose such complex problems of interpretation as to be of little or no value for climatological purposes. For example, while it is obvious that grain harvest prices were in the past very substantially influenced by weather conditions, the fact that so many other factors operated to affect price levels makes it very difficult to interpret price series in climatic terms. And as Le Roy Ladurie (1972) has stressed, phenomena even more remotely affected by climate such as revolts and migrations are in themselves virtually useless as climatic indicators. Nevertheless, it must be recognised that some forms of indirect information can be interpreted meteorologically with an acceptable degree of confidence – see, for example, the discussion of wine harvest dates and tithe auction dates by Pfister (this volume).

The second basic factor conditioning the usefulness of historical data is the extent to which the material forms a highly systematic series which is continuous, based on a fixed observational time unit and otherwise homogeneous, and can be readily expressed in numerical form.

Certain sources, unfortunately rather few in number and relatively recent (post AD 1500), provide series which are both highly systematic in this sense and closely correlated with standard meteorological variables. Such series thus offer the best substitute for instrumental data. They can in fact be readily calibrated in terms of a standard variable by correlating the series (or a modern-day equivalent) with instrumental data, provided that there is a sufficient period of overlap between the two forms of record.

One of the first examples of this type of analysis was the winter temperature reconstruction derived by Flohn (1949) from the meteorological journals of Tycho Brahe for Hven, Denmark, 1582–97 (La Cour, 1876), and of Haller for Zürich 1546–76 (Wolf, 1885). Flohn found from modern instrumental observations that the ratio of days of snowfall to days of rainfall in winter was well correlated with winter temperature. Using the snow and rainfall records of Brahe and Haller, Flohn was able to show that winters in Zürich were about 1.3 deg C colder in 1561–76 than in 1546–61, whereas winter temperatures on Hven from 1582 to 1597 were about 1.5 deg C below those around the turn of the twentieth century. The implied severity of the latter part of the sixteenth century accords with the wind direction observations given by Brahe and analysed by La Cour (1876).

As far as parameteorological series are concerned, de Vries (1977) and van den Dool, Krijnen & Schuurmans (1978) have calibrated data on the freezing of Dutch canals in terms of winter temperatures to extend the record of Dutch winter temperatures back to 1634; and Pfister (1977) has used data on snow-falls and the duration of snow cover from various documentary sources to estimate winter temperatures near Zürich back to 1683. Moodie & Catchpole (1975) have analysed Hudson's Bay

Company post journals to establish a series of dates for the break up and freezing of ice in the estuaries of various rivers in North America. Possibly these data could, like the European material already described, be calibrated with modern meteorological data to construct a temperature record back to 1760, though in fact this step has not so far been accomplished.

The outstanding example of the use of phenological or indeed any other form of climatically indirect evidence from documentary sources is the work of Le Roy Ladurie and others on wine harvest dates in France. The dates for the beginning of harvest were fixed and registered annually by municipal authorities, and can be abstracted from the archives to form continuous series for various localities which in some cases extend back to the end of the fifteenth century. Essentially the dates reflect when the grapes reached maturity, which was largely dependent on climatic conditions during the immediately preceding spring and summer. Precipitation and insolation were among the factors which affected how quickly the grapes matured, but the most important variable was the temperature levels prevailing during the growing season from March/April to September/October. For the nineteenth century, Le Roy Ladurie has demonstrated high correlation coefficients between wine harvest dates and temperature levels derived from instrumental series. It is true that some doubts have been cast on the homogeneity of wine harvest series and the strength of the correlation with temperatures in the eighteenth century and earlier (de Vries, 1977), while the work of Pfister (1980) indicates that wine harvest dates are more closely correlated with temperatures from April to June than with those from July to September. But the overall conclusion must be that wine harvest dates do appear to provide a reasonably good index of spring/summer temperatures over long periods when instrumental data are lacking (see the summary of data and analysis in Le Roy Ladurie & Baulant, 1980). So far, however, Le Roy Ladurie has not ventured to calibrate the entire series in terms of absolute temperature values, and it is probably true (as Le Roy Ladurie has himself stressed) that the series are more reliable as indicators of inter-annual temperature fluctuations over relatively short time scales than as an index of secular trends.

The impressive series described in the previous paragraphs are exceptional. Unfortunately, most of the information available from documentary sources does not form a systematic series nor can it be easily converted to standard meteorological variables. On the contrary, as will be clear from the survey of sources presented earlier, the bulk of the available data, especially for periods before the late sixteenth century, is markedly discontinuous, non-homogeneous, and biased towards the record of extreme events.

When the available data are very fragmentary, it may be possible only to use the material in a very impressionistic way to provide some indication of possible major trends. Purists may denounce such an unsystematic approach, but in fact the method is a perfectly permissible one so long as conclusions are presented with appropriate signals of caution. A good example of this kind of work is found in the earlier sections of Chu Ko-Chen's pioneering survey (1973) of climatic fluctuations in China over the last 5000 years. Freely admitting the imperfections of the evidence, Chu impressionistically but skilfully deployed a wide range of fragmentary materials to present a coherent framework of climatic change in remote periods.

More systematic approaches to the available data are, however, often feasible. One form of analysis of fragmentary non-homogeneous sets of information is to transform the material into a numerical index prior to interpreting it in terms of standard meteorological variables. One of the first to do this was Brooks (1926) who derived 50-year indices of wetness and winter severity for Europe using data from a number of compilations. Lamb (1963, 1977) refined this work by calculating decadal indices for Europe near 50° N at these longitudes, 0° E (England), 25° E (central Europe) and 50° E (the Russian Plain), back to AD 1100. Lamb's data, derived indirectly via compilations from chronicles, annals, diaries and similar sources, consisted of fragmentary information indicating extremely mild or severe weather in the winter months (December, January and February) and extremes of raininess or drought in high summer (July and August). The indices he used may be defined as follows. For 'winter severity', the index is C–M, where C is the number of unmistakably cold winter months in a decade, and M the number of unmistakably mild months; and for 'summer wetness', W–D, where W is the number of summer months with evidence of frequent rains, and D is the number of months with evidence of drought per decade.

Two factors are critical to the validity of such indexing procedures. One is the rigorous interpretation of the data, and here the principles of historical interpretation and the content analysis techniques described above are relevant. The second is the adequacy of the data set. Lamb worked on the assumption that, despite the manifest imperfections of the sources he used by modern scientific standards, they did, for the areas and periods he studied, provide a complete record of major extremes.

This assumption was based on the following test. Lamb counted the overall number of extreme months per decade for the most recent part of the record (where he could be sure that there were very few gaps) and then tested the earlier data to confirm that the total number of extremes was approximately the same. This is a valid test provided there have been no changes in either the perception of extremes, or the actual frequency of extremes, but (as Lamb himself recognises) both these propositions are open to doubt. In any case the value of Lamb's test has been questioned on the grounds that the materials he used, being derived from unreliable compilations, include too much spurious information to give a true impression of the completeness of the record (Ingram *et al.*, 1978). However, further work will be necessary to discover just how serious, in terms of final results, this problem actually is.

It is methodologically important to try to establish, in general, how many gaps are permissible before decadal values become unrepresentative, and to devise techniques to minimise the problem. Testing reveals that different types of index show differing sensitivity to missing data. 'Difference' indices, as used by Lamb, are quite sensitive. 'Ratio' indices, on the other hand, show less sensitivity. (An example of a ratio index for winter severity is $(C-M)/(C+M)$, where C and M are the number of cold and mild months respectively.) The sensitivity of indices can be tested by a Monte Carlo type of simulation where one starts with an artificially constructed (but realistic) complete record and progressively introduces more and more random gaps until there is a significant change in the average index value for an ensemble of decades. Such simulations show that ratio indices can tolerate small gaps in the record. (Equally importantly, they indicate that ratio indices are robust to uncertainties in the perception and/or actual frequency of extremes.)

The second step in indexing is to convert from crude index to meteorological parameter. Here the method is to correlate the index values with instrumental temperature or precipitation data over an overlap period. If a significant correlation cannot be established then the index itself is of doubtful value, but if there is a correlation then it can be used to interpret index values prior to the overlap period directly in terms of meteorological variables. Lamb (1965) found that the correlation coefficient between summer wetness and late summer (July plus August) precipitation was 0.91, and the correlation between winter severity and mean winter temperature (December, January, February) was -0.87 (see Lamb, this volume: Fig. 11.2). However, neither index shows any significant correlation with summer temperatures.

The third step in producing meteorological time series from fragmentary historical data is that of adjustment to take account of long-term fluctuations. It has already been noted that historical data (excluding the most easily quantifiable items discussed initially) tends to be biased towards the reporting of extreme events. But 'extreme' here is not absolute, and is only relative to the observer's own idea of what is normal, which in turn is largely determined by his own experience. Long-term trends in the norm go unobserved, particularly because they are small relative to year-to-year fluctuations. As a crude meteorological instrument, man tends to act as a high pass filter and his subjective observations show only short-term fluctuations about an ever-changing norm. The response of this filter is uncertain: for example, a trend may show up as a series of extreme years giving the appearance of a step-function change in climate.

This problem has long been realised by climatologists (see, for example, Brooks, 1926), and the solution suggested by Brooks still applies. This is to use either proxy data, or records which are not subject to filtering, to derive long-term trends independently, and to use these as the background for shorter-term fluctuations. Many possibilities present themselves: glacier fluctuations, pollen records, lake levels, tree-line changes, vegetation limits and others are suitable, Some of these items may themselves derive from historical documents and so require care in compilation and analysis, while others are imperfectly understood as climatic proxies and can only be used after detailed individual studies of their responses to climate.

A more ambitious use of fragmentary climatic data from documentary sources is to construct a time series of maps of seasonal or intra-seasonal circulation patterns based on qualitative data from different locations. Such maps can provide a synthesis of the diverse types of data which are available for Europe and the North Atlantic. Starting with a well-distributed set of climatic descriptions pertaining to a particular season or month, putative isobars can be drawn so that the implied wind pattern and the positions of centres of cyclonic activity are consistent with the descriptive evidence. The principle involved is not unlike that used by twentieth-century meteorologists to extend the surface pressure patterns on synoptic maps over regions of scarce data. Qualitative accounts of temperature and precipitation patterns and other kinds of meteorological and parameteorological phenomena are all of value for this kind of analysis, wind strength and direction being particularly useful.

An example of this work is shown in Fig. 7.2: a reconstruction for the severe winter of 1683/84, arguably the most severe winter in England since 1659, and one of the most severe for Europe as a whole (possibly surpassed by 1708/9 (Lenke, 1964)). The pressure map for January 1963 is shown for comparison: the similarity is striking and reinforces earlier comments by Manley (1975) on these two winters.

The results of this technique are highly dependent on the skill and experience of the analyst in making subjective interpretations. However,

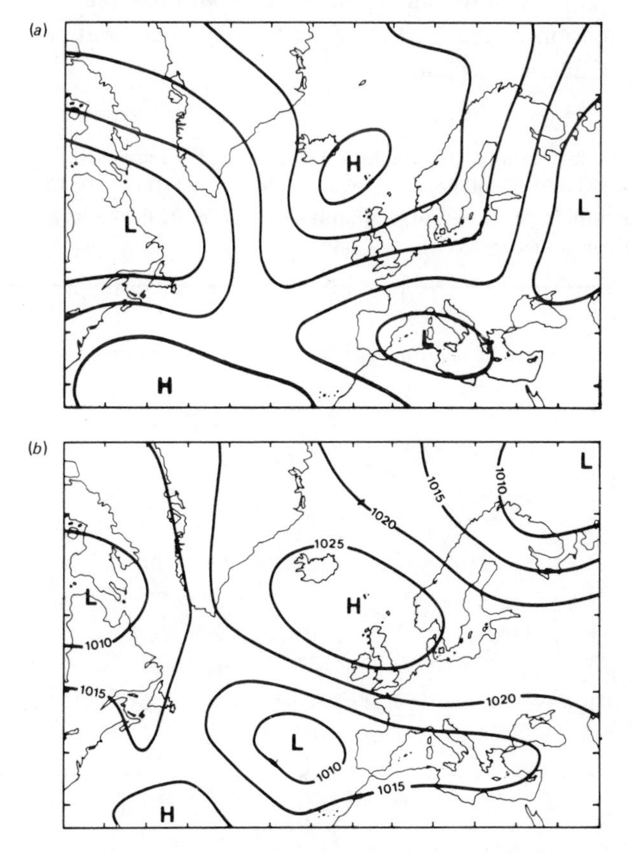

Fig. 7.2. (*a*) Preliminary reconstruction for the winter 1683–84 circulation pattern, and (*b*) the pressure map for January 1963. (1683-84 reconstruction supplied by H. H. Lamb.) Both seasons exhibit strong blocking patterns with persistent E to NE winds over England. The normal winter pattern shows a belt of low pressure at the latitude of Iceland, and southwesterly winds over England. The patterns here are highly anomalous.

the subjective element may be substantially reduced in the following ways. Initially the analyst must be trained in the method using qualitative data from the instrumental period, so that the results of subjective analysis can be tested against results objectively derived from instrumental data. Moreover, spatial correlation techniques similar to those used by Blasing (1975), which group together years of dominant circulation patterns, may be employed as guidelines. Monthly and seasonal data for the period 1881–1978 have been studied in this way with encouraging results (Farmer, 1979). Further guidelines for analysis may be obtained from correlation work for individual stations using anomaly data of rainfall, temperature and sea level pressure. The general associations of pressure with rainfall and, to a lesser extent, temperature, are already well known. Fig. 7.3 shows this kind of study applied at Valentia, Eire, with monthly data for 1900–74. Anomaly correlations of this type illustrate how much confidence may be placed in an analyst's association of, for example, a 'very rainy' climatic description with low pressure. It is an obvious extension of the work to derive anomaly correlation maps using data from different locations.

Fig. 7.3. Correlations between monthly pressure (p), temperature (T) and rainfall (R) anomalies (i.e. deviations from the 1900 to 1974 mean) for Valentia, Eire (1900–74): p–R, dotted line; T–R, dashed line; p–T, full line. (From Farmer, 1979.)

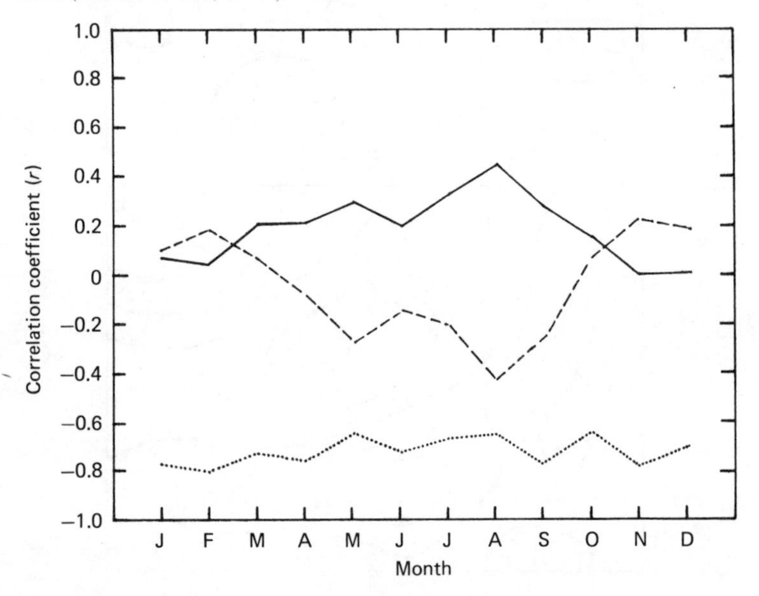

Clearly all these techniques for the scientific analysis of documentary sources have their individual limits. But for certain favoured areas and periods it is possible to combine virtually all of them to produce a richly textured picture of climatic change. This has been achieved, for example, by Pfister (this volume), whose work may be taken as a good illustration of much of what has been said in the preceding pages. Such a comprehensive and methodologically sophisticated study amply vindicates the use of documentary sources for climatic reconstruction and demonstrates how they can add greater detail and definition to the outlines of climatic shift derived from field data. Documentary climatology is still at a relatively infant stage, and undoubtedly some work produced earlier has been to some extent marred by the methodological weaknesses which are bound to afflict pioneering efforts. But a greater awareness of the problems involved in the scientific use of documentary sources is now current, and the future prospects of this field of study seem excellent.

Acknowledgements

We should like to acknowledge the assistance of H. H. Lamb, T. M. L. Wigley, A. E. J. Ogilvie and other members of the Climatic Research Unit, University of East Anglia. The chapter was written when M. J. Ingram and G. Farmer were supported by a grant from the Rockefeller Foundation, and D. J. Underhill was supported by funds from the US National Science Foundation.

References

Alexandre, P. (1974), Histoire du climat et sources narratives du moyen âge. *Le Moyen Age*, **80**, 101–16.

Alexandre, P. (1976). *Le Climat au Moyen Age en Belgique et dans les Régions Voisines (Rhénanie, Nord de la France)*. Liège/Louvain: Centre Belge d'Histoire Rurale, Publication No. 50.

Alexandre, P. (1977). Les variations climatiques au moyen âge (Belgique, Rhénanie, Nord de la France). *Annales: Economies, Sociétés, Civilisations*, **32**, 183–97.

del Badia, I. (1927). *A Florentine Diary from 1450 to 1516 by Luca Landucci*, transl. Alice de Rosen Jervis. London: J. M. Dent & Sons. 308 pp.

Baker, J. N. L. (1932). The climate of England in the seventeenth century. *Quarterly Journal of the Royal Meteorological Society*, **58**, 421–39.

Bell, B. (1970). The oldest records of the Nile floods. *Geographical Journal*, **136**, 569–73.

Bell, B. (1971). The Dark Ages in ancient history. I. The First Dark Age in Egypt. *American Journal of Archaeology*, **75**, 1–26.

Bell, W. T. & Ogilvie, A. E. J. (1978). Weather compilations as a source of data for the reconstruction of European climate during the medieval period. *Climatic Change*, **1**, 331–48.

Berelson, B. R. (1952). *Content Analysis in Communication Research.* Glencoe, Ill.: Free Press. 220 pp.

Bergeron, T., Fries, M., Moberg, C-A. & Ström, F. (1956). Fimbulvinter. *Fornvännen. Tidskrift för Svensk Antikvarisk Forskning*, **51**, 1–18.

Bernheim, E. (1889). *Lehrbuch der Historischen Methode.* Leipzig: Duncker & Humblot. 530 pp.

Blasing, T. J. (1975). Methods for Analyzing Climatic Variations in the North Pacific Sector and Western North America for the last few Centuries. Unpublished PhD thesis, University of Wisconsin. 177 pp.

Brandon, P. F. (1971). Late-medieval weather in Sussex and its agricultural significance. *Transactions of the Institute of British Geographers*, No. 54, 1–17.

Brooks, C. E. P. (1926). *Climate through the Ages.* London: Benn. 439 pp.

Butzer, K. W. (1958). Studien zum vor- und frühgeschichtlichen Landschaftswandel der Sahara. *Abhandlungen der Mathematisch-Naturwissenschaftlichen Klasse, Akademie der Wissenschaften und der Literatur in Mainz*, unnumbered vol., 1–49.

Castañeda y Alcover, V., ed. (1934). *Coses Evengudes en la Civtat y Regne de Valencia: Dietario de Mosén Juan Porcar, Capellán de San Martín (1589–1629).* Madrid: Cuerpo Facultativo de Archiveros, Bibliotecarios y Arqueólogos. 2 vols., 333, 341 pp.

Catchpole, A. J. W. & Moodie, D. W. (1978). Archives and the environmental scientist. *Archivaria*, **6**, 113–36.

Chu Ko-Chen (1973). A preliminary study on the climatic fluctuations during the last 5000 years in China. *Scientia Sinica*, **16**, 226–56.

Chu Ping-Lai (1968). *Climate of China*, English language version, 2nd printing. Washington, DC : US Department of Commerce. 621 pp.

Craddock, J. M. (1976). Annual Rainfall in England since 1725. *Quarterly Journal of the Royal Meteorological Society*, **102**, 823–40.

van den Dool, H. M., Krijnen, H. J., & Schuurmans, C. J. E. (1978). Average winter temperatures at de Bilt (The Netherlands) : 1634–1977. *Climatic Change*, **1**, 319–30.

Farmer, G. (1979). Typing of North Atlantic January surface pressure patterns. *Climate Monitor*, **8**, 105–9.

Flohn, H. (1949). Klima und Witterungsablauf in Zürich im 16. Jahrhundert. *Vierteljahrsschrift der Naturforschungs–Gesellschaft in Zürich*, **95**, 28–41.

Flohn, H. (1950). Klimaschwankung im Mittelalter und ihre historisch–geographische Bedeutung. *Berichte zur deutschen Landeskunde*, **7**, 347–57.

Folland, C. K. & Wales-Smith, B. G. (1977). Richard Towneley and 300 years of regular rainfall measurement. *Weather*, **32**, 438–45.

Gottschalk, M. K. E. (1971). *Stormvloeden en Rivieroverstromingen in Nederland. I: De periode vóór 1400.* Assen: van Gorcum. 581 pp.

Gottschalk, M. K. E. (1975). *Stormvloeden en Rivieroverstromingen in Nederland. II: De periode 1400–1600.* Assen: van Gorcum. 896 pp.

Gottschalk, M. K. E. (1977). *Stormvloeden en Rivieroverstromingen in Nederland. III: De periode 1600–1700.* Assen: van Gorcum. 474 pp.

Graves, E. B. (1975). *A Bibliography of English History to 1485.* Oxford: Clarendon Press. 1103 pp.

Hellman, G. (1916). Über die ägyptischen Witterungsangaben im Kalender von Claudius Ptolemaeus. *Sitzungsberichte der Königlich Preussischen Akademie der Wissenschaften, Berlin*, unnumbered volume, 332–41.

Hennig, R. (1904). Katalog bemerkenswerter Witterungsereignisse von den ältesten Zeiten bis zum Jahre 1800. *Abhandlungen des Königlich Preussischen Meteorologischen Instituts*, **2**, No. 4. 93 pp.

Hockliffe, E., ed. (1908). *The Diary of the Rev. Ralph Josselin 1616–1683*. London: Camden Society, 3rd series, 15. 192 pp.

Huxham, J. (1739). *Observations de Aëre, et Morbis Epidemicis, ab Anno 1728 ad Finem Anni 1737 Plymuthi factae*. London.

Ingram, M. J., Underhill, D. J. & Wigley, T. M. L. (1978). Historical climatology. *Nature*, **276**, 329–34.

La Cour, P. (1876). Tyge Brahes Meteorologiske Dagbog, holdt paa Uranienborg for Aarene 1582–1597. Appendix to *Collectanea Meteorologica*, Copenhagen: Kgl. Danske Videnskabernes Selskab.

Lamb, H. H. (1961). Climatic change within historical times as seen in circulation maps and diagrams. *Annals of the New York Academy of Sciences*, **95**, 124–61.

Lamb, H. H. (1963). On the Nature of Certain Climatic Epochs which differed from the Modern (1900–39) Normal. In *Proceedings of the WMO/UNESCO Rome Symposium on Changes of Climate*, pp. 125–50. Paris: UNESCO, Arid Zone Research Series, 20. Reprinted in H. H. Lamb, *The Changing Climate*, London: Methuen, 1966, pp. 58–112.

Lamb, H. H. (1965). The early medieval warm epoch and its sequel, *Palaeogeography, Palaeoclimatology, Palaeoecology*, **1**, 13–37.

Lamb, H. H. (1967). Britain's changing climate. *Geographical Journal*, **133**, 445–68.

Lamb, H. H. (1977). *Climate: Present, Past and Future*, vol. 2. London: Methuen. 835 pp.

Lamb, H. H. & Johnson, A. I. (1966). Secular variations of the atmospheric circulation since 1750. *Geophysical Memoirs, Meteorological Office, London*, **110**. 125 pp.

Landsberg, H. E., Yu, C. S. & Huang, L. (1968). *Preliminary Reconstruction of a Long Time Series of Climatic Data for the Eastern United States*. University of Maryland: Institute for Fluid Dynamics and Applied Mathematics, Technical Note BN-571.

Langlois, C. V. & Seignobos, C. (1898). *Introduction to the Study of History*, transl. G. G. Berry. London: Duckworth & Co. 350 pp.

Le Roy Ladurie, E. (1972). *Times of Feast, Times of Famine: a History of Climate since the Year 1000*, transl. B. Bray. London: George Allen & Unwin. 428 pp.

Le Roy Ladurie, E. & Baulant, M. (1980). Grape harvests from the fifteenth through the nineteenth centuries. *Journal of Interdisciplinary History*, **10**, 839–49.

Lenke, W. (1964). Untersuchung der ältesten Temperaturmessungen mit Hilfe des strengen Winters 1708–1709. *Berichte des Deutschen Wetterdienstes*, **13**, No. 92.

Lisle, E. (1757). *Observations in Husbandry*, 2nd edn, 2 vols. London.

McClure, N. E., ed. (1939). *The Letters of John Chamberlain*. Philadelphia: The American Philosophical Society, Memoirs, 12. 2 vols, 627, 694 pp.

Macfarlane, A. (1970). *The Family Life of Ralph Josselin.* Cambridge: Cambridge University Press. 241 pp.

Macfarlane, A., ed. (1976). *The Diary of Ralph Josselin, 1616–83.* London: British Academy, Records of Social & Economic History, new series, 3.

Manley, G. (1952). Thomas Barker's meteorological journals, 1748–1763 and 1777–1789. *Quarterly Journal of the Royal Meteorological Society*, **78**, 255–9.

Manley, G. (1953). The mean temperature of Central England, 1698–1952. *Quarterly Journal of the Royal Meteorological Society*, **79**, 242–61.

Manley, G. (1974). Central England temperatures: monthly means 1659 to 1973. *Quarterly Journal of the Royal Meteorological Society*, **100**, 389–405.

Manley, G. (1975). 1684: The coldest winter in the English instrumental record. *Weather*, **30**, 382–8.

Messerli, B., Messerli, P., Pfister, C. & Zumbühl, H. J. (1978). Fluctuations of climate and glaciers in the Bernese Oberland, Switzerland, and their geoecological significance, 1600 to 1975. *Arctic and Alpine Research*, **10**, 247–60.

Moodie, D. W. & Catchpole, A. J. W. (1975). *Environmental Data from Historical Documents by Content Analysis: Freeze-up and Break-up of Estuaries on Hudson Bay 1714–1871.* Winnipeg: Manitoba Geographical Studies, 5. 119 pp.

Moodie, D. W. & Catchpole, A. J. W. (1976). Valid climatological data from historical sources by content analysis, *Science*, **193**, 51–3.

Oliver, J. & Kington, J. A. (1970). The usefulness of ships' log-books in the synoptic analysis of past climates. *Weather*, **25**, 520–8.

Pearson, M. G. (1973). Snowstorms in Scotland, 1782–1786. *Weather*, **28**, 195–201.

Pfister, C. (1977). Zum Klima des Raumes Zürich im späten 17. und frühen 18. Jahrhundert. *Vierteljahrsschrift der Naturforschungs-Gesellschaft in Zürich*, **122**, 447–71.

Pfister, C. (1980). The Little Ice Age: thermal and wetness indices for Central Europe. *Journal of Interdisciplinary History*, **10**, 665–96.

Plummer, C. & Earle, J., eds. (1952). *Two of the Saxon Chronicles Parallel*, 2nd edn. Oxford: Clarendon Press. 2 vols., 420, 463 pp.

von Rudloff, H. (1967). *Die Schwankungen und Pendelungen des Klimas in Europa seit dem Beginn der regelmässigen Instrumenten-Beobachtungen* (*1670*). Braunschweig: Vieweg & Sohn, Die Wissenschaft series, 122. 370 pp.

Samaran, C., ed. (1961). *L'Histoire et ses Méthodes.* [Paris]: Gallimard, Bibliothèque de la Pléiade, Encyclopédie de la Pléiade, 11. 1771 pp.

Schove, D. J. (1949). Chinese 'raininess' through the centuries. *Meteorological Magazine, London*, **78**, 11–16.

Shaaber, M. A. (1929). *Some Forerunners of the Newspaper in England, 1476–1622.* Philadelphia: University of Pennsylvania Press. 368 pp.

Shirley, J., ed. & transl. (1968). *A Parisian Journal 1405–1449.* Oxford: Clarendon Press. 418 pp.

Stern, D. V. (1978). A Hertfordshire Manor of Westminister Abbey: An Examination of Demesne Profits, Corn Yields, and Weather Evidence. Unpublished PhD thesis, Kings College, University of London. 419 pp.

Symons, G. J., ed. & transl. (1891). *Consideraciones Temperiei pro 7 annis per magistrum Willelmum Merle ... 1337–1344.* London: Edward Stanford.

Thornes, J. E. (1978). *The Accurate Dating of certain of John Constable's Cloud Studies 1821–22 using Historical Weather Records*. University College, London: Department of Geography, Occasional Papers, No. 34.

Titow, J. Z. (1960). Evidence of weather in the account rolls of the Bishopric of Winchester, 1209–1350. *Economic History Review*, 2nd series, **12**, 360–407.

Titow, J. Z. (1970). Le climat à travers les rôles de comptabilité de l'Evêché de Winchester, 1350–1450. *Annales: Economies, Sociétés, Civilisations*, **25**, 312–50.

Tout, T. F. (1934). Mediaeval forgers and forgeries. In *The Collected Papers of Thomas Frederick Tout*, vol. 3, pp. 117–43. Manchester: Manchester University Press.

de Vries, J. (1977). Histoire du climat et Économie: des Faits Nouveaux, une interprétation différente. *Annales: Économies, Sociétés, Civilisations*, **32**, 198–226.

Weikinn, C. (1958). *Quellentexte sur Witterungsgeschichte Europas von der Zeitwende bis zum Jahre 1850. I: Hydrographie*, Part 1. Berlin: Deutsche Akademie der Wissenschaft zu Berlin, Institut für Physikalische Hydrographie.

Wintringham, C. (1727). *Commentarium Nosologicum Morbos Epidemicos et Aeris Variations in Urbe Eboracensi Locisque Vicinis, ab Anno 1715 usque ad finem Anni 1725 Grassantes Complectens*. London.

Wolf, R. (1885). *Meteorologische Beobachtungen*. Schweizerische Meteorologische Beobachtungen, Suppl. vol. 180.

8

An analysis of the Little Ice Age climate in Switzerland and its consequences for agricultural production*

CHRISTIAN PFISTER

Abstract

A careful search across a number of Swiss archives has yielded an unexpected wealth of documentary information on climate. Based on these data the main spells of heat and cold, wetness and drought can be traced back to the early sixteenth century at the level of individual months. In the category of 'documentary data' a variety of indicators has been found which permits the estimation of seasonal or monthly temperature patterns. As well as direct weather observations the data include phenological and paraphenological records, such as vine harvest dates, and vine yield data and parameteorological records such as snow phenomena and freezing lakes. Before using any such documentary data the sources need to be carefully verified. Dating must be corrected and standardised. Every type of documentary proxy record needs to be carefully calibrated with temperature measurements. The relative importance of individual months should be assessed by using multivariate statistics: vine harvest dates are, for example, taken as proxies for the temperature pattern for the whole growing season by many authors, but analysis has revealed that the crucial months are from April to June or July. The quality and accuracy of results can be improved through a combined interpretation of historical proxy data and non-instrumental descriptive weather observations. The magnitude of a temperature deviation from the mean can be estimated from proxy data while the weather observations detail the timing of the event. A codebook and a set of programs for coding, sorting and decoding data have been devised, which allow the pool of information to be conveniently stored as a data bank from which information can be extracted in a readable and understandable form. In order to assess the impact of the Little Ice Age (1525–1825) climate upon agricultural production the fluctuations of the major staple foods (grain, wine, dairy products) are considered and compared with documentary climate evidence. It appears that the very poor crops were always the result of extreme meteorological conditions, sometimes far beyond the maxima of the present century.

* Preparatory work for this article has been made possible by a post-doctoral fellowship from the Swiss National Science Foundation.

Introduction: survey of data types

The term 'documentary data' is used here as a blanket term applied to all types of written and illustrated sources created prior to the establishment of modern meteorological networks in the individual countries. Three subgroups of documentary data may be distinguished: *instrumental measurements*; *non-instrumental observations* with specific references to individual weather factors (temperature, precipitation, wind, etc.); and *documentary proxy data*, a variety of information which reflects the combined effect of several weather factors during a period of several months. Like the field proxy data of the scientist, documentary proxy data must be calibrated with instrumental measurements before they can be used to estimate specific meteorological variables (Pfister, 1980a). The other basic data source useful in studies of recent climatic change is field proxy data (or field data) a term which covers various parameters which vary as a result of changes in climate; such as pollen, glacial deposits, tree rings, oxygen isotopes in ice sheets, etc., discussed at length in other chapters in this volume.

The contribution of Ingram, Underhill & Farmer (this volume) has revealed that documentary data need to be carefully examined and verified before they can be taken as valid climatic evidence. Such data are also limited in coverage and are influenced by subjective biases and the fallibility of the untrained observer. In this chapter the possibilities for overcoming some of the shortcomings of documentary data are discussed; in particular by pooling and correlating the information from instrumental measurements, non-instrumental observations, documentary proxy data and field proxy data. Data from Switzerland are used to illustrate the approach, and, in so doing, details of the climate of Switzerland between AD 1525 and 1825 are presented. In a final section the impact of the severe climate of the Little Ice Age on agriculture and society is discussed.

Basic steps of the analysis

Fig. 8.1 gives an outline of the five basic steps used in constructing the Swiss Historical Weather Documentation: data compilation, data verification, data management, data combination results (first and second level), and data interpretation results (third level).

The compilation of data sources

Instrumental measurements and non-instrumental observations
The number of reliable records in these two categories increases

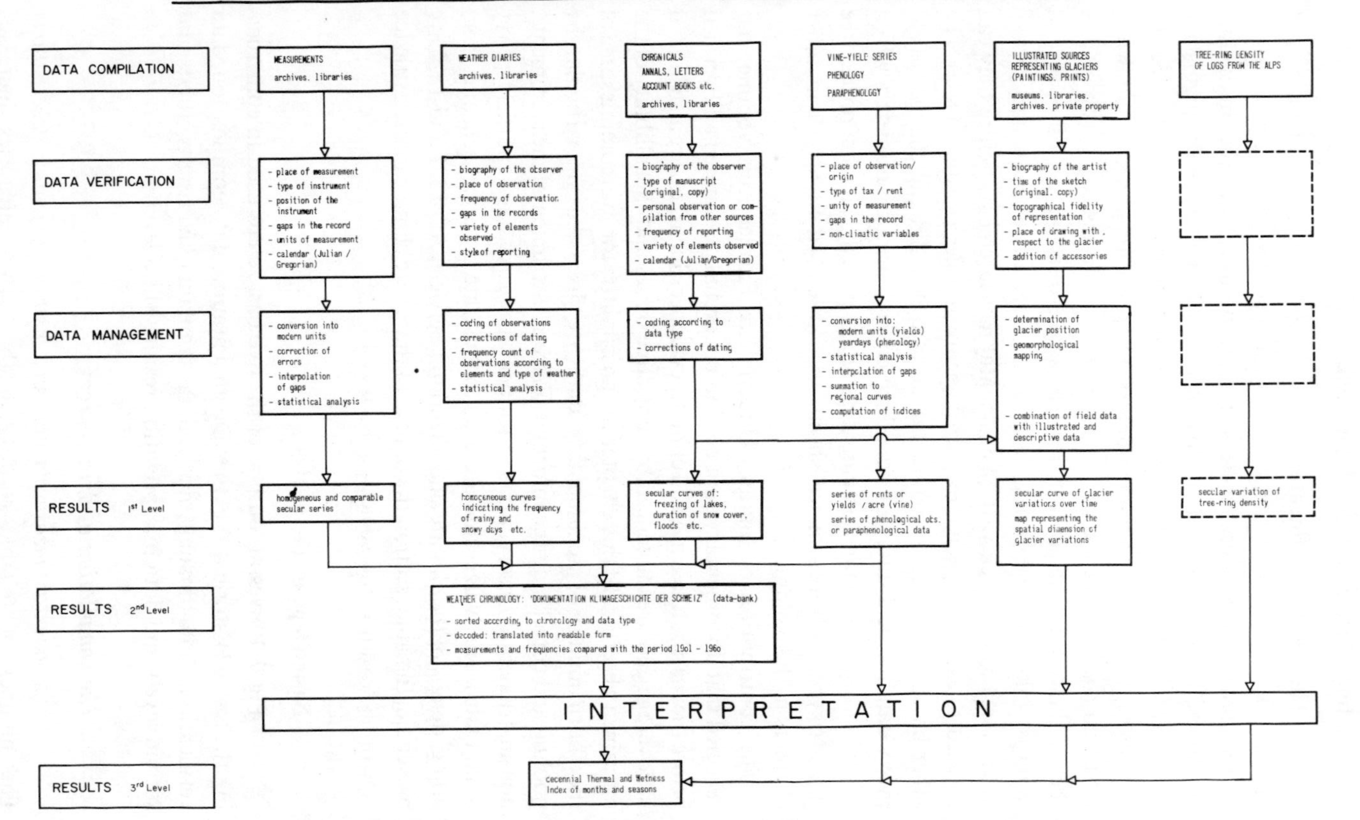

Fig. 8.1. Basic steps of the procedure used in the analysis of the different data types.

Table 8.1. *Percentage of missing monthly observations*

Period	Jan.	Feb.	Mar.	Apr.	May	Jun.	Jul.	Aug.	Sep.	Oct.	Nov.	Dec.
1525–1549	54	66	75	79	75	67	71	54	83	88	92	95
1550–1658	8	10	7	4	8	4	2	9	8	20	8	16
1659–1825	no missing observations											
1525–1825	8	9	8	8	9	7	6	7	9	14	10	13

from an average of 30 observations per year from 1525 to 1550 up to more than 150 observations per year in the first half of the eighteenth century. Before 1550 observations at the level of individual months are rare, especially in autumn and early winter: most reports refer to the weather pattern of entire seasons. From 1550 to 1657 there are very few months for which we have no information. After 1659, descriptive information is available for every month (Table 8.1).

The relative abundance of the different data types changes through time. In the sixteenth and seventeenth centuries, annals, chronicles and personal papers from the backbone of the documentation. They provide a mixture of weather observations interspersed between reports of the daily activity of the observer, fires, political events, religious festivities, military campaigns, food prices, and so on. The emphasis given to a certain type of reports varies. During periods of political crises or civil wars weather reports were often neglected. Nevertheless the percentage of weather reports in many of these documents is very large. The abundance of lakes, the vicinity of the Alps, and the variety of agricultural products (grain and wine in the lowlands, dairy produce in the pre-Alps and Alps) provides a broad spectrum of observations from different altitudes which can be combined to give a coherent picture of climatic patterns.

From 1550 to 1750, 82 out of 200 years are covered by daily non-instrumental observations. The diary of the parson Wolfgang Haller kept near and in Zürich in the sixteenth century (1545–46; 1550–76) has already been analysed by Flohn (1949) to provide estimates of winter temperatures for the period. Haller describes the daily weather with one, sometimes two expressions. Obviously his number of rainy and snowy days is too small to be directly compared to modern measurements. Nevertheless, their relative frequencies have provided sufficient information to give a rough idea of changes in climate during the period.

Much more detailed are the diaries from the period 1680–1750. One of them, kept by the baker Hans Rudolf Rieter in the town of Winterthur near Zürich from 1721 to 1738 gives an account of the weather situation almost hour by hour, not only during the day, but also during considerable parts of the night (Pfister, 1977).

In 1708 the period of systematic meteorological measurement begins with Johann Jakob Scheuchzer. His precipitation series for the years 1708–12 and 1717–31 have been published (Pfister, 1978b), whereas his measurements of pressure and temperature still await scientific analysis. The secular temperature and pressure series from Basel begin in 1755 (Bider *et al.*, 1959), followed by similar series from Geneva beginning in 1768 (Bider & Schüepp, 1961; Schüepp, 1961). From the second half of the eighteenth century the overabundance of information has called for restrictions on data abstracted for analysis. Thus, only instrumental measurements, the monthly number of rainy and snowy days from weather diaries (Riggenbach, 1891; Pfister, 1975, 1978b), and observations regarding phenology, snow cover, snowfalls on the alpine pastures during summer, floods and the freezing of lakes were included in the documentation.

Documentary proxy data Most of the written evidence in this category is provided by records of public administration and government. The vine harvest dates are well documented, since they can be drawn from local police registers. The Swiss scientist Dufour (1870) was the first to discover this type of climatic evidence. He was followed by Angot (1885), Garnier (1955), Le Roy Ladurie (1971) and Le Roy Ladurie & Baulant (1980). Although vine harvest dates are a good proxy climate indicator, climate information based on the quality of wines (Weger, 1952; Müller, 1953) is less reliable. To a large degree this is because much of the original documentary data is unverified.

The pioneering work in the analysis of illustrated sources has been carried out by Zumbühl (1980). His search through the most important museums, archives, libraries and galleries throughout Europe has brought to light an unexpectedly large number of previously unknown paintings, prints and photographs of Swiss glaciers. Using these data he has been able to retrace the fluctuations and dimensions of the two Grindelwald glaciers from 1750 onwards with a precision approaching the standard of modern measurements.

Phenological observations are frequently found in chronicles, annals and diaries, but at present very few climatic historians have used this type of data as a substitute for instrumental measurements (Margary,

1926; Pfister, 1972, 1980a; Kington, 1974). Comparison of the vegetative stages within a year and between years is a way of obtaining a relatively objective idea of prevailing temperatures prior to instrumental observations. Most phenological observations concern different stages in the development and maturity of fruit trees, grapes or crops.

Another type of proxy data has recently been discovered: Swiss tithe records. In some series of records the date on which the tithes (paid in grain) were sold by auction is regularly listed year after year. The oldest series date from the early seventeenth century. It has been shown that the date of the auction was closely related to the time of maturation of the earliest winter grain, mainly rye. In most cases the rye harvest immediately followed the auction of the standing crops. An analysis of both modern phenological observations of the beginning of the rye harvest and the historical series of tithe auction dates has revealed that both reflect the temperature during May and June (Pfister, 1979). Tithe auction dates may therefore be used as substitutes for phenological observations (such as vine harvest dates). In order to distinguish such data types from true phenological observations, they are called *para-phenological data.*

In contrast to grain harvests very little research has been devoted to fluctuations in wine production in historical times. Most of the secular series found in Swiss archives list the product of vineyard share-cropping agreements (Teilreben) received by public institutions and authorities. This means that the production was divided between the tenant and the landlord, the latter's part being listed in the document. In addition, several records give the acreage of the vineyard. Four regional series for different parts of the Swiss plateau, compiled from a variety of local series for which the acreage is *not* known, show correlations between 0.56 and 0.76; the main series compiled from these four regional series is highly correlated ($r = 0.87$) with another series compiled from several local series for which the acreage *is* known (Pfister, 1981).

Modern viticulture theory shows that the yearly fluctuations in vine yields are related to the weather pattern during the summer. High yields can be expected when the months of July and August are hot and dry; a collapse in yields is mainly the result of a cool rainy summer (Primault, 1969). This same relationship is also evident in the historical vine yield curve (Pfister, 1980a).

Field proxy data Field proxy data may be used to supplement other data for late summer. The use of X-ray densitometry for measuring tree rings makes it possible to consider the relation between climate and tree rings in a new light. Investigations so far have shown that it is the maximum density of the annual rings that provides the best climatological information (Schweingruber, Bräker & Schär, 1978). Maximum density is primarily influenced by the temperature in late summer. By cross-dating the annual tree-ring sequences of living trees with samples of timber taken from ancient houses a 700-year long chronology has been developed for the Northern pre-Alps of Switzerland (Schweingruber, Braeker & Schaer, 1979). The series discussed in this chapter originates from the Bernese Oberland (Lauenen, Saanen) and has been kindly provided by Dr Schweingruber.

Guidelines for data verification

Bell & Ogilvie (1978), Ingram, Underhill & Wigley (1978), Zumbühl (1980) and Ingram (this volume) have convincingly shown that a considerable amount of previously accepted historical climate material is unreliable or wrongly dated. Historical data require careful examination and verification before use. Fig. 8.1 shows the basic steps in data verification, and the subject is discussed further by Ingram, Underhill & Farmer (this volume). Accurate dating was one of the major problems in the verification of the Swiss data. Between 1583 and the end of the eighteenth century the Gregorian and Julian calendars were simultaneously in use and the calendar was a constant matter of debate between catholic and protestant cantons. The catholic cantons adopted the Gregorian style in 1583. Most protestant cantons did so in 1701, but others followed at different times during the eighteenth century. In this situation it is essential that every dating correction is made explicit either by a special note or by giving both dates with each climate reference.

Data management

Because of the unexpectedly large amount of data which were collected it was necessary to use a computer for data handling. The following procedure was applied:

(*a*) devising the framework of a codebook suitable for many different types of documentary data,

(*b*) coding the sources and refinement of the codebook,

(*c*) devising a sequence of four programs for checking, interpreting, sorting and decoding the information,

(*d*) correction of coding errors based on the decoded printout.

A crucial factor was the framework of the codebook, which had to be

Table 8.2. *Organisation of the basic record (punch card) in the data bank on historical weather observations in Switzerland*

Data category	Subject		Details	Code	Columns[a]
All	Chronology		NNN = year − 1000	NNN	1–3
			Month	1–12	
			Seasons	13–16	4–5
			Entire year	17	
	Data category:	A, B:	10-day intervals	1–3	6
		A, B:	entire month	4	
		C:	precip. measure	5	
		D:	frequency counts	6	
		E:	temperature measure	7	
	Region			1–15	8–9
	Source		Tied to author, place and altitude of observation. NNNN = reference number	NNNN	11–14
A	Daily observations (to be processed by the program MET1K)		10 three-column observations separated by blanks (winds, thermal character, sky/precip.)	M[b]	16–54
A, B	reliability			M	56
	Prevailing weather in a 10-day interval, month or season (see columns 4,5)		Winds	M	57–58
			Thermal character	M	59–60
			Sky/precip.	M	62–63
			Snow cover	M	64
	Impact of weather upon the lithosphere and hydrosphere		Landslides, glacier fluctuations, river levels, freezing of lakes	M	66–67
	Phenological observations		Day within 10-day period	0–9	6–8
			58 phenophases[d]	M	69–70

Table 8.2. (*cont.*)

Data category	Subject	Details	Code	Columns[a]
A,B(ctd)	Impact of weather on the biosphere	Droughts, floods, frosts, snow etc. affecting crops or livestock	M	71–72
	Impact of weather on Man	Epidemics, quality and quantity of harvests, fluctuations of prices, dearths, famines	M	73–74
	Type of calendar/inclusion of footnotes	—	M	75
C	Monthly precipitation sums (NNN = actual amount in mm)		NNN	16–18
D	Monthly frequency counts based on A[c]		M	16–71
E	Monthly mean temperatures (NNNN = temperature in °C and tenths)		NNNN	16–19

[a] Columns 15, 55 and 65 are blank (except possibly for data category D).
[b] M is used to denote a miscellaneous numerical code.
[c] This is the output produced by the program MET1K. It mainly contains monthly frequency counts of the items given in the daily observations.
[d] 58 phenophases: 58 different phases of flowering and ripening for different crops and plants.

compatible with the logic of computer programming. In particular, the maximum record length was restricted to 80 characters, the length of a standard punch card. Furthermore, a certain number of blank columns facilitated readability and the detection of errors. An outline of the organisation of the codebook is given in Table 8.2. The organisation has been tailored to the management of five different data categories:

A – daily non-instrumental weather observations
B – miscellaneous observations from chronicles and annals
C – precipitation measurements
D – monthly frequency counts of various observations
E – temperature measurements

The basic principle is to sort the data chronologically and thematically and to identify the data source using a reference number keyed to a file giving the author, the region and precise location and altitude of the observation point.

An interval of ten days (i.e. 36 intervals per year) was taken as the smallest reporting unit (except in the case of phenological data where subdivision down to the day was allowed for). Working with 10-day periods has the advantage that it allows a dating resolution of less than a month. The inclusion of more than 70 000 individual daily observations on the other hand would have excessively inflated the data set; as it is the total number of items exceeds 27 000.

Through coding, the message of the historical document is converted into a sequence of numbers which are then transferred to the computer. Most weather reports consist of a number of repetitive elements, which are not greatly distorted through coding. Sometimes, however, reports are so unusual that no codebook can be refined enough to express the content fully. These reports are quoted verbatim as a footnote in the documentation (see column 75 in Table 8.1). In order to distinguish between items which partly depend on the interpretation of the researcher and those which are precise reports from first class observers, a reliability code has been incorporated in the codebook (column 56).

The main advantages of this computerised procedure can be summarised as follows.

1. The coding forces the researcher to focus upon the essence of the source.

2. After a while the researcher knows most of his codes by heart and becomes able to code information very rapidly, almost as quickly as he reads it. Coding avoids the slow and cumbersome work of transcribing documents.

3. The homogeneous sequence and representation of the data facilitates interpretation.

4. Records containing quantified information (measurements, numbers of rainy days, etc.) are preinterpreted by the program, in the sense that a comparison is automatically made with the quartile statistics of the 1901–60 measurements at the same place in order to determine the character of the month in question.

5. A large number of attributes (author, quality, place, altitude of observation, etc.) can be attached to every single record without error.

6. Coding errors, even if they are detected in the final stage of the

procedure through inconsistencies in the decoded text, can be corrected easily, rapidly and inexpensively.

7. As the information is stored in numerical form it could be translated into languages other than German by including additional decoding loops in the program.

8. This would allow the content of the Swiss data bank to be incorporated easily into any future international climatic-history data bank.

Results

Secular series of specific climatic features (first level results)

One way of abstracting information from the data bank is to sort and decode it according to specific climatic features (such as floods, droughts, the freezing of lakes, the frequency of rainy days, or the duration of snow cover) and to represent these data as a time series or a frequency distribution over time. This has already been done for features such as snow cover (Pfister, 1978a) and the freezing of lakes (Pfister, 1980b). However, this type of analysis has the disadvantage that the series produced are often short or not entirely homogeneous and comparable. For a thorough analysis of the pattern of changes in climate another approach must be used.

The weather chronology 'Dokumentation Klimageschichte der Schweiz' (second level results)

The data bank assembled to date is very extensive. A small section of it is shown in Table 8.3 which lists some observations during the winter 1572/3 (in Switzerland this was the most severe winter of the last half millennium). The information listed on lines 2–5 is for various places for the entire month (line 1) of December 1572. The climate information is followed by a region number (Reg: given to facilitate the location of small, little-known places), the place, the altitude (in metres), the name of the observer and a reference number. If the name of the observer is missing there is doubt that the observation is contemporary. The bottom line for December 1572 (line 6) gives the interpretation of the daily observations of Wolfgang Haller: 12 days with precipitation, with a snow frequency of 99 per cent (i.e. 100 per cent) (indicating a temperature below the lower quartile for the period 1901–60). This interpretation is consistent with the freezing of Lake Geneva observed by Savion (Geisendorf, 1942; cf. Table 8.3, line 4 and information for 1573).

Table 8.3. *Swiss Historical Weather Documentation: an example of the decoded printout. The top line gives the place (Schaffhausen), altitude (403 m), name of observer and the reference number. The information listed below is for various places for the entire month of December 1572. For 1573 the two lists of information are for the first and second 10-day periods in January (i.e., 1–10 January and 11–20 January).*

```
GESAMTER MONAT-

 SEHR KALT.  REG: 7,GENEVE: 375 M(SAVION, 167)(    )
 DAUERNDE SCHNEEDECKE.  REG: 6,ZUERICH: 430 M(W.HALLER, 125)
 ANZEICHEN VON STARKER KAELTE ALLE GEWAESSER ZUGEFROREN UND BEGEHBAR.  REG: 7,GENEVE: 375 M( 207)**
 ,GENEVE: 375 M(SAVION, 167)(    )

VERAENDERLICH 12 NIEDERSCHLAGSTAGE VORWIEGEND SCHNEE:99% KALT.  REG: 6,ZUERICH: 430 M(W HALLER, 170)#


  1   5   7   3
  *************

  J A N U A R

1.DEKADE-


 KALT.  VORW.BEWOELKT(VEREINZELT SCHNEE OD.SCHNEE+REGEN).   REG: 6,ZUERICH: 430 M(W HALLER, 170)
 ANZEICHEN VON STARKER KAELTE ALLE GEWAESSER ZUGEFROREN UND BEGEHBAR.  REG: 6,ZUERICH: 430 M(W.HALLER, 125)
 ,ZUERICH: 430 M(BULLINGER, 152)REG: 7,GENEVE: 375 M(  64),GENEVE: 375 M(  68)

2.DEKADE-
 SEHR KALT/MILD.  REG: 5,BERN: 540 M(J HALLER,  17)(    )
 KALT/WARM.  VORW.SONNIG(VEREINZELT SCHNEE OD.SCHNEE+REGEN).   REG: 6,ZUERICH: 430 M(W HALLER, 170)
 AUSGEAPERT.  REG: 6,ZUERICH: 430 M( 128)
 UEBERSCHWEMMUNG.  REG: 6,ZUERICH: 430 M(W.HALLER, 125)
 SEE AUFGETAUT.  REG: 7,GENEVE: 375 M(SAVION, 167)
 ANZEICHEN VON STARKER KAELTE ALLE GEWAESSER ZUGEFROREN UND BEGEHBAR.  REG: 5,BERN: 540 M(J HALLER,  17)(    )REG: 6
 ,ZUERICH: 430 M(W.HALLER, 125)(    )REG: 7,RM.GENF-HOCHSAVOYEN (W.HALLER, 125)(    )
```

For 1573 two lists of information are shown covering the first and second 10-day periods in January (i.e. 1–10 January and 11–20 January). The first two lines of the second list refer to 5-day intervals: 'sehr kalt/mild' and 'kalt/warm' means that the cold spell was broken between 16 January and 20 January. Note again the excellent correlation of this information from Bern and Zürich with Savion's observation that Lake Geneva became ice-free between 11 January and 20 January. The computer listing makes such mutual confirmation of independent data readily apparent and, equally, exposes reports which are incompatible.

The large amount of very detailed weather information contained in the data bank may in some cases be needed by historians interested in, for example, military campaigns, the history of transportation, or harvest failures. The climatologist, however, may be more interested in other aspects of information contained in the data bank, such as quantitative data, and material pertaining to the monthly or seasonal time scale.

Data interpretation (third level results)

An attempt has been made to derive quantitative estimates of wetness and coldness for individual months. Two types of monthly index, an unweighted and a weighted one have been derived. For the unweighted index the month is given a value of −1 (cold or dry), 0 (unremarkable), or +1 (warm or wet). Months without observations score zero.

In constructing weighted indices, weight factors ranging from 1 to 3 are used. The greatest weight factors have been applied to those cases which can unmistakably be considered as 'extreme' by twentieth-century standards. The weights for the individual months have then been summed up to seasonal values. The value of the weighted seasonal index may therefore range from −9 to +9.

Decadal indices have also been calculated. The *unweighted decadal thermal index* resembles Lamb's winter severity index (Lamb, 1977): it gives the excess number per decade of unmistakably mild months (*M*) over unmistakably cold months (*C*) (i.e. *M-C*) for the individual months and for entire seasons. The *unweighted decadal wetness index* is *W-D*, where *W* is the number of months with evidence of frequent rains and *D* is the number of months with evidence of drought per decade. Because there are very few missing data, a 'difference' index has been used instead of a 'ratio' index, even though the latter is less sensitive to uncertainties resulting from missing data (Ingram *et al.*, 1978).

For the weighted thermal indices we cannot rely solely upon the emphasis given by the observer (like 'bitter cold' or 'extremely hot') as we know that these statements are biased by the subjective perception of the individual. We can, however, estimate the degree of severity by comparison with suitably chosen proxy data. To take an example: if one observer describes the month of May in a particular year as being very dry and hot, while another reports the opening of the first vine flowers towards the end of the month, the qualitative statement of the first observer is confirmed by the quantifiable phenological observation of the second. Based upon the analysis of modern phenological data it can be estimated that such a month may have had a heat excess of at least 1–1.5 deg C.

If, however, the statement is based upon phenological evidence alone, the bias can be considerable: due to exceptional warmth in early spring (February to April) the vegetation can be so advanced that an early vine flower is no longer a safe indicator of a warm May. Thus we need the mutual correspondence of observation and proxy data in order to make a safe estimate. While the magnitude of a deviation can be estimated through phenology, the qualitative observation determines its timing. The precise timing is of special importance when proxy data which reflect the thermal conditions over a period of several months are used.

Table 8.4 gives a survey of several types of proxy data which have been

Table 8.4. *Survey of indicators used for the determination of the thermal character of individual months*

Month	Cold	Warm
Dec., Jan., Feb.	Uninterrupted snow cover, freezing of lakes	Scarcity of snow cover, signs of vegetation
Mar.	Long snow cover, high snow-frequency	Sweet cherry first flower (±1.3 deg.)
Apr.	Snow cover and snow frequency, beech tree leaf emergence, sweet cherry flower in May	beech tree leaf emergence, (tithe auction date)
May	Tithe auction dates (±0.6 deg) vine first flower (±1.2 deg) barley harvest beginning	
Jun.	tithe auction dates (±0.6 deg) vine full flower (±1.2 deg) vine last flower colouration of first grapes	
Jul.	vine yields (±0.6 deg) colouration/maturity of first grapes	
Apr.–Jul.	wine harvest dates (±0.6 deg)	
Aug.	wine yields (±0.6 deg) tree-ring density (±0.8 deg)	
Sept.	vine quality tree-ring density (±0.8 deg)	
Oct.	Snow cover, snow-frequency	Reappearance of spring vegetation (cherry flower, etc.)
Nov.	Long snow cover, high snow-frequency, freezing of lakes	No snowfall, cattle in pastures

The figures give the standard error of estimate in degrees C.

used for weighting the thermal index. The figures give the standard error of estimate. The interpretation of the character of winter months is mainly based upon observations of snow and ice phenomena. Flohn (1949) has shown that the ratio of the number of snowy days to the number of rainy days can be used to estimate winter temperatures. Pfister (1978a) has combined observations of snow frequency with observations of snow cover in order to attempt to estimate temperature for the winter months from 1683 to 1738.

During the vegetative period, a variety of phenological, paraphenological and field proxy data (e.g., tree-ring densities) can be used. Most of the biological proxy data used reflect the temperature patterns over a period of several months. The longer the period involved, the higher the precision of the estimate. On the other hand, longer periods result in less precision of timing and therefore lower the quality of monthly estimates. For every type of proxy data stepwise multiple linear regression has been used to determine in which months the temperatures are most decisive for determining the proxy data fluctuations. The data type taken as the best indicator for the temperature patterns in an individual month is underlined in Table 8.4.

The results for the vine harvest dates require a comment. According to the prevailing theory these dates can be taken as indicators of the mean temperatures from April to September (Garnier, 1955; Le Roy Ladurie, 1971; Le Roy Ladurie & Baulant, 1980). The regression analysis made by Pfister, however, revealed that the dates were much more determined by temperatures in spring and early summer than by the warmth of August and September. This holds for nineteenth- and twentieth-century data also. Vine harvest dates are also much more strongly correlated ($r = +0.74$) with early summer indicators such as tithe auction dates, than with the tree-ring densities ($r = +0.47$), which reflect late summer temperatures. These results give strong support to the importance of spring and early summer warmth.

The weighted wetness index is based mainly upon the number of rainy days counted in weather diaries, or occasionally given in chronicles. Descriptions of heavy rainfall events which caused floods of the major rivers in the Swiss plateau (Rhône, Rhine, Aare, Reuss, Limmat) and descriptions of hydrological drought have also been used.

Seasonal values of the weighted thermal and wetness indices (see Appendix) show correlations which are similar to those between instrumental values of temperature and precipitation. These correlations are shown in Table 8.5. A strong negative correlation exists between the thermal and wetness indices for summer, with weaker (yet still highly significant) correlations for spring and autumn. No significant correlation exists for winter. Comparisons with similar coefficients based on instrumental evidence for the period 1931–60 reveal striking similarities. The correlations in each season show no significant differences at the 0.05 level. These results strongly support the accuracy and realism of the index values.

Table 8.5. *Correlation coefficients between seasonal weighted thermal and wetness indices (1525–1825: $n=300$) compared with those between measured temperature and precipitation*[a] *(1931–60: $n=30$).*

		Winter	Spring	Summer	Autumn
Indices	Correlation coefficient	+0.01	−0.37[b]	−0.68[b]	−0.32[b]
	95% confidence limits	−0.10 to 0.14	−0.26 to −0.47	−0.61 to −0.73	−0.20 to −0.42
Measurements	Correlation coefficient	+0.30	−0.40	−0.65	−0.16
	95% confidence limits	−0.05 to 0.58	−0.04 to −0.64	−0.39 to −0.80	−0.47 to 0.23

[a] Average values for a sample of stations from the Swiss plateau and the pre-Alps taken from Fliri (1970).
[b] Significant at the 0.001 level.

Examples of data synthesis

In order to compare the sensitivity of a variety of data types, two significant yet different climatic events will be considered; the short-term (time scale less than 10 years) climatic fluctuation of 1812–17, and the long-term (time scale 10–100 years) deterioration of the late sixteenth century. The latter can be considered as the onset of the maximum phase of the Little Ice Age. The main evidence for a marked climatic deterioration during this time is the advance of the alpine glaciers, which reached their historical maxima around 1600 (Le Roy Ladurie, 1971; Zumbühl, 1980). Around 1820 a similar widespread glacier advance occurred, the result of a climatic fluctuation from 1812 to 1817. In contrast to the advance of the sixteenth century, this particular advance is fully documented by meteorological measurements in the lowlands and

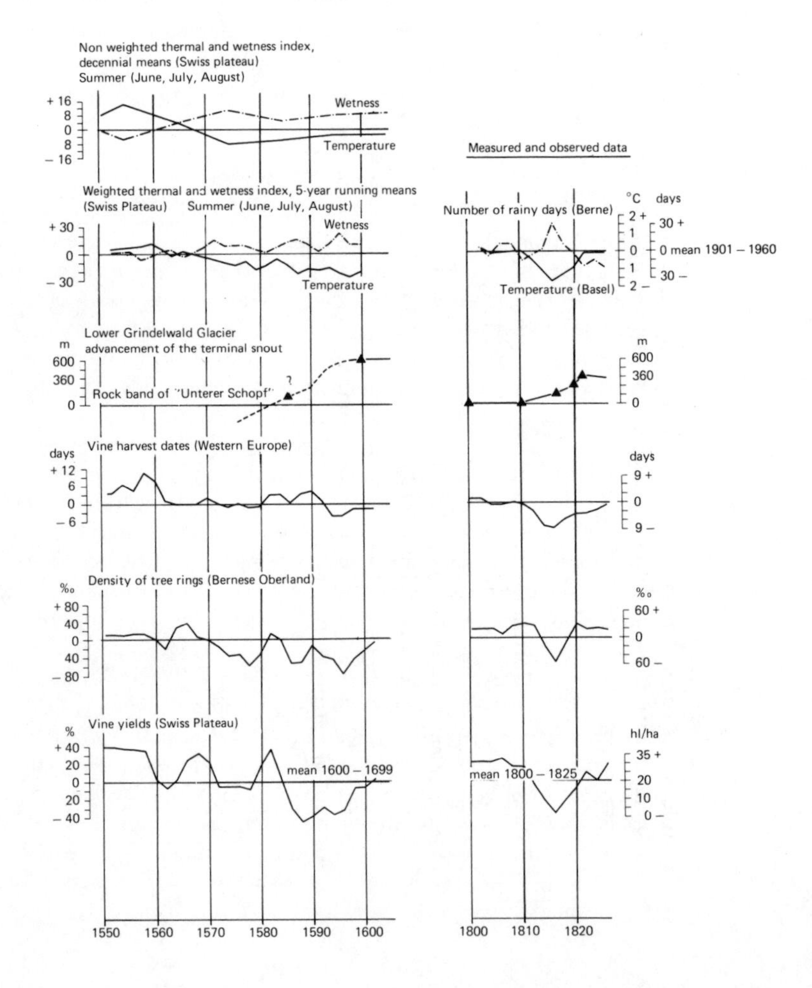

detailed observations of the weather patterns (summer snowfalls, time of thaw) in the Alps. The climatic fluctuation can be detected in many parts of Europe and even in the United States (von Rudloff, 1967; Post, 1977; Messerli, Messerli, Pfister & Zumbühl, 1978).

The short-term fluctuation of 1812–17

The fluctuations of all data types are remarkably well co-ordinated. From 1812 to 1817 an uninterrupted succession of cold and wet summers occurred, during which snow fell more frequently and reached lower altitudes than in 'normal' years. In addition a fall in the temporary snowline was observed. Tree-ring densities in the Alps show a marked regression, vine harvests across Europe were delayed, vine production in Switzerland (and probably also in other countries) collapsed. The snout of the lower Grindelwald glacier advanced rapidly after a time lag of 3 years. All of these items are shown graphically in Fig. 8.2 where the correspondences between different data types are readily apparent.

Fig. 8.2. Climate fluctuations 1570–1600 and 1812–17 as determined by various types of historical and proxy data. The indices are as described in the text. *Temperature* (1803–26) are 5-year running means of summer temperature expressed as departures from the 1901–60 mean (from Bider *et al.*, 1959). *Number of rainy days* are 5-year running means of the number of rainy days in the summer months (June, July, August), based on the frequency counts in weather diaries from Basel (1800–03) and Berne (1804–25). Values are expressed as deviations from the mean of the number of days with 0.3 mm or more during the period 1901–60. (Data for Basel from Riggenbach (1891) and for Berne from Pfister, in Messerli *et al.* (1978).) *Lower Grindelwald glacier:* data from Zumbühl (1980). *Vine harvest dates* are 5-year running means from 102 series from north-eastern France, western Switzerland and Rhenania, expressed as deviations from the mean, 1484–1879 (data from Baulant & Le Roy Ladurie, 1978). *Tree-ring density:* 5-year running means for logs from Lauenen and Saanen (Bernese Oberland) expressed as deviations from the mean 1600–1975. (By kind permission of Dr F. H. Schweingruber, Federal Institute of Forestry Research, Switzerland.) *Vine yields* (1550–1600) are 5-year running means of vine yields (half and third part of harvests from the lakes of Geneva, Neuchatel, Biel, Zurich and the Canton of Argovia). The data are means of the deviations from the trend, expressed as a percentage of the serial means from 1600–99. *Vine yields* (1800–20) are 5-year running means of yields (hl/ha) from five vineyards in the Canton of Zurich and two vineyards in the Canton of Berne.

Table 8.6. *Number of 'cold', 'warm', 'wet' and 'dry' months (unweighted index) and seasonal and annual mean values of the weighted index: 1525–69 contrasted with 1570–1600*

	Unweighted index: number of months which were undoubtedly				Weighted index: seasonal and annual mean values	
	warm	cold	dry	wet	Thermal index	Wetness index
Summer						
1570–1600	26 (28%)	44 (47%)	11 (12%)	35 (38%)	−1.4	1.8
1525–1569	49 (36%)	24 (18%)	25 (18%)	32 (24%)	0.7	0.1
Difference	−8%	+29%	−6%	+14%	−2.1	1.7
Autumn						
1570–1600	22 (24%)	18 (19%)	12 (13%)	17 (18%)	−0.1	0.1
1525–1569	33 (24%)	9 (7%)	27 (20%)	15 (11%)	0.6	−0.4
Difference	0%	+12%	−7%	+7%	−0.7	0.5
Winter						
1570–1600	11 (12%)	41 (44%)	15 (16%)	17 (18%)	−1.9	0.0
1525–1569	20 (15%)	33 (24%)	12 (9%)	8 (6%)	−0.5	−0.2
Difference	−3%	+20%	+7%	+12%	−1.4	0.2
Spring						
1570–1600	17 (18%)	31 (33%)	15 (16%)	16 (17%)	−1.1	−0.1
1525–1569	27 (20%)	34 (24%)	21 (16%)	24 (16%)	−0.2	−0.2
Difference	−2%	+9%	0%	+1%	−0.9	0.1
Year						
1570–1600	76 (20%)	134 (36%)	53 (14%)	85 (23%)	−4.4	1.9
1525–1569	129 (24%)	99 (18%)	84 (16%)	76 (15%)	0.6	−0.7
Difference	−4%	+18%	−2%	+8%	−5.0	2.6

The climatic deterioration in the sixteenth century

The second half of the sixteenth century saw a significant deterioration in European climate. The details of this decline in Switzerland are shown in Table 8.6 and Fig. 8.2. From Table 8.6 it can be seen that, in all seasons, the period 1570–1600 was significantly cooler and wetter than the period 1525–69. The cooling trend can be seen in more detail by reference to the data given in the Appendix. The decrease in the number of warm and dry months was less pronounced than the increase in the number of cold and wet months. The largest changes occurred in summer. These results are corroborated by the evidence shown in Fig. 8.2, where vine yields and tree-ring densities show strong cooling trends over the period 1550–1600. For 1525–69 there were twice as many warm months compared with cold months in summer. In contrast, in the period 1570–1600 the cold months dominated the warm, to the extent that almost half of the summer months were cold. There were similar changes in wetness, with a considerable increase in the percentage of wet months. In winter the percentage of cold months doubled and the percentage of wet months tripled between 1525–69 and 1570–1600. (These changes are similar to (but more detailed than) those noted by Flohn (1949) in his analysis of the diary of Haller. Flohn found that, in Zürich, the proportion of snow compared with rain increased from 44 per cent to 63 per cent from 1550–63 to 1564–76). In autumn and spring, changes were in the same direction as those in summer and winter, but the magnitudes of change were somewhat smaller. Warm dry autumns were frequent in the first half of the sixteenth century; consistent with the observation of Müller (1953) of the large number of years with good quality wine.

Together with the increase of coldness and wetness the number of extreme months also increased. The frequency of extremes was measured for proxy data and in the number of rainy days by counting the number of occurrences beyond the first or the eleventh duodecile defined using data for the present century. The results are shown in Fig. 8.3. More extremes meant an enhanced risk of a bad harvest. Accompanying the increase in the number of extreme months was the tendency for the extremes to be more extreme. As an example, consider the case of snow

Fig. 8.3. The number of extreme months per decade for the period 1530–1610. Extremes are defined as occurring when temperature or wetness conditions are in the first or eleventh duodecile (based on the period 1901–60).

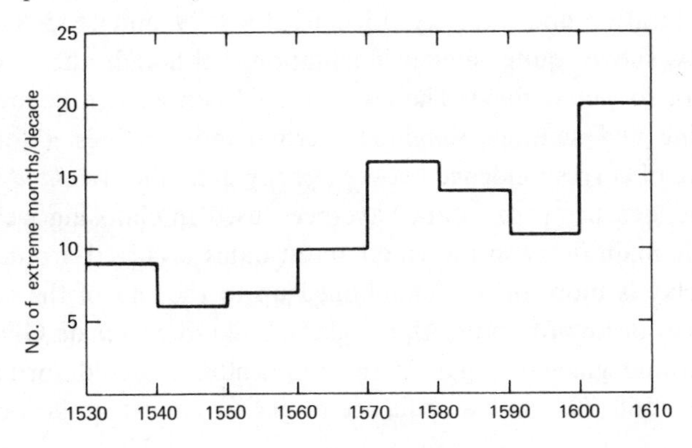

cover (Pfister, 1978b). From 1525 to 1564 there is no evidence of a winter in which snow cover in the lowlands lay more than 80 days. Then, in a single decade (1565–74) four extremely long and cold winters occurred, one of them (1573) probably being the coldest (though not the longest) in the last half-millennium. Over the period 1565–1830, more than 20 winters apparently had a longer duration of snow cover than the most extreme case recorded in the last 100 years (85 days, Zürich) (Uttinger, 1962). The famous summer of 1588, fatal for the Spanish Armada, was by far the rainiest ever recorded in the Swiss Plateau (Lucerne): from June to August it rained 77 out of 92 days (84 per cent). The summer of 1596 had the same number of days of rain in Lucerne, but was somewhat drier in Basel. For comparison; in the 'year without a summer' (1816) it rained on only 54 days in Berne, while the maximum since 1900 is 60 days with more than 0.3 mm in 1910 (Schüepp, 1976).

In addition to the long-term trend evident from the figures in Table 8.6, there were superimposed short-term fluctuations and interesting differences between seasons in the timing of the deterioration. Identifying the beginning of the deterioration is somewhat subjective, but the following details can be obtained from the data in the Appendix. The main sequence of cold winters began in the 1550s with two main phases between 1561–73 and 1582–1603. Cold springs began later, around 1568, and show a brief period of amelioration between 1579 and 1585. For the summers, 1573 is the first really bad summer, and no subsequent summer compares with the good seasons of the 1530s and 1550s. 1579 and 1588 are particularly bad. Autumns are less variable and show no well-defined starting point for the deterioration.

Fluctuations in proxy data are shown in Fig. 8.2. Vine yields show three distinct cold (and probably wet because of the demonstrated correlation between summer temperatures and precipitation) phases of increasing duration and intensity: 1560–64, 1572–79, and 1585–97. Tree-ring density shows quite similar fluctuations, although the first cold period is not so well defined. The last two cold periods can be identified in the spring and summer weighted thermal index values. (Note that some degree of correspondence between proxy data and *weighted* indices is inevitable, because proxy data have been used in choosing weighting factors.) The main delay in the vine harvest dates occurred around 1560 and this delay is more or less maintained up to the end of the century. The advances of low-reaching Alpine glaciers like the Mer de Glace and the Grindelwald glaciers began in the 1580s after the cold springs and summers of 1568–79, while a sequence of glacier catastrophies occurred in the 1590s with a time lag of 5–7 years after the series of extremely wet and cold summers in the second half of the 1580s (see Fig. 8.2, but also Le Roy Ladurie, 1971; Furrer, Gamper-Schollenberger & Suter, 1980; Zumbühl, 1980).

The impact of the Little Ice Age on agricultural production

Agriculture is an important aspect of the man–environment relationship. Most scholars agree that year-to-year fluctuations in agricultural output are mainly controlled by the weather. The consequences of this fact for pre-industrial economies and societies are, however, controversial. Hoskins (1968) believes that the (weather-controlled) annual fluctuations in harvest yields had an important influence on demography, mortality and diseases, agrarian legislation, and social unrest; and perhaps even the fundamental process of economic growth. De Vries (1980), on the other hand, believes that, in most areas of early modern Europe, 'the level of economic integration was sufficient – including trade, markets, inventory formation and even future trading – to loosen greatly the asserted links between weather and harvests and between harvests and economic life more generally'.

Anyone who is familiar with agricultural theory and plant physiology knows that the response of plants to weather conditions cannot be described in terms of simple equations which would fit into an econometric model; vine grapes are an exception as the yield correlates very well ($r^2 = 0.65$) with midsummer temperatures (Pfister, 1981). In the case of winter grains, however, the weather conditions affecting growth are more complicated, as they fall into eight different phases and extend over a 12-month period (Slicher van Bath, 1977). In addition, attention must be given to ecological interactions; i.e., the interrelationships between fluctuations in the output of basic foodstuffs such as grain, potatoes, fruit, vegetables and dairy products. These factors may either mitigate or magnify the fluctuations in specific commodity processes and available food resources (Pfister, 1978c). In order to model properly the effect of climate on an agricultural economy a climate response model should be developed for every major crop and all the different crop models should then be combined into one large model. This is, of course, impossible – an adequate data basis for such an elaborate model does not exist even for the present century, let alone earlier periods of history. Nevertheless, as will be demonstrated below, there is strong evidence of climatic impact on past society in Switzerland.

In order to obtain information of the levels and fluctuations of agricultural production in Switzerland from 1525 to 1825, the major staple foods (grain, wine, dairy products) were considered. Grain production was estimated from aggregate tithes (regional curves for western, central and eastern Switzerland and the Uplands). For wine production, local series of tithes or yields per acre from the major vine-growing counties were aggregated to form a single series (Pfister, 1980a). Dairy production, although it is an important branch of agriculture, is, unfortunately, not covered by quantitative evidence and had to be documented with qualitative data. Only grain production and dairy production are considered here.

To examine the sensitivity of crops to weather, results obtained from the agricultural sciences are compared with historical climate evidence. Wheat was affected by heavy rainfall in autumn and winter, which presumably washed nitrogen out of the soil (e.g., 1570/1, 1629/30, 1768/9). Spelt and rye suffered from extended snow cover in spring, which favoured infections like the fungus 'snow mould', *fusarium nivale* (e.g., 1587, 1614, 1785); see Pfister (1978c).

Dairy production in summer depended on the length and character of the grazing season on alpine pastures (i.e., the date of thaw and the frequency of fresh snow layers). Hay was of prime importance to dairy production since hay was needed for fodder during winter. Hay-making in the valleys during summer suffered particularly from wet spells (which lowered the protein content of the hay) and, to a lesser degree, from long droughts (which diminished the quantity of hay produced). The amount eaten in the winter period varied according to the length of the grazing period in autumn and spring.

It is believed that during the Little Ice Age, or at least during the most severe parts of it, there was an enhanced variability of climate (see above and Fig. 8.3). From the agricultural sciences it is well known that the yield of a crop is vulnerable to large-scale deviations from the long-term means during critical phases of the vegetative period. It is probable, therefore, that the more variable climate of the Little Ice Age was worse for agricultural production than the climate of the late nineteenth and twentieth centuries. Grain production in the Upland Zone of Switzerland supports this assertion. Most of the very bad harvests – 1565, 1566, 1571, 1573, 1587, 1614, 1644, 1681, 1716, 1731, 1770, 1785, 1789, 1816 and 1824 – occurred after an unusually long period of spring snow cover, giving conditions which favoured massive outbreaks of *fusarium nivale*. This parasite, so significant for the harvests in the Little Ice Age period, is hardly mentioned in Swiss agronomical textbooks today. The reason for this is simple: none of the winters in the last 120 years has been long enough to allow massive outbreaks of *fusarium nivale* to occur. Of the 15 poor harvests of the Little Ice Age noted above, at least 10 were the direct consequence of climatic extremes never seen in the present century. The consequences of a poor grain harvest were mostly felt in less than one year's time, when grain stocks ran dry before the ripening of the next harvest.

These long winters and snowy springs were also damaging to dairy production. As long as the snow lasted the cattle had to be fed from hay harvested in the previous year. When the hay ran out the cattle either had to be fed with branches of pine, or slaughtered. The stressful conditions during winter and spring meant that most cows which survived were dry. Hence a popular dictum, which came out of the experience of the Little Ice Age climate: 'If the cattle are hungry, human beings will also become hungry.'

In spite of these links between severe seasons and harvest and dairy production, the relationships are far from simple. For instance, a comparison between agricultural production, grain prices and demographic data for the period 1755–97 (Pfister, 1978c) has revealed that even a severe harvest failure accompanied by a heavy loss in herds sometimes only caused a very moderate price peak and had no demographic consequences. The year 1785 is a good example. In this severe year, the winter grains were heavily damaged but there was a normal harvest of summer grains, potatoes, fruit, vegetables and wine and sufficient supplementary grain could be imported from abroad to lessen the impact.

Subsistence crises only occurred whenever crops failed repeatedly and when several staple foods were affected at the same time. This was always the result of the cumulative impact of a number of unfavourable weather conditions (wet autumns, long winters, cold springs, cool wet summers), stretching over several years and involving large areas. A good example is the cluster of bad years from 1812 to 1817, which caused the last great subsistence crisis of the western world (Post, 1977). Similar periods of overall price peaks in western Europe (1570–74, 1585–89, 1592–96, 1627–30 and 1769–71) are, in Switzerland, clearly connected to slumps in agricultural production (grain, wine, dairy products) and an exceptional amount of meteorological stress.

This leads one to the conclusion that subsistence crises were not primarily the result of weak social and economic structures, nor were they only the result of climatic stress. In fact, it is astonishing how efficiently past societies managed to overcome the impact of a *limited* climatic stress. The evidence for Switzerland does, however, strongly implicate climate as a causative factor, but a great amount of stress was needed in order to initiate a subsistence crisis; far greater than we have experienced in our own century.

Appendix

The following tables give data for seasonal weighted thermal and wetness indices for Switzerland for the period 1550–1820. ND is used to indicate insufficient data: i.e. when information is available for less than two of the three months.

Weighted thermal index for winter (DJF[a])

	0	1	2	3	4	5	6	7	8	9	Decade
1550	0	−6	2	0	−2	0	−1	−2	−1	2	−8
1560	0	−4	5	0	−1	−6	−5	−3	1	1	−12
1570	0	−5	0	−9	1	−2	−1	1	0	1	−14
1580	ND	ND	0	ND	3	−2	ND	−6	−1	−3	−9
1590	1	−4	3	−5	−2	−9	−1	0	−6	0	−23
1600	−4	−8	2	−5	3	ND	1	5	−7	2	−11
1610	1	0	−1	8	ND	1	−6	5	−2	0	6
1620	1	−3	−1	−6	−6	7	2	−4	3	3	−4
1630	ND	0	ND	1	−4	−2	0	−3	−3	5	−6
1640	2	−1	−2	−3	−2	−3	ND	2	ND	−6	−13
1650	7	−1	0	−7	−3	ND	−3	0	2	2	−3
1660	−7	6	2	−7	−2	−6	−3	−3	0	−1	−21
1670	−3	7	−3	2	−3	−2	−2	−2	2	−4	−8
1680	5	−8	5	1	−5	−5	1	−6	−6	−2	−20
1690	1	−7	−6	−1	−4	−8	1	−7	−3	−3	−37
1700	−4	−1	3	−1	−6	−2	1	4	1	−8	−13
1710	−1	2	0	2	0	−1	−8	−4	−5	3	−12
1720	0	−1	2	−4	4	−1	−3	−1	1	−4	−7
1730	0	−7	−1	4	6	−2	−1	0	−4	0	−5
1740	−5	5	0	0	−3	−3	2	2	1	4	3
1750	0	1	1	−3	0	−6	0	−6	−6	3	−16
1760	−2	−1	−4	−4	4	−1	−7	−2	−4	−2	−23
1770	−4	0	2	−2	2	0	−3	−6	−4	−1	−16
1780	−1	−3	1	1	−6	−4	0	−2	4	−3	−13
1790	0	2	0	−1	3	−5	4	−1	0	−3	−1
1800	−1	0	−2	−4	2	−4	2	1	−4	0	−10
1810	−5	0	−4	−4	−7	2	−4	2	−2	−1	−23
1820	−2	−4	4	0	2	2	0	−4	6	−3	1

[a]Year given is that for the January

Weighted wetness index for winter (DJF)

	0	1	2	3	4	5	6	7	8	9	Decade
1550	0	0	1	0	0	1	3	−1	−1	−1	2
1560	0	0	−1	0	0	0	0	−3	−1	1	− 4
1570	1	5	1	0	2	−1	0	−1	−1	3	9
1580	ND	ND	0	ND	0	−4	ND	4	−1	−5	− 6
1590	2	−3	0	0	1	−2	0	0	6	0	4
1600	−2	0	0	0	0	ND	0	0	6	−4	0
1610	−1	−3	1	−2	ND	0	0	0	−2	0	− 7
1620	−1	0	−1	0	0	0	−1	4	1	0	2
1630	ND	−1	ND	−2	−2	−1	0	0	−1	−1	− 8
1640	1	0	1	0	0	0	ND	1	ND	0	3
1650	−2	2	−1	−1	−1	ND	0	0	2	0	− 1
1660	0	0	−1	−2	−3	1	−1	0	0	−1	− 7
1670	−1	2	−8	1	−2	−3	−2	0	0	0	−13
1680	−2	1	−6	4	0	0	−1	4	−2	0	− 2
1690	0	−4	0	1	−1	0	−3	0	−2	2	− 7
1700	0	0	−1	0	0	−2	−3	0	2	0	− 4
1710	−5	3	0	0	−5	−2	−3	0	2	6	− 4
1720	6	2	−1	3	0	0	4	0	0	1	15
1730	−3	2	0	−4	−2	0	−1	2	1	1	− 4
1740	−2	5	0	−2	−6	0	0	1	0	0	− 4
1750	−3	0	−2	−1	−1	−6	0	−4	1	−2	−18
1760	4	−1	−1	−3	4	−1	−3	−1	0	0	− 2
1770	0	2	3	0	−1	0	2	0	−2	−5	− 1
1780	2	4	0	1	2	2	0	−3	4	3	15
1790	−3	2	1	−1	−1	−2	2	−4	1	2	− 3
1800	−1	0	3	−1	2	2	1	2	−2	2	8
1810	−3	1	−2	−3	1	1	4	0	0	−3	− 4
1820	−2	−4	−2	1	−3	0	−3	0	−2	−2	−17

Weighted thermal index for spring (MAM)

	0	1	2	3	4	5	6	7	8	9	Decade
1550	−1	−2	2	−2	1	1	0	−1	3	6	7
1560	3	1	2	0	−1	0	1	3	−4	−4	−1
1570	−7	3	1	−5	−2	−1	−2	0	−4	1	−16
1580	1	0	ND	3	0	0	−3	−5	0	−1	−5
1590	−1	1	−2	1	−3	−5	0	−2	−1	1	−11
1600	−2	−7	1	9	2	2	ND	0	−2	0	3
1610	−1	6	−1	−2	−8	0	1	2	0	0	−3
1620	1	−1	0	−1	2	−2	−1	−8	−5	−4	−17
1630	−2	2	0	1	ND	−1	6	4	ND	1	11
1640	−4	−1	−3	−5	−1	−3	1	2	−1	−5	−20
1650	−2	−2	−3	3	0	−1	0	−1	−2	−1	−9
1660	2	3	1	3	2	−3	−2	−3	−1	0	2
1670	−1	2	0	0	−2	−2	5	−1	−1	−5	−5
1680	3	−5	2	−3	1	0	4	−3	−6	−4	−11
1690	−7	−1	−5	−5	−2	−5	−3	−3	−3	−8	−42
1700	−7	−8	0	0	1	−3	1	−4	2	0	−18
1710	1	0	−3	−2	−7	2	−7	0	1	1	−14
1720	−2	−5	0	2	5	−2	−1	2	6	−4	1
1730	0	−5	3	−1	1	−3	0	0	−2	−3	−10
1740	−9	−2	−4	0	1	1	0	−1	−6	1	−19
1750	4	−3	1	5	−2	−2	−6	−2	0	−1	−6
1760	0	2	1	−4	−3	0	−2	−5	−2	−1	−14
1770	−5	−1	0	−2	4	−1	−2	−1	−1	−1	−10
1780	1	4	−3	−4	−2	−5	0	−1	1	0	−9
1790	1	2	1	−1	4	0	−3	1	0	−6	−1
1800	0	0	0	−2	−1	−6	0	−4	−3	−3	−19
1810	2	6	−5	−2	−2	3	−4	−7	−1	2	−8
1820	0	0	4	−1	−4	−1	−2	4	0	−1	−1

Weighted wetness index for spring (MAM)

	0	1	2	3	4	5	6	7	8	9	Decade
1550	4	2	−1	2	−1	0	0	0	0	0	6
1560	−1	−2	0	−1	1	−2	1	−2	0	1	−5
1570	0	−1	−1	3	−1	3	2	−1	0	0	4
1580	−1	0	ND	−6	1	0	2	0	−4	−4	−12
1590	−2	−1	1	1	1	1	2	2	1	0	6
1600	0	0	−3	−6	0	0	ND	−4	−2	2	−13
1610	−2	−8	−3	−3	0	0	−1	−1	0	0	−18
1620	−1	0	−1	0	−2	2	1	5	2	0	6
1630	2	−2	−1	−1	ND	1	−6	−3	ND	−2	−12
1640	2	1	1	0	−1	−1	−2	−2	0	3	1
1650	1	2	2	−4	−2	1	0	−1	0	−2	−3
1660	0	−3	−2	−2	−3	−1	0	1	0	0	−10
1670	0	0	−2	1	1	0	−4	0	0	1	−3
1680	−1	0	−1	−1	−2	−3	−2	0	−2	1	−11
1690	−2	0	−1	6	−2	0	2	0	2	0	5
1700	−2	0	0	0	1	2	−2	1	−2	1	−1
1710	−2	2	0	0	0	0	−2	−2	2	−2	−4
1720	3	4	0	−3	2	−1	−4	2	−3	−1	−1
1730	1	−4	−2	0	2	1	−2	0	1	0	−3
1740	−2	−3	0	4	0	0	0	−1	0	1	−1
1750	2	6	−1	−5	−3	−2	3	0	0	−1	−1
1760	−3	2	−4	0	1	3	1	0	−2	0	−2
1770	1	−2	2	−3	0	−1	1	1	0	−2	−3
1780	1	−1	5	2	1	0	3	3	−3	−1	10
1790	−1	−2	1	−1	1	0	−3	−2	−2	4	−5
1800	1	0	−1	−2	−1	0	0	−4	−3	2	−8
1810	−1	−5	−1	−1	−5	−1	2	1	3	−3	−11
1820	2	1	−3	0	−2	−4	−4	2	−4	2	−14

Weighted thermal index for summer (JJA)

	0	1	2	3	4	5	6	7	8	9	Decade
1550	−2	2	3	1	2	−3	6	3	2	5	19
1560	−4	0	1	1	0	2	3	0	0	−1	2
1570	−1	0	2	−5	0	2	−3	−2	0	−8	−15
1580	0	2	0	2	−2	−4	1	−2	−6	−4	−13
1590	3	1	2	−4	−4	1	−4	−4	−1	3	− 7
1600	ND	−4	−1	4	2	2	−3	1	−5	−2	− 6
1610	1	3	0	−3	ND	1	8	ND	−5	−1	4
1620	0	−4	1	5	2	−3	1	−4	−7	3	− 6
1630	3	1	−2	−1	−1	0	1	2	1	−3	0
1640	−2	−1	0	2	2	4	1	0	−3	−1	2
1650	−1	2	0	1	0	0	0	0	−1	1	2
1660	1	0	−1	−3	−1	2	4	−1	1	5	7
1670	1	0	1	−1	0	−5	1	−2	1	−2	− 6
1680	2	2	−3	3	3	−4	2	1	−2	−2	2
1690	0	2	−5	1	1	−1	0	0	−4	−1	− 7
1700	−1	−1	−1	−1	2	2	4	0	1	−1	4
1710	0	2	2	−4	−3	0	−4	0	6	4	3
1720	−2	0	0	0	3	−3	3	2	4	2	9
1730	0	1	−3	0	−1	0	0	−2	0	1	− 4
1740	0	1	2	0	1	−1	4	−1	0	1	7
1750	0	−1	0	2	2	0	2	2	−3	2	6
1760	1	1	1	2	1	1	0	0	0	−3	4
1770	−2	−1	2	−2	2	0	0	−1	3	−1	0
1780	3	2	2	2	0	−2	0	−1	1	−3	4
1790	−2	1	1	2	2	−2	−2	0	1	−2	− 1
1800	−1	−1	2	3	1	−6	0	5	1	−2	2
1810	−3	3	−4	−6	−2	−3	−9	−4	−1	0	−29
1820	−1	−4	3	−1	−1	−2	4	2	0	−2	−2

Weighted wetness index for summer (JJA)

	0	1	2	3	4	5	6	7	8	9	Decade
1550	2	0	−1	0	2	3	−2	−1	0	−6	−3
1560	4	0	3	2	0	0	0	−4	1	2	8
1570	4	2	1	4	2	−3	3	−1	2	6	20
1580	0	−2	0	−2	7	5	−1	1	8	0	16
1590	−2	3	−2	2	4	0	8	4	0	−4	13
1600	ND	3	1	0	0	0	2	0	4	7	17
1610	0	0	1	3	ND	0	−7	ND	3	0	0
1620	1	5	1	−7	2	1	2	4	5	−1	13
1630	−3	−2	1	1	1	−1	0	−1	0	2	−2
1640	1	0	0	−2	−1	−4	−3	2	3	1	−3
1650	1	−1	1	−2	0	0	3	0	0	0	2
1660	−2	2	0	3	1	−2	−2	0	1	−3	−2
1670	0	0	−1	4	1	4	−1	3	0	2	12
1680	2	−4	3	−3	−1	3	0	0	3	4	7
1690	4	0	5	0	2	2	2	2	−1	0	16
1700	3	0	−1	1	0	−1	−3	6	1	1	7
1710	0	−3	−1	6	0	1	1	2	−3	−4	−1
1720	8	1	0	1	−2	6	1	3	0	−2	16
1730	1	1	2	4	6	3	1	6	4	0	28
1740	2	1	−2	0	−2	0	−4	5	1	−1	0
1750	1	0	2	−3	−2	−1	0	−1	3	0	−1
1760	2	0	−1	1	3	2	0	2	3	3	15
1770	2	1	0	1	0	3	1	1	−1	0	8
1780	−5	2	−2	1	0	2	4	2	2	3	9
1790	1	−1	1	−4	−4	1	1	0	1	1	−3
1800	−3	−1	−2	−4	0	1	0	−4	0	−2	−15
1810	0	−4	1	3	2	0	5	2	−5	0	4
1820	1	0	−2	−1	0	−2	−5	−3	−2	−2	−12

Weighted thermal index for autumn (SON)

	0	1	2	3	4	5	6	7	8	9	Decade
1550	0	−3	1	1	1	−1	0	0	1	−1	− 1
1560	2	2	2	−1	0	−3	1	0	0	4	7
1570	0	2	−2	−3	2	1	0	ND	ND	ND	0
1580	ND	ND	ND	ND	ND	ND	ND	−5	2	1	− 2
1590	3	1	1	−2	−2	0	0	−2	−1	5	3
1600	−1	−2	0	4	0	2	−2	1	−3	−2	− 3
1610	2	−2	0	−3	−1	ND	0	0	0	−1	− 5
1620	1	1	1	2	1	−1	2	−3	−5	0	− 1
1630	1	2	−3	−2	3	3	1	0	−1	2	6
1640	−2	−1	2	1	1	−2	−1	2	−3	0	− 3
1650	−2	0	6	1	1	−1	0	0	2	−1	6
1660	1	3	3	2	−2	−3	1	−3	2	2	6
1670	1	−2	1	−1	−1	−3	−4	−2	−1	−1	−13
1680	6	3	0	0	−4	2	1	−2	−3	0	3
1690	0	−2	−3	−1	−2	−1	0	−2	−3	0	−14
1700	−1	0	−4	−2	2	0	3	−1	−1	−1	− 5
1710	1	1	0	−4	−3	0	−2	0	1	1	− 5
1720	−1	−2	4	3	−2	1	−1	−1	1	0	2
1730	−2	2	−2	−3	−4	−1	1	−1	−1	−2	−13
1740	−4	0	−1	0	0	0	−1	2	−1	−1	− 6
1750	0	−2	1	2	0	−4	−4	−4	−6	−1	−18
1760	0	−1	−2	−4	−4	−6	−2	2	0	−1	−18
1770	4	−2	6	2	−4	2	−2	2	−2	5	−11
1780	2	0	−5	2	0	2	−6	2	−3	−2	− 8
1790	0	−2	0	−2	0	−2	2	0	0	−2	− 6
1800	−2	0	2	−4	0	−5	2	0	−3	−6	−16
1810	3	2	−4	−3	−3	−2	−4	0	−2	−2	−15
1820	−7	0	0	0	0	2	3	−1	−1	−5	− 9

Weighted wetness index for autumn (SON)

	0	1	2	3	4	5	6	7	8	9	Decade
1550	0	2	0	2	−2	6	4	0	−2	0	10
1560	2	2	0	−2	2	−1	−2	−6	−2	0	− 7
1570	1	2	2	4	2	−2	0	ND	ND	ND	9
1580	−4	ND	ND	ND	−2	ND	ND	2	1	0	− 3
1590	−1	0	−1	2	3	1	0	0	1	−4	1
1600	ND	0	ND	−2	0	−1	−3	0	2	−1	− 5
1610	0	0	1	−1	0	ND	ND	ND	−1	−1	− 2
1620	−4	ND	0	ND	−2	0	−7	1	4	1	− 7
1630	ND	−2	0	ND	−4	0	−1	0	ND	ND	− 7
1640	0	0	−1	−1	−1	ND	2	−2	−2	−1	− 6
1650	2	3	−7	−1	−1	0	ND	0	0	5	1
1660	1	−3	−1	−1	1	2	−3	0	−3	−6	−13
1670	1	−1	1	−1	−1	1	0	−2	0	0	− 2
1680	−5	−2	−1	2	3	0	−2	3	2	2	2
1690	2	1	2	0	2	−2	0	−1	4	−2	6
1700	2	−2	0	1	0	3	−2	0	−2	4	4
1710	−1	6	0	2	0	0	1	−2	−1	4	9
1720	4	2	0	−2	−6	−2	3	2	−2	2	1
1730	0	−3	−1	2	1	−3	−3	1	−1	2	− 5
1740	2	2	2	0	0	−3	−2	−2	−2	−6	− 9
1750	−1	−3	−7	−3	−3	2	−4	0	−1	−2	−22
1760	3	2	0	1	0	2	−2	0	4	1	11
1770	1	−1	−1	2	1	−1	0	−2	6	−1	4
1780	4	1	1	−2	−2	1	3	1	−4	0	3
1790	0	−1	3	3	5	0	−1	0	4	4	17
1800	1	6	−1	0	1	−1	−2	3	3	−1	9
1810	−1	0	1	0	−3	−2	2	−3	−2	−2	−10
1820	−2	−3	−1	−4	4	1	0	−3	0	3	− 5

References

Angot, A. (1885). Étude sur les vendanges en France. *Annales du Bureau Central Météorologique de France* for 1883.

Baulant, M. & Le Roy Ladurie, E. (1978). Une synthèse provisoire: les vendanges du XVe siècle au XIXe siècle. *Annales Économies Sociétés Civilisations*, **33**, 763–71.

Bell, W. T. & Ogilvie, A. E. J. (1978). Weather compilations as a source of data for the reconstruction of European climate during the Medieval period. *Climatic Change*, **1**, 331–48.

Bider, M., Schüepp, M. & von Rudloff, H. (1959). Die Reduktion der 200 jährigen Basler Temperaturreihe. *Archiv für Meteorologie, Geophysik und Bioklimatologie*, series B, **9**, 360–412.

Bider, M. & Schüepp, M. (1961). Luftdruckreihen der letzten zwei Jahrhunderte von Basel und Genf. *Archiv für Meteorologie, Geophysik und Bioklimatologie*, Series B, **11**, 1–36.

Dufour, M. L. (1870). Notes sur le problème de la variation du climat. *Bulletin de la Société vaudoise des Sciences naturelles*, **10**, 359–436.

Fliri, F. (1970). Probleme und Methoden einer gesamtalpinen Klimatographie. *Jahresbericht der Geographische Gesellschaft Bern*, **49**, 113–27.

Flohn, H. (1949). Klima und Witterungsablauf in Zürich im 16 Jahrhundert. *Viertel-jahrsschrift der Naturforschungs-Gesellschaft in Zürich*, **95**, 28–41.

Furrer, G., Gamper-Schollenberger, B. & Suter, J. (1980). Zur Geschichte unserer Gletscher in der Nacheiszeit. In: *Das Klima. Auf dem Wege zum Verständnis der Klima-mechanismen und ihrer Beeinflussung durch den Menschen*. New York: Springer.

Garnier, M. (1955). Contribution de la phénologie à l'étude des variations climatiques. *La Météorologie*, **40**, 291–300.

Geisendorf, P. (1942). *Les Annalistes Genevois du Début du Dix-septième Siècle: Savion-Piaget-Perrin. Études et Textes*. Memoires et Documents publiés par la Société d'Histoire et d'Archéologie de Genève, Geneva.

Hoskins, W. G. (1968). Harvest fluctuations and English economic history, 1620–1759. *Agricultural History Review*, **16**, 15–31.

Ingram, M. J., Underhill, D. J. & Wigley, T. M. L. (1978). Historical climatology. *Nature*, **276**, 329–34.

Kington, J. A. (1974). An application of phenological data to historical climatology. *Weather*, **29**, 320–28.

Lamb, H. H. (1977). *Climate: Present, Past and Future*, vol. 2. London: Methuen. 835 pp.

Le Roy Ladurie, E. (1971). *Times of Feast, Times of Famine. A history of climate since the year 1000*, transl. B. Bray. New York: Doubleday.

Le Roy Ladurie, E. & Baulant, M. (1980). Grape harvests from the fifteenth through the nineteenth centuries. *Journal of Interdisciplinary History*, **10**, 839–49.

Margary, I. D. (1926). The Marsham phenological record in Norfolk, 1736–1925, and some others. *Quarterly Journal of the Royal Meteorological Society*, **52**, 27–54.

Messerli, B., Messerli, P., Pfister, C. & Zumbühl, H. T. (1978). Fluctuations of climate and glaciers in the Bernese Oberland, Switzerland, and their geoecological significance, 1600 to 1975. *Arctic and Alpine Research*, **10**, 247–60.

Müller, K. (1953). *Geschichte des Badischen Weinbaus*, 2nd edn. Lahr: Baden Schauenburg Verlag.

Pfister, C. (1972). Phänologische Beobachtungen in der Schweiz der Aufklärung. *Informationen und Beiträge zur Klimaforachung*, **9**, 15–30, edited by the Geographical Institute, University of Berne, Switzerland.

Pfister, C. (1975). *Agrarkonjunktur und Witterungsverlauf im westlichen Schweizer Mittelland*. Geographica Bernensia, G2. Bern.

Pfister, C. (1977). Zum Klima des Raumes Zürich im späten 17. und frühen 18. Jahrhundert. *Vierteljahrsschrift der Naturforschenden Gesellschaft Zurich*, **122**, 447–71.

Pfister, C. (1978a). Fluctuations in the duration of snow cover in Switzerland since the late seventeenth century. In *Proceedings of the Nordic symposium on climatic changes and related problems*, ed. K. Frydendahl, pp. 1–6. Copenhagen: Danish Meteorological Institute, Climatological Paper No. 4.

Pfister, C. (1978b). Die älteste Niederschlagsreihe Mitteleuropas; Zürich 1708–1754. *Meteorologische Rundschau*, **31**, 56–62.

Pfister, C. (1978c). Climate and economy in eighteenth century Switzerland. *Journal of Interdisciplinary History*, **9**, 223–43.

Pfister, C. (1979). Getreide-Erntebeginn und Frühsommertemperaturen im schweizerischen-Mittelland seit dem frühen 17. Jahrhundert. *Geographica Helvetica*, **34**, 23–35.

Pfister, C. (1980a). The Little Ice Age: thermal and wetness indices for Central Europe. *Journal of Interdisciplinary History*, **10**, 665–96.

Pfister, C. (1980b). Klimaschwankungen und Witterungsverläufe im schweizerischen Alpenraum und Alpenvorland zur Zeit des Little Ice Age. Die Aussage der historischen Quellen. In *Das Klima. Auf dem Weg zum Verständnis der Klima-mechanismen und ihrer Beeinflussung durch den Menschen*. New York: Springer.

Pfister, C. (1981). Die Fluktuationen der Weinmosterträge im Schweizerischen 'Weinland' vom 16. bis ins frühe 19. Jahrhundert. Klimatische Ursachen und sozioökonomische. *Bedentung Schweizerische Zeitschrift für Geschichte* (in press).

Post, J. D. (1977). *The last Great Subsistence Crisis in the Western World*. Baltimore: Johns Hopkins.

Primault, B. (1969). Le climat et la viticulture. *International Journal of Biometeorology*, **13**, 7–24.

Riggenbach, A. (1891). Die Niederschlags – Verhältnisse von Basel. *Denkschriften der Schweizerische Naturforschenden Gesellschaft, Zürich*, **32**, No. 2.

von Rudloff, H. (1967). *Die Schwankungen und Pendelungen des Klimas in Europa seit dem Beginn der regelmässigen Instrumenten-Beobachtungen* (*1670*). Braunschweig: Vieweg & Sohn, Die Wissenschaft series, 122, 370 pp.

Schüepp, M. (1961). Lufttemperatur. Section c of *Klimatologie der Schweiz*, published as *Beiheft zu den Annalen der Schweizerischen Meteorologischen Zentralanstalt, Zürich.*

Schüepp, M. (1976). Niederschlag, parts 9–12. Section I/E/16 of *Klimatologie der Schweize*, published as *Beiheft zu den Annalen der Schweizerischen Meteorologischen Zentralanstalt, Zürich*, Jahrgang, 1976.

Schweingruber, F. H., Bräker, O. U. & Schär, E. (1978). Dendroclimatic studies in Great Britain and in the Alps. In *Evolution of Planetary Atmospheres and Climatology of the Earth.* Toulouse: Centre National d'Études Spatiales.

Schweingruber, F. H., Braeker, O. U. & Schaer, E. (1979). *Boreas*, **8**, 427–52.

Slicher van Bath, B. H. (1977). Agriculture in the Vital Revolution. *Cambridge Economic History of Europe*, vol. 5, pp. 42–132. Cambridge University Press.

de Vries, J. (1980). Measuring the impact of climate on history: the search for appropriate methodologies. *Journal of Interdisciplinary History*, **10**, 599–630.

Uttinger, H. (1962). Die Dauer der Schneedecke in Zürich. *Archiv für Meteorologie, Geophysik und Bioklimatologie*, Series B, **12**, 404–12.

Weger, N. (1952). Weinernten und sonnenflecken. *Berichte des Deutschen Wetterdienst in der US-Zone*, **38**, 229–37.

Zumbühl, H. J. (1980). Die Schwankungen der Grindelwaldgletscher in den historischen Bild- und Schriftquellen des 12. bis 19. Jahrhunderts. *Denkschriften der Schweitzerischen Naturforschenden Gesellschaft, Zürich*, **92**.

9

The historical climatology of Africa

S. E. NICHOLSON

Abstract

This chapter discusses the methodology of historical climatology as developed for Africa, presents the results of recent historical climate studies of southern Africa, and compares this with the chronology previously developed for sub-Saharan Africa.

While for sub-Saharan Africa local chronicles were valuable sources, they are replaced in southern Africa by archival material and historical texts. Particularly important in all regions are travellers' journals, European settlers' diaries, as well as older geographical sources. Changes of lakes, rivers and landscape; references to climate and seasons; occasional meteorological observations; and references to famine, droughts and floods are usual types of information. Information is particularly rich for nearly all areas of the Cape, for the southeastern Kalahari, for northern Botswana and Ngamiland and for parts of Namibia.

Results of the investigation for southern Africa are only tentative, but seem to confirm that very anomalous periods in sub-Saharan Africa and times of climatic discontinuity there are also reflected in the southern African record. This includes a major rainfall change toward 1800, and anomalous weather patterns in the 1680s and early nineteenth century (especially the 1830s). There is some evidence of a change of rainfall seasonality in certain areas and of greater snowfall in the nineteenth century than at present.

Introduction

This chapter, while relating to the entire African continent, focusses on three particular regions: the Sahelo–Soudan zone of northern Africa and the Kalahari and Karroo of southern Africa (Fig. 9.1). These regions have several features in common which make them particularly suitable for such a study of climatic history. All are semi-arid, drought-prone regions sensitive to climatic fluctuations and human mismanagement; rainfall variability is extremely high; all three have undergone significant ecological changes within the past few centuries; and in each case it is not yet possible to distinguish adequately between climatic and human factors in the observed ecological degradation. The comparison between the Kalahari and Sahel goes further: both receive between 100 and 500 mm of rainfall annually during a brief 'summer' rainy season, determined in part by the advance and retreat of the Inter-Tropical Convergence Zone or ITCZ (Fig. 9.1).

The Sahel, Kalahari and Karroo are obvious choices for a study of climatic history. Their climatic sensitivity and aridity magnify the effects of small magnitude but large-scale climatic changes. As a result, the area's inhabitants and their histories are frequently and markedly affected by climate. Furthermore, the traces left by climatic fluctuations in

Fig. 9.1. Map of African rainfall, the ITCZ and described locations.

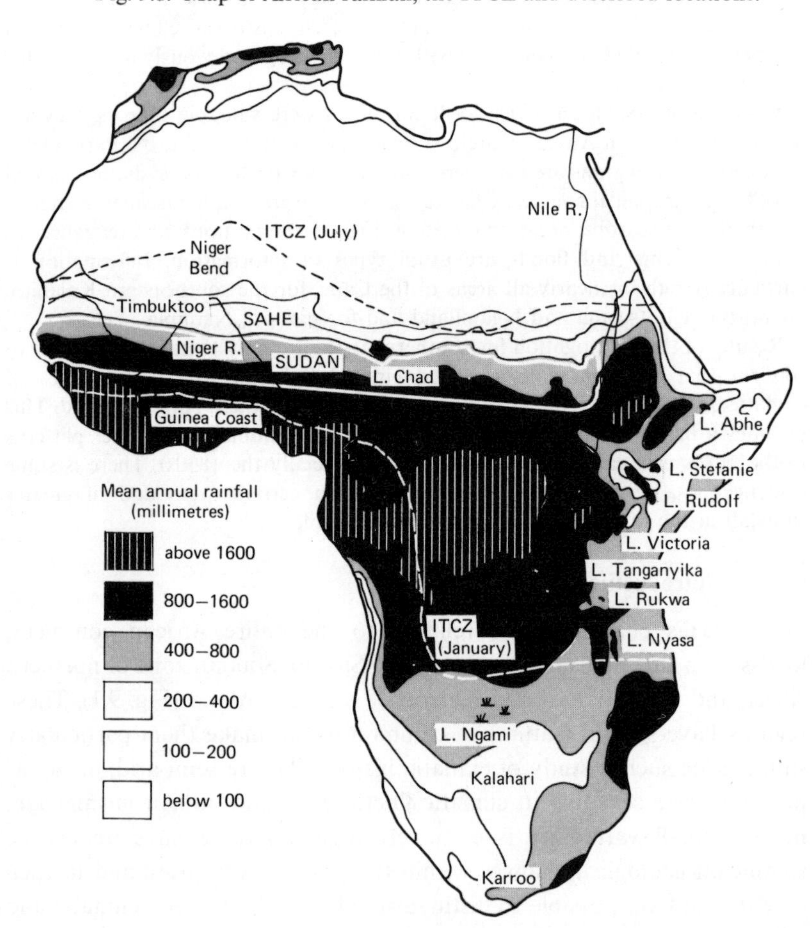

such semi-arid regions – for example, a change in rainfall seasonality or the desiccation of a lake – are highly visible. Finally, the relationship of the ITCZ to the Sahel and Kalahari means that climatic fluctuations in these areas represent a large part of the tropics and subtropics as well as the climatic interaction of the two hemispheres. Thus, knowledge of these areas is crucial in understanding global climatic history.

This discussion deals first with methodology. The results of a previous study of the Sahel are then briefly summarised. Next, the results of the current southern Africa project are presented and compared with those for the Sahel. Finally, three climatically anomalous periods are considered.

Methodology

Material for reconstructing past African climates derives from several disciplines. Archaeological studies, such as those of Munson (1971) and Toupet (1973) in Mauritania, may indicate occupation of now uninhabitable areas, or subsistence on resources (such as fishing, wet cultivation or snails) which are either no longer available or no longer used in these areas. Geological studies, such as those of Maley (1973) and Servant (1973) in Chad and Daveau (1965) in Mauritania, provide information on lake-level variation, former lakes and waterways, dunes and evaporites and pollen. Tree-ring studies have also been carried out in parts of the continent and nearby areas: Morocco, Tunisia, Algeria, the Sinai, Kenya, Namibia, Natal and other parts of South Africa (see, for example, Ginestous, 1927; Walter, 1936; Fahn, Wachs & Ginzburg, 1963; Waisel & Lipschitz, 1968; Hall, 1976; Lilly, 1977). Historical and geographical sources provide the most valuable data for the present study. They include local chronicles (such as those of Bornu and Timbuktoo, which record at least a millennium), reports of travellers and settlers, historical reports, archives, maps, early geographical journals and oral histories. These sources and the methodology used to evaluate the data they contain are considered in greater detail elsewhere (Nicholson, 1976, 1979a, 1980a). The types of information they provide are summarised in Table 9.1. For some regions, certain indicators are particularly sensitive. In the Sahel, the key indicator is the length of the rainy season. In various parts of southern Africa, the critical indicators include the seasonality of rainfall, the wind regime, and the occurrence of snow.

There are special problems involved in reconstructing the historical climatology of Africa (Nicholson, 1979a). Information is scarce, especially continuous records, so the researcher must often rely on types of

Table 9.1. *Types of data useful for historical climatic reconstructions, particularly with respect to Africa*

I Landscape descriptions

1 Forests and vegetation: are they as today?
2 Conditions of lakes and rivers:
  (*a*) height of the annual flood, month of maximum flow of the river
  (*b*) villages directly along lakeshores
  (*c*) size of the lake (e.g. as indicated on map)
  (*d*) navigability of rivers
  (*e*) desiccation of present-day lakes or appearance of lakes no longer existing
  (*f*) floods
  (*g*) seasonality of flow; condition in wet and dry seasons
3 Wells, oases, bogs in presently dry areas
4 Flow of wadis
5 Measured height of lake surfaces (frequently given in travel journals, but optimally some instrumental calibration or standard should accompany this)

II Drought, famine and other agricultural information

1 References to famine or drought, preferably accompanied by the following:
  (*a*) where occurred and when occurred: as precisely as possible
  (*b*) who reported it; whether the information is second hand
  (*c*) severity of the famine or drought; local or widespread?
  (*d*) cause of the famine
2 Agricultural prosperity:
  (*a*) condition of harvest
  (*b*) what produced this condition
  (*c*) months of harvests, in both bad years and good years
  (*d*) what crops grown
3 Wet cultivation in regions presently too arid

III Climate and meteorology

1 Measurements of temperature, rainfall, etc.
2 Weather diaries
3. Descriptions of climate and the rainy season: when do the rains occur, what winds prevail?
4 References to occurrence of rain, tornado, storm
5 Seasonality and frequency of tornadoes, storms
6 Snowfalls: is this clearly snow or may the reporter be mistakenly reporting frost, etc?
7 Freezing temperatures, frost, hail
8 Duration of snow cover on mountains (or absence)
9 References to dry or wet years, severe or mild winters, other unusual seasons

information less commonly used in European reconstructions (see Table 9.1). Furthermore, the useful indicators tend to be qualitative and often are not directly comparable with present data. Finally, interpretation tends to be difficult; it involves unravelling the factors controlling regional integrators (such as lakes and rivers) and differentiating between human and natural factors. The subjective nature of much of the information further complicates interpretation.

These problems are not as great as might at first sight appear (indeed many of them are common to historical climatology in general (see Ingram, Underhill & Farmer, this volume, and Pfister, this volume)). First, even information which is insufficient to provide continuous chronologies can be used when it relates to markedly anomalous periods in the past, such as major drought episodes. In these circumstances fragmentary evidence becomes useful and, even without a direct comparison with the present, information about such periods is useful for climatologists as well as for historians and geographers. Secondly, convergent (but independent) items of evidence, though individually weak may often be combined to provide reliable information about a given event. To take a hypothetical example, a satisfactory combination of evidence would be an oral tradition of a drought in several regions, concurrent geological evidence of lake desiccation, coupled with a firmly dated observation from an explorer's weather diary. Furthermore, the few indicators that are continuous, such as the record of Lake Chad, can sometimes be used to calibrate qualitative indicators such as drought or floods, thus providing a direct comparison with the present.

Information derived from a number of different areas which can be linked with known global or large-scale events also lends credibility to the reconstructions. This can also help to differentiate between climatic and human factors in observed environmental change. If non-random spatial rainfall anomaly (or departure) patterns are obtained which are physically meaningful, this also increases the credibility of historical climate reconstructions.

The rainfall anomaly pattern for *c.* 1738–56 (Fig. 9.2) illustrates the use of this method. Severe and extensive drought severely affected areas across nearly the entire Sahel–Soudan zone; reportedly half the population of Timbuktoo and other parts of the Niger Bend perished during the resulting famine (Curtin, 1975). Chronicles and tradition report this event in most areas but the dating is only approximate (within about one decade); however, archives and historical documents from Senegal also describe the drought and fix its date precisely. There is also geological evidence of a near desiccation of Lake Chad at the time dated by both historical and radiocarbon methods. The period also appears as extremely anomalous in the quantitative Nile record, which provides semi-annual measurements of the river's flood. This record indicates well above average rainfall over the Ethiopian highlands (from maximum flood height) and (from the summer minimum levels) low rainfall in

Fig. 9.2. African rainfall anomalies *c.* 1738–56; minus signs denote evidence of drier conditions; plus signs denote evidence of above-average rainfall; circled symbols denote regional integrators, e.g. lakes and rivers. (From Nicholson, 1978)

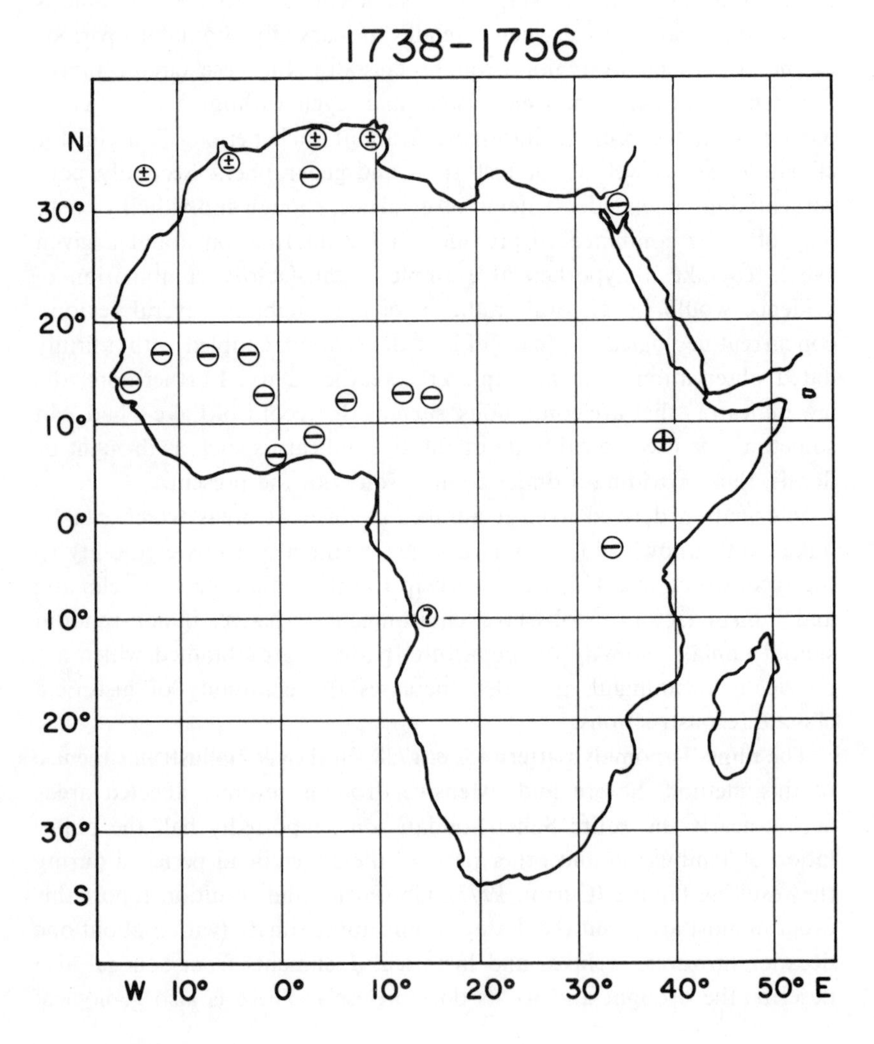

equatorial East Africa. Moreover, the record of a rain gauge operating in Funchal, Madeira, indicates not only increased rainfall but also a longer than average rainy season at Funchal, pointing to an increased equatorial penetration of the mid-latitude cyclones in winter. The suggested rainfall increase in North Africa derives some support from isolated tree-ring studies. For further details of this example see Nicholson (1978).

Summary of climatic fluctuations

General results for the Sahel and southern Africa

Fig. 9.3 summarises the climatic chronology for the semi-arid Sahel and Soudan, along the southern margin of the Sahara (Nicholson, 1978, 1979a). In general, conditions wetter than the present prevailed from about the sixteenth to and including the eighteenth centuries, declining after the late nineteenth century. However, within this wetter episode, major droughts occurred in about the 1680s, *c.* 1738–56, in the 1830s and in the early 1900s and 1910s, but at the end of the last century relatively good rains prevailed, and the Sahel's 'normal' rainfall was then perhaps 20–40 per cent greater than today. Most of these episodes are also present in the southern African record.

The results for southern Africa have been summarised for ten of the eleven regions shown in Fig. 9.4. Chronologies of droughts and wetter years for each of these regions, including the earliest quantitative rainfall records, are shown in Fig. 9.5; the top five relate to the Kalahari and nearby areas, while northeastern-Cape/Griqualand and central Cape chronologies relate to the Karroo and other semi-arid regions. In

Fig. 9.3. Summary of long-term rainfall changes in the sub-Saharan Sahelo–Soudan zones.

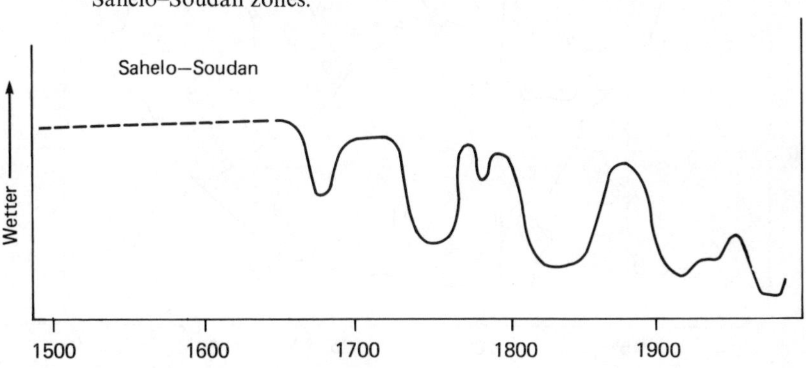

addition to being a catalogue of historical events, these chronologies also underline the anomalous conditions of the 1830s, of the late nineteenth century and of the early twentieth century. These anomalous periods will be discussed later in greater detail. Other results include reconstructions (still tentative) of the historical fluctuations of Lakes Ngami, Rukwa and Nyasa, which are here compared with lakes in other parts of the continent (Fig. 9.6).

In addition to these chronologies, there exists much general information not yet thoroughly analysed. (Nicholson, 1980b). A preliminary analysis suggests the following:

1. In earlier centuries (i.e., pre-1850) snow and freezing temperatures were more common in South Africa. Deep and lasting snow persisted through the winter on the Roggeveld, and was frequent and lasting on higher parts of central Namibia.

2. The area just north of the Cape was within the winter rainfall province of the Cape before about 1800, but after that time the summer rainfall regime prevailing further north extended southwards (to, for example, the Roggeveld).

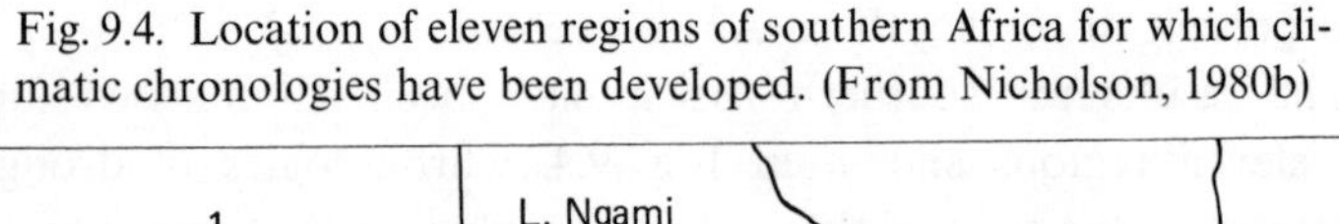

Fig. 9.4. Location of eleven regions of southern Africa for which climatic chronologies have been developed. (From Nicholson, 1980b)

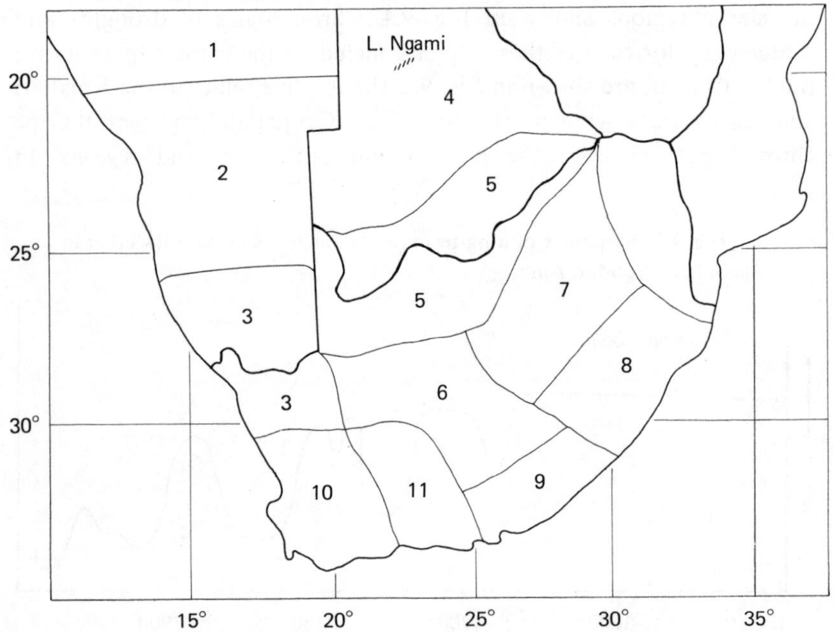

3. The wind and rainfall regime at the Cape has been relatively constant throughout the last three centuries.

4. Landscape descriptions and faunal evidence suggest that before about 1800 parts of the Karroo, the dry northwestern Cape, much of the Kalahari and other dry regions of South Africa were considerably less arid than subsequently.

Early nineteenth century

Consider now the 1830s. In southern Africa there are numerous reports of a continual decline in rainfall and progressive desiccation of lakes, rivers and springs from about 1800 to 1830, and numerous droughts in the 1820s and 1830s. This complements evidence based on landscape descriptions and fauna summarised by Schwarz (1920). In 1797, for example, Barrow (cited by Schwarz) described the Beer Vley, north of Willowmore, as a periodical stream running through a vast

Fig. 9.5. Climatic chronologies for ten regions of southern Africa. (From Nicholson, 1980b)

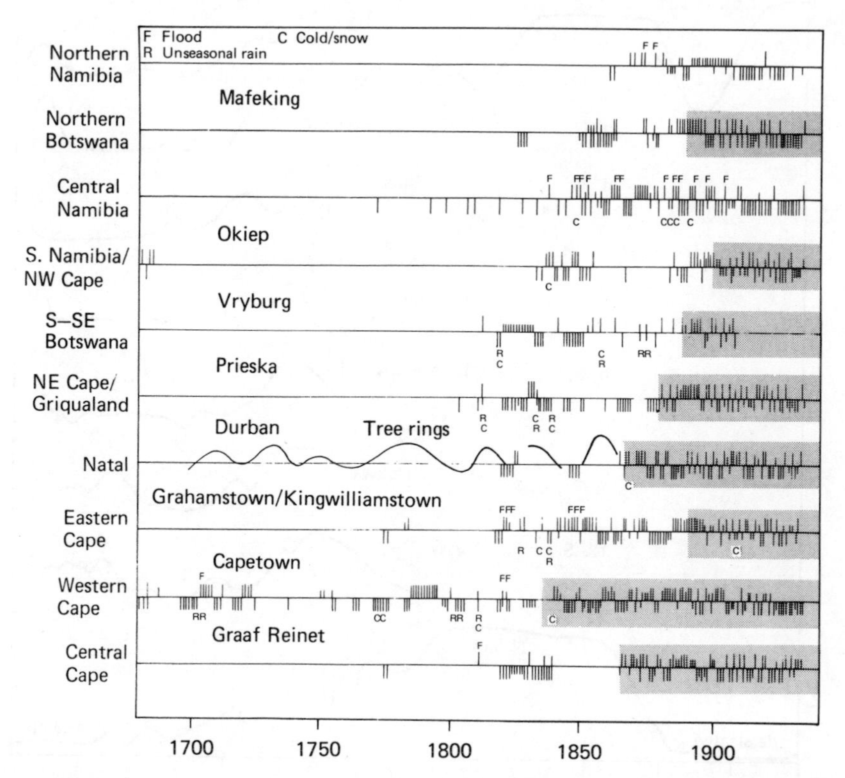

plain of rushy grasses, swamps, springs and periodical rivers. In Schwarz's time, the vley was completely dry, its surroundings dry, barren and desolate. Similarly, le Vaillant (again, from Schwarz) in about 1780, described a fauna reminiscent of East Africa in the Great Fish River area near Craddock; lions, buffalo and numerous types of antelope roamed the plains, while hippopotomi thrived in the rivers there. In Schwarz's time, as well as today, this was typically barren Karroo landscape.

Schwarz (1920) claims that a major change took place toward 1820, when Lake Ngami dried up and many marginal areas, where woods previously lined river banks turned into arid Karroo. A number of reports substantiate Schwarz's claim (Nicholson, 1980b). Early in the nineteenth century Lake Ngami was so deep that waves formed which

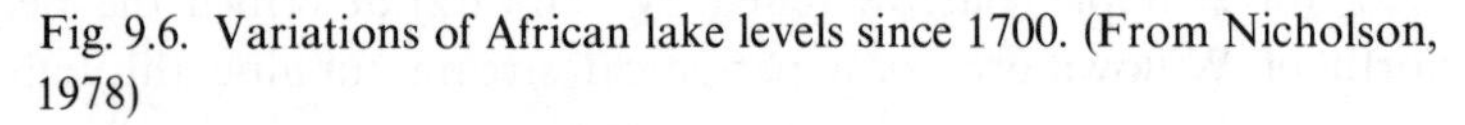
Fig. 9.6. Variations of African lake levels since 1700. (From Nicholson, 1978)

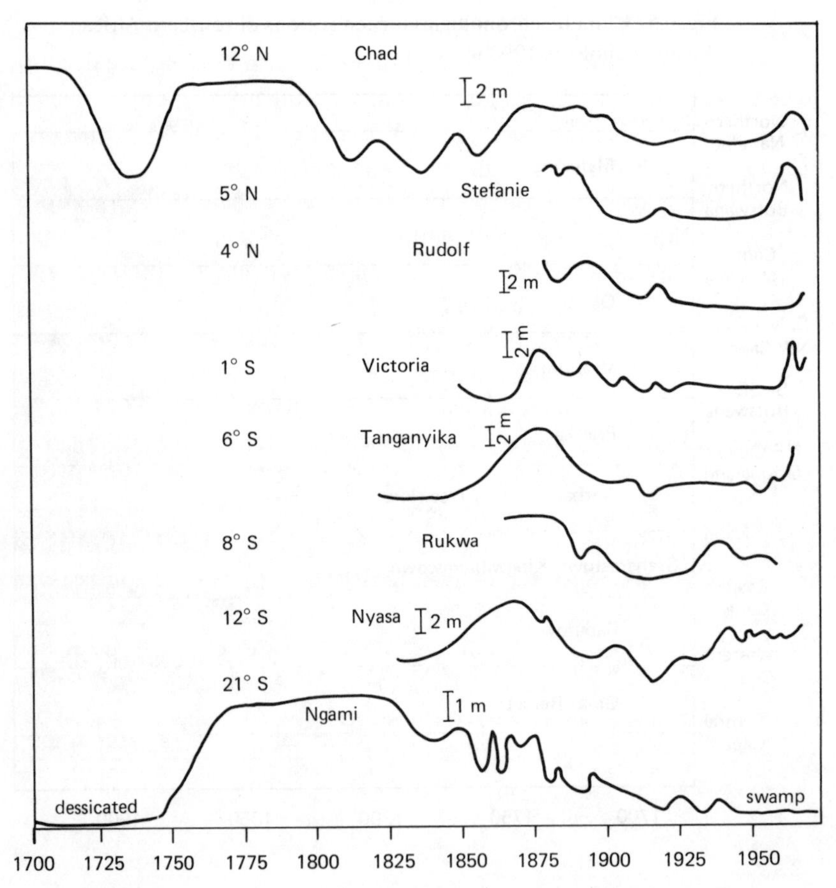

were powerful enough to throw hippopotomi and fish to shore. The Shua River had been navigable and the Makarikari pans, today only occasionally flooded, never dried up. After about 1820 the pans received little flood-water from the Botletlie River. Both the Damaras and Namaquas and numerous white settlers claimed toward 1850 that the rains had been more abundant earlier in the century in northern Namibia and Botswana. The situation was similar in the southern Kalahari. Before the 1830s the Matlaring River ran at Kuruman, and rains fell more abundantly; the Kuruman and Maclare Rivers were deep and strong and the former ran across the desert. In southern Namibia, the flow of the spring at Copper Berg continually decreased from *c.* 1800 to *c.* 1830. These descriptions are summarised in Fig. 9.7. The suggested desiccation toward the 1830s is evident also in the chronologies in Fig. 9.5. The central Cape, Natal, the northeastern Cape and Griqualand, south and southeastern Botswana and the northwestern Cape and southern Namibia suffered from frequent and severe droughts.

Reviewing these data in a broader African context, we see that this period is climatically anomalous and a similar arid episode occurred in most areas. The levels of Lakes Nyasa, Tanganyika, and probably Victoria (Fig. 9.6) were very low, while Lake Chad was partially desiccated. Both the flood and annual minimum levels of the Nile were very low, indicating decreased rainfall in large areas of equatorial East Africa and in the Ethiopian highlands. Drought affected a number of areas in southern and eastern Africa, and devastated parts of the Sahel (Nicholson, 1976, 1978). Chad and parts of the Sudan experienced about twelve years of drought between about 1828 and 1839. Other areas affected included Senegal and parts of Mauritania, Mali and Upper Volta. The information for all areas is summarised in Fig. 9.8; more detail for areas in the northern hemisphere is found in Nicholson (1976, 1978).

In summary, a pattern of drought and below-average rainfall covers most of the continent while increased rainfall is restricted primarily to the north African coast, the Guinea Coast (Accra), and the southeast of South Africa.

There is evidence that conditions similar to those of the 1820s and 1830s probably prevailed until about 1850. Apparently synchronous anomalous conditions occurred in other parts of the world. While drought was frequent over much of Africa, rapidly variable conditions, severe winters, glacial advances and aberrant circulation patterns occurred in Europe and severe and highly variable winters plagued parts of North America (Lamb, 1968, 1977).

Fig. 9.7. Comparison of climate and environment in southern Africa (*a*) *c*. 1790–1810 and (*b*) *c*. 1820–40. (From Nicholson, 1980b)

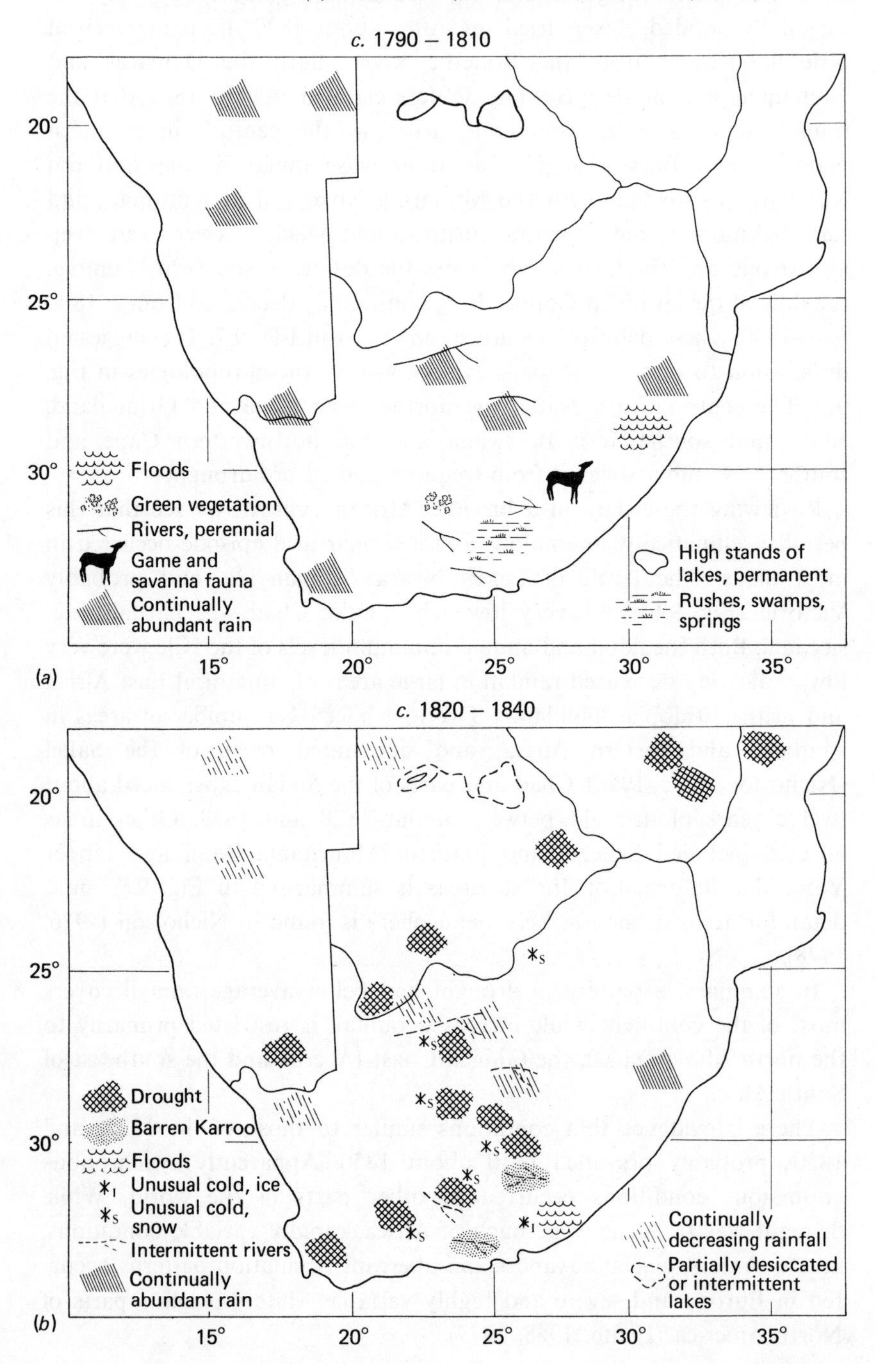

Late nineteenth century

The situation in Africa changed rapidly in the late nineteenth century. Information for this period is plentiful enough for detailed rainfall anomaly maps to be constructed for each five-year period from 1870 to 1900 (Nicholson, 1976). Fig. 9.9 summarises the overall character

Fig. 9.8. African rainfall anomalies *c.* 1820–40; minus signs denote evidence of drier conditions; plus signs denote evidence of above-average rainfall; small circles denote average conditions; circled symbols denote regional integrators, e.g. lakes and rivers. (From Nicholson, 1978)

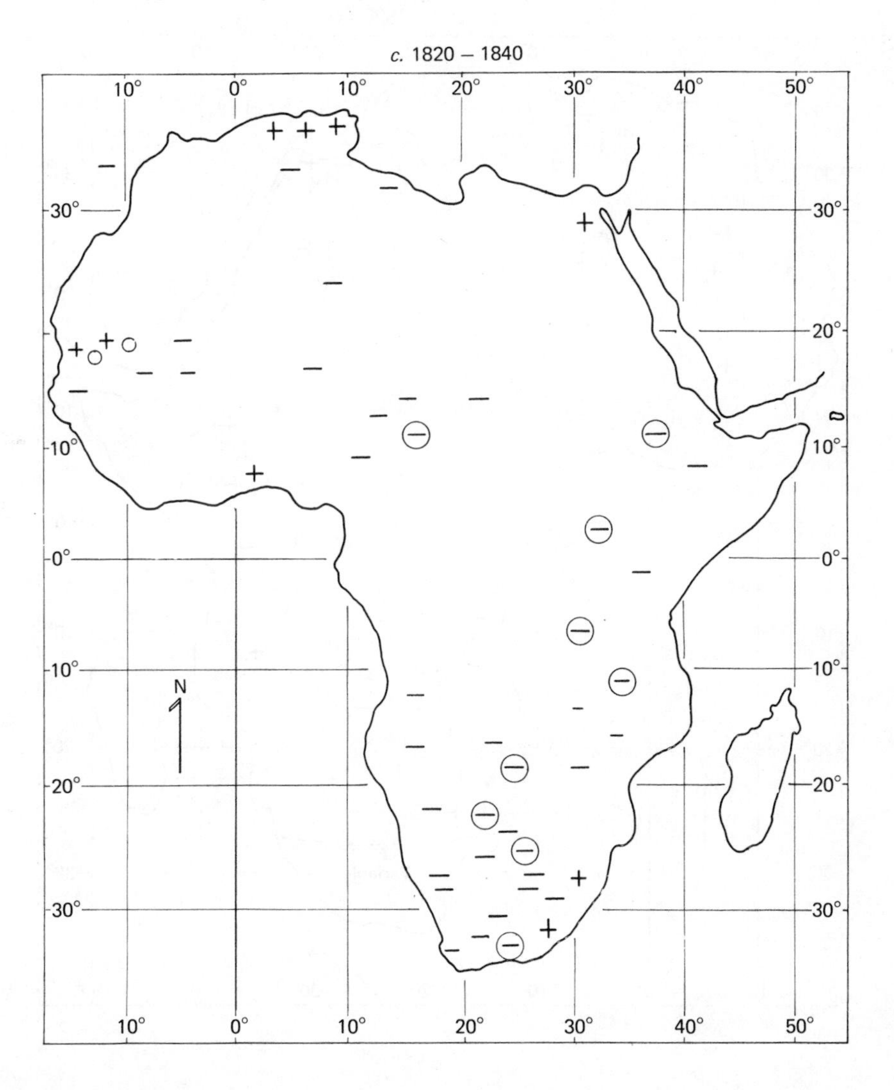

of the period 1870–95 using these maps as a basis. Increased rainfall characterised much of southern Africa, as well as east Africa and both the northern and southern margins of the Sahara. The levels of lakes throughout southern and eastern Africa and Chad (Fig. 9.6), and the discharge of the Nile, Niger and Senegal Rivers, and of streams in

Fig. 9.9. African rainfall anomalies *c.* 1870–95; minus signs denote evidence of drier conditions; plus signs denote evidence of above-average rainfall; small circles denote average conditions. (From Nicholson, 1978)

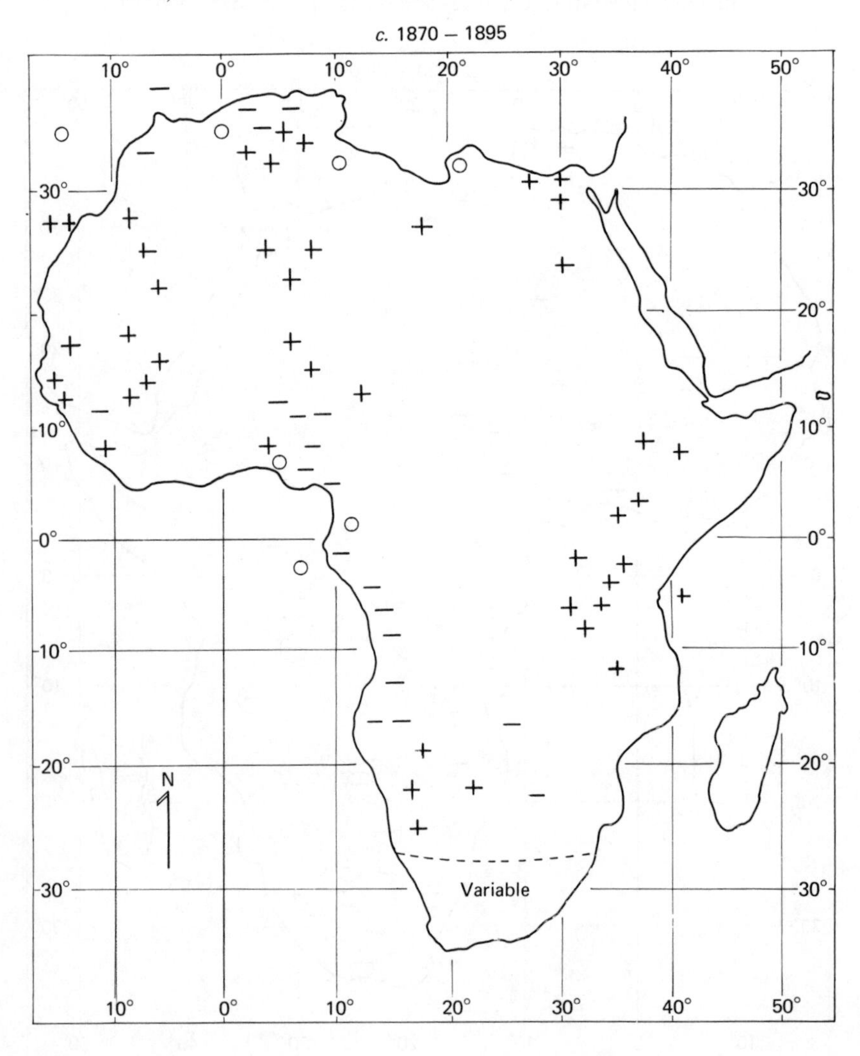

southern Africa, far exceeded their twentieth-century values. Consistently good harvests and sufficient rainfall benefited the inhabitants of the Niger Bend, southern Algeria (Nicholson, 1976, 1978), Namibia, Botswana, and the areas in South Africa, including Griqualand and most of the Cape Colony (Fig. 9.5). The scattered quantitative rainfall records

Fig. 9.10. African rainfall anomalies *c.* 1895–1920; minus signs denote evidence of drier conditions; plus signs denote evidence of above-average rainfall; small circles denote average conditions. (From Nicholson, 1978)

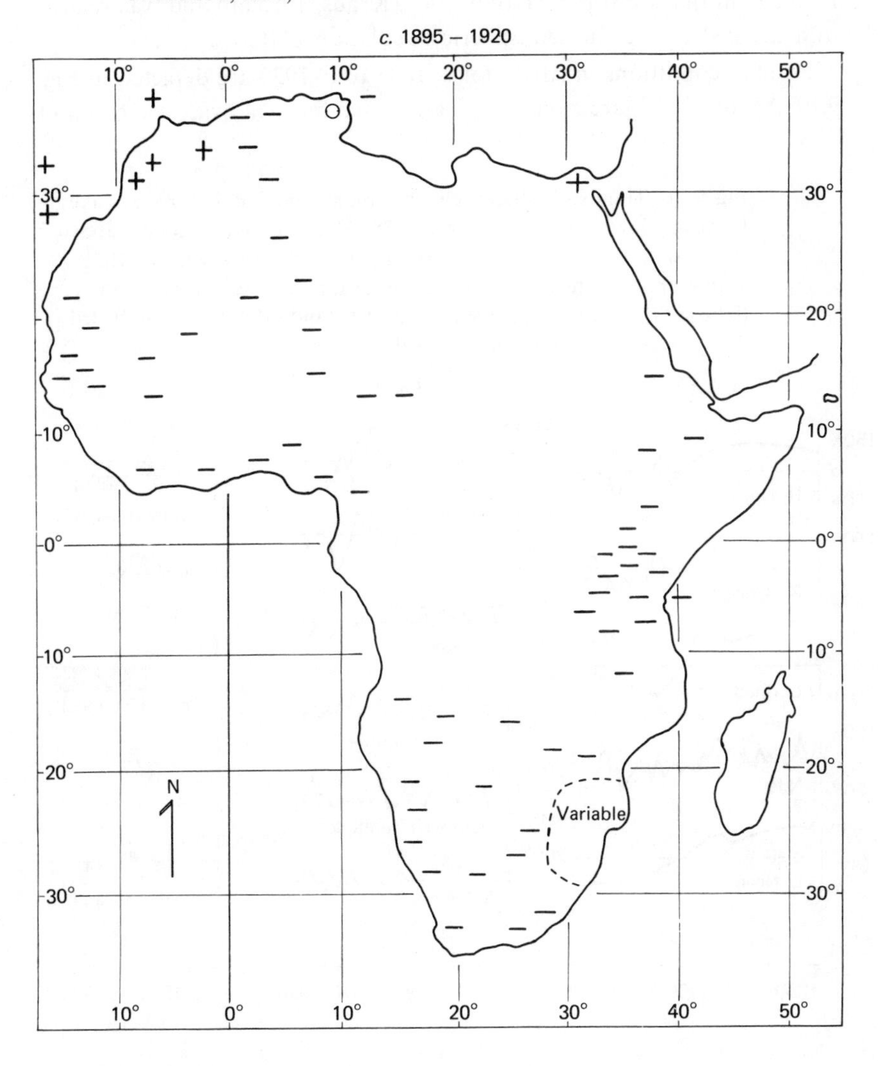

provide supporting evidence. Decreased rainfall characterised only the north African coast and the equatorial regions (*c*. 10 °N to 20 °S) of the more western parts of the continent.

The twentieth century

After 1895 the situation rapidly reversed; a general and progressive trend to drier conditions culminated in severe drought toward 1913/14 in most tropical and subtropical regions of Africa. This occurred synchronously with a major global change which was particularly pronounced in the subtropics (Lamb, 1966, Kraus, 1955a,b) and was related to a marked change in hemispheric circulation patterns.

Rainfall conditions in Africa for *c*. 1895 to *c*. 1920 are depicted in Fig. 9.10. Again, this figure summarises a massive body of evidence, much of

Fig. 9.11. Trends of African climatic indicators (rainfall, rivers, lakes, harvests), 1880–1920 (Nicholson, 1978). Harvest quality: good = above the axis, poor = below the axis. Rainfall (*r*) and river discharge (*d*): graphs show changes expressed either as a percentage of the mean (labelled $\bar{r}$ or $\bar{d}$) or as a percentage of the standard deviation (labelled $\Delta r/\sigma_r$). Lake levels: annual mean values in metres.

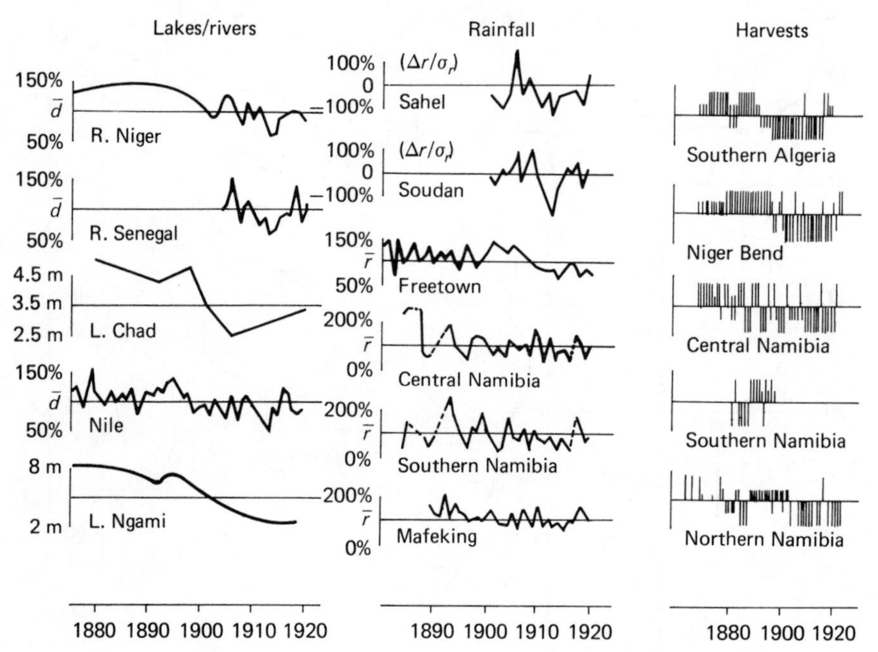

which is still unpublished; the anomaly pattern in the map is representative of the general character of these three decades although short-term regional variations occur. Droughts ended the periods of good harvests in Algeria, west Africa, and Namibia (Fig. 9.5); rainfall decreased markedly in Botswana and numerous areas of South Africa (Fig. 9.5). Lake levels fell sharply and progressively (Fig. 9.6) throughout the continent, as did the discharge of rivers: the Senegal, Niger, Nile and streams in Namibia are some examples. This desiccation was most severe toward 1913/14.

A comparison of a few quantitative records (e.g. Fig. 9.11) for the two periods *c.* 1880–95 and 1910–40 underlines the magnitude of the change between these periods. Mean rainfall for Freetown in Sierra Leone was 30–35 per cent greater for the earlier period, the mean depth of Lake Chad was about 50 per cent greater and the discharge of the Nile was about 35 per cent greater for the earlier period.

Fig. 9.12. African rainfall anomalies 1968–73: blank areas, no data available; black areas, well above normal rainfall; heavy stippling, above normal rainfall; light stippling, normal rainfall; horizontal shading, below normal rainfall; cross-hatching, well below normal rainfall. Inset map shows the broad features of the rainfall departure pattern (stippling indicates no data).

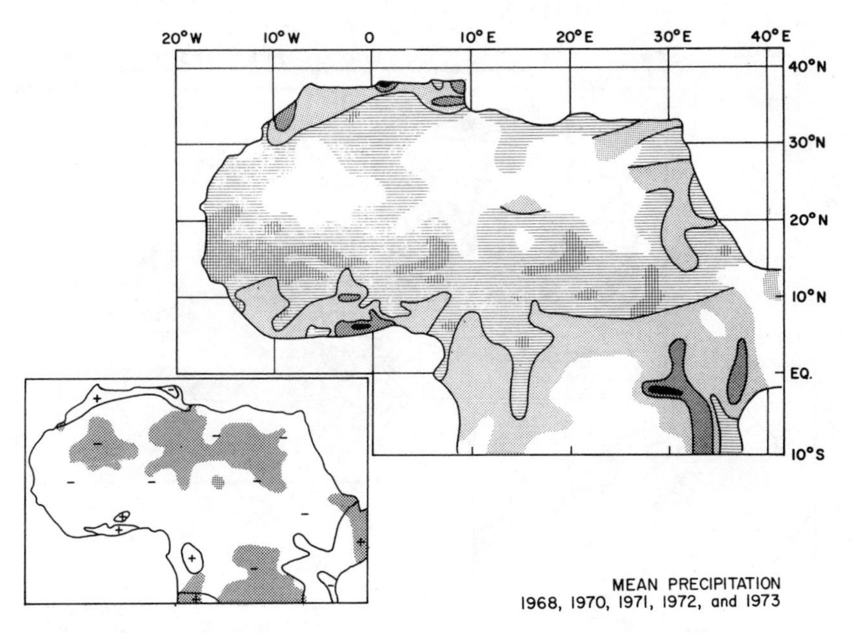

Conclusion

In conclusion and summary I would like to focus on two of the periods discussed, the 1830s and the period 1870–95, representing two realistic climatic extremes. Parallel changes seem to occur in the sub-Sahara or Sahel and southern Africa with respect to anomalous climatic episodes and fluctuations in historical times: the nineteenth century provides two dramatic illustrations of this. This synchroneity suggests an important teleconnection between the semi-arid subtropics of the two hemispheres, a result which has important implications concerning anomalous movements of the ITCZ. The agreement of the sub-Saharan and southern African chronologies, as well as the spatially coherent and physically meaningful anomaly patterns described, supports the validity of the climatic reconstructions presented in this chapter. The physically realistic nature of the rainfall anomalies is indicated by comparison of

Fig. 9.13. African rainfall anomalies characteristic of the 1950s (composite of 1946, 1950, 1952, 1956 and 1958): blank areas, no data available; black areas, well above normal rainfall; heavy stippling, above normal rainfall; light stippling, normal rainfall; horizontal shading, below normal rainfall; cross-hatching, well below normal rainfall. Inset map shows the broad features of the rainfall departure pattern (stippling indicates no data).

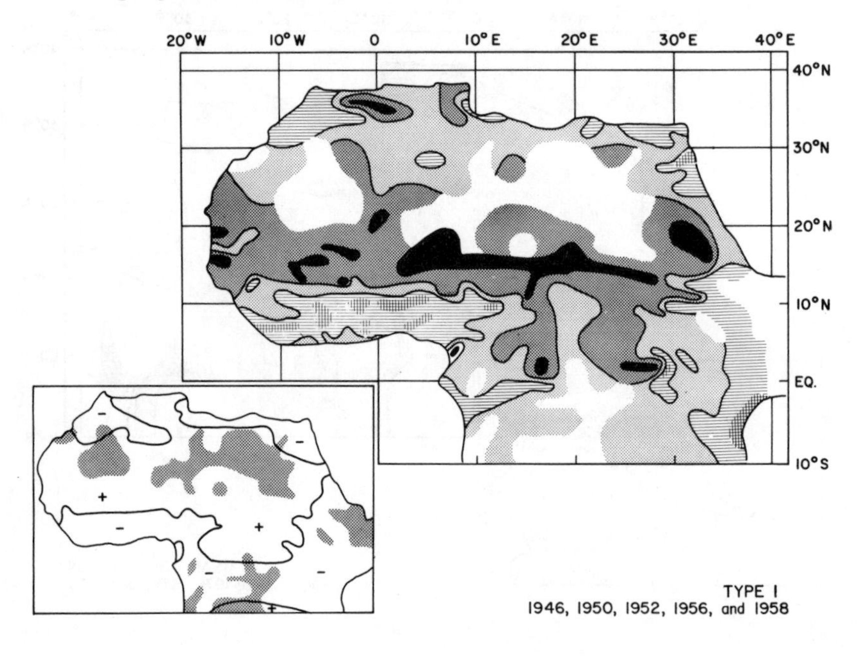

the patterns with anomaly patterns based on instrumental data, and is confirmed by theoretical considerations of displacement of the ITCZ. In areas north of 10 °S, the 1830s pattern is analogous to the recent drought years *c.* 1968–73 (Fig. 9.12), and the 1870–95 periods finds its analog, though an incomplete one, in the 1950s (Fig. 9.13), when Sahelian rainfall was as much as 40 per cent above the normal (Nicholson, 1979b). It will be important to find out if these analogies hold for southern Africa.

Finally, both of these nineteenth-century periods resemble two climatically important periods of Africa's geological past. Continental desiccation, like that of the 1830s, occurred at the peak of the last Ice Age,

Fig. 9.14. Summary of African climatic conditions *c.* 18 000 BP. (After Nicholson, 1978)

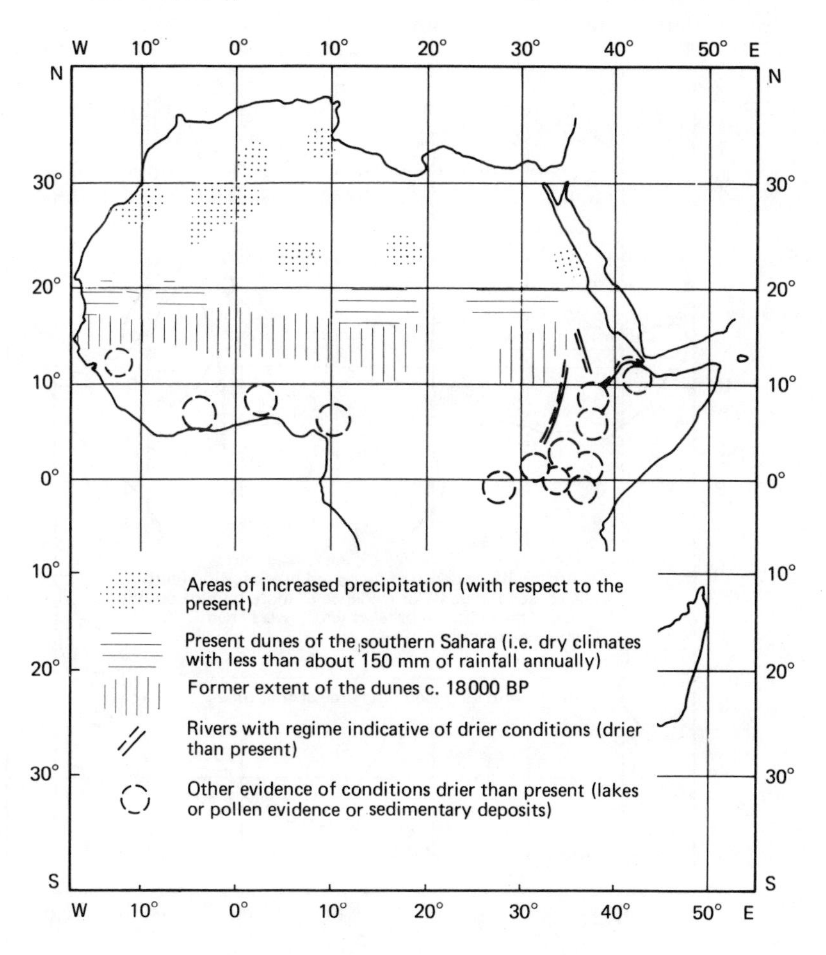

c. 18 000 BP. At this time the Sahara expanded southward as increased aridity prevailed throughout the African tropics (Fig. 9.14; see also Nicholson & Flohn, 1980) and the semi-arid Kalahari (Heine, 1978) penetrated equatorwards. During the Neolithic or Atlantic period, 5000–6000 years ago, the Sahara contracted along both its tropical and temperate margins, giving a rainfall pattern similar to that for the late nineteenth century (Fig. 9.15; see also Nicholson & Flohn, 1980). Civilisations flourished throughout the Sahara (Gabriel, 1977) during this early period. At the same time, more humid conditions than at present probably prevailed throughout much of eastern and southern

Fig. 9.15. Summary of African climatic conditions *c.* 5000 BP (After Nicholson, 1978)

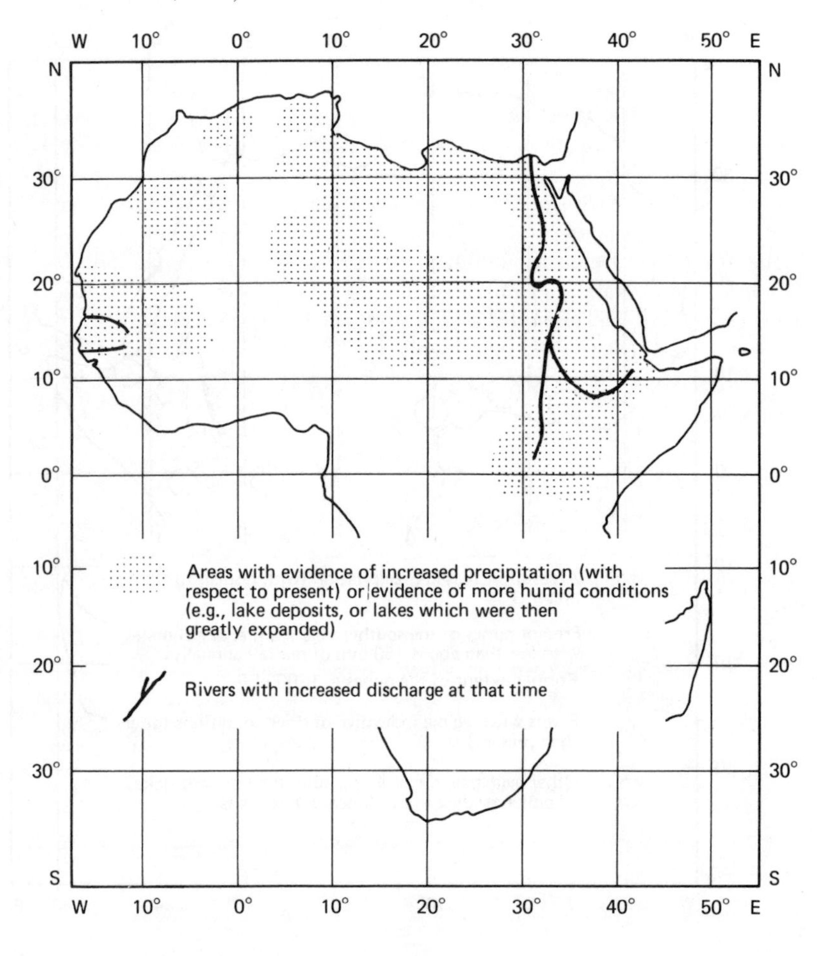

Africa (Nicholson & Flohn, 1980), as they did also in the late nineteenth century. Thus, study of the two extreme historical episodes might greatly increase our understanding of both African and global climatic fluctuations on all time scales.

Acknowledgement

This work was funded in part by the Atmospheric Research Section of the National Science Foundation under Grant ATM77-21547 to the University of Virginia.

References

Curtin, P. D. (1975). *Economic Change in Pre-Colonial Africa: Supplementary Evidence.* Madison: University of Wisconsin Press. 307 pp.

Daveau, S. (1965). Dunes ravinées, et dépôts du Quaternaire récent dans le Sahel mauritanien. *Révue Géographie d'Afrique occidentale-Dakar*, **1**, 7–48.

Fahn, A., Wachs, N. & Ginzburg, C. (1963). Dendrochronological studies in the Negev. *Israel Exploration Journal*, **13**, 291–9.

Gabriel, B. (1977). Zum oekologischen Wandel in der oestlichen Zentralsahara. *Berliner Geographische Abhandlungen*, **27**.

Ginestous, G. (1927). Le chêne zeen d'ain Draham. *Bulletin. Direction générale de l'Agriculture Commerce et Colonisation*, pp. 3–12. Tunis.

Hall, M. (1976). Dendroclimatology, rainfall and human adaptation in the Later Iron Age of Natal and Zululand. *Annals Natal Museum*, **22**, 693–703.

Heine, K. (1978). Jungquartäre Pluviale and Interpluviale in der Kalahari (südliches Afrika). *Palaeoecology of Africa*, vol. 10, ed. E. M. van Zinderen Bakker and J. A. Coetzee, pp. 31–40. Rotterdam: Balkema.

Kraus, E. B. (1955a). Secular changes of tropical rainfall regimes. *Quarterly Journal of the Royal Meteorological Society*, **81**, 198–210.

Kraus, E. B. (1955b). Secular changes of east-coast rainfall regimes. *Quarterly Journal of the Royal Meteorological Society*, **81**, 430–9.

Lamb, H. H. (1966). Climate in the 1960's. *Geographical Journal*, **132**, 183–212.

Lamb, H. H. (1968). *The Changing Climate.* London: Methuen, 236 pp.

Lamb, H. H. (1977). *Climate: Present, Past and Future*, vol. 2. London: Methuen. 835 pp.

Lilly, M. A. (1977). An assessment of the dendrochronological potential of indigenous species of trees in South Africa. *Environmental Studies*, Occasional Paper 18, Department of Geography and Environmental Studies, University of the Witwatersrand, Johannesburg. 96 pp.

Maley, J. (1973). Mécanisme des changements climatiques aux basses latitudes. *Palaeogeography, Palaeoclimatology, Palaeoecology*, **14**, 193–227.

Munson, P. J. (1971). The Tichitt Tradition: A Late Prehistoric Occupation of the Southwestern Sahara. Unpublished PhD dissertation, University of Illinois, Urbana. 393 pp.

Nicholson, S. E. (1976). A Climatic Chronology for Africa: Synthesis of

Geological, Historical, and Meteorological Information and Data. Unpublished PhD dissertation, University of Wisconsin, Madison. 324 pp.

Nicholson, S. E. (1978). Climatic variations in the Sahel and other African regions during the past five centuries. *Journal of Arid Environments*, **1**, 3–24.

Nicholson, S. E. (1979a). The methodology of historical climate reconstruction and its application to Africa. *Journal of African History*, **20**, 31–49.

Nicholson, S. E. (1979b). Revised rainfall series for the West African subtropics. *Monthly Weather Review*, **107**, 620–3.

Nicholson, S. E. (1980a). Saharan climates in historic times. In *The Sahara and the Nile*, ed. M. A. J. Williams & H. Faure. Rotterdam: Balkema.

Nicholson, S. E. (1980b). Study of Environmental and Climatic Changes in Africa during the Past Five Centuries, NSF Final Report, ATM 77-21547. *c.* 200 pp.

Nicholson, S. E. & Flohn, H. (1980). African environmental and climatic changes and the general atmospheric circulation in Late Pleistocene and Holocene. *Climatic Change*, **2**, 313–48.

Schwarz, E. H. L. (1920). *The Kalahari or Thirstland Redemption.* Cape Town.

Servant, M. (1973). Séquences continentales et variations climatiques: évolution du bassin du Tchad du Cenozoïque supérieur. Unpublished PhD thesis, Université Paris VI. 348 pp.

Toupet, C. (1973). L'évolution du climat de la Mauritanie du Moyen-Age jusqu'à nos jours. Colloque de Nouakchott sur les problèms de la désertification. *c.* 20 pp.

Waisel, Y. & Lipschitz, N. (1968). Dendrochronological studies in Israel: *Juniperus phoenica* of north and central Sinai. *La-Yaaran*, 18, 1–22.

Walter, H. (1936). Die Periodizitaet von Trocken und Regenzeiten in Deutsch-Suedwestafrika auf Grund von Jahresringmessungen an Baeumen. *Berichten der Deutschen Botanischen Gesellschaft*, **54**, 608–11.

10

Droughts and floods in China, 1470–1979

WANG SHAO-WU AND ZHAO ZONG-CI

Abstract

Documentary historical data from yearbooks, diaries and gazettes have been used to construct drought and flood charts for China for the period AD 1470–1979. The characteristics of spatial and temporal variations of drought and flood have been examined by means of empirical orthogonal function analysis and power spectrum analysis. The eigenvectors based on historical data are similar to those obtained from an analysis of instrumental records for the period 1951–74. This provides strong support for the reliability of the historical data and for the methods of interpretation of these data. Six distinct types of drought and flood distribution can be distinguished, and a chronology of types from 1470 to 1979 is given. The relationship between the variations of drought and flood and fluctuations of the atmospheric circulation is also discussed.

Introduction

In recent years abnormal climate events have caused widespread concern among meteorologists and climatologists. At the present time drought and flood have such serious influences on agriculture in China that it is of prime importance to look for regularities in their variations and to understand the causes and the mechanism of their occurrence. Meteorological data for the last 30 years or so are not sufficient for investigations of long-term variations of climate. Fortunately, in China, there are numerous historical writings dating from as early as 2000 years ago. These works (yearbooks, diaries, county annals, gazettes, Ming and Qing records, etc.) frequently refer to abnormal events and disasters. These descriptions provide important information about climatic variations.

During 1975 and 1977, studies on droughts and floods in the past 500 years were made by a co-operative team consisting of members from the Research Institute of the Central Meteorological Service, provincial

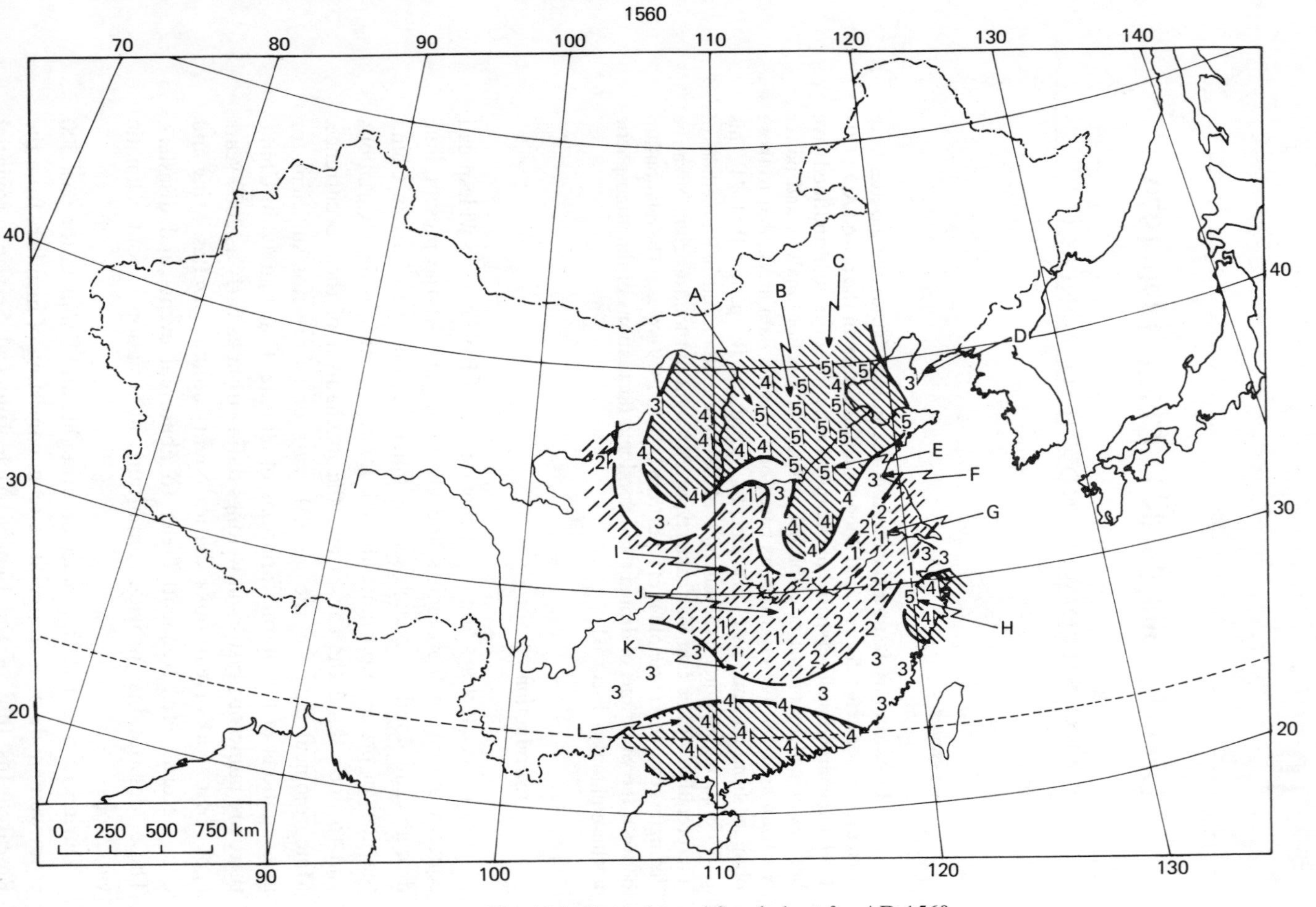

Fig. 10.1 Drought and flood chart for AD 1560.

Table 10.1. *Some data for AD 1560*

	Location	Description	Grade
A	Datong	Men eating men	5
B	Shijiazhuang	No rain in spring and summer	5
C	Peking	Locust pest	5
D	Dalian	A rich harvest	3
E	Hezhei	Famine	5
F	Linyi	No historical records	3
G	Nanking	Flood	1
H	Jinhua	No rain in June and July	5
I	Yichang	The town overwhelmed by flood	1
J	Yueyang	Flood	1
K	Shaoyang	The town flooded	1
L	Liuzhou	Drought in late summer	4

Meteorological Services, the University of Nanking and the University of Peking. The outcome of these studies is the reconstruction of a series of drought and flood charts from 1470 to 1977 inclusive (Central Meteorological Institute, *et al.* 1981). This chapter presents a preliminary analysis of these charts.

Drought and flood data

The original data, which consist of numerous written descriptions, are not immediately suitable for quantitative analysis. They have, therefore, been classified into five grades: very wet, wet, normal, dry and very dry, denoted by the numbers 1, 2, 3, 4 and 5 respectively. As an example, the spatial distribution of grades in the year AD 1560 is given in Fig. 10.1. In Table 10.1 a brief description of the data for 12 stations for the year 1560 is given: their positions are shown in Fig. 10.1 by the letters A, B, C, . . . , L respectively. Each station in the figure represents a region including dozens of counties. It is usually possible, therefore, to find some information on the occurrence of drought or flood for each station and for each year. Occasionally there is an abundance of data, although this does not happen for early years. Overall it has been possible to build up a series of drought and flood grades for 118 stations covering almost the whole east part of China from 1470 to 1977 with only a few intermissions.

Spatial patterns

In studying the spatial patterns of droughts and floods over China, an empirical orthogonal function analysis has been carried out by using data from 25 stations for 1470–1977. Fig. 10.2(*a*)–(*c*) shows the first three eigenvectors expressed by $X1$, $X2$ and $X3$. The positive areas are shaded.

Fig. 10.2(*a*) shows the simultaneous occurrence of drought or flood over the Yang-tze and the Yellow Rivers: i.e. both regions experience the same extreme in the same year. The second eigenvector (Fig. 10.2(*b*)) shows reversed character in the distribution of droughts and floods: i.e. there was drought over the Yellow River and flood over the Yang-tze, or vice versa. There are three belts of drought or flood in the third eigenvector (Fig. 10.2(*c*)) with drought or flood occurring simultaneously to the south of the Yang-tze River and north of the Yellow River. This analysis has been performed for different time periods. Table 10.2 shows how the percentage variance accounted for by the first ten eigenvectors has changed with time.

The first three eigenvectors combine to yield a picture of zonal distribution of drought and flood. The first eigenvector may be related to low-frequency variations in the Southern Oscillation Index (SOI) which are linked with the intensity of the western Pacific subtropical high. The second eigenvector also appears to be linked to the position of the subtropical high: specifically whether or not it extended north of the Yang-tze River. The third eigenvector may be related to high-frequency variations in the East Asian Circulation Index (EAI). These relationships will be discussed further below.

The high-order eigenvectors (figures omitted) give a rather meridional picture. The fourth eigenvector shows a reversed feature between the east and west of North China, while the fifth eigenvector shows opposite extremes in the eastern and western parts of South China. The former may be associated with the presence or absence of high pressure over the Sea of Japan, while the latter may be associated with the frequency of occurrence of tropical cyclones over the southeast coast of China. The variance accounted for by the higher order eigenvectors is by no means negligible.

One of the reasons for the subdivision into 100-year periods was to investigate the stability of the eigenvector patterns. The patterns of the first five eigenvectors for any one 100-year period were very similar to those for other 100-year periods. The stability of the eigenvector patterns with time, and the fact that they have physically realistic interpretations,

Table 10.2. *Percentage variance accounted for by the first ten eigenvectors*

Period	$X1$	$X2$	$X3$	$X4$	$X5$	$X6$	$X7$	$X8$	$X9$	$X10$	Data source	Sum, $X1$ to $X10$
1471–1570	14.4	13.0	8.1	6.7	6.3	5.8	5.0	4.6	4.2	3.7	Historical data	71.8
1571–1670	19.2	11.6	7.6	6.7	6.5	5.3	4.0	4.0	3.7	3.5	Historical data	72.1
1671–1770	13.8	10.5	8.1	7.1	6.6	6.4	5.4	4.8	4.2	4.1	Historical data	71.0
1771–1870	14.8	13.4	8.0	7.7	5.9	4.9	4.8	4.2	4.0	3.9	Historical data	71.6
1871–1970	17.0	13.0	9.2	6.8	5.4	5.2	4.6	4.5	4.3	3.0	Historical data	73.0
1470–1977	14.8	11.0	7.2	6.2	5.2	5.2	4.2	4.0	3.8	3.7	Historical data	65.3
1951–1974	18.2	12.5	11.0	8.3	6.5	5.6	4.6	3.9	3.5	3.2	Instrumental data	77.3

Fig. 10.2. Eigenvectors of drought and flood using historical data from AD 1470 to AD 1977.

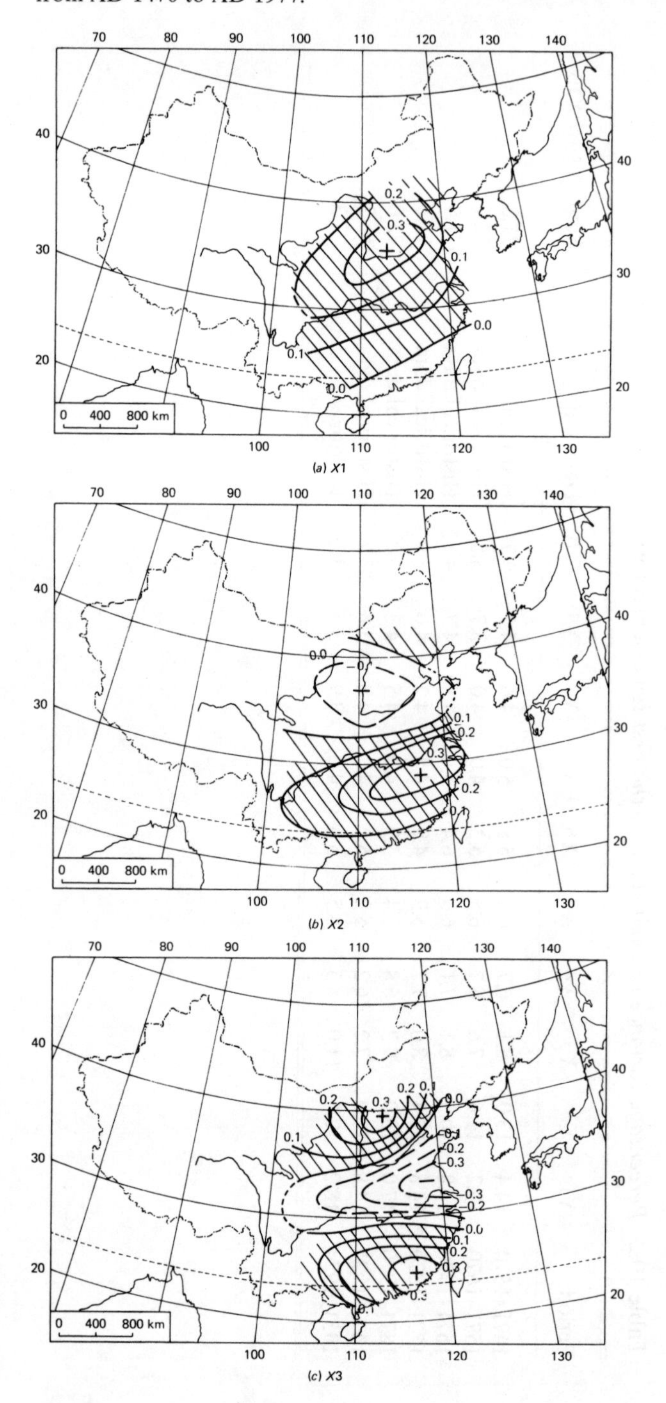

Fig. 10.3. Eigenvectors of rainfall using instrumental data from AD 1951 to AD 1974.

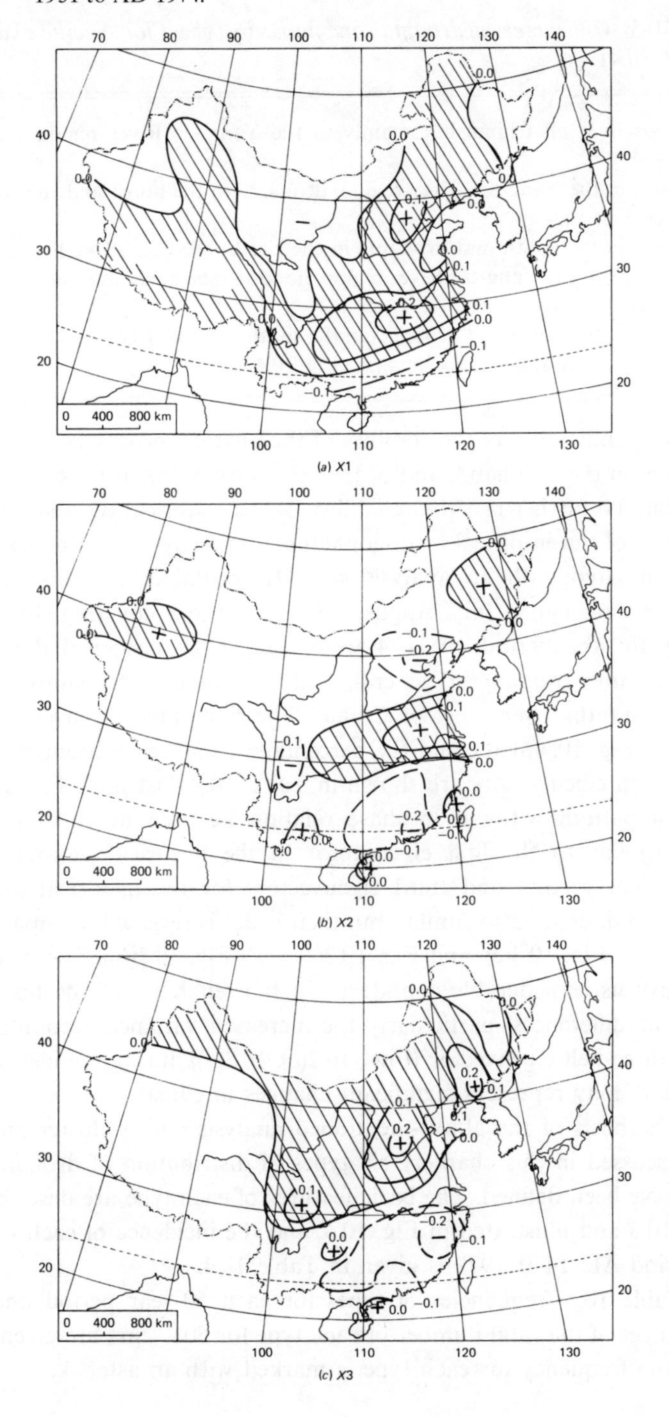

Table 10.3. *Characters of droughts and floods by types (for specific examples see Fig. 10.4)*

1a. Floods over all China, but mainly in the Yang-tze River region (see Fig. 10.4(*a*))
1b. Floods in the Yang-tze River region, droughts to the north and the south of it (see Fig. 10.4(*b*))
2. Floods in the south and droughts in the north (see Fig. 10.4(*c*))
3. Droughts in the Yang-tze River region, floods to the north and the south of it (see Fig. 10.4(*e*))
4. Floods in the north and droughts in the south (see Fig. 10.4(*f*)).
5. Droughts over almost all China (see Fig. 10.4(*d*)).

is a strong indication of the stability of the physical processes of climatic variation on the one hand, and of the reliability of the historical data on the other. To further test the reliability of the historical reconstructions, the results of the empirical orthogonal function analysis of historical data were compared with an analysis of instrumental data. Summer precipitation data (June, July, August) for 100 stations covering the whole country for the period 1951–74 were used. (The historical data also reflect summer rainfall in general, and rainfall occurs mainly in the summer months over most of China.) The first three eigenvectors are shown in Fig. 10.3(*a*)–(*c*), and the percentage variances accounted for by the first ten eigenvectors are shown in Table 10.2 (last line). Fig. 10.3(*a*) shows a pattern with an in-phase relation between north and south China similar to the first eigenvector in the historical analysis (Fig. 10.2(*a*)). The second and third eigenvectors for the historical and instrumental data are also similar, but their order is reversed (compare Fig. 10.2(*b*) with Fig. 10.3(*c*), and Fig. 10.2(*c*) with Fig. 10.3(*b*)). These similarities give us considerable confidence in the reliability of the historical data. The differences, particularly the increased variance accounted for by the three-belt eigenvector (Figs. 10.2(*c*), 10.3(*b*)) during the instrumental period, may represent significant changes in climate.

On the basis of the above-mentioned analysis and a cluster analysis (not discussed in this chapter), six types of distribution of drought and flood have been defined. The basic features of each type are described in Table 10.3 and illustrated in Fig. 10.4, and the incidence of each type in the period AD 1470–1979 is given in Table 10.4.

In Table 10.5 frequencies of types for each 50-year period and the percentages of the total number of each type for 508 years are given. The maximum frequency for each type is marked with an asterisk.

Table 10.4. *Chronology of types for AD 1470–1979*

	0	1	2	3	4	5	6	7	8	9		0	1	2	3	4	5	6	7	8	9
1470	2	1a	2	2	1a	1a	4	3	3	2	1730	1a	1a	1a	3	1a	3	1a	3	5	4
1480	2	2	3	2	2*	2	1b	5	5	3	1740	3	1b	1b	1b	3	4	2	4	5	1a
1490	2	1b	2	2	1a	2	1b	2	1b	1b	1750	3	4	3	4	1a	1a	1a	4	1b	2
1500	1b	1a	1a	3	2	2	1b	5	3	5	1760	1b	1a*	1b	1b	2	2	1a	4	3	1b
1510	1a	1a	5	5	4	1b	2	1a	1a	1a	1770	3	3	3	1a	2	3	1a	1b	5	4
1520	3	2	2	3	2	5	4	2	5	5	1780	3	4	3	2	2	5	5	1b	1b	4
1530	3	1b	3	2	3	3	4	1a	4	2	1790	4	1a	2	2	2	1a	3	3	4	1a
1540	2	5	3	4	4	5	3	3	1a	3	1800	1a	1a	5	1b	2	1b	3	5	4	3
1550	1b	3	1a	3	3	1b	3	1a	1b	3	1810	4	2	2	2	5	4	4	2	3	3
1560	1b*	1b	1a	1a	1a	2	1b	1a	2	1a	1820	4	3	4	1a	3	2	1b	1b	4	3
1570	1a	1a	2	2	1a	1a	3	1b	1b	1b	1830	4	1a	1b	1b	2	5	5	2	1b	2
1580	1a	2	2	3	2	5	2	2	5	4	1840	1a	1a	2	4	1a	3	2	2	1a	1a
1590	5	1b	4	4	4	4	2	1a	4	2	1850	1a	4	4	1a	3	3	2	5	2	2
1600	2	1b	1a	4	4	1b	3	1a	1b	2	1860	2	2	2	3	3	1b	1b	5	4	2
1610	1b	2	1b	1a	1a	2	2	2	5	2	1870	1b	3	3	3	3	1b	2	2	2	3
1620	2	3	3	4	4	5	4	2	5	5	1880	2	2	1b	4	4	1a	1a	4	3	1a
1630	3	4	3	2	2	2	5	2	2	2	1890	3	3	3	4	4	4*	4	4	4	2
1640	2	2	5	5	3	4	4	3	4	1b	1900	5	1a	5	1a	3	1b	1a	5	2	1b
1650	2	1a	3*	4	4	1b	1a	4	1a	4	1910	4	1a	1b	3	3	1a	1b	4	3	2
1660	4	2	4	1a	4	5	5	3	1a	1b	1920	2	4	3	4	2	5	1b	5	5	5
1670	1b	3	3	1b	2	1a	1a	3	3	3	1930	5	1b	3	4	5	2	2	3	1b	2
1680	1b	4	1b	1b	4	4	3	1b	3	2	1940	4	5	2	2	3	5	2	3	2	3
1690	2	5	3	3	3	1a	1b	4	1b	3	1950	3	5	2	3	1a	2	4	1b	4	3
1700	4	3	4	4	2	1a	2	3	1b	1a	1960	5	5	1a	4	4	5	5	4	2	1b
1710	1b	5	2	2	2	1a	4	3	1a	2	1970	2	4	5*	3	5	2	3	3	3	1a
1720	2	5	2	2	4	1a	1a	1a	4	4	1980										

* Examples used in Fig. 10.4.

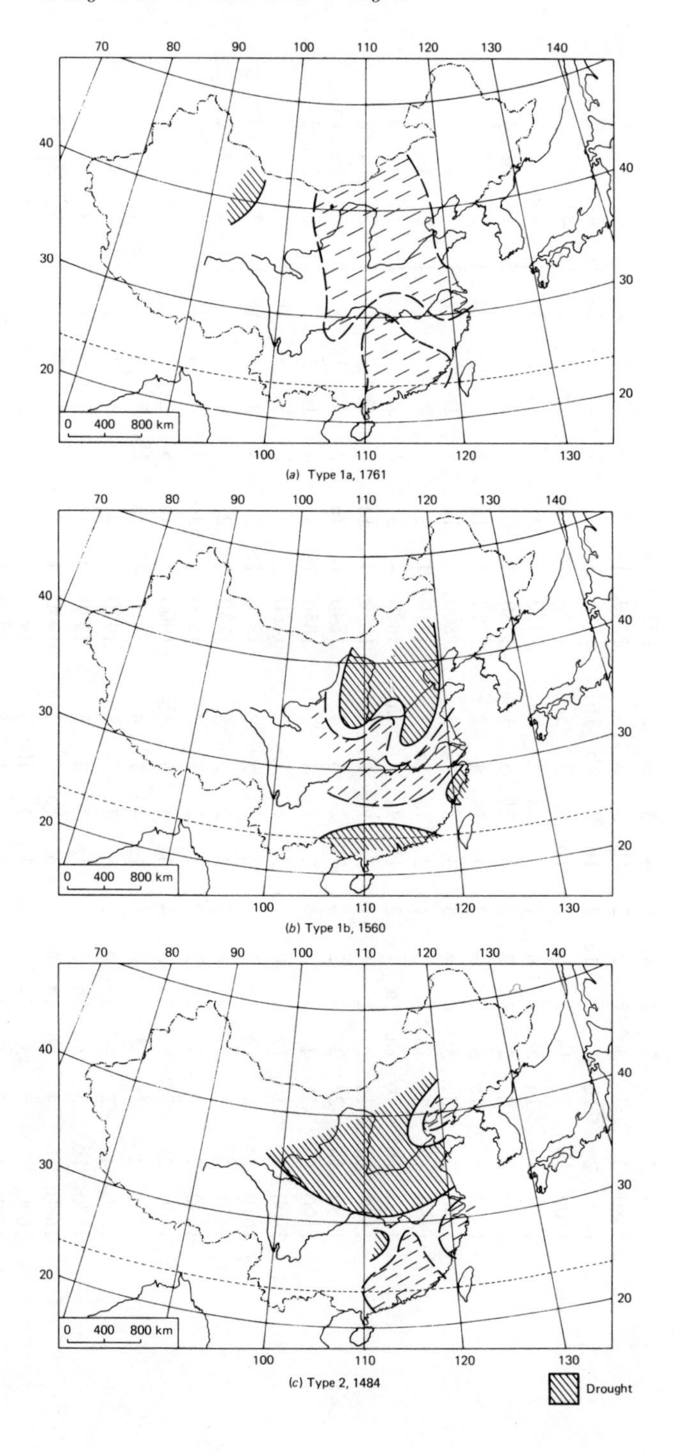

(*a*) Type 1a, 1761

(*b*) Type 1b, 1560

(*c*) Type 2, 1484

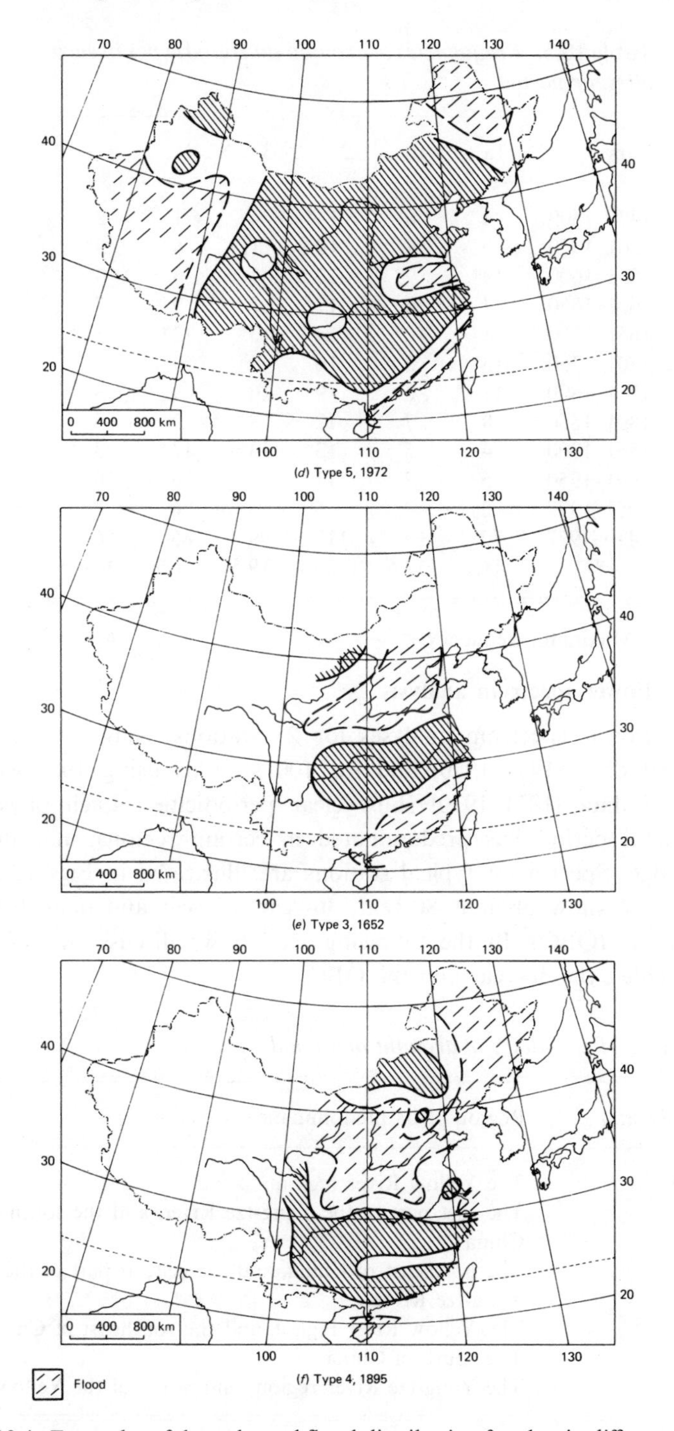

Fig. 10.4. Examples of drought and flood distribution for the six different types.

Table 10.5. *Frequencies of drought and flood types in each 50-year period*

Type	1a	1b	2	3	4	5
1470–1500	4	6	14*	4	1	2
1501–1550	9	4	10	12*	6	9*
1551–1600	13*	9*	12	7	6	3
1601–1650	4	6	17*	7	9	7
1651–1700	8	10*	4	13*	12*	3
1701–1750	13*	5	10	9	9	4
1751–1800	11	8	9	10	9	3
1801–1850	9	7	12	8	9	5
1851–1900	4	5	13*	13*	12*	3
1901–1950	5	7	11	11	6	10*
1951–1977	2	2	5	5	6	7
1470–1977	82	69	117	99	85	56
%	16.1	13.6	23.1	19.5	16.7	11.0

* Maximum frequency.

Power spectrum analysis

Power spectrum analyses for 25 stations using 500 years of historical data (1471–1970) and for 100 stations using 100 years of historical data (1871–1970) show clear periodicities. Each physically meaningful period has predominated in certain regions; indicated in Table 10.6. Spectra for typical stations are illustrated in Figs 10.5 and 10.6. These show obvious 80-year, 36-year, 22-year and quasi-biennial oscillations (QBO). In the following section we discuss two of these periodicities, the 36-year and the QBO.

Table 10.6. *Periodicity of drought and flood*

Period (years)	Region of its predominance
80–100	The Yellow River region
36	The east part of the Yang-tze River, and the southwest of China
22	The northeast of China, and the central part of the Yang-tze River
11	The Yellow River region, and the southeast of China
5–6	The south of China
QBO	The Yang-tze River region, and north of the Yellow River

Mechanisms for the 36-year period and the QBO

Figs. 10.7(*a*) and (*b*) show the correlation coefficients between drought and flood grade at Shanghai and sea-level pressure for high and low frequencies respectively. These were constructed by the following procedure. Each series was smoothed by taking weighted 3-year running means with weighting factors 0.23, 0.54 and 0.23, in order to filter out the QBO and to obtain a series in which the low-frequency variations were dominant. Then the low-frequency series was subtracted from the origonal series to obtain a high-frequency series in which the QBO is dominant. After that the correlation coefficients between drought and flood grade at Shanghai and sea-level pressure were calculated for each series. Areas where the coefficients are significantly different from zero at the 5 per cent level are shaded.

From Fig. 10.7 it is clear that the circulation mechanism of drought and flood formation depends on the time scale of the climatic fluctuation. We have also found that the QBO rainfall cycle is well correlated with the EAI (expressed by the sum of pressure at intersections 50° N, 130° E and 50° N, 140° E) and that the 36-year cycle is closely linked to the SOI (expressed by the pressure difference between intersections 10° N, 160° W, and 10° S, 140° E). Further details of the pressure charts used in this analysis are given by Wang (1964, 1965) and Wang & Zhao (1981). The

Fig. 10.5. Power spectra for Peking and Nanking, using historical data from 1471 to 1970 (500 years). Abscissa is frequency in cycles per 160 years. Figures above peaks give period in years. Ordinate is relative power.

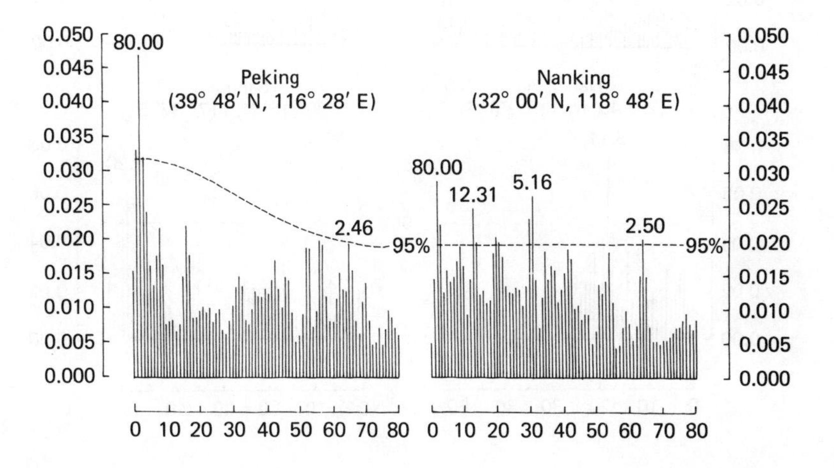

charts used cover the region 70° N to 50° S, and the months January, April, July and October for the period 1873–1970.

Power spectra of the EAI and the SOI are given in Fig. 10.8. In Fig. 10.9(*a*) the correlation coefficients between the EAI and the high frequency component of drought and flood grades of 100 stations are shown. Fig. 10.9(*b*) shows the distribution of correlation coefficients

Fig. 10.6. Power spectra for selected stations, using historical data from 1871 to 1970 (100 years). Abscissa is frequency in cycles per 110 years. Figures above peaks give period in years. Ordinate is relative power.

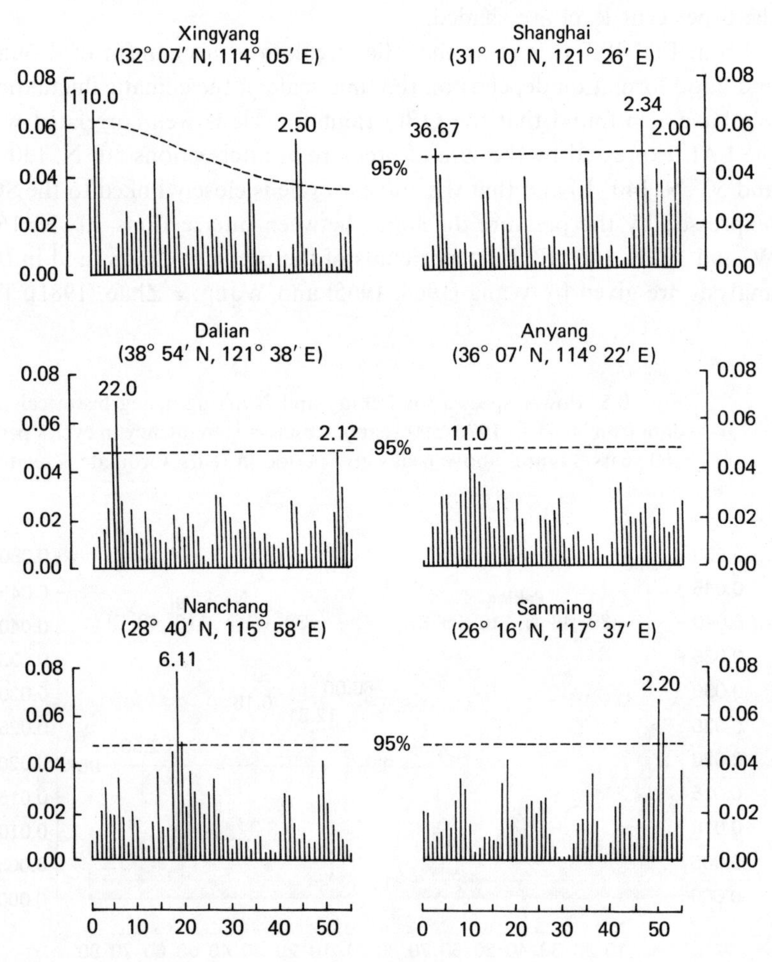

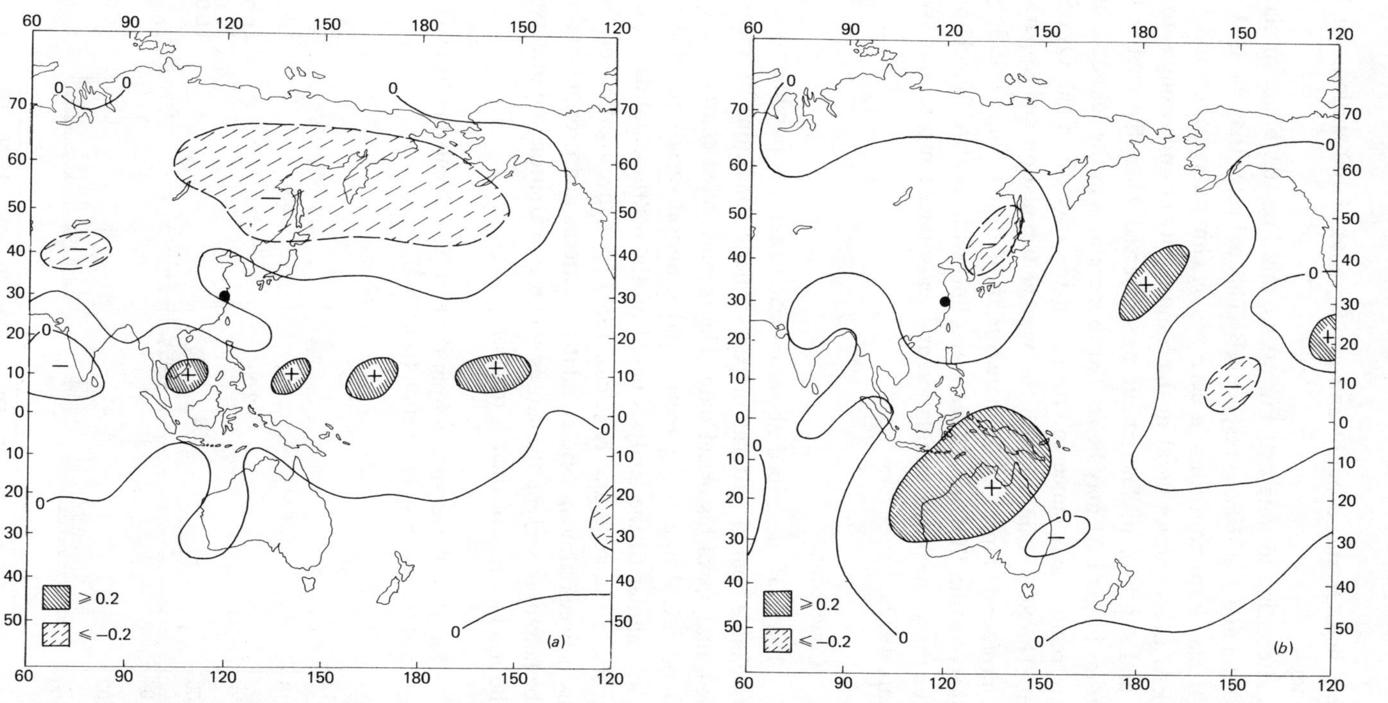

Fig. 10.7. Correlation coefficients between drought and flood data for Shanghai and sea-level pressure data from 1871 to 1970 for (*a*) *high-frequency* series, (*b*) *low-frequency* series. Shanghai is marked by a dot.

between the SOI and drought and flood grades of 100 stations for low frequency.

It is interesting to note that Fig. 10.9(*a*) and Fig. 10.9(*b*) are similar to Fig. 10.2(*c*) and Fig. 10.2(*a*) respectively. This indicates that the EAI and the SOI play important roles in determining climatic variation in China. High EAI accompanies flood in the Yang-tze River and drought to the north and south of it. The situation is reversed when the EAI is low. Variations in the EAI may be related to variations in the Okhotsk high-pressure region, and hence to changes in the strength of the Oya Shio current. The SOI is linked with the Walker Circulation and apparently has a profound impact on climate variation over China. The 36-year perodicity in the SOI may have some link with feedback mechanisms between the atmospheric circulation and sea-surface temperatures (Wang & Zhao, 1979a, b).

Conclusion

The work described above indicates that it is possible to reconstruct reliable spatial patterns of past climatic conditions by means of detailed analysis of historical data. The reconstructed charts of drought and flood reveal interesting spatial and temporal variations that in the main are similar to the analysis based on the instrumental data (Wang, 1962a, b, 1963a, b; Wang & Zhao, 1981). The investigation of the relations between atmospheric circulation patterns and drought and flood may be helpful in seeking the causes of climate variation and understanding the mechanisms of their formation.

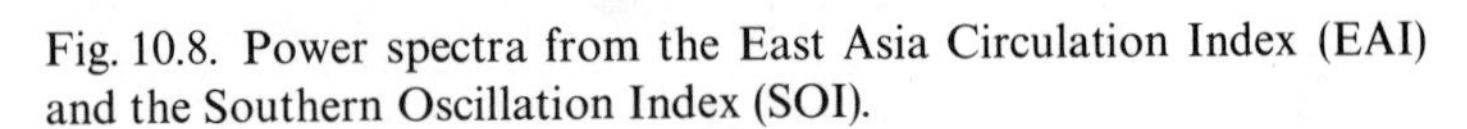

Fig. 10.8. Power spectra from the East Asia Circulation Index (EAI) and the Southern Oscillation Index (SOI).

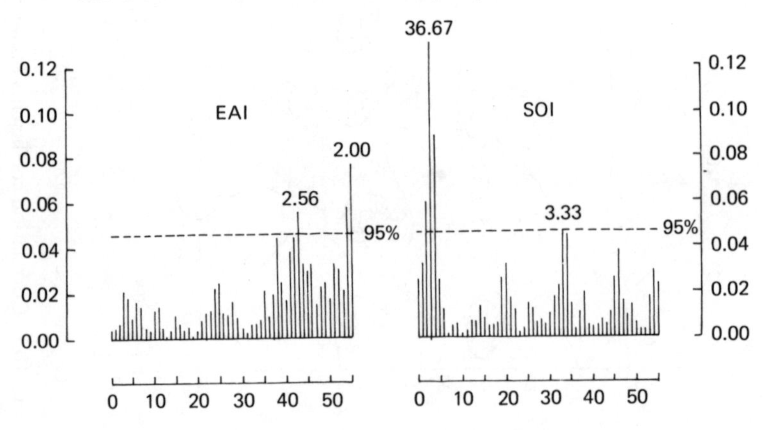

Fig. 10.9. Correlation coefficients between drought and flood data over China from 1871 to 1970 and: (*a*) EAI for high-frequency series, (*b*) SOI for low-frequency series.

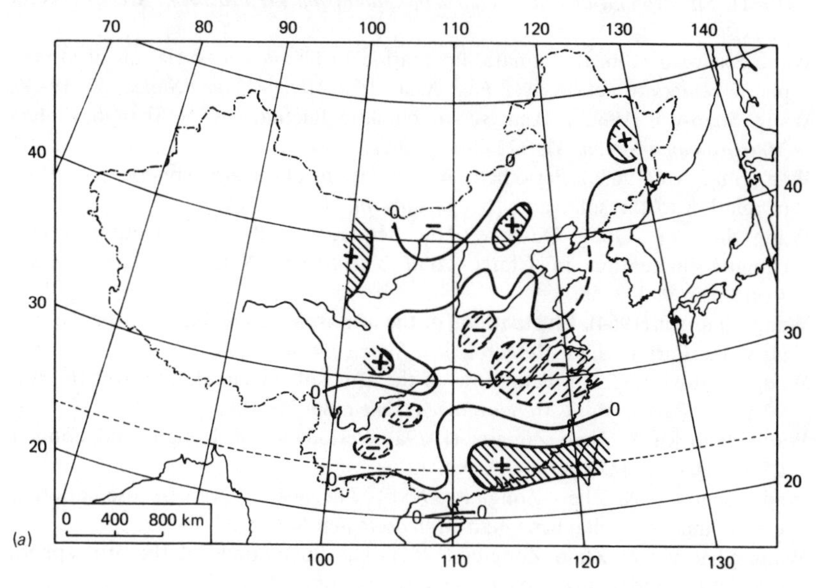

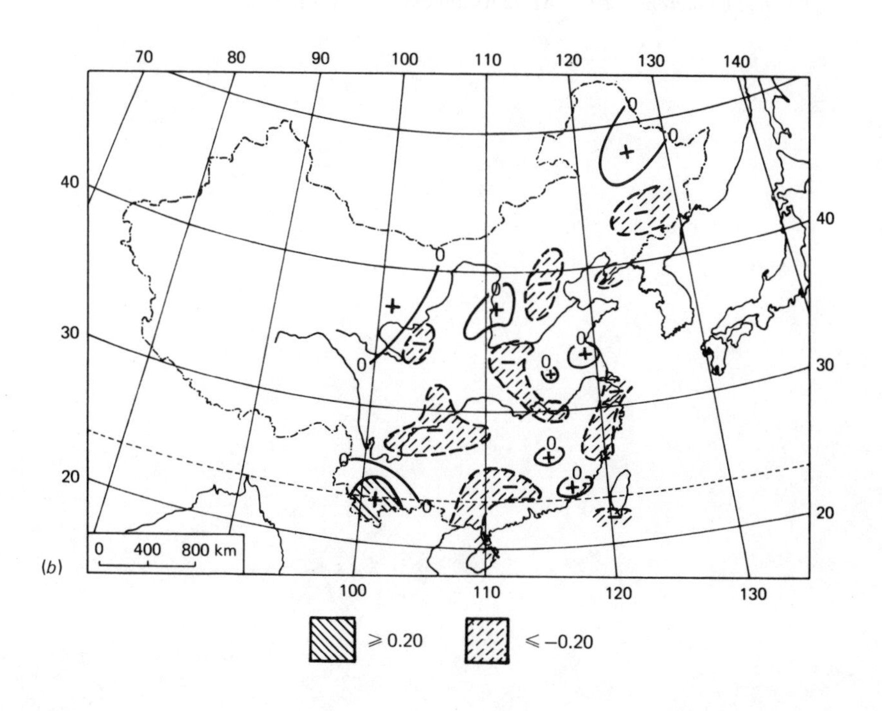

References

Central Meteorological Institute, Peking University, Nanking University, etc. (1981). *The drought and flood charts of China from 1470 to 1977*. Atlas Press (in press).

Wang Shao-wu (1962a). Climatic fluctuation in China and variation of atmospheric centers of action over East Asia. *Acta Meteorologica Sinica*, **32**, 19–36.

Wang Shao-wu (1962b). Analysis of climatic fluctuations of Shanghai. *Acta Meteorologica Sinica*, **32**, 322–36.

Wang Shao-wu (1963a). Periodicity of fluctuations of the atmospheric circulation in relation to the solar activity. *Advance in Meteorology*, 48–67.

Wang Shao-wu, Gong Qu-er & Niu Fen-lau (1963b). Investigation on the climatic fluctuation in China. *Acta Scientiarum Naturalium Universitatis Pekinensis*, **9**, 343–58.

Wang Shao-wu (1964). Fluctuations of the atmospheric circulation over the past 90 years, Part 1. *Acta Meteorologica Sinica*, **34**, 486–506.

Wang Shao-wu (1965). Fluctuations of the atmospheric circulation over the past 90 years, Part 2. *Acta Meteorologica Sinica*, **35**, 200–14.

Wang Shao-wu & Zhao Zong-ci (1979a). Periodicity of drought and flood in China. *Qixiang* (*Meteorology*), No. 1.

Wang Shao-wu & Zhao Zong-ci (1979b). The 36-year wetness oscillation in China and its mechanism. *Acta Meteorologica Sinica*, **37**, 64–73.

Wang Shao-wu & Zhao Zong-ci (1981). The fluctuations of the atmospheric circulation and climate over the past century, *Proceedings of the Climatic Change Conference, 1978*. Meteorological Press (in press).

PART III

TOWARDS A THEORY OF CLIMATE–HISTORY INTERACTIONS

11

An approach to the study of the development of climate and its impact in human affairs

HUBERT H. LAMB

Abstract

This chapter first reviews the reasons why climatic change came to be conventionally ignored for many years, until quite recently, in planning for the future just as much as in the analysis of human history and archaeology. The prevailing attitude seems now to have changed, although much confusion remains: more knowledge of the behaviour of climate is greatly needed. The next section introduces the great variety of types of data and analysis which can now be used to reconstruct a very long past record of climate and the possibilities of corroboration and assessment of margins of uncertainty. The possibilities of reconstruction, particularly in Europe, North America and the Far East, are quite detailed for recent centuries; but outline reconstruction of the successive global regimes and their times of change can go back far beyond the dawn of history. Next, the results of study of the course of climatic development over about the last thousand years, chiefly in Europe, through the Medieval Warm Epoch and the Little Ice Age, are presented and certain apparent impacts on human history and the environment are mentioned. The transition between these extreme climatic periods produced many severe events. A final section mentions some probable outstanding cases of climatic change which affected human history in other times, including our own.

Introduction

We live in a time when climatic change, or the threat of it, is forever being discussed. Those concerned with practical affairs and planning in government, industry and trade are confronted with a confusing and alarming range of conflicting views of the probable future development of climate from the scientists working in meteorology and climatology. How are the conflicting views to be resolved so that we can improve the layman's situation?

Ever since the first national weather services were established in the late nineteenth century, meteorology has been largely occupied with the scientifically very difficult problem of forecasting the development of the flow of the atmosphere, and hence development of the weather, over periods ranging from some hours to some days ahead. In the past 30 years the greatest effort has gone into reducing this task more and more to a matter of calculation, employing highly developed mathematical expressions of the atmospheric processes and requiring the most advanced computers.

Around 1880–1900 it was apparent from the first 100-year series of weather observations that the climatic averages of that time were very similar to those of a century earlier, and it became conventional to treat climate as essentially constant. Accordingly, study of the development and possible changes of climate was given a very low priority or more generally neglected altogether. This was the attitude to climate adopted in textbooks, and illustrated by the tables of climatic statistics published for planning purposes seldom mentioning which years' data they were based on.

Given these circumstances, it was natural that the general public and workers in other disciplines also ignored the possibility that climate might not always be so stable and climatic change might affect their affairs. Of course, this did not apply to occasional extreme events; but the incidence of these, it was generally supposed, could be treated as random. Hence the idea that special pleading must be involved in any attempt to invoke climate as a cause in human history, except where some isolated extreme event – or short run of extreme years – upset a society which was already in some unstable state.

By 1950, however, it was obvious that the climate had been changing significantly during the twentieth century, though in ways which made things easier for most human activities in most parts of the world. There had been a general rise of prevailing temperatures (Fig. 11.1), with recession of the ice on the Arctic seas and increasingly rapid recession of the glaciers almost everywhere. There were fewer failures of the monsoons in India and West Africa, and rainfall had increased in continental interiors (except in the Americas, where the 'rain-shadow' of the Rockies and the Andes produced an opposite effect). A scientific explanation of global warming in terms of the effects on the (solar incoming versus terrestrial outgoing) radiation balance by carbon dioxide which Man was adding to the atmosphere – an end-product of all the fossil fuels burnt – won wide acceptance (Plass, 1956). This left it possible to believe that climate had been essentially constant until the industrial revolution. But then it was discovered that the trends had changed, so that the most careful assessments indicate global cooling from around 1945 to 1965, since when – despite a recent renewal of interest in the carbon dioxide question and a very proper concern with the possible effects of other

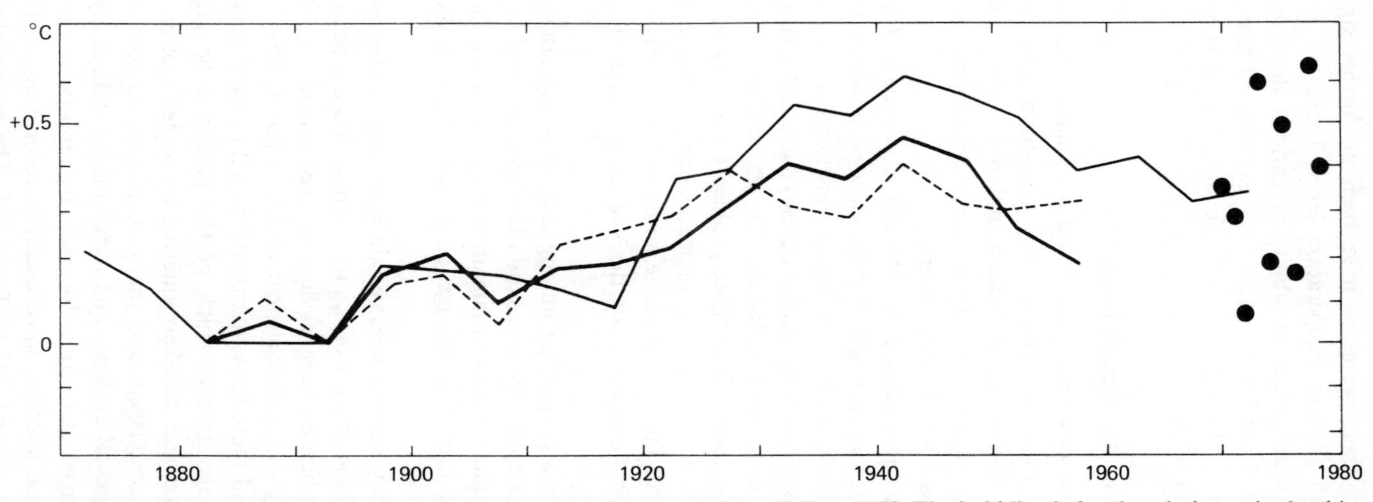

Fig. 11.1. Changes in the average surface air temperature of the earth from 1870 to 1978. The bold line is for the whole earth, the thin line is for the latitude band 0°–80° N, and the broken line is for the latitude band 0°–60° S. The data are from Mitchell (1961) and Brinkmann (1976) and are 5-year averages. The dots are annual average values for the northern hemisphere based on data given by Jones & Wigley (1980; see also Ingram, Farmer & Wigley, this volume, Fig. 1.1).

intrusions by Man – there seems to have been no further significant change of the overall average temperature level. There are, however, indications that the year-to-year variability of various elements of the climate has increased, involving a frequency of extremes of temperature and rainfall (droughts and floods), and of gale situations in some areas, not seen since 1900 or 1920.

Investigation of the climatic record

Reconstruction of a long, past record of climate is necessary (*a*) for meteorology to trace and identify any long-term and recurrent processes of climatic change and fluctuation, and (*b*) to enable historians to see when and where significant shifts of climatic behaviour may have introduced vital stresses into human affairs.

Historians, while acknowledging the obvious importance of harvests and a sound economy, have probably for the most part preferred to seek explanations for the phenomena of human history from within the realms of human personality and societal development and relationships with which they are more familiar. Indeed, this attitude has its counterpart in the physical sciences where those problems are most easily and satisfactorily analysed where external influences can be treated as constant and so eliminated. Thus, for example, some meteorologists prefer to discount any possible fluctuations of the sun as a cause of climatic change.

There has also been a proper fear among historians and others of the danger of arguing in a circle: the accusation has been heard of inventing climatic change to explain events in human history which are then held to prove the occurrence of a change of climate. This danger can, however, be avoided.

Because climate provides or denies certain necessary conditions of life to all constituents of the living world, and controls basic aspects of the physical environment besides, fossil evidence of past climatic regimes and events exists on every side and in a great variety of forms. These can be used by ever more and more techniques, each to tell its own story of the past record of climate. Hence, much of the past can be gradually surveyed without preconceived ideas entering in; margins of error and uncertainty can be established; and many items can be corroborated from independent types of evidence and techniques of data analysis.

The types of systematic data from documentary reports, and from the physical and biological sciences, most useful in reconstructing the past record of climate are summarised in Table 11,1. The potentialities and

Table 11.1. *Data for reconstruction of the past record of climate*

Type of data	Climatic elements concerned	Time resolution of the observations	Lags in response	Beginning of record	Areas covered
Standard meteorological instrument observations: barometer, thermometer rain-gauge, etc.	Surface pressure, temperature, rainfall, windflow, etc. (humidity observations start later)	Virtually instantaneous	Insignificant	Mid seventeenth century for temperature, late seventeenth century for pressure and precipitation; some fragmentary rainfall measurements from earlier times	Gradual growth of world network beginning in Europe
Upper air instrument measurements	Upper air temperature, humidity, pressure values, winds	Virtually instantaneous	Insignificant for climatological purposes	1930s in parts of Europe and North America; fragments much earlier from mountain top stations in Europe, and from balloon flights	Northern hemisphere from 1949; southern hemisphere from 1957
Ship-borne instruments	Sea temperatures (salinity and ocean current observations start later)	Virtually instantaneous	Insignificant for climatological purposes	1850s; fragments from 1780	Mainly Atlantic Ocean for first 20–30 years

Table 11.1. (*cont.*)

Type of data	Climatic elements concerned	Time resolution of the observations	Lags in response	Beginning of record	Areas covered
Descriptive weather registers, weather diaries	Wind, weather, rain and snow frequencies, etc.	Daily	—	Earliest significant example is Merle (1337–44), but mainly mid-sixteenth century onwards	Many parts of Europe; scattered data from early expeditions in the eastern half of North America and elsewhere from the late 1500s onwards
Ships' logs (mainly useful in port or patrolling short sections of coast)	Wind, weather, rain and snow frequencies, etc.	Once or several times a day	—	*c.* 1670–1700; isolated much earlier reports of exploration voyages	Europe, China, Japan: similar middle eastern and Indian records probably exist
Annals, chronicles, account books, state and local documents, farm and estate management reports, accounts of military campaigns, etc.	Weather, especially extremes and long spells of weather, droughts, floods, frost, snow, great heat, great cold, harvest results, etc.	Month or season, sometimes to the specific day	—	About AD 1100 in Europe; occasional reports much earlier, seventh century AD or earlier in Japan; 1000 BC in China	Europe, China, Japan: similar middle eastern and Indian records probably exist

	River flood levels (rainfall and evaporation, snowmelt)		Varies from a few hours to half a year (Nile)	AD 622; fragments much earlier, from 3100 BC	Earliest records are for the River Nile; otherwise as above
	Lake levels (rainfall and evaporation)		1–20(?) years	*c.* 400 BC	Earliest records are for the Caspian Sea
Glaciers (advances and retreats reported, old moraines dated, etc.)	Temperature, sunshine and cloudiness, snowfall	Minimum 1 year	About 1–30 years characteristic, depending on size of glacier	10 000 BP or earlier	Mountainous regions in many latitudes
Varves (year layers in lake-bed and a few river estuary and sea-bed sediments)	Stream-flow, rainfall, temperature – but interpretation difficult	1 year: much more difficulty than with tree rings in eliminating dating errors and uncertainties	Days or weeks	*c.* 8000 BC	Sweden and northern United States, also Japan
Tree rings	Temperature, rainfall	Ring width 1 year, cell structure 1–5 weeks	Ring width depends on up to 15 months previous weather	4000–6000 BC in the southwestern United States	Much of Europe and North America – potentially most temperature regions in both hemispheres
Pollen analysis	Temperature, rainfall	About 50 years; some deposits dateable to order 1 year	Quick response to adverse conditions, up to 5000 years lag in recolonisation of northern Europe after ice age	>100 000 BP	All the world's land areas and some ocean bed deposits

Table 11.1. (*cont.*)

Type of data	Climatic elements concerned	Time resolution of the observations	Lags in response	Beginning of record	Areas covered
Insect faunas	Temperature	About 100 years	Probably never more than a few decades or at most centuries	> 100 000 BP	Limited coverage so far: most work in England
Marine micro-fauna (foramin-ifera, etc.)	Surface and deep-water temperatures (according to species habitat)	Ranges from 100 to 2500 years depending on rate of de-position on the ocean bed at the site	—	> 1 000 000 BP	Samples available from every ocean and all latitudes
Isotope measurements: (*a*) on ice sheets (^{18}O)	Temperature, snowfall	At best a few days or weeks, aids recognition of year layers	—	> 1 000 000 BP: around 100 000 BP for high reso-lution data	Mainly Greenland; Canadian Arctic and Antarctica

(*b*) on tree rings (^{18}O, ^{2}H, ^{13}C)	Temperature, humidity, rainfall (interpretation problems still controversial)	Generally 1 year or more, but seasonal data is possible	—	AD 200 AD 1350	California Central Europe
(*c*) $CaCO_3$ sediments	Temperatures, amount of H_2O in ice (i.e. removed from the oceans)	Ranges from about 100 to 2500 years depending on rate of deposition at the site	—	>1 000 000 BP	Samples available from every ocean and all latitudes
(*d*) speleothems	Temperature, precipitation, ice volume (interpretation problems still exist)	Of order 1000 years	—	>1 000 000 BP	Most land areas (but still untapped)

problems of handling these different types of evidence have been described by Lamb (1977). For a remarkably full guide and critique to the meteorological instrument data available for Europe see Rudloff (1967).

Because of the adaptability of Man and the larger animals, only in rare cases (as when the regular traversing of high mountain passes or deserts begins and ends or lake settlements and routes across marshes are abandoned) can deductions about climate be tentatively made directly from their limitations: and, even in these cases, corroboration from other indicators of environmental history is sought.

On the other hand, the value of the numerous reports in many types of document, particularly from Europe and the Far East, describing the weather of individual seasons, and the extent to which these reports can be used to map the weather and diagnose the prevailing atmospheric circulation conditions, is only now beginning to be explored and exploited. Great care is needed in examining the reliability of the historical sources used, but it is not inconceivable that every season since AD 1200 containing any kind of severity of weather in Europe can be established.

The reliability of the general indications and broad trends derived from this kind of documentary research is indicated by comparisons between the works of Lamb, Titow and Alexandre. A comparison between series of 15 decadal values from AD 1200 to 1350 of the Summer Wetness index for England defined by Lamb (1965) on the basis of counts of wet and dry months reported in collections of manuscript data and Titow's (1970) rating of the summer and autumn weather (June through to October) derived from the manorial accounts of the bishopric of Winchester yields a correlation coefficient of 0.77, significant at the 1 per cent level. Successive 50-year mean values of Lamb's Winter Severity index for England from AD 1100 to 1600 (derived from counts of cold and mild months) and values derived in a similar fashion from records from southeast Belgium and northeast France by Alexandre (1976) gives a correlation of 0.85, significant at the 1 per cent level. For summer wetness, however, the correlation coefficient is 0.22, which is not statistically significant. Alexandre's *decadal* indices can be compared with Lamb's for the period 1100–1400 (data from Alexandre, 1977) giving the following results – winter severity, $r=0.22$, not statistically significant; summer wetness, $r=0.44$, significant at the 5 per cent level. The overall picture is one of encouraging agreement on both long (50-year) and short (10-year) time scales, with some interesting discrepancies on the short time scale.

I have not presumed to derive prevailing temperature values for less than 50-year time steps from these documentary data. (This is a precaution to allow for the possibility of some misleading reports and errors in transcription in the sources of data then known.) However, circumspect use of combinations of special types of information such as phenological data (flowering dates of several plant species, fruit and grain ripening dates, dates of leaf-fall), wood densities and tree-ring widths, dates and durations of ice cover on canals, rivers and lakes, oxygen-isotope measurements, etc. may open the way to useful estimates of temperatures and perhaps other items of the weather in individual years. Ways must always be sought, however, of assessing margins of error and uncertainty or of the degree of internal consistency and confidence in the evidence.

The climatic record of the last thousand years

If we will demonstrate the influence of the climatic development (and shocks produced in the course of this) on human history, we must look first at extreme cases.

For this purpose, the record of the last thousand years is well suited because it seems to have included a notably wide range of climatic regimes. The High Middle Ages evidently saw a persistently warm climatic epoch which lasted until about 1300–10 in Europe and affected about two-thirds of the northern hemisphere in the previous 100–200 years. A peak of warmth seems to have been attained a few centuries earlier in Greenland and much more of the Arctic (Dansgaard *et al.*, 1975), and cooling in those regions may have been responsible for an increase of storminess affecting the North Sea and perhaps much of the Atlantic and Europe after AD 1200 (Britton, 1937; but see also Gottschalk, 1971). The following centuries brought a series of changes, some of them abrupt, leading to the so-called Little Ice Age between about 1550 and 1700 or later, when the extent of ice on the Arctic seas and of ice and snow on land seems to have been greater than at any time since the last major glaciation. This Little Ice Age period probably ranks with the warmest Postglacial times around 6000 years ago as the only times when the mean climatic conditions seem to have departed in just one direction everywhere from those of the present century.

Even so, there was undoubtedly a strong geographical pattern of regions where the anomaly was particularly intense and others where it was much weaker. There was also a continual variation with time, and indeed the variability of the weather from year to year and from one 5-year period to the next (at least in Europe) seems to have been particularly great during the early onset phases of the Little Ice Age between AD 1300 and about 1450 and in the climax stages in the sixteenth and seventeenth centuries.

The general course of climatic development is indicated by the sequence of half-century mean values of winter temperature and summer rainfall for typical low-lying sites in central England in Fig. 11.2. These estimates were derived from regression equations relating prevailing

Fig. 11.2. Fifty-year averages of winter temperature (top) and summer rainfall (bottom expressed as percentages of the 1916–50 average) for England. The continuous line gives unadjusted values based solely on regression analysis and Winter Severity and Summer Wetness index values prior to 1700 for temperature and prior to 1750 for rainfall. 95 per cent confidence limits based on the regression equations and using the *t*-distribution are shown. Values shown after 1700 (temperature) and 1750 (rainfall) are observed instrumental data from Manley's central England temperature series (Manley, 1974) and the England and Wales rainfall series. The long dashed lines project the unadjusted estimates back in time, but only 100- to 200-year means are given based on sparse data. The short dashed line in the temperature curve gives adjusted values, accounting for botanical evidence. The dotted curve is the analyst's opinion based on a wide variety of evidence.

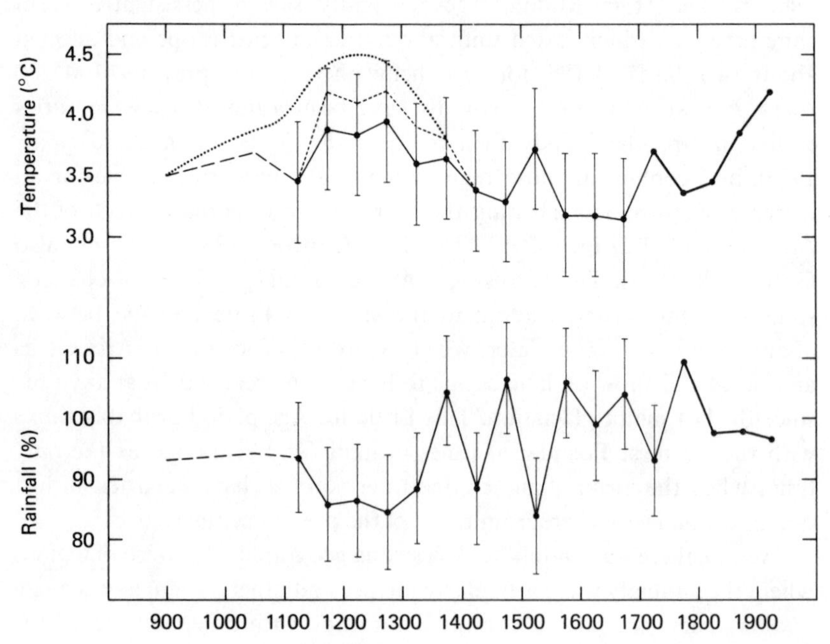

temperature and rainfall values to the winter severity and summer wetness indices discussed above (see Lamb, 1965 and 1977). Error bars mark the 95 per cent confidence limits based on the statistical relationships (but not accounting for possible uncertainties in the historical data; see, for example, Bell & Ogilvie, 1978; Ingram, Underhill & Wigley, 1978). Estimates for other seasons are given by Lamb (1965, 1977).

The changes of winter temperature shown in Fig. 11.2, amounting to a fall of the 50-year means of 1.0–1.5 deg C between the thirteenth and seventeenth centuries, appear to have been accompanied by similar changes in other seasons. These changes imply a shortening of the average growing season by about a month and a decline of almost 200 metres in the upper limit of productive cultivation of crops in England (and, presumably, in other places at about the same latitude in Europe). The latter aspect seems to be confirmed by the history of the upper limit of cultivation on the Lammermuir Hills in southeast Scotland over about the same period traced by Parry (1975, and this volume). Firbas & Losert (1949) estimated also that the upper tree line on the mountains of central Europe fell by about 200 metres over this time. The recovery of the 50-year averages from 1700 to the first half of the present century amounted to about 0.8 deg C and seems to have lengthened the mean growing season by about 3 weeks.

The frequency of frosts and the duration of snow cover were clearly affected similarly. In the case of England this can be traced in various papers by Manley (e.g. 1969). Pfister's studies of daily weather records near Bern and Zürich, in Switzerland, from 1683 onwards show that there were eight winters in the first 56 years (between 1683 and 1738) when the length of snow cover exceeded the six most extreme cases (80 to 86 days) observed in the last 100 years: in two of those winters (1684/5 and 1730/1) the duration was 112 and 110 days. There were more cases later in the eighteenth century (1769/70 and 1788/9) of about the same duration; and in one earlier case in 1613/14 and again in 1784/5 the snow lay in Bern for over 150 days. A common feature of those times in the Swiss records is the bitter winter weather which lasted through March, and as Pfister has pointed out (1975, 1978) the years with longest snow cover and delayed springs produced very low grain yields (recorded in the tithe statistics) and necessitated great slaughtering of dairy cattle when the hay ran out.

The effects of the climate on vine cultivation in Europe can be traced in the data tabulated by Le Roy Ladurie (1967, 1971) and Müller (1953) and include a southward shift of the limit of vineyard cultivation which lasted until our own times. Similarly, in the Kiangsi province of China the cultivation of oranges which had been practised for centuries was given up as a result of frequent frosts between 1654 and 1676 (Chu Ko-Chen, 1961).

The fossil evidence and many actual reports of advancing glaciers in most parts of the world in the Little Ice Age period have been catalogued by many authors and are summarised in Lamb (1977; Ch. 17). For example, in Iceland, Norway and the Alps the records include the overrunning of farms and farmland by the glaciers as well as disasters from avalanches, landslides and the bursting of glacier-dammed lakes. Even in Ethiopia there is evidence from Portugese travellers that the snowline came down on the highest mountains, and in eastern equatorial Africa the glaciers on Mt Kenya and Kilimanjaro were in more advanced positions than their present limits.

The advance of the Arctic sea ice had already by the fifteenth century isolated the old Viking colony in Greenland from all contact with the outside world, and the European population there died out (see McGovern, this volume). In the late seventeenth century a further advance of the ice caused Iceland to be surrounded and cut off in the extreme year 1695, while a recent study (Lamb, 1979) indicated that in that year the polar water reached the entire coast of Norway and for 30 years (1675–1704) it dominated the ocean surface between Iceland and the Faeroe Islands (61° N). This meant that the ocean surface in that area was about 5 deg C colder than in the present century (with corresponding southward displacement of the fish populations) presumably also explaining the reports of permanent snow on the highest hills in northeast Scotland where one or two high-level lochs bore permanent ice (see, for example, Pennant, 1771). Another probable consequence would be a greater lapse of the prevailing temperatures with height, particularly with winds from the northwest quadrant, affecting the hills and mountains in Scandinavia and central Europe.

It is quite clear that there was in the 1580s and 1690s an enhanced north–south gradient of temperature between about latitude 50° and 60° N in this part of the world. And this would provide a basis for occasional windstorms of a severity not likely to be experienced in warmer climatic periods. The greater storms resulted in coastal changes by flooding and shifting sands, accompanied by disasters to settlements near the coasts and to ships at sea (Lamb 1977: Ch. 17; Gottschalk, 1975, 1977). The wind situation has been at least partially verified in the case of the

Spanish Armada summer of 1588 by a recent study (Douglas *et al.*, 1978).

Many studies – for example of early thermometer records in Sweden by Wallén (1953), of the descriptive data on snow cover in Switzerland by Pfister (1978, and this volume) and freezing of the Dutch canals by de Vries (1977; see also van den Dool, Krijnen & Schuurmans, 1978), as well as of tree rings in Germany by Fürst (1963) – appear to indicate a significantly wider range of year-to-year variability in the Little Ice Age, which seems to have been associated with a reduced frequency of westerly winds in Europe and increases in the frequency of winds from most other directions. Periods around the climax of this climatic regime in the seventeenth and eighteenth centuries, as well as, to judge from tree-ring evidence (Fürst, 1963) the earliest onset phases around AD 1300, seem to have been particularly affected. Thus, there were some great 'heat waves', as in England in 1665 and 1666, 1718 and 1719, and some exceptionally mild winters, in the midst of the period of generally colder climate. And particularly in the fourteenth, sixteenth and eighteenth centureis there seem to have been notable runs of flood years and drought years both in England and elsewhere in Europe (Bull, 1925; Buchinsky, 1957). In Africa north of the equator, in Algeria and in the Sahel zone, records exist which show heavy concentrations of drought years at times around 1590–1650 and 1690–1780 (Nicholson, 1980 and this volume). These alternations of extreme seasons may have been the most damaging aspects of the behaviour of the climate to the economies of those times, with severe effects on farming routines and harvests and on the health of people and animals.

Other periods of history

Space hardly permits reference here to examples of the correspondence of earlier well-established times of climatic stress and change with major events in human history and times of cultural change. Mention must, however, be made of the following.

1. The upland farm village of Hoset about 350 metres above sea level at latitude 63° N in central Norway can be shown to have been abandoned to the forest three times – in about the seventh century AD, in 1435, and in the 1690s – each time in a period of marked deterioration of the climate known elsewhere in Europe (Salvesen, Sandnes, Farbregd & Halvorsen, 1977).

2. The storm floods of the North Sea between AD 1240 and 1362 caused sixty Danish parishes that produced over half the revenues of the

diocese of Slesvig (now Schleswig) to be 'swallowed by the salt sea' (Steenstrup, 1907), doubtless with much loss of life. And this presumably also occurred in northwest Germany and the Netherlands where the Jadebusen and the Zuyder Zee were formed at about that time. (For further details see Gottschalk, 1971.)

3. Advancing Alpine glaciers in or about the last millennium BC closed the prehistoric gold mines in the Hohe Tauern, Austria (Gams, 1937), an event which occurred again after mining had been resumed in the warm period of the High Middle Ages.

4. Archaeologically dated former shorelines of the Caspian Sea confirm a remarkable, long drought period around AD 300 (Leontev & Federov, 1953), when Huntington long ago (1907) suggested that drying up of the pastures used by the nomads in central Asia may have started the *Völkerwanderungen* which led to the barbarian invasions of the Roman Empire.

In all these cases it is likely that a long-continued climatic shift played a critical part in the patterns of human life.

The studies with which this chapter is concerned have a strong relevance to our world today. After a period in which advancing technology greatly increased agricultural yields and enormously reduced the risks of harvest losses due to weather, our vulnerability to climatic shifts and individual bad years is probably increasing again, as was shown by the doubling of the world price of wheat in 1972 after harvest failures in Soviet Asia, India and the Sahel. Vulnerability is surely increased by the rationalisation of world agriculture and trade in ways that have returned large areas to concentrating mainly on a single crop (or just one or two crops). It is also increased by the growth of population to a point where, in years with harvest shortfall in any one of the major grain-producing regions of the world, total stocks are run down and even major countries are forced – regardless of political considerations – to buy food from another major producer.

In these circumstances, climatological research must urgently seek better knowledge and understanding of the time scale and ranges of natural climatic fluctuations, particularly the incidence of periods of enhanced variability, as well as considering the possible side-effects of Man's various activities and their increasing scale and variety. The latter demand both theoretical and experimental studies and continual monitoring. Climatic research is a field which calls urgently for collaboration between scientists and workers skilled in the disciplines of the humanities.

References

Alexandre, P. (1976). Le climat dans le sud de la Belgique et en Rhénanie de 1400 à 1600. Prélude au 'Petit Age Glaciaire' de l'époque moderne. *Annales du XLIV-ième Congrès,* Huy, 18–22 Août 1976. Federation des Cercles d'Archaeologie et d'Histoire de Belgique, ASBL.

Alexandre, P. (1977). Variations climatiques au Moyen Age, *Annales: Économies, Sociétés, Civilisations,* **32**, 183–97.

Bell, W. T. & Ogilvie, A. E. J. (1978). Weather compilations as a source of data for the reconstruction of European climate during the Medieval Period. *Climatic Change,* **1**, 331–48.

Brinkman, W. A. R. (1976). Surface temperature trend for the northern hemisphere – updated. *Quaternary Research,* **6** (3), 355–8.

Britton, C. E. (1937). Meteorological chronology to AD 1450, Meteorological Office *Geophysical Memoir,* 70. London: HMSO.

Buchinsky, I. E. (1957). *The Past Climate of the Russian Plain,* 2nd edn. Leningrad: Gidrometeoizdat. (In Russian: English transl. in Meteorological Office Library.)

Bull, E. (1925). Klimaskifte og nedgang i Noreg i sein mellomalderen. *Syn. og Segn,* **31**, 12–19.

Chu Ko-Chen (1961). The pulsation of world climate during historical times. *New China Monthly,* **6**, Peking (in Chinese, English translation in Meteorological. Office Library).

Dansgaard, W., Johnsen, S. J., Reah, N., Gundestrup, N., Clausen, H. B. & Hammer, C. U. (1975). Climatic changes, Norsemen and modern man. *Nature,* **255**, 24–8.

van den Dool, H. M., Krijnen, H. J. & Schuurmans, C. J. E. (1978). Average winter temperatures at De Bilt, Netherlands, 1634–1977. *Climatic Change,* **1**, 319–30.

Douglas, K. S., Lamb, H. H. & Loader, C. (1978). A meteorological study of July to October 1588: the Spanish Armada storms, Norwich (Climatic Research Unit, University of East Anglia – *Research Publication CRU RP 6*). 76 pp.

Firbas, F. & Losert, H. (1949). Untersuchungen über die Entstehung der heutigen Waldstufen in den Sudegen, *Planta,* **36**, 478–506.

Fürst, O. (1963). Vergleichende Untersuchungen über räumliche und Zeitliche Unterschiede interannueller Jahrringbreiterschwankungen und ihre klimatologische Auswertung. *Flora,* **153**, 469–508.

Gams, H. (1937). Aus der Geschichte der Alpenwälder. *Zeitschrift des deutschen und österreichischen Alpenvereins,* **68**, 157–70.

Gottschalk, M. K. E. (1971). Stormvloeden en rivieroverstromingen in Nederland. *Part I: De periode voor 1400.* Assen: van Gorcum.

Gottschalk, M. K. E. (1975). Stormvloeden en rivieroverstromingen in Nederland. *Part II: De periode 1400–1600.* Assen: van Gorcum.

Gottschalk, M. K. E. (1977). Stormvloeden en rivieroverstromingen in Nederland. *Part III: De periode 1600–1700.* Assen: van Gorcum.

Huntington, E. (1907). *The Pulse of Asia.* Boston, Mass: Houghton Mifflin. 415 pp.

Ingram, M. J., Underhill, D. J. & Wigley, T. M. L. (1978). Historical climatology. *Nature*, **276**, 329–34.

Jones, P. D. & Wigley, T. M. L. (1980). Northern Hemisphere temperatures, 1881–1979. *Climate Monitor*, **9**, 43–7.

Leontev, O. K. & Federov, P. V. (1953). History of the Caspian Sea in late- and post-Hvalynsk time. *Izvestia Ser Geogr*. No. 4, pp. 64–74. Moscow: Akademii Nauk. In Russian.

Lamb, H. H. (1965). The early medieval warm epoch and its sequel. *Palaeogeography, Palaeoclimatology, Palaeoecology*, **1** (1), 13–37.

Lamb, H. H. (1977). *Climate: Present, Past and Future*, vol. 2, *Climatic History and the Future*. London: Methuen. 835 pp.

Lamb, H. H. (1979). Climatic variation and changes in the wind and ocean circulation: the Little Ice Age in the northeast Atlantic. *Quaternary Research*, **11**, 1–20.

Le Roy Ladurie, E. (1967). *Histoire du climat depuis l'an mil*. Paris: Flammarion. 317 pp.

Le Roy Ladurie, E. (1971). *Times of Feast, Times of Famine*, transl. B. Bray. New York: Doubleday. 426 pp.

Manley, G. (1969). Snowfall in Britain over the past 300 years. *Weather*, **24**, 428–37.

Manley, G. (1974). Central England temperatures: monthly means 1659 to 1973. *Quarterly Journal of the Royal Meteorological Society*, **100**, 389–405.

Mitchell, J. M. (1961). Recent secular changes of global temperature. *Annals of the New York Academy of Science*, **95**(1), 235–50.

Müller, K. (1953). *Geschichte des Badischen Weinbaus*. Lahr in Baden: Moritz Schauenburg. 283 pp.

Nicholson, S. E. (1980). Saharan climates in historic times, Ch. 8 in *The Sahara and the Nile*, ed. M. A. J. Williams & H. Faure. Rotterdam: A. A. Balkema.

Parry, M. J. (1975). Secular climatic change and marginal agriculture, *Transactions of the Institute of British Geographers*, Publication No. 64, pp. 1–13, London.

Pennant, T. (1771). *A tour of Scotland, 1769*. Chester.

Pfister, C. (1975). *Agrarkonjunktur und Witterungsverlauf im westlichen Schweizer Mittelland 1755–1797*, Bern (Geographisches Institut der Universität.) 289 pp.

Pfister, C. (1978). Fluctuations in the duration of snow cover in Switzerland since the late seventeenth century, *Proceedings of the Nordic Symposium on Climatic Changes and related problems*, pp. 1–6 (organised by the Danish Natural History Society and Danish Meteorological Institute, ed. K. Frydendahl). Copenhagen (Danish Meteorological Institute – *Climatological Papers No. 4*). 260 pp.

Plass, G. N. (1956). The carbon dioxide theory of climatic change. *Tellus*, **8**, 140–54.

von Rudloff, H. (1967). *Die Schwankungen und Pendelungen des Klimas in Europa seit dem Beginn der regelmässigen Instrumenten-Beobachtungen* (*1670*), Braunschweig (Vieweg: *Die Wissenschaft*, *122*).

Salvesen, H., Sandnes, J., Farbregd, O. & Halvorsen, A.-M. (1977). The Hoset Project: an interdisciplinary study of a marginal settlement. *Norwegian Archaeological Review*, **10**, 107–54.

Steenstrup, J. (1907). Danmarks Tab til Havet i den Historiske Tid. *Historisk Tidsskrift*, **8**, 153–66.

Titow, J. Z. (1970). Le climat à travers les rôles de comptabilité de l'évêche de Winchester (1350–1450). *Annales: Économies, Sociétés, Civilisations*, **25**, 312–50. (Fold-out diagram gives data over a longer period.)

de Vries, J. (1977). Histoire du climat et économie: des faits nouveaux, une interprétation différente. *Annales: Économies, Sociétés, Civilisations*, **32**, 198–226.

Wallén, C. C. (1953). The variability of summer temperature in Sweden and its connection with changes in the general circulation. *Tellus*, **5** (2), 157–78.

12

Short-term climatic fluctuations and their economic role

H. FLOHN

Abstract

This chapter is an introductory study of the socio-economic impact of short-term interannual fluctuations in climate. Minor long-term climatic fluctuations are generally associated with changes in the number of extreme events. Extreme anomalies on the seasonal time scale often occur in clusters (for example, the three severe winters, 1977–79, in the USA, or the Sahel drought years, 1969–73), and such clustering greatly magnifies the impact. Relationships between climatic anomalies and harvest production are summarised, with emphasis given to the importance of long-distance climatic spatial correlations (teleconnections) and their effects on modern agricultural economies.

Looking back at the fourteenth–eighteenth centuries, vine harvest dates and rye harvest prices are suggested as useful proxy data by their fairly good correlations with climatic data. Weather diaries are identified as a valuable source of information, not only for local studies, but also for international comparisons.

The role of extreme climatic anomalies

From the point of view of the layman, the comparatively small magnitude of variations of climate (for example, the small changes in 30-year averages), are mostly felt by varying frequencies and intensities of extreme climate anomalies of individual seasons and even individual weather events. Rare events remain in the memory: examples are the droughts of 1976 and 1921 in western and central Europe, or the severe winter of 1962/3. The economic role of such extreme events reinforces this impression: for example, the sudden rise of coffee prices after some days of severe frost in Brazil; and the rise of cereal prices, by a factor near 3, correlated with the bad harvests in 1972 in the USSR and in India. These examples illustrate the sensitivity of prices to relatively minor fluctuations in yield in recent times. Similar sensitivity apparently existed in the past (see, for example, the medieval Winchester record: Fig. 12.1). *Short-term interannual fluctuations of climate* play a much greater role in the economy than the weak long-term fluctuations (i.e., those on a time scale of decades and centuries). Unfortunately, such anomalies have been of only limited interest to meteorologists. In long-range forecasting their prediction seems to be, as yet, out of reach, while in classical climatology they are conceived simply as part of the statistical noise.

Disregarding the statistical properties of extreme events, they do tend to cluster in adjacent years. Recent examples of clustering are the three severe winters, 1977–79, in the USA, the Sahel drought years, 1969–73, and the severe winters, 1939–42, in Europe. Sometimes anomalies of different sign occur in the same group, for example the sequence of the wettest and driest year, 1917/18, in India, and the sequence 1916–18 in the equatorial Pacific. The special danger of such clustering to sub-

Fig. 12.1. Wheat yield and prices, Winchester 1211–1448. (Data from Titow, 1960, 1970)

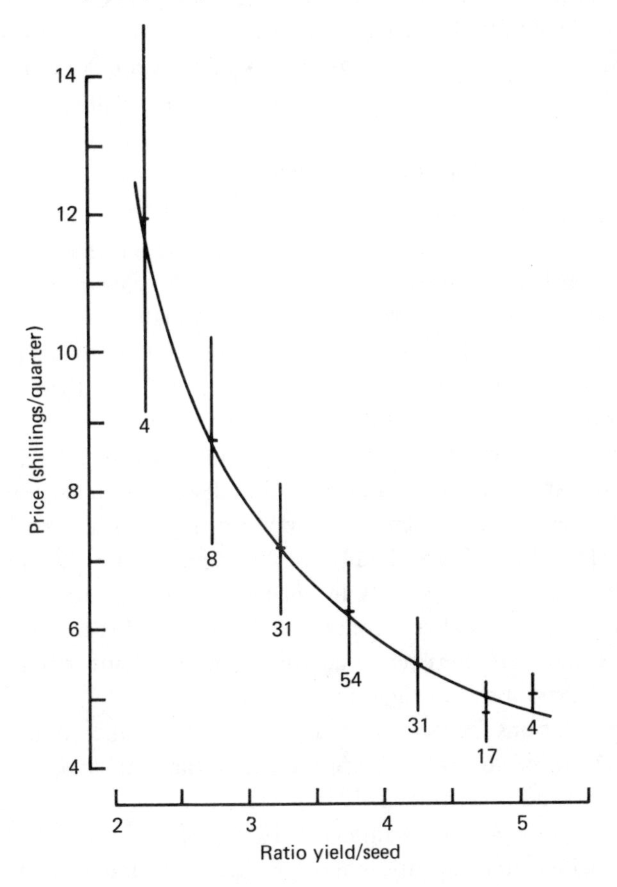

sistence agriculture, at least in marginal situations, has been stressed by Parry (this volume).

The case of the anomalous year, 1972, has drawn attention to the spatial distribution of such climatic extremes. Extremes often occur at the same time in widely separated parts of the globe: these 'links' are called *teleconnections* and have been investigated long ago in relation to the Indian summer Monsoon (Walker, 1923/4). Some climatic events are related to characteristic flow patterns in the upper troposphere westerlies. These quasi-stationary Rossby waves frequently penetrate deep into the tropics and are associated with blocking anticyclones. Such blocking situations are part of the physical background of teleconnections.

In long records of climatological parameters the time variation of running 30-year averages is usually much smaller than that of similar 30-year sequences of interannual variability. While the averages rarely change by more than 20 per cent, the variability (as measured, for example, by the standard deviation), changes by a factor of up to 2. A world-wide survey of time variations of variability is much needed. This would rely heavily on the availability of data before 1900.

Climatic anomalies and cereal harvests

Looking at the food problem, the sign of climatic anomalies producing bad harvests – or famines in the case of the subsistence agriculture of earlier centuries or more recently, in developing countries – depends on the crop type and on the regional climate.

In semi-arid or semi-humid areas moist and cool summers produce excellent harvests, while droughts are most feared. In humid western and central Europe, however, droughts only rarely cause harvest failures. In these regions bad harvests are mostly (but not exclusively) related to prolonged cold spells in spring and to cool and wet summers, expecially during the harvest period, when such conditions may cause the grains to rot in the fields. (For further details see, for example, Le Roy Ladurie, 1971: 288–92.) Such cold periods in spring or summer are known to occur (but not regularly) after a cold winter. This would seem to indicate a persistence of characteristic circulation patterns, but comparative statistical and synoptic investigations are needed.

Spatial correlations between distant climatic anomalies (teleconnections), believed to be controlled by properties of the atmospheric general circulation, may be of considerable importance to the world economy. The coincidence of a wet cool summer in 1976 in the central USSR (with high grain yields) and the drought in western Europe is a noteable example. From the agricultural point of view, these teleconnections deserve much more attention, especially since about 90 per cent of the world's grain export derives from the North American mid-western states with their imminent drought risk. A recent paper by Fleer (1978) has given ample evidence for teleconnections in the rainfall belt of the tropics and subtropics. These particular teleconnections are related to the Walker circulation (Bjerknes, 1969).

Regarding climatic teleconnections and their impact on agriculture and fisheries, much statistical work remains to be done, as numerical climate models do not yet take into account all details of topography and air–sea–ice interactions. The relationship between droughts in the American mid-west and conditions in the grain-producing area of the USSR and the humid European countries is one of the most urgent problems (Rauner, 1979). Another is the relationship between European circulation anomalies and tropical summer rains in India and in the Sahel/Sudan belt (Flohn & Nicholson, 1980).

Climatic data, vine harvests and cereal prices, fourteenth–eighteenth centuries

In addition to statistical studies based on recent instrumental data (which rarely extend back beyond the mid-nineteenth century), one of the most essential tasks is a comparative investigation of European historical records of temperature and rainfall. This covers at first early instrumental records, some of which (because of the unknown scale of early thermometers) require calibration using qualitative observations from documentary sources (Brumme, 1981).

The value of early weather diaries can hardly be overestimated (see Pfister, this volume). For the years 1508–18 and 1513–31 two partly overlapping diaries from adjacent towns are now under investigation: from Ingolstadt and from the Rebdorf monastery near Eichstätt, locations separated by not more than 30 km (Flohn, 1979). The latter is, for the last six years, practically complete. The learned prior of this monastery used the observations for a critical check of weather folklore rules, mostly with negative results. Although the evaluation of these diaries is not yet finished, preliminary analysis shows that the climatic data relate fairly well to combined rye prices (from Elsas, 1936–40) for Munich, Augsburg and Würzburg. Correlation coefficients between these three price series range from 0.82 to 0.86 for the period 1501–50.

During this period isolated peaks of prices occurred in 1517, possibly related to the cold winter of 1516/17 and the wet summer of 1517, and in

1529 and 1530, after a relatively wet summer in 1529. Immediately before the German peasants' revolt of 1524/5, however, no extraordinary weather events (or price changes) could be found in Bavaria.

The beginning of the vine harvest has frequently been used to indicate the average summer temperature at least before recent times (see, for example, Pfister, this volume, who points out some uncertainties in this relationship; and Piuz, 1974). In addition to the records collected by Le Roy Ladurie (1971), two other long series are available: Geneva 1560–1789 (covering 205 out of 230 years: Piuz, 1974) and Vienna 1523–1779 (226 out of 257 years: Pribram, 1938). While the Geneva record is highly correlated with France (1560–1619 complete: +0.748), Geneva and Vienna (separated by 800 km) naturally show greater deviations. Nevertheless, all available data from two consecutive subperiods between 1560 and 1779 (85 and 87 pairs) are significantly correlated (0.515 and 0.543) at the 0.1 per cent level. Piuz (1974) has correlated the wine harvest data (1760–90) with the temperatures during the vegetative period (April–September) with a resulting correlation of −0.68 which is significant at the 1 per cent level. This indicates that such data can (cautiously) be used for estimating individual summer temperature anomalies, especially when used in conjunction with records of wine quality.

Another useful source of data on past climate and climatic impact might be cereal prices if unrepresentative values (due to local events) can be eliminated. The derivation of area averages may be useful in eliminating events of purely local significance. Area averaging is, however, meaningful only if positive spatial correlations exist on a year-to-year time scale. As an example of such spatial correlations, grain prices from Frankfurt/Main (Elsas, 1936–40), Brügge, (Verlinden, 1959, 1965, 1972, 1973), Winchester, (Titow, 1970) and Exeter (Beveridge, 1929) for the periods 1350/1–1399/1400 and (in brackets) 1400/1–1450/1, are compared. The following correlation coefficients have been found:

Frankfurt–Brügge	0.435	(0.597)
Brügge–Winchester	0.355	(0.681)
Frankfurt–Winchester	0.378	(0.637)
Winchester–Exeter	0.768	(0.846)

The low values during the first period are caused by a one-year shift of some peaks between two records. Another cause is the high prices at Brügge between 1385 and 1389 possibly due to a local distortion. The number of degrees of freedom of the data is certainly smaller than 50 because of rather high autocorrelations (with one-year lag). For the first period these vary in the four series between 0.36 and 0.43. In the second period the autocorrelations are between 0.60 and 0.68. After 1401 it appears that the extremes of the marked famine period of 1437–39 contribute significantly to the high correlations; the prices rose by a factor of 2–4 (cf. Abel, 1966:63, for Germany).

A critical comparative investigation using all available data from western and central Europe – I have collected more than 25 long series between 1348 and about 1790 from England and the Netherlands, Germany and Austria – together with independent weather data from diaries and original chronicles promises to give insight into the possible role of the vagaries of weather (via harvest failures) in socio-economic events. The largest known volcanic eruption (Tambora in spring 1815, preceded by two other heavy eruptions) coincided with anomalous climate and cereal prices (Abel, 1966) in 1816 and 1817: see also Pfister (this volume). In many areas of Europe the 1816/17 cereal prices were higher than in any other year before 1950; only the catastrophic years 1315/16 may be a comparable case. Similar effects were observed in Japan, while in the eastern USA, 1816 was the year 'without a summer' (Landsberg & Albert, 1974; Post, 1977).

Another example of a severe famine triggered by climatic anomalies happened in 1770–72 (Fig. 12.2), with minor effects in the Lower Rhine region and in Belgium, but with a drastic increase in grain prices (by a factor of up to 4) in the Alps. During these years quite unusually severe cold periods in spring and summer occurred (Pfister, 1975; see also this volume, pp. 240 and 242).

Available meteorological data indicate that circulation anomalies between January and July were characterised by frequent blocking anticyclones over the British Isles (where 1771 was a dry year and 1770 was dry during the summer), together with frequent troughs and/or cold cyclonic centres in the Alps – similar to July 1954 with its abundant floods. This was the last severe famine in many countries of Europe (Abel, 1974). It was particularly felt in Scandinavia, in the mountain areas of Switzerland (Pfister, 1975), and in Saxony, where the number of inhabitants (1.63 million) dropped by about 6 per cent. In the mountains (Erzgebirge und Vogtland, mostly between 500 and 900 m.a.s.l.) the decline was greater, 8–9 per cent during 1770–72 (for further details see Abel, 1974: 200ff, especially 253).

Conclusions

These few examples allow the following conclusions:

1. Minor long-term fluctuations of temperature or rainfall are as-

Fig. 12.2. Selected rye prices 1701–1800 (harvest years, silver standard: g Ag/100 kg) at Xanten (Elsas-Archiv), Köln (Ebeling & Irrsigler, 1976), Namur (Ruwet, Hélin, Ladrier & van Buyten, 1966), Würzburg and Augsburg (Elsas, 1936–40). Note differences in relative height of peaks: e.g. 1770–2 very high in southern Germany, low in Belgium and in the lower Rhine region.

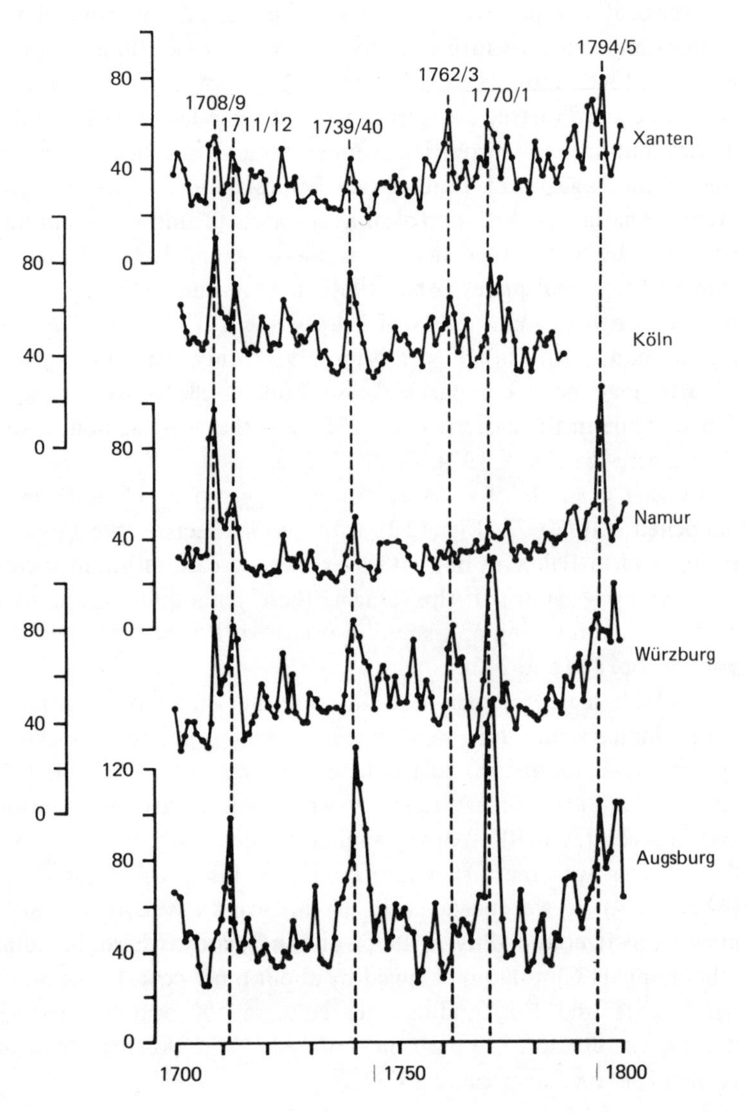

sociated with changes in the frequency of extreme weather events. Observations show that such extreme events often occur in groups of consecutive years. Until very recently their role in economic and social development has been greatly underrated.

2. Such extreme events often show long-distance spatial correlations as a consequence of atmospheric motion scales, with similar or opposite anomalies in far-distant areas of the globe (teleconnections). Their relation to regional anomalies of air–sea interaction (e.g. oceanic upwelling) or sea–ice should be investigated extensively using available instrumental data.

3. Historical sources (in European countries mainly after AD 1300) such as weather diaries and harvest data should be investigated not only at a local scale, but also for international comparisons. This facilitates the distinction between weather-related variations in harvest data and those caused by social and economic factors.

References

Abel, W. (1966). *Agrarkrisen und Agrarkonjunktur*, 2nd edn. Hamburg/Berlin: Parey.

Abel, W. (1974). *Massenarmut und Hungerkrisen im vorindustriellen Europa.* Hamburg/Berlin: Parey.

Beveridge, W. H. (1929). A statistical crime of the seventeenth century. *Journal of Economic and Business History*, **1**, 503–33.

Bjerknes, J. (1969). Atmospheric teleconnections from the equatorial Pacific. *Monthly Weather Review*, **97**, 163–72.

Brumme, B. (1981). Methoden zur Bearbeitung historischer Meß-und Beobachtungsdaten (Berlin und Mitteldeutschland 1683–1770). *Archiv für Meteorologie, Geophysik und Bioklimatologie.* (in press).

Ebeling, D. & Irrsigler, F. (1976). Getreideumsatz, Getreide und Brotpreise in Köln 1368–1797. *Mitteilungen aus dem Stadtarchiv von Köln*, vol. 65. Heft, Köln-Wien, Böhlau Verlag.

Elsas, J. (1936–40). *Umrisse einer Geschichte der Preise und Löhne in Deutschland.* (3 vols.) Leiden.

Elsas-Archiv, University of Göttingen: unpublished data, courtesy of Dr H. Saalfeld.

Fleer, H. (1978). Statistische Analysen und Telekonnektionen der Niederschlagsschwankungen in der Tropenzone. PhD dissertation, University of Bonn, (see also *Bonner Meteorologischen Abhandhlungen*, **26** (1981).

Flohn, H. (1979). Zwei bayrische Wetterkalender aus der Reformationszeit. *Mitteilungen aus dem Meteorologischen Institut der Universität München*, **35**, 173–7.

Flohn, H. & Nicholson, S. (1980). Climatic fluctuations in the arid belt of the 'Old World' since the last glacial maximum; possible causes and future implications. *Palaeoecology of Africa*, **12**, 3–21.

Le Roy Ladurie, E. (1971). *Times of feast, times of famine. A history of climate since the year 1000*, transl. B. Bray. New York: Doubleday. 426 pp.

Landsberg, H. E. & Albert, J. M. (1974). The summer of 1816 and volcanism. *Weatherwise*, **27**, 63–66.

Pfister, C. (1975). Agrarkonjunktur und Witterungsverlauf im westlichen Schweizer Mittelland zur Zeit der ökonomischen Patrioten 1755–97. *Jahrbuch der Geographischen Gesellschaft von Bern*, 2.

Piuz, A. M. (1974). Climat, récoltes et vie des hommes à Genève, XVIe–XVIIIe siécle. *Annales: Economies, Sociétés, Civilisations*, **29**, 599–618.

Post, J. D. (1977). *The Last Great Subsistence Crisis in the Western World*. Baltimore: The Johns Hopkins University Press.

Pribram, A. F. (1938). *Materialien zur Geschichte der Preise und Löhne in Österreich*, vol. 1, Wien.

Rauner, Y. L. (1979). Synchronous recurrence of droughts in the northern hemisphere and crop yield fluctuations. Lecture at USSR–USA Symposium on Climate Modelling, Climatic Change and Statistics, Tbilissi.

Ruwet, J., Hélin, E., Ladrier, F. & van Buyten, L. (1966). *Marché des Céréales a Ruremonde, Luxemburg, Namur et Diest aux XVIII siècles*. Louvain: University of Louvain. 492 pp.

Titow, J. (1960). Evidence of weather in the account rolls of the Bishopric of Winchester 1209–1350. *Economic History Review*, **12**, 360–407.

Titow, J. (1970). Le climat á travers les rôles de compatibilité de l'évèche de Winchester (1350–1450). *Annales: Economies, Sociétés, Civilisations*, **25**, 312–50.

Verlinden, C. (1973). *Dokumenten voor de Geschiedenis van Prijzen en Lonen in Vlaanderen en Brabant*, Deel I (1959), II (1965), III (1972), VI. Brügge: DeTempel.

Walker, J. T. (1923/4). Correlations in seasonal variation of weather. *Memoirs of the Indian Meteorological Department*, **24**, 75–131 and 275–332.

13

Climatic change and the agricultural frontier: a research strategy

M. L. PARRY

Abstract

There has been a tendency to evaluate the historical significance of climatic change by inductive studies of the spatial and temporal coincidence of economic and climatic events in the past. Such studies incorporate the weakness of inductivism: they lack an unbroken logic which establishes the relationships between the various propositions, and they have not sought to test the propositions either empirically or theoretically. There is, therefore, little evidence that the (supposedly related) economic event is contingent to the climatic event. One way of establishing more precisely the contingent relationships between climate and economy in the past is to adopt a 'retrodictive' strategy. This needs to be based on an understanding of the processes linking climate, agriculture and settlement derived from analogue models of contemporary crop–climate links, from studies of the interaction between weather and farming decisions, and from the analysis of the different effects on agriculture of long- and short-term climatic change. Retrodictions (or rational reconstructions of history) made on the basis of this understanding, can be tested against historical actuality, and their confirmation can point to a causal relationship. The retrodictive strategy is demonstrated with examples which focus on the connection between permanent land abandonment and climatic cooling in late medieval Britain. Retrodicted areas of farmland which may have become marginal to cereal cropping as a result of possible changes in climate over the period AD 1200–1700 correlate closely with extensive areas of farmland and numerous settlements abandoned during this period. The retrodictive strategy is advocated as a technique which would be useful in a variety of climate-history studies.

Introduction

The task of evaluating the impact of changes of climate on the path of economic history is an extraordinarily difficult one. It embraces two disciplines which have traditionally adopted very different paths of enquiry and which require from their disciples very different realms of expertise. Perhaps we should not be surprised that, in general, climatologists make poor historians and that historians make poor climatologists, but this simple fact probably accounts both for the curious ambivalence of economic historians to climatic change and for the lack of success that climatologists have had in their attempts to measure the economic implications of climatic change. Part of the solution to this problem is, of course, to be found in collaboration between historians and climatologists, but the contention of this chapter is that a solution of the problem also requires the adoption of a particular research strategy.

The ambivalence, sometimes outright hostility, of economic historians to studies of the role of climate in history may be explained in several ways. First, some historians have an incomplete or outdated understanding of the ways in which climates change, and how much and at what times in the past they have changed. For example, the chronology proposed by Brooks in 1949 (Brooks, 1949) – and now superseded – is still quoted by some historians as evidence that the early medieval period was 'one of Europe's cold cycles' (Postan, 1975). This had led to the assertion that village desertions in England had little to do with climate because few occurred during or shortly after the supposed cold period (Beresford, 1971). In fact, there is a broad coincidence between the phase of most widespread desertion, AD 1370–1500, and a phase of more frequently cool and wet summers detected by Lamb in fifteenth-century England (Lamb, 1977; Parry, 1978). This coincidence should not, of itself, encourage us to draw a connection between the two events, but it at least suggests that the enquiry should be re-opened.

Secondly, historians may not always appreciate that climatic variations have very different effects on different regimes of climate. For example, the growing conditions of summer in maritime climates are much more seriously diminished by overall cooling than those in more continental climates. It may be misleading, therefore, to describe a long-term climatic fluctuation of 1 deg C over western Europe as 'slight, almost intangible' (Le Roy Ladurie, 1972:244). In areas of marginal farming under a maritime regime, such changes would have a pronounced effect on the length of the growing season (Parry, 1978).

Thirdly, there is not always a complete understanding of the very different biological and agronomic effects produced by short-term and long-term changes of climate. Short-term 'runs' of unseasonable weather which warranted the remarks of contemporary observers may, over the long term, have balanced themselves out. The ambivalence of such historians as Beresford and Postan to climatic change seems to have resulted in part from disappointing comparisons of short-term climatic variations with long-term changes in agriculture and rural settlement, but we should not be surprised to find little connection between the two.

Most important amongst those factors affecting the historian's attitude to the role of climate in history has been his low opinion of the performance of climatologists in evaluating the economic and social impact of climatic changes. In particular, climatologists have sometimes failed to grasp the peculiar constraints imposed by historical problems on the deductive strategy which they have traditionally employed. There are two realms of misunderstanding. The first relates to the fact that, while historical explanation is frequently the product of what can broadly be termed a deductive approach, historians cannot state all the premises behind their explanations (which are simply too numerous), and thus historical explanations are *probabilistic*: historians seek to mention only some of the 'contingent conditions' for occurrences (Nagel, 1961). The probability of a given statement thus measures the confidence the historian has in the truth of the statement, and this probability he frequently describes by use of such terms as 'primary cause', 'secondary cause', etc. The statement thus has an uneliminable subjective component, and is not open to analysis in the strictly deductive form which is most familiar to the climatologist. Perhaps because of the difficulty in defining a rigorous framework for hypothesis testing, there has been an unfortunate tendency among climatologists to assume that the testing of predictions is inappropriate in historical investigations of this kind. Despairing of the viability of a more rigorous approach, climatic historians have sometimes resorted merely to an investigation of the synchroneity of climatic and economic events – a search for a temporal or 'phased' fit between two sets of events and a search which has involved little scrutiny of causality. This has been the second failing: to assume that events occurring both synchronously and in the same place are events which necessarily have a causal connection – yet all that can be established by this argument is a space–time coincidence. In several studies which have adopted this line of reasoning there is little evidence that climatic change was a *contingently necessary condition* of coincident economic change.

An example may illustrate this point. A recent investigation of the role of climate in Mycenaean decline has focused on the possibility that the incidence of intense drought, analogous to conditions which occurred in the eastern Mediterranean for a few weeks during the winter of 1954/55 but more prolonged, could have induced famine and hence out-migration by Mycenaean farmers around 1000 BC (Bryson, Lamb & Donley, 1974). A number of problems are raised by this analysis, but the point to emphasise here is that the premise of a connection between reduced rainfall, famine, and out-migration is not examined. Precisely what agrometeorological effect would the 1955 type of drought have had on the farming systems of 1000 BC? The processes linking rainfall, yield and farming decisions need to be traced before it is reasonable to venture a connection between a reduction in one and a change in the other. Without a study of such processes, the conclusion relating cultural decline to climatic change, whether it be southern Greece, northwest India or Roman north Africa, is derived from an inductive form of reasoning and incorporates the weaknesses of inductivism; that is, it lacks an unbroken logic which establishes the relationships between the various propositions and it has not sought to test the propositions either empirically or theoretically.

Now, the historian is faced daily with myriad events that are connected by time and space. He seeks to identify *only* those events that are contingent conditions (i.e. explanations) of his 'problem' event, and would rightly be alarmed at the historical climatologist's apparent willingness to rest content with a space–time coincidence as evidence of a causal link. There is, therefore, a particular need for a research strategy that seeks to establish the contingent relationships between climate and economy in the past.

The retrodictive strategy

There are two elements in the research strategy that require special attention. First, there is the need to formulate strategies which incorporate the powerful predictive tool of the deductive method. One way in which the climatic historian can do this is to predict (or 'retrodict') a rational reconstruction of history and then test this against historical actuality (Fig. 13.1). Use of the retrodictive method is limited in historical studies because historical problems are not often composed of two sets of dichotomous data, and because empirical studies of present-day processes linking the data are not generally feasible owing to the complex and changeable nature of socio-economic processes. However, *both* these requirements are satisfied in many problems facing the climatic historian. For example, it is possible to 'retrodict' the effect of climatic fluctuations on crop yields in the past by using analogous data of the effect of weather on yields in the present, allowing for differences in these relationships with developments in crop strains over time. This makes it possible to fulfil the second element in the strategy: the modelling of processes linking economic activity and weather in the past, so that confirmation of a retrodiction based on an understanding of these process links is more than a statement of a space–time coincidence but is a statement about a functional relationship. If the retrodiction is refuted, then the model requires re-examining; if it is confirmed, then this evidence can be compared with data from independent sources. If there is a convergence of evidence to support the conclusion then the hypothesis can be said to be confirmed (Fig. 13.1).

The retrodictive strategy outlined above is valuable in a wide range of problems met at the interface of climate and agricultural history. This may be illustrated by reference to a specific example in which the problem is to explain the extensive and permanent abandonment of high-level farmland (250 m OD and above) in northern Europe (Fig. 13.2). The range of alternative explanations may include an error of farming judgement, change in farming objectives (for example, the change in tolerance of different levels of yield), change in farming systems (parti-

Fig. 13.1. A 'retrodictive' strategy in historical research.

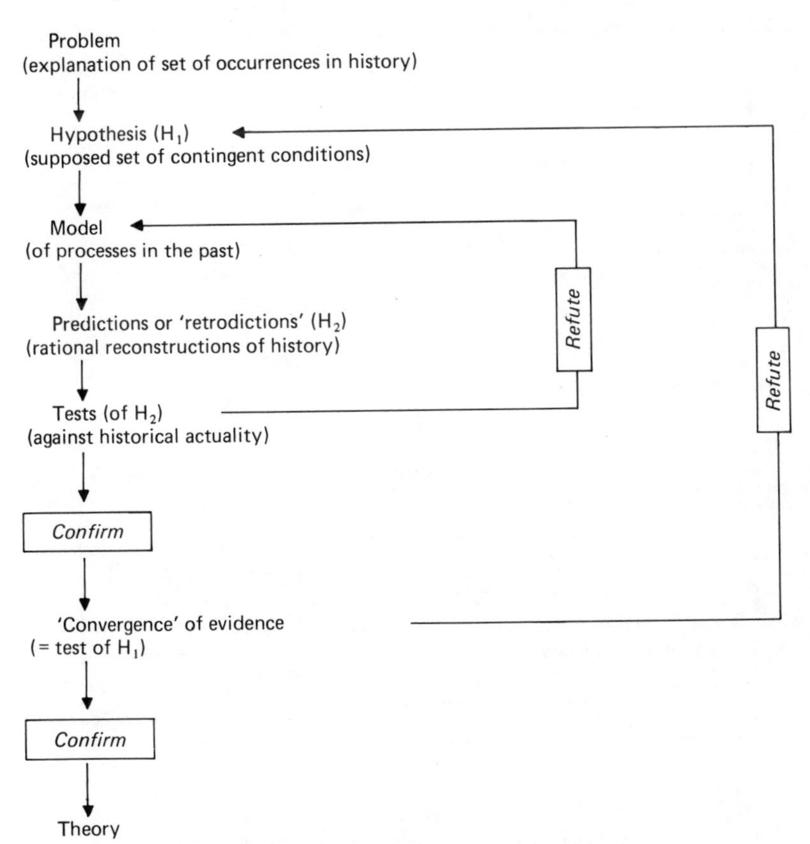

cularly changes in technology enabling an increase of yield or productivity elsewhere), and climatic change leading to changes in the physiological viability of farming and to changes in productivity. In the case under review the hypothesis is that reduced warmth and increased wetness are factors which partly explain the high-level abandonment. In

Fig. 13.2. The retrodictive strategy: climatic change and land abandonment. Note: d/d = summer warmth in day-degrees above 4.4 °C; PWS = end-of-season potential water surplus.

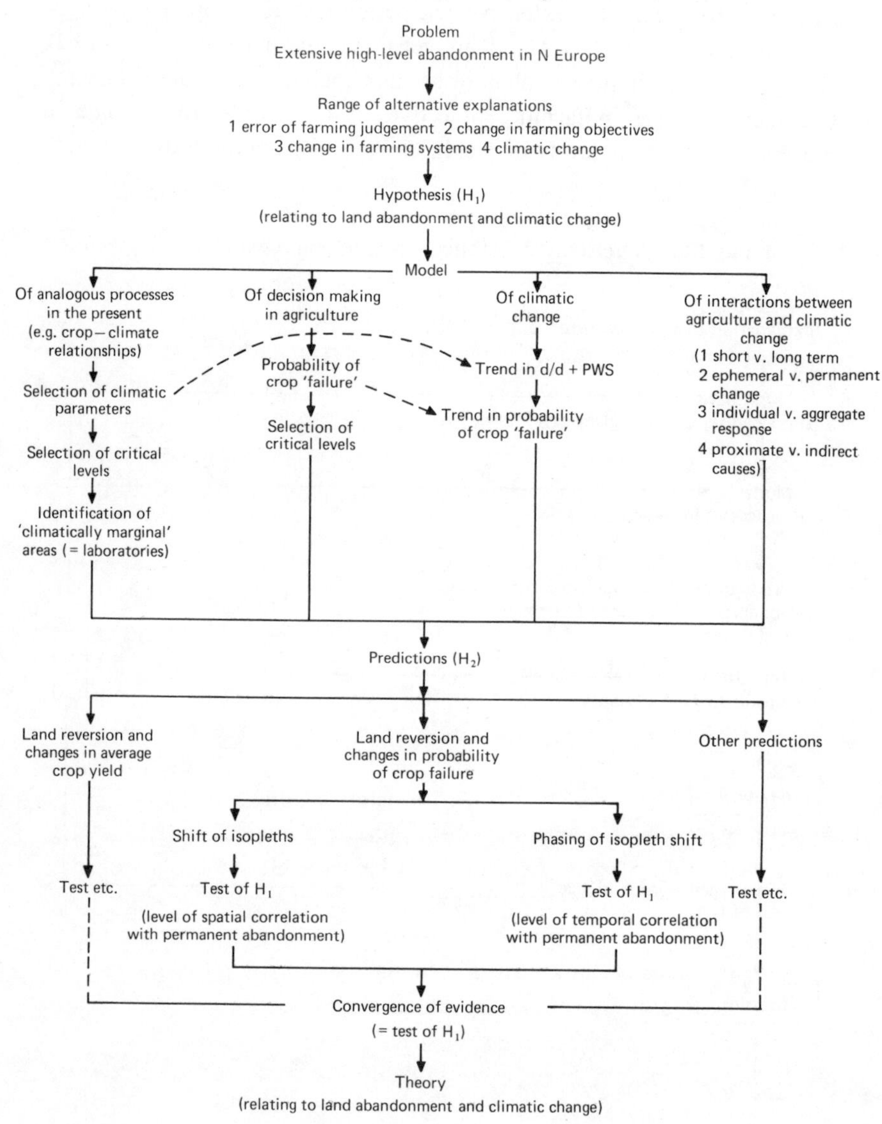

the research strategy outlined here, emphasis is placed on two areas of activity: the precise modelling of processes linking climate and farming decisions, and the testing of predictions made on the basis of such modelling.

Modelling the processes

Our understanding of the relationships between climate and agriculture in the past can be increased by attention to four sets of processes: (*a*) analogous crop–climate processes in the present, (*b*) the interaction between uncertain weather and farming decisions, (*c*) changes of agro-climates and (*d*) the interaction between agriculture and short- and long-term changes of climate (Fig. 13.2).

Analogous crop–climate processes It is possible to comprehend the effects of past changes of climate on crop yields by adopting present-day processes as analogues of those in the past. This can provide a degree of detail to our understanding of crop–climate processes so that the more difficult task of understanding processes at work in the past stands a reasonable chance of success. In particular, it is important to establish those parameters of climate to which the main agricultural activity was most sensitive. In much of northwest Europe extensive high-level cultivation was largely based on the cropping of oats. A study of both modern and early varieties of oats reveals that oats cultivation above 250 m OD in Britain is particularly sensitive to three climatic factors: exposure, summer warmth and summer wetness. The most effective measures of these restraints are average windspeed, accumulated temperature and an 'end-of-summer' potential water surplus (PWS) (Parry, 1975). The relative role of these parameters in constraining the upward advance of oats cultivation will have varied throughout northern Europe according to regional differences in the regime of summer warmth and wetness. But for any particular region it it possible to establish minimum levels of these parameters required for critical phenological developments in the oats crop and, by drawing isopleths of these levels, to establish the zone which is climatically marginal for oats-cropping.

In southern Scotland minimum levels of summer warmth for the ripening of oats were established as being 1050 day-degrees above a base of 4.4 °C. Maximum levels of PWS and exposure were established at 60 mm PWS and 6.2 m/s average windspeed (Parry, 1975). At the same time it was also possible to establish the approximate levels of warmth, wetness and exposure critical to commercial cereal cultivation: 1200 day-degrees, 20 mm PWS and 5.0 m/s average windspeed. Combined isopleths of these three parameters, drawn around the hillsides of southern Scotland, indicate the zone that is climatically marginal to the cropping of oats – the zone in which oats-cropping is particularly sensitive to changes of climate.

In isolating such zones we are, in a sense, establishing a 'crop ecotone' which provides us with a suitable laboratory in which the effects of climatic changes can more readily be studied because their effects are likely to have been particularly marked. Climatically marginal zones of this kind can be established, in a similar way, for a variety of cropping systems and for a range of different areas. Fig. 13.3 indicates the

Fig. 13.3. Climatically marginal land in the British Isles. (After Parry, 1978).

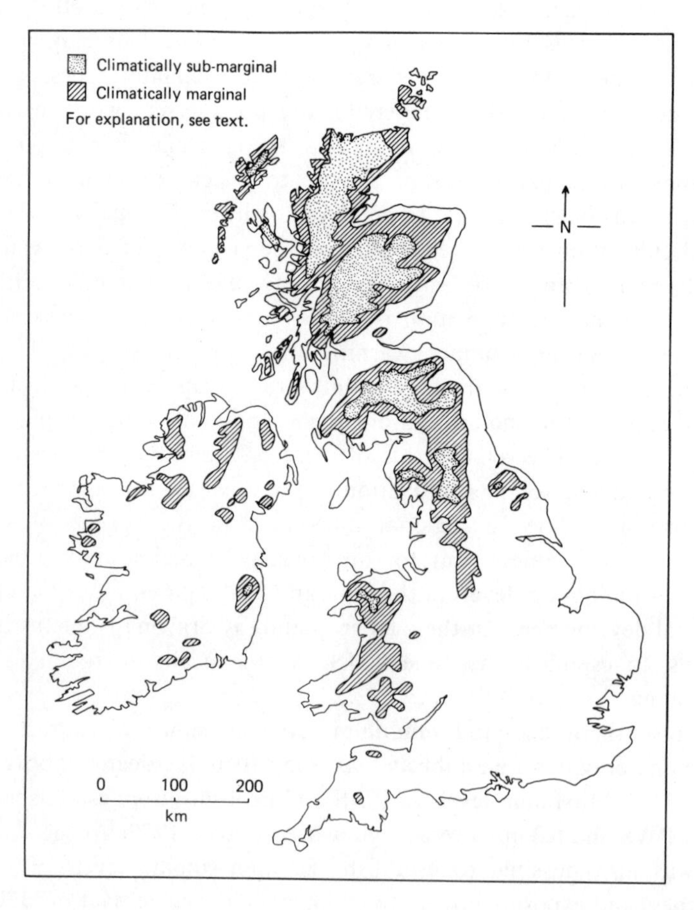

approximate distribution of climatically marginal land in the British Isles mapped from published data of accumulated temperature alone. The decision to reclaim or abandon farmland in this zone is likely to have been influenced strongly by the farmer's perception of the climatic viability for cultivation.

Farming decisions and uncertain weather Any evaluation of the connection between climatic change and farmland abandonment ultimately focuses on those factors which the farmer considered in weighing his decision to abandon farmland. It cannot be assumed that long- or medium-term changes in yield, which would be important to the modern commercial farmer, necessarily played a similar role in the past. In marginal areas of semi-commercial or subsistence cropping an important factor in the viability of farmland in pre-modern times was its ability to sustain the farming population from one year to another. The size of yield above this minimum was probably less significant than the likelihood of achieving the minimum in a given year. Failure of the harvest, particularly in two successive years, was disastrous since it led to the consumption of seed grain and the foreclosing of farming options for the future. The probability of outright failure of a harvest may thus have been at least as important as average yield over a run of years. This is vital to any evaluation of the effect of climatic variability on farming viability because, given the low seasonal range of marginal maritime climates such as those in upland Britain, the probability of accumulated temperatures falling below a critical level required for the ripening of oats increases with elevation, not linearly, but in a characteristically S-shaped curve. Fig. 13.4 illustrates the curve of 'probability of harvest failure' (i.e. probability of occurrence of a growing season of only 970 day-degrees above 4.4 °C) for the Lammermuir Hills in southeast Scotland. At the lower end of this curve, which broadly coincides with marginal foothill cropping in Britain, the probability of harvest failure increases quasi-exponentially with elevation. At the limit of cultivation in southern Scotland (*c.* 300 m OD) the frequency of harvest failure is about one year in five and doubles with a 50 m increase in elevation. Changes in the frequency of two consecutive failures are even more marked, increasing 100-fold over 150 m (Parry, 1976).

Changes of agro-climate This discussion of climatic constraints on cereal cropping and of their influence on farming decisions leads to the following conclusions.

If changes in climate are to be evaluated in farming terms, then these changes need to be expressed, first, in measures appropriate to the constraint they impose on cropping (i.e. in accumulated temperature, PWS and exposure) and, secondly, in terms appropriate to the effect they may have exerted on farming decisions (i.e. in frequency of crop failure).

Trends of accumulated temperature and PWS for AD 1100–1950 have been estimated from Lamb's data for central England (Parry, 1975). If Lamb's data are correct, the scale of fluctuation was substantial. During the first phase of cooling, about AD 1250–1450, average summer warmth fell by 175 day-degrees in southeast Scotland and PWS increased by about 70 mm. While mean maximum temperatures may have fallen by

Fig. 13.4. Probability of 'harvest failure' in southeast Scotland. (After Parry, 1976).

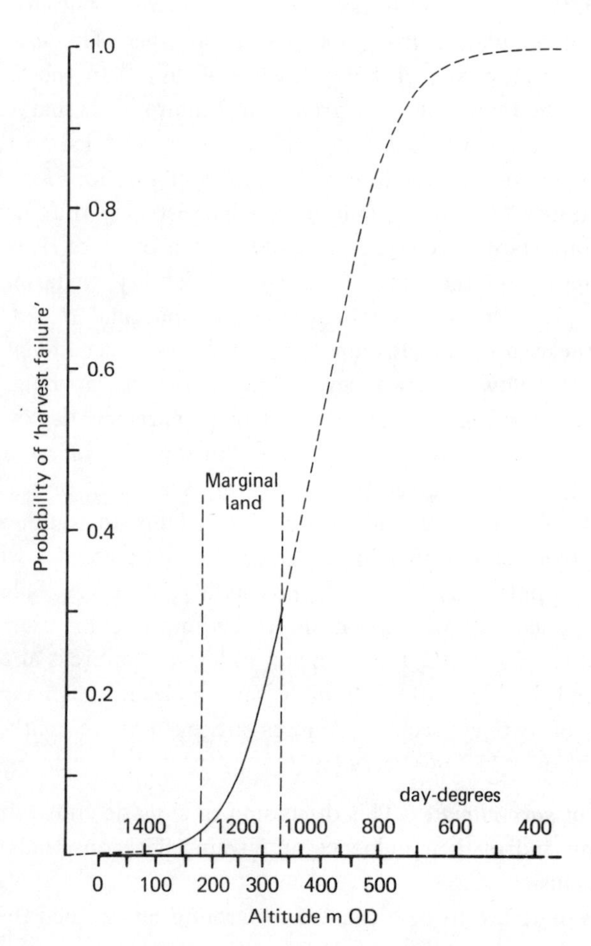

less than 1 deg C, summer warmth was thus reduced by almost 15 per cent and the frequency of harvest failure at 300 m OD increased from about one year in twenty to one year in three. In the subsequent two centuries mean maximum temperatures may have fallen by less than 0.5 deg C, summer warmth by about 7 per cent and the average frequency of failed harvests increased to two years in five. It should be emphasised that these are broad estimates, their accuracy being constrained by the precision of existing data (Parry, 1978).

There are two implications that arise from these climate-induced trends in growth potential. First, there may have been substantial changes in the viability of upland farming over time. Some high-lying farmland, near the present cultivation limit, has at one time been climatically submarginal and at another supra-marginal. Secondly, there may have been substantial shifts of the theoretical limit to cultivation, which has risen and fallen with the fluctuation of the potential for upland plant growth over the last millennium. Isopleths of the climatic limits to cropping, drawn for selected intervals between AD 1100 (the peak of the supposed medieval 'optimum') and AD 1700 (the nadir of the subsequent cold phase) would thus display an overall 'fall' in theoretical climatic limits to cultivation.

The nature of short- and long-term interactions These findings may be used to formulate a prediction about the agricultural response to climatic change which may be tested against the historical actuality. But the nature of such a prediction depends on certain assumptions and distinctions which must first be made explicit and defended.

Our basic premise is that *permanent* land abandonment in northern Europe is somehow connected with *long-term* climatic change. The assumptions underlying this contention are as follows (Fig. 13.2). First, a distinction is assumed between the very different effects of long- and short-term changes of climate. Short-term climatic changes might be expected to have a short-term (but not a long-term) agricultural effect, unless they trigger a response that has been pre-conditioned by other long-term factors. (The latter concept is necessary to explain such phenomena as the failure to re-settle an area initially deserted in response to a run of harsh years, once more benign climatic conditions had returned). Conversely, it is reasonable to assume that long-term climatic changes could have long-term agricultural effects.

Secondly, a distinction is assumed between ephemeral (or temporary) agricultural response and permanent agricultural response; and between single events which are the products of individual decisions (e.g. the desertion of single farms) and aggregate events which constitute the actions of many men (the desertion of entire farming regions).

Finally, we assume a distinction between hypotheses of climatic change as proximate or precipitating causes of agricultural change and those which are underlying or indirect causes of agricultural change.

The branches of these dichotomies (short-term/long-term; ephemeral/permanent; individual/aggregate; proximate/indirect) are interlocked. For example, if the concern is to explain the long-term (or permanent) desertion of farmland, which was the aggregate product of the decisions of many farmers, then the concern will be with long-term or step-wise climatic changes (since wealthy farmers may survive two or three hard years but will go down with the rest over the long term), and with climatic changes as underlying or indirect causes of farm desertion. On the other hand, if the concern is to explain ephemeral events such as the temporary desertion of a few farms, then the interest will focus on short-term climatic change as a proximate or precipitating factor.

The distinction has thus been made between brief climatic fluctuations as possible proximate causes of short-term economic events, and long-term climatic change as a factor that might induce long-term changes in the environmental resource base. Upon this shifting resource base, rather like a moving stage, play the range of proximate factors that affect the viability of agriculture: fluctuations of market prices, changes in farming technology, the passage of epidemics and wars – even short-term 'runs' of good or bad weather (Fig. 13.5). Marginal agriculture and settlement

Fig. 13.5. Proximate and indirect factors behind land abandonment or settlement desertion. $M_{1...3}$ = marginal tolerance of resource scarcity (e.g. highest frequency of crop failure, lowest average yield tolerated, etc.); ta = temporary land abandonment; pa = permanent land abandonment.

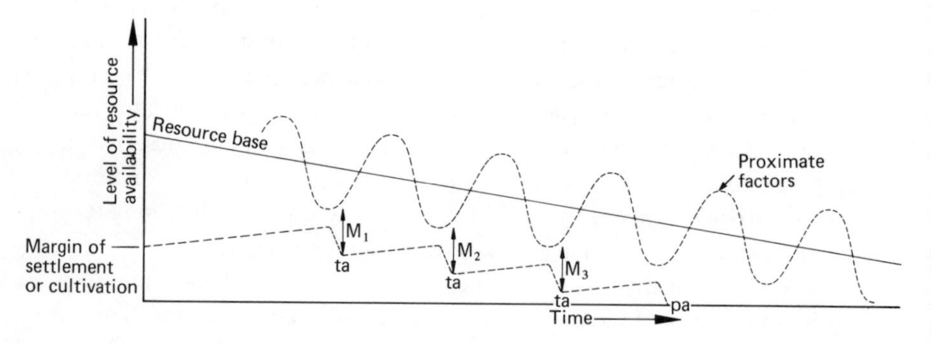

may respond to such forces in a limited and ephemeral way: there may occur the short-term retreat of such margins where the marginal tolerance of resource scarcity has been temporarily exceeded. But extensive and permanent retreat is likely to occur only where long-term changes in the resource base, perhaps (though by no means always) induced by climatic change, have reduced the stage upon which the proximate factors play. It may be argued that without this secular decline the proximate factors would have promoted a less widespread and a less permanent response.

In the schema that has been drawn, the proximate factors are the stuff of orthodox history – political, economic and social forces which were perceived by individuals and were responded to in their individual and collective actions; and these surely account for the great proportion of events that historians have sought to explain. But the degree to which agriculture and settlement were responsive to such factors depended, in part, upon their marginality. A decline in their resource base would make them particularly sensitive to any social and economic changes that touched upon their viability. An example of the complementary role of resource base and proximate factors is given in Fig. 13.6. It portrays, in schematic form, the possible response of the upper limit of cultivation in Scotland over the period AD 1200–1700 to short-term political, economic and social events which were played out on a stage shifting with trends in summer warmth and summer wetness. The diagram emphasises the contention that, while political upheaval, the Black Death, the decline of the monasteries, or even runs of disastrously poor summers in the 1590s and 1690s, may have *triggered* the retreat of high-level

Fig. 13.6. Proximate and indirect factors behind farmland abandonment in upland Scotland: a schematic chronology.

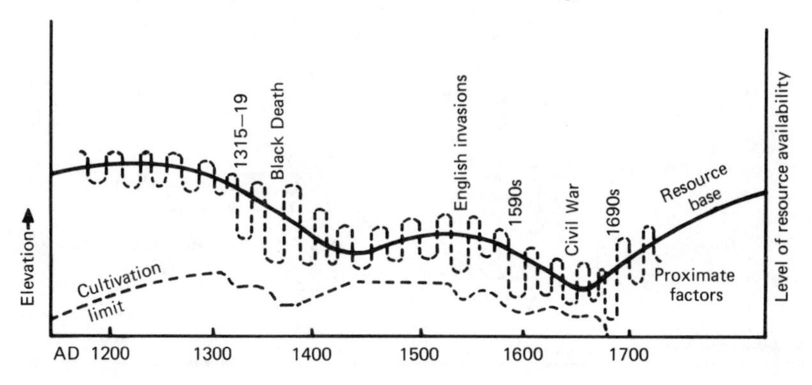

cultivation, it was the operation of these forces on a marginality enhanced by climatic deterioration in the fourteenth and seventeenth centuries that produced such a marked and lasting effect.

Testing retrodictions

Given an understanding of the relative roles of proximate and indirect factors, of the chronology of climatic change, and of the functional links between climate, crop yield and the frequency of crop failure it is possible to make a retrodiction about changes in the upper limit of cultivation (Fig. 13.2). In this case the retrodiction is that the shift of isopleths of critical levels of the probability of crop failure is correlated with areas in which the permanent abandonment of farmland most frequently occurred. Had there been a perfect response in shift of *real* cultivation limits to shift in *theoretical* limits to cultivation then permanent reversion would be found exactly in the area defined by shifting theoretical limits, and would have occurred in phase with this shift. In fact a perfect response is not to be expected (owing to the many

Fig. 13.7. Abandoned farmland and lowered climatic limits to cultivation in southeast Scotland, AD 1300–1600. (After Parry, 1975)

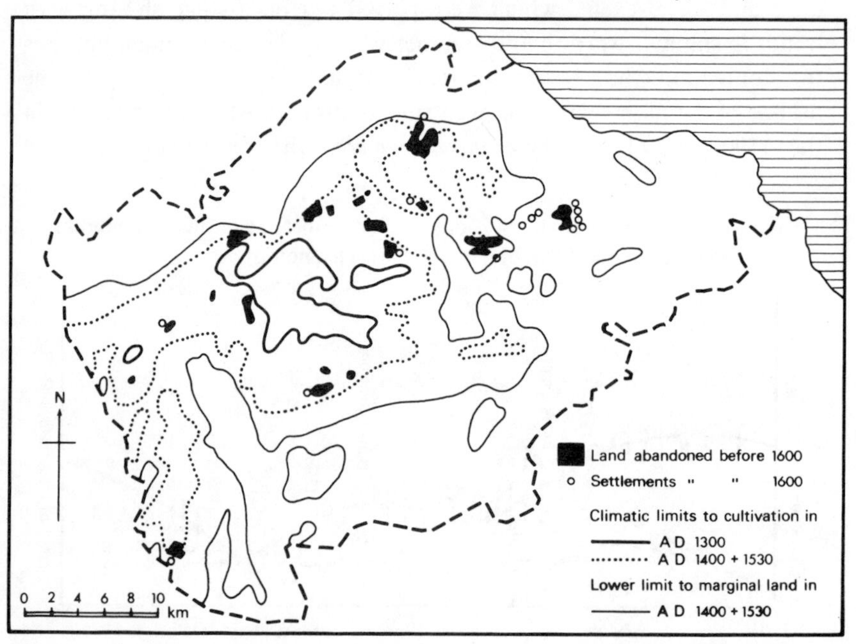

proximate and indirect factors operating), but we can predict that reversion will be focused in this area.

This prediction (or rather retrodiction) can now be tested. In the Lammermuir Hills in southeast Scotland there is a close temporal and spatial fit between the distribution of permanently abandoned farmland mapped from aerial photographs and the zone over which theoretical limits to cultivation fell between AD 1300 and 1600, and AD 1600 and 1800 (Figs. 13.7 and 13.8). Between AD 1600 and 1750 three out of an existing total of fifteen farmsteads were abandoned within the zone that probably became submarginal by about 1530. A further eight were deserted in the zone which became submarginal between 1530 and 1600, and all but one of the remainder was in the marginal zone. Moreover, of 2990 ha of cultivation which had reverted to permanent moorland between AD 1600 and 1800, about three-quarters had become climatically submarginal before 1600. The retrodiction is thus confirmed. Since the predicted space–time correlation was derived theoretically and empirically from an understanding of process links it is logical to assume

Fig. 13.8. Abandoned farmland and lowered climatic limits to cultivation in southeast Scotland, AD 1600–1800. (After Parry, 1975)

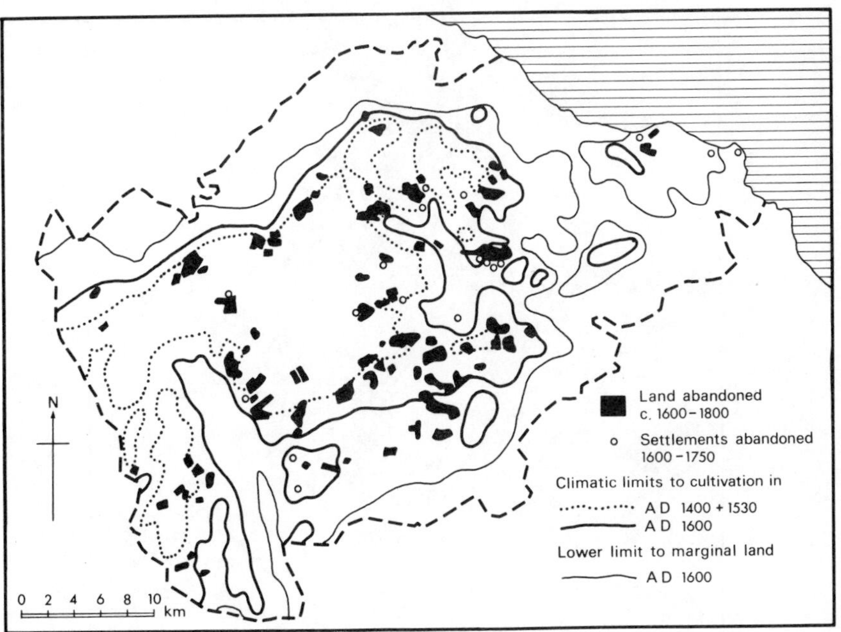

that the observed correlation reflects a causal relationship between climatic change and land abandonment. Of course we can never be certain of this causality – the evidence simply does not exist in sufficient detail; yet it would have been remarkable if the highest cultivation had not been abandoned, because it would have experienced an average frequency of two failed harvests in five towards the end of the seventeenth century. Nevertheless, a note of caution is due. The estimates of temperature and precipitation used in our calculations necessarily have a wide margin of error and, more importantly, are partly based upon data from agricultural diaries. We should therefore be wary of circular

Fig. 13.9. 'Retrodicted' shift of cultivation limits in the British Isles, AD 1300–1600. The 'recurrently marginal' zone is that over which theoretical climatic limits to successful cereal cultivation have shifted during this period. (After Parry, 1978)

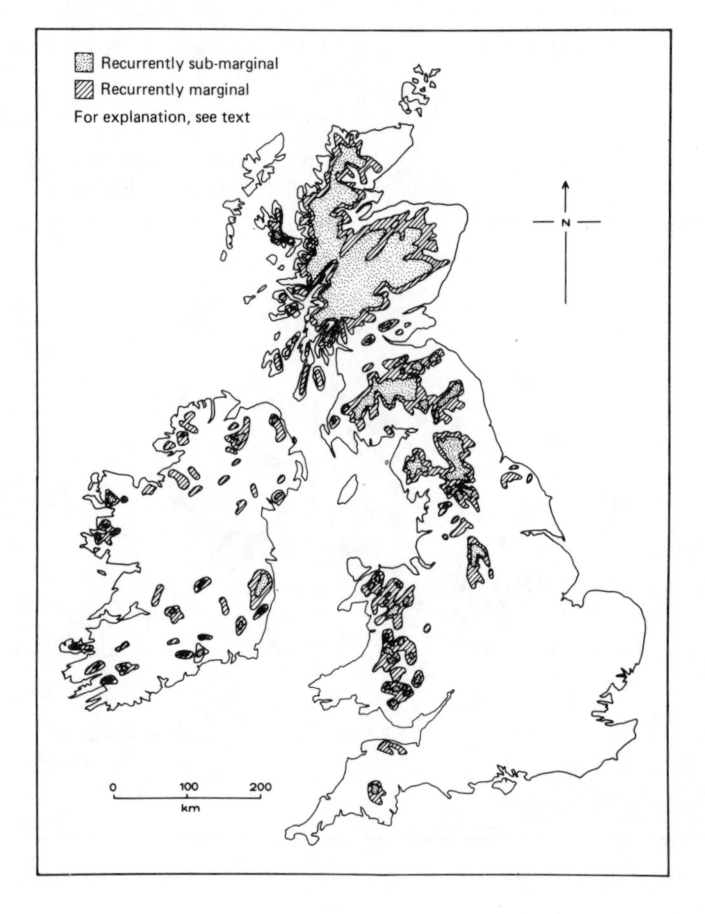

argument. Further discussion of this problem will be found in Parry (1978). Moreover, it should be emphasised that this chapter has focused on only one of a range of alternative explanations for land abandonment. Other explanations hypothesised in Fig. 13.1, such as errors of farming judgement, changes in farming objectives and changes in farming systems, remain to be tested. Any theory of land abandonment must draw upon a mix of these explanations, and I have emphasised elsewhere that, when balanced against the alternatives, climate change proves to be very much the subordinate factor (Parry, 1978).

It is also possible to retrodict the distribution of permanent land abandonment in the same way for a similar period, AD 1300–1600, for the British Isles as a whole (Fig. 13.9). This is based solely on temperature accumulation and does not include regional and local variations in PWS and exposure. Nevertheless, it points to the area where the economic historian should look for the full economic consequences of climatic cooling. This 'recurrently marginal' land covers more than two million hectares, or about one-third of Britain's presently unimproved moorland; and if the pattern displayed in the Lammermuir Hills is characteristic of Britain as a whole, a retreat of agriculture and settlement in these areas may have occurred in the late Middle Ages. This is a 'retrodiction' which remains to be tested and awaits the attention of the economic historian.

Conclusion

There is a need to follow a more rigorously deductive path in evaluating the economic significance of changes of climate. The coincidence of climatic and economic events in space and time cannot be taken as an adequate indication of a causal connection, but causality can be evaluated by use of a retrodictive strategy based upon the careful modelling of processes linking weather and agriculture. The retrodictive technique is a valuable research strategy which should be useful in a wide range of studies designed to investigate the economic impact of long-term climatic fluctuations.

References

Beresford, M. W. (1971). A review of historical research (to 1968). In *Deserted Medieval Villages*, ed. M. W. Beresford & J. G. Hurst, pp. 21–75. London: Lutterworth Press.

Brooks, C. E. P. (1949). *Climate through the Ages*, 2nd edn. London: Benn.

Bryson, R. A., Lamb, H. H. & Donley, D. L. (1974). Drought and the decline of the Mycenae. *Antiquity*, **48**, 46–50.

Gregory, S. (1954). Accumulated temperature maps of the British Isles. *Transactions of the Institute of British Geographers*, **20**, 59–73.

Lamb, H. H. (1977). *Climate: Present, Past and Future*, vol. 2, *Climatic History and the Future*. London: Methuen.

Le Roy Ladurie, E. (1972). *Times of Feast, Times of Famine*. A History of climate since the year 1000, transl. B. Bray. London: Allen & Unwin.

Nagel, E. (1961). *The Structure of Science*. London: Routledge.

Parry, M. L. (1975). Secular climatic change and marginal land. *Transactions of the Institute of British Geographers*, **64**, 1–13.

Parry, M. L. (1976). The significance of the variability of summer warmth in upland Britain. *Weather*, **31**, 212–17.

Parry, M. L. (1978). *Climatic Change, Agriculture and Settlement*. Folkestone: Dawson.

Postan, M. M. (1975). *The Medieval Economy and Society*, 2nd edn. Harmondsworth: Penguin.

14

History and climate: some economic models

J.L.ANDERSON

Abstract

The trends of certain economic indicators in Europe show broad correspondences with that of annual average temperature over the last millennium. However, examination of models of economic change in the period reveals no general causal relationship.

A number of conceptual difficulties are involved in attempts to demonstrate a relationship between climatic and social change. The principal problem is that a given change in climate may have a variety of effects, and a particular historical event may be the product of either climatic amelioration or deterioration. Between a change in the circulation pattern and a change in historical behaviour lie numerous physical, economic, and social factors which influence any process of causation by their nature, and by their variation and interaction independent of climate.

An indication of the uncertainty surrounding the nature and effects of climatic change in early times is that both amelioration and deterioration have been argued to be principal elements in the Near Eastern origins of agriculture, and in an expansion of Scandinavian settlement about the tenth century AD.

The climatic pattern in historical times has been adequately established. The economic 'depressions' of the fifteenth and seventeenth centuries coincided with periods of unfavourable climate, but the effects of the climate are not discernible in the long-run movements of agricultural prices, incomes, and yields. Spells of bad weather precipitated subsistence crises, notably in the early fourteenth and late sixteenth centuries when growth of population had been exceeding that of resources.

Evidence provides little support for a climatic interpretation of the general course of European history. Research may more usefully be directed to the analysis of the effects of changing climate upon particular regions, and of the nature and costs of weather shocks.

Introduction

This essay outlines the difficulties involved in any endeavour to identify links between climatic and economic change. Consideration is given to general problems of analysis and assessment in this field, and to explanations of the major events in European history for which climatic change has been claimed to have been at least partly responsible. These are the Viking irruption, the crisis of the late Middle Ages, and the crisis of Early Modern times. The aim of this brief survey obviously cannot be to present a comprehensive, interpretative history of all the relevant periods, or even of any one of them. Rather, the objective is to test some of the arguments, or models, that have been used to link climatic change with historical change.[1]

Correlation invites investigation, and in view of the correspondence over the last millennium between the long-term variation of annual average temperature and some economic indicators, the question of the influence of climatic change on human affairs in Europe cannot reasonably be ignored. The expansion of the medieval economy which culminated in the thirteenth century, and its subsequent contraction, are paralleled by the rise and fall of the long-run temperature trend, while in more recent times both temperature series and an index of the purchasing power of wage rates show peaks around AD 1500, 1750 and in the twentieth century, with troughs in the seventeenth century and about 1800 (Fig. 14.1). However, suggestive as the correlations may be, it will be argued here that the relationships are without demonstrable historical significance.

Between the warmth of the medieval 'optimum' and the chill of the Little Ice Age, temperature in Europe fell by about 1 deg C, and rainfall

Fig. 14.1. Climatic and economic indicators. Bold line: annual temperature, central England, 50-year averages (after Lamb, 1966: 186). Light line: index of purchasing power of builders' wage rates, southern England, general trend (after Phelps Brown & Hopkins, 1956: 302).

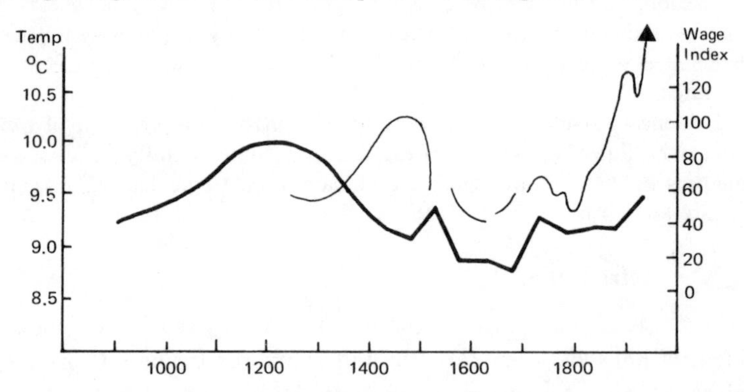

by about 10 per cent in terms of annual averages. In addition, there was some shift in the seasonality of rainfall, and an increased variability of the weather (Lamb, 1977: 412, 447–8). It may be accepted that these changes were such as would have affected plant growth to some extent, and therefore agrarian output. This in turn would have affected the harvest-sensitive, pre-industrial economy and society. However, the general problem examined in this chapter is not whether climatic change had *any* effect, but whether the effects can be isolated analytically in a study of long-run economic change, and can convincingly be shown to have been an important determinant of the course of European history. The following analysis suggests that these questions can be answered only in the negative, and that the effects of climatic change can be incorporated in historical explanation only in the most general and impressionistic terms.

Conceptual problems

In historical explanations which involve changes in climate, models based on climatic deterioration have been the most popular. However, the response of a social group to the challenge of a climatic deterioration is unpredictable. In different circumstances, it may be extinction, endurance, migration, adaptation, or technical development. With no unique link between climatic change and social action, 'explanation' on the basis of correlation is simply rationalisation, unless in any particular case evidential and theoretical links exist. Amelioration models are even more vague in their implications, as amelioration, being permissive rather than mandatory, may have as much effect as pushing a piece of string. Further, the effects of a slight easing of environmental constraints over a period of centuries, as occurred in the Middle Ages (Duby, 1974: 6–11), generally would be inextricably mixed with those of progress in the organisation and techniques of production. Yet these models represent perhaps the most promising areas for further investigation, as amelioration may lead, either directly or through social change, to long-run population growth, which in turn can demonstrably influence the economy. Population change is central to a number of explanations of economic change in Europe and elsewhere (Chambers, 1972; North & Thomas, 1970, 1977; Wilkinson, 1973).

The fact that a given change in climate may have a variety of effects, and that a given historical event – such as a migration – may result from either climatic amelioration or deterioration (or neither), indicates the central difficulty in analysing the role of climatic change in history. While the nature of climate at a given time may in some instances be inferred from its effects, the converse does not hold; the historical effect cannot be inferred from the climate. Historians of climate regularly and reasonably infer the nature of the climate from its apparent manifestations in aspects of the natural and cultural landscape. For example, the *existence* of vineyards in medieval England indicates a certain climatic regime, which is supported by independent, convergent evidence (Lamb, 1977:435–6). But the climate was not necessarily a causal factor in either their proliferation or in the reduction of their number. Meteorological parameters merely indicate the degree of success that may be expected, all else constant, to attend upon attempts to grow a specific plant. Whether a particular crop is cultivated will depend not only upon the climate, but also upon a complex interaction between the cost, quality, and availability of co-operating factors of production, the range of perceived and acceptable alternative uses for those factors, and the sets of available and expected market opportunities.

In sum, the historian of climate, and the historian endeavouring to use climatic change as an explanatory variable, may both examine the same data, but do so with quite separate and distinct objectives. Failure to appreciate this can lead at best to confusion, at worst to circularity of argument (Le Roy Ladurie, 1971:17).

Conditioning factors

The inferring of a climatic cause from a given event is generally precluded by the number and nature of other elements which exist in the chain of causation linking change in the climate with social change. These elements, designated as conditioning factors in the accompanying diagram (Fig. 14.2), will affect the nature of a climatically induced change at each stage in the process of causation. In addition, they may themselves be changed as a consequence of such induced change, and may also change autonomously over time. Finally, elements intrude which must be considered to be wholly or partly exogenous to this process, such as the effects of war and, in the present state of knowledge, of a number of demographically important epidemic diseases, including plague (Appleby, 1980; cf. Post, 1980). These complicate to the point of impossibility the task of isolating analytically the climatic component of long-run historical change.

In the short run, when the conditioning factors may be treated as being constant, responsibility for variations in output of a crop can be apportioned between the various inputs – including weather elements –

by statistical analysis based on the production function. This is a model, in algebraic form, relating output of a product to inputs of factors of production, broadly classified as land, labour, and capital. But there remains the problem of determining the effect on the economy of a more general change in output, such as a harvest deficiency. Certainly, society will be worse off in total; but there are secondary effects. Farmers producing for the market may benefit from a poor season. The sensitivity of demand for grain to price changes being relatively low, the percentage fall in quantity sold may be less than the percentage increase in price, resulting in an increase in farmers' total revenue. The effects of this change in income distribution on the economy will depend on the relative investing, spending, and importing habits of gainers and losers in the agricultural and non-agricultural sectors, and on the reaction of labour to changes in food prices (Gould, 1962). Uncertainty in this area is reflected in the debate that has surrounded the assessment of the economic consequences of harvest failure (Post, 1977:143–4); but a recent contribution has clarified one aspect of the problem. Mokyr (1977:984) has tested the hypothesis that a shift in income distribution in favour of farmers, consequent upon a rise in grain prices (contemporaneous with a deterioration in the weather), was a source of increased demand for the products of the nascent manufacturing industries in Britain in the second

Fig. 14.2.

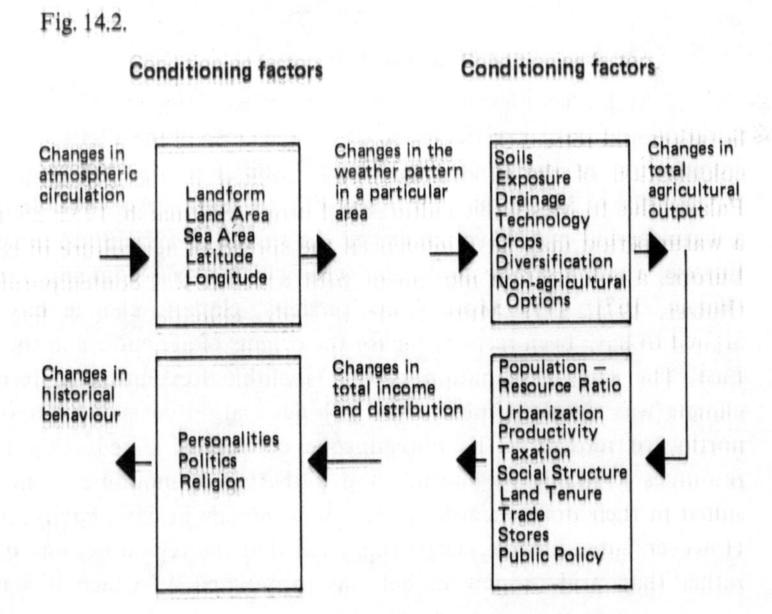

half of the eighteenth century. It was shown that a net rise in the demand for non-agricultural goods in these circumstances was unlikely, as this could occur only if the demand for agricultural products – of all kinds – were of such a nature that fewer of them were consumed when income rose.

Analysis of the long-run is still more complex. In that time perspective the conditioning factors can vary autonomously. In the present state of knowledge, the long-run aggregate production function is an organising concept, not an operational one. Its elements – technology, population, resources, and the social and institutional structure – each represent groups of variables which are interacting, and may or may not be quantifiable (Cameron, 1976: 13ff). The difficulties are such that the 'goal of partitioning the impact of causal forces underlying long-term economic change – for many, the most exciting goal of economic history – would seem as yet beyond our grasp' (McClelland, 1975: 240).

Long-term effects

Given that the effect of climatic change on historical processes cannot be specified rigorously, can it be identified in more general terms? To find a clear association, it would seem appropriate to consider social groups subjected to a change in climate substantial in terms of the degree of marginality of their environments and the level of their technologies.

The beginnings of agriculture

Major ecological changes were effected by the climatic amelioration and retreat of the ice sheets at the close of the Pleistocene. The colonisation of the tundra by forest resulted in the transition from Palaeolithic to Mesolithic cultures in Europe (Saltmarsh, 1952: 269); and a warm period may have influenced the spread of agriculture in central Europe, a colonisation movement with which it was contemporaneous (Butzer, 1971: 577). More controversially, climatic change has been argued to have been responsible for the origins of agriculture in the Near East. The early explanation of the Neolithic Revolution in terms of climate was simple (Childe, 1954: 48): desiccation consequent upon the northward retreat of the pluvial zone occasioned a reduction in the resources available to hunting and gathering communities, which resulted in their domesticating plants and animals in oasis environments. However, subsequent research suggested that the region became humid, rather than arid. A new model was formulated, in which it was hypothesised that the more bountiful environment encouraged sedentism, and the consequent growth of population led to attempts to augment the increasingly inadequate supply of naturally available food by the practice of agriculture (Smith & Young, 1972: 31–4). The explanation is plausible, internally and externally consistent, and may be useful; but it encompasses neither necessary nor sufficient conditions. Even if population growth were central to the process of domestication, it may have been autonomous, climatic change being merely coincidental (Cohen, 1977). More importantly, because food production developed in many microenvironments (Bender, 1975: 211–12), any general model would appear to be unsatisfactory.

The Viking irruption

A similar amelioration model has been used to account for the Scandinavian migrations of the Viking era (Claiborne, 1973: 349ff). The problem may conveniently be considered in terms of the 'pull' and 'push' complexes of motivations governing migratory movements. Climatic amelioration undoubtedly occasioned a 'pull' by creating relatively attractive areas for settlement in Iceland and Greenland, but it can scarcely have had relevance to the perennial attractiveness of the warmer lands to the south. Also fundamental to the model is a 'push' effect, generated by an increase in population, occasioned by a climatic amelioration. Certainly, population growth is an important element in interpretations of the Viking period; and it is probable that an increase in the number of people in northern lands would have been permitted or encouraged by the relative warmth which glaciological evidence suggests was enjoyed for over a millennium in Scandinavia from about the second century AD to the beginning of the Little Ice Age (Ahlmann, 1953: 38).[2] However, the effect of the onset of the medieval optimum, upon which the model is based, is much more uncertain, owing to the inadequacy of the physical and historical data, and to the nature of the Viking episode itself, which was only one of a number of comparable movements.

No clear consensus has emerged in the literature concerning climatic patterns during the Dark Ages. It seems that in Europe some climatic instability occurred in the eighth century, and cool summers in the ninth (Lamb, 1977: 426–7). If these conditions had been experienced in northern areas the climate would have tended to constrain rather than encourage a rise in population – but perhaps by this negative effect it influenced some to venture overseas. Indeed, the fragility of the data base is such that an explanation for an expansion of agriculture in Norway, and by implication for the Viking ventures, has been offered in terms of a progressive climatic decline from the eighth, or even sixth, century AD (Lunden, 1974).

The 'deterioration' model rests on the assumption that peasants endeavour to maintain a 'target' income, i.e., if their income is reduced, in order to restore it to the desired level peasants will increase effort and extend the cultivated area. From this, the expansion of farming in Norway at the close of the first millennium AD emerges as an indication not of an easing of climatic constraints, but of a general reduction in agricultural productivity consequent upon climatic deterioration. Apart from the dubious assumption of climatic decline based explicitly on Russell's views (1969: 51–2), the model does not appear to be applicable in the Scandinavian context. It assumes that the economy was closed with no alternatives to agriculture, and that intra-marginal land was readily available. Neither assumption appears to fit the circumstances in Scandinavia: seafaring pursuits and migration offered obvious alternatives, and in northern and mountainous lands climatic deterioration would be likely to reduce the area suitable for arable husbandry, and so tend to frustrate rather than initiate an internal colonisation movement. More crucially, the model deals with an increase in cultivation, not settlement, and it is accepted by Lunden that both increased in the period under review.

In an appropriate historical perspective, the Viking irruption may be seen as an event in a series, rather than as a unique occurrence explicable solely or largely by a natural force exogenous to the development of European society. For the previous thousand years there had been a persistent if somewhat episodic migration from the Scandinavian area to more climatically favoured and economically attractive lands. In the ninth century a fresh impetus was given to southward incursions. The accumulated wealth, and the military weakness of western Europe during the decline and dissolution of the Carolingian empire, were such as to attract predators from all parts of its periphery – the Saracens and Magyars, as well as the Northmen (Duby, 1974: 113–15). Also, at about that time, the process of technical evolution had made available to Norse adventurers the longship, and the less celebrated but more seaworthy *knörr* (Jones, 1968: 183ff).

The factors involved in any migration movement are numerous and complex. However, the experience of later and vastly more extensive maritime empires, begun in pillage or conquest, continued in trade, and culminating in colonisation, suggests that the motive, means and opportunity outlined above represent sufficient and probably necessary conditions for the initiation and success of the Viking intrusions.

Subsistence crises and economic recessions

If the role of climatic change cannot be unequivocally identified in very long-term demographic and economic changes, then an alternative avenue of inquiry is the investigation of notable historical episodes – subsistence crises, and subsequent economic recessions – to which a deteriorating climate may have perceptibly contributed by reducing production.

Crises of a spectacular kind occurred in the early fourteenth century and at the close of the sixteenth century. At those times the climate was demonstrably deteriorating in terms of its suitability for grain growing in Europe. But the crisis decades, in each of those cases, followed long periods of population growth, without corresponding increases in productivity, particularly in agriculture. Society had been rendered increasingly crisis-prone by the increased demand pressing upon resources, the expansion of which was limited by the constraints of the physical environment and the institutional structure. With both supply and demand changing, the contribution of each to the observed scarcity cannot be rigorously apportioned. However, as will be shown, the circumstances in each period tend to suggest that the effect of climatic change would have been to aggravate, rather than to initiate, problems the origins of which lay in the demographic, technological, and political conditions of those periods.[3]

The Middle Ages: 1300–1450

In England during the thirteenth century, there was a rise in prices and in what may broadly be termed rent and profits, while the standard of living of the peasantry declined (Miller & Hatcher, 1978: 45–53). Similar conditions obtained in other areas of Europe. The intensifying problem of subsistence, which culminated in the crisis of the early decades of the fourteenth century, has conventionally been interpreted in terms of the effects of population growth on the availability and productivity of land (Postan, 1966; Postan & Hatcher, 1978).[4] The model can be expressed diagrammatically (Fig. 14.3). It is assumed that recourse to less fertile land, and intensification of cultivation on existing plots, will encounter diminishing returns; i.e., although total output increases, average and marginal returns will diminish with each successive application of labour to land. For this reason, the marginal product curve (MP) has a negative slope. In these circumstances, an increasing population will reduce the average wage through competition in the augmented labour market; while total output and the surplus – rent and profit –

will increase. This is illustrated by the fall in average wage OW to OW_1, and increase in the surplus $\Delta\ ABW$ to $\Delta\ AB_1W_1$.

Following this analysis, a fall in productivity occasioned by a deterioration of the climate could be represented by a downward shift in the marginal product curve. If, as a consequence of the reduction, the surplus tended to remain constant or fall, the inconsistency with historical data would indicate that climatic change was not the principal cause of the scarcity reflected in rising prices. Unfortunately, the results of this analysis would be sensitive to assumptions concerning the climatically induced changes in the parameters of the production function. Accordingly, even if the complication of changes in the product price were ignored, it is not possible to discriminate between the competing hypotheses on these theoretical grounds.

The interpretation in which population growth, rather than climatic change, was the principal variable is the more consistent with available evidence of agricultural yields in this period (Titow, 1972). Wheat yields to seed sown were generally steady, and although there was a tendency for yields of barley and oats to decline in the closing decades of the thirteenth century, it does not appear to have been the consequence of climatic deterioration. Historical and experimental evidence suggest it to have been largely the consequence of an over-extended colonisation, resulting in the cultivation of less fertile land, and the destruction of the balance between pasture and arable which was necessary for maintenance of soil fertility (Titow, 1972: 24, 30–3; Cooter, 1978).

It is possible that a changing climate contributed something to the problems of the half-century centred upon AD 1300, and the wet years of the second decade of the fourteenth century certainly intensified them. However, the effects of climatic change are not clearly discernible in the general course of economic change over that period.

Fig. 14.3. Population and productivity.

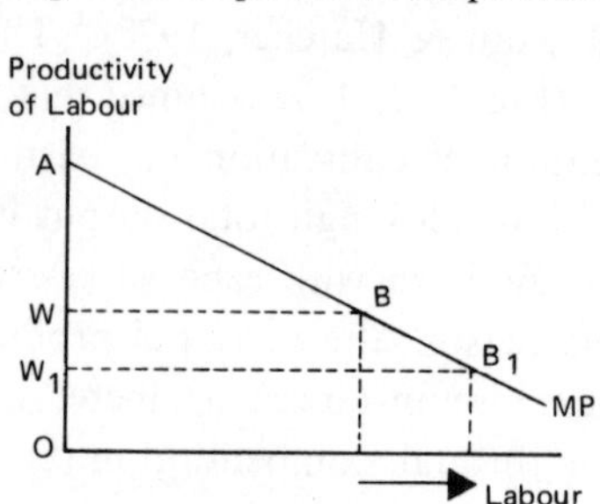

More generally, the secular agrarian depression experienced in Europe toward the close of the Middle Ages was a protracted crisis of profitability, not of subsistence. The century following the Black Death was characterised by population decline. Real wages rose, while rents and grain prices fell, the latter both absolutely and in relation to other rural products (Slicher van Bath, 1963: 140). The material well-being of many in Europe who were fortunate enough to avoid the ravages of plague and the horrors of war seems to have stood higher in the fifteenth century than at any time until the late nineteenth century. The subsistence crises which nevertheless punctuated this relatively prosperous but disturbed period are readily explicable in terms of the limited responsiveness of agricultural production to short-run variation in the price of food-stuffs occasioned by poor seasons. Higher real incomes meant dearer labour, and also an increased demand for luxuries such as wool and meat. Not only did animal products tend to hold their value, but grazing was a much less labour-intensive form of husbandry than cropping. The consequence was a conversion of some of the arable to pasture, less biologically efficient in the production of food but generally more profitable given the change in relative prices both of products and of factors of production. This in turn meant that although the area of fertile soil available per head had increased, the relationship between population and the short-term capacity of the land to increase production of the staple cereals had changed but little (Helleiner, 1967: 69).

In aggregate, agricultural land was converted to pasture or abandoned in this period not because of an intensifying climatic constraint, but because there was no longer profit or point in its continued cultivation in the face of a declining demand for its product. Even in areas demonstrably marginal in climatic terms, caution must be exercised in assessing the fundamental reasons for land abandonment (Parry, 1978: 117). Climate, soil, conditions of tenure, and economic and military aspects of location would all have influenced the selection of land to be abandoned; and it may be presumed that, whatever the reason for agricultural contraction, land which was marginal in any of those senses would generally be the first to have been converted to pasture or relinquished to waste.

The late sixteenth and the seventeenth century

The European subsistence crises of the last decade of the sixteenth century were clearly precipitated by bad weather. However, as in the thirteenth century, rising rents and falling wages (Davis, 1973: 103–4) suggest that the growing population was placing increasing pressure upon an insufficiently responsive resource base. The consequent sensitivity of society to variations in the food supply would have been further increased by urbanisation, by the increased number of horses required for the expansion of commerce and manufacturing, and by the revenue requirements of the developing absolutist states.

By contrast, the lengthy period of social and economic disturbances which followed, frequently though debatably termed the 'general crisis' of the seventeenth century, has been characterised in its economic aspect as a crisis of distribution, occasioned by the weight and incidence of taxation, rather than one of production (Steensgaard, 1978: 40–2). As the century progressed, and population growth slowed or ceased, grain prices declined relative to those of other goods and wages (Slicher van Bath, 1977: 53), indicating that the scarcity of basic food-stuffs was being eased. Further, the nature and intensity of the problems differed between sectors, and between the states of Europe, in ways which suggest that the differences may more reasonably be ascribed to differing institutional and economic structures than to the effects of the general climatic deterioration termed the Little Ice Age. Broadly, England and the United Provinces of the Netherlands suffered little if any retardation of their economic progress, while France stagnated, and Germany, Spain, and Italy declined.

Certainly, it may be expected that a change in the general circulation of the atmosphere will express itself in terms of changed weather patterns differently in different regions, particularly in an area of such diverse landform and latitudinal extent as Europe. Therefore diversity of economic experience does not *ipso facto* exclude climatic change from consideration as a prime mover of historical events. However, any case for the operation of climatic change as the general cause of a 'general crisis', based simply on the idea of simultaneity implicit in that term, would in such circumstances be correspondingly weakened (cf. Braudel, 1974: 18–19).

In this context it is illuminating to consider the sharp contrast between the fortunes of France and England in the seventeenth century. During this period France suffered successive subsistence crises, while across the channel fears of oversupply led to the liberalisation of restrictions upon the export of grain, and finally to export bounties (John, 1976: 47). Wheat prices moved synchronously in both countries, suggesting that over the long term the behaviour of the weather, and hence the climatic pattern, in each was comparable. However, the differential movement in England of the prices of grains other than wheat indicates that organisational change in agriculture had there reduced the sensitivity of total agricultural output to vagaries of the weather, and had correspondingly reduced the vulnerability of the economy (Appleby, 1979).

From a more generalised comparison between the economies of Poland, Silesia, and Pomerania, Topolski (1974:137–40) argues that the decline of Poland in the seventeenth century was attributable partly to the ravages of war, but principally to the effects on productivity of the social and tenurial structure. Unfavourable climatic change is explicitly rejected as an explanatory variable.

Climatic change has been emphasised in connection with the decline of southern Europe in this period – excess rainfall in the Italian region (Utterström, 1955:40), and drought in Spain (Vives, 1970:126). However, the relative decline of these states can be explained adequately in economic and social terms. The long-term problems of southern Italian agriculture may be attributed to a rising demand for grain, which encouraged imprudent husbandry and financing (de Vries, 1976: 51–2), and therefore would appear to have been the consequence rather than the cause of rising prices. More importantly, the decline in commerce and urban industries, upon which Italian prosperity had been based, was the result of a disinclination or inability to adjust to fundamental changes in the nature and pattern of international trade (Cipolla, 1970). Spanish difficulties in the sixteenth and seventeenth centuries can be accounted for essentially by reference to the familiar, dismal catalogue of the costs of empire, the effects of disease, the economic implications of social attitudes, and ill-conceived financial, commercial, and agricultural policies. More specifically, the interests of agriculture were subordinated to those of fiscal and political expediency (de Vries, 1976: 48–51). To take but one example, the imposition of grain price ceilings, while of benefit to the consumer in the short term, would reduce the responsiveness of an agricultural sector to the needs of a growing population by adversely affecting incentives (North & Thomas, 1973: 130). This would have increased the vulnerability of the society to any variations in the weather pattern, random or otherwise.

In relation to the economic recessions of the late Middle Ages and of the Early Modern period in Europe, there are no explanatory lacunae which need to be filled by reference to the presumed effects of climatic deterioration. However, the intensity and timing of particular crises may well have been influenced by weather shocks, themselves possibly a product of the variability associated with a deterioration of the climate toward a more continental regime.

Climate, grain yields and prices

In an attempt to test for evidence of the effect of climatic change, English grain price series for the period 1480–1619 have been analysed in terms of the frequency and magnitude of deviations from the trend (Harrison, 1971). Although the relative frequency of harvests designated poor on price criteria was found to have increased, this does not necessarily indicate an increased occurrence of bad weather. Given a constant relationship between the quantity harvested and weather variation, an oscillation of constant amplitude and frequency between favourable and unfavourable weather could produce an apparent increase in harvest failure. If population were increasing, as it was (Chambers, 1972: 14), variations in output of a given amount might be expected eventually to occasion greater price rises than falls. With a society being pushed to the edge of subsistence by its increasing demand for food pressing upon a relatively fixed normal or average supply, the sensitivity of price to any reduction in output would increase. This is illustrated in Fig. 14.4, a diagramatic representation of the grain market. Average supply in the short-run is represented by the curve *S*. Normal variations in supply from one harvest to the next are represented by the curves S_v. These variations result in fluctuations of price to P_v about a normal price *P*. The kinked form of demand curve for grain (*D*) rests on the assumption that when available supply becomes barely adequate for subsistence, any reductions, by occasioning speculation and consumer desperation, would cause sharper increases in price than in a period when adequate surpluses or reserves existed.[5] As the demand curve shifts outward from *D* to D_1 under the influence of the requirements of an increasing population, the 'kink' in the curve would enter the zone of

Fig. 14.4. Foodstuffs: demand and supply. (*a*) Period in which supply can readily satisfy needs at levels above subsistence. (*b*) Period in which normal supply just meets subsistence requirements. (After Anderson, 1981)

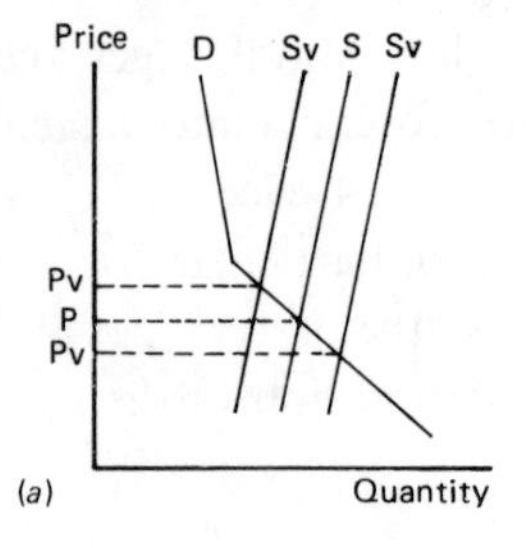

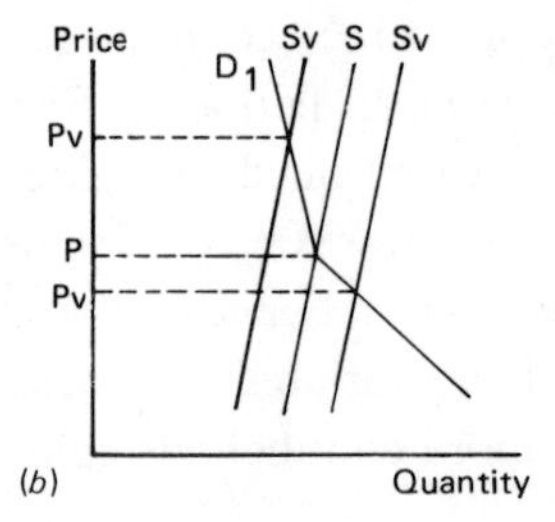

'normal' short-run fluctuation in supply. In these circumstances, a normal deficiency in supply would result in the price rising toward famine levels. Conversely, equivalent movements of output in the opposite direction would not generate a price fall of corresponding magnitude, the price being supported by the numerous non-food sources of demand for grain which would be largely unsatisfied if output were near subsistence level.

It is of course not necessary to establish the possibility of an asymmetrical reaction of prices to variations in output of a given magnitude to demonstrate that models in which prices are used as a proxy for production are unsatisfactory. Even in the short term, prices are affected by monetary factors, the extent of trade, the quantities in public and private stores, and the possibility of drawing produce from extraordinary sources. Further, although rising prices indicate an increasing 'scarcity', this does not necessarily imply a diminution of welfare. In a region open to trade, a rise in the price of locally produced commodities could as well indicate an improvement in the terms of trade with a consequent increase in local prosperity. Indeed, models based on price movements appear in the literature only for want of a better alternative; historical data for productivity and production in agriculture are generally exiguous, representative of only specific localities, and often difficult to interpret (Slicher van Bath, 1977: 80). However, those available show a long-run correspondence between trends of yields, prices, and population in Europe; indicating that population was the independent variable, influencing prices and through them productivity (Steensgaard, 1978: 30–1). If change in the climatic constraint were the principal factor affecting long-run average output, then an inverse relationship between yield and price would be expected.

Conclusion

The somewhat negative thrust of this chapter can be summed up by a counterfactual conditional proposition: if there had been no climatic change, then the history of Europe would not have been different in any general sense. Marginal areas by definition would have been affected, perhaps profoundly but, except for the Scandinavians, peoples at the margins of cultivation in Europe were of little political or economic significance to the mainstream of European history. Disease, war, technology, and economic and social institutions must continue to hold central place in general analyses of European history concerned with long-term change.

By contrast local, and even regional, histories could well have been very different in the absence of changes in climate. Shocks occasioned by changing weather patterns, if not diffused by the operation of the market or actions of state agencies, may entail economic and social disaster for a particular area; while adjustment to changed physical parameters involves costs which are often unevenly borne. The possibility of weather variations of opposite signs within national boundaries (Manley, 1962: 236–7), and of a complexity of response of agrarian output to given changes in the weather (Jones, 1964: 46–8, 56–8), makes the nation, although an entity convenient from a statistical point of view, less useful as a primary unit for examining the effects of climatic change than the geographic region, the industry, the economic class – or even the demographic cohort (Manley, 1958; Sutherland, this volume).

Similarly, and paradoxically, the most fruitful approach to the study of the historical effects of climatic change would appear to be the investigation of the impact, and short-run implications, of climatic shocks experienced by the productive system. Studies of this sort, focussing on factors which have determined the degree of economic sensitivity to changes in weather patterns, and on the problems and costs of adjustment that such changes have entailed (de Vries, 1980), are likely to be not only intellectually satisfying, but of most value as a background for contemporary economic planning.

Notes

1. A number of the issues raised in this outline, and associated problems, are examined in more detail in Anderson (1981).
2. The trend of movement of Swedish glaciers suggests a slight increase in warmth over the first millennium AD, while that of Norwegian glaciers indicates a slight cooling from Roman times to 1600, with a warm period about AD 1000. Both show a pronounced glacial maximum about the middle of the first millennium BC (Ahlmann, 1953: 38).
3. Glantz (1978) cites more recent examples to caution against the assumption of a climatic cause when the weather had merely precipitated crises which were fundamentally economic, social, or political in origin.
4. The principal challenge to this view is offered by those who consider resources to have been limited essentially by the implications of the institutional structure, rather than by physical constraints. See Brenner (1976), and contributors to the debate in subsequent issues of *Past & Present*.
5. The principles underlying the use of a kinked demand curve are outlined, in a different context, by Reutlinger (1976: 3). I owe this reference to Dr J. Kennedy.

References

Ahlmann, H. W. (1953), *Glacier Variations and Climatic Fluctuations*. New York: The American Geographical Society, 51 pp.

Anderson, J. L. (1981). Climatic change in European economic history. *Research in Economic History, A Research Annual*, ed. Paul Uselding, vol. 6. Greenwich, Connecticut: JAI Press (in press).

Appleby, Andrew B. (1979). Grain prices and subsistence crises in England and France, 1590–1740. *Journal of Economic History*, **XXXIX**, 4, 865–87.

Appleby, Andrew B. (1980). Epidemics and famine in the Little Ice Age. *Journal of Interdisciplinary History*, **X**, 4, 643–63.

Bender, Barbara (1975). *Farming in Pre-history. From Hunter–gatherer to Food-producer.* London: John Baker. 268 pp.

Braudel, Fernand (1974). *Capitalism and Material Life 1400–1800*, trans. Miriam Kochan. Glasgow: Fontana/Collins. 462 pp.

Brenner, Robert (1976). Agrarian class structure and economic development in pre-industrial Europe. *Past & Present*, **70**, 30–75.

Butzer, Karl W. (1971). *Environment and Archaeology. An Ecological Approach to Prehistory*, 2nd edn. Chicago: Aldine. 703 pp.

Cameron, Rondo (1976). Economic history, pure and applied. *Journal of Economic History*, **XXXVI**, 1–27.

Chambers, J. D. (1972). *Population, Economy and Society in Pre-Industrial England.* London: Oxford University Press. 162 pp.

Childe, Gordon (1954). *What Happened in History.* Harmondsworth: Penguin. 288 pp.

Cipolla, Carlo M. (1970). The economic decline of Italy. In *The Economic Decline of Empires*, ed. C. M. Cipolla, pp. 196–214. London: Methuen.

Claiborne, Robert (1973). *Climate, Man, and History.* London: Angus & Robertson. 444 pp.

Cohen, Mark Nathan (1977). *The Food Crisis in Prehistory. Overpopulation and the Origins of Agriculture.* New Haven: Yale University Press. 341 pp.

Cooter, William S. (1978). Ecological dimensions of medieval agrarian systems. *Agricultural History*, **52**, 458–77.

Davis, Ralph (1973). *The Rise of the Atlantic Economies.* London: Weidenfeld & Nicolson. 352 pp.

de Vries, Jan (1976). *Economy of Europe in an Age of Crisis 1600–1750.* Cambridge: Cambridge University Press. 284 pp.

de Vries, Jan (1980). Measuring the impact of climate on the economy: the search for appropriate methodologies. *Journal of Interdisciplinary History*, **X**, 599–630.

Duby, Georges (1974). *The Early Growth of the European Economy. Warriors and Peasants from the Seventh to the Twelfth Century*, transl. Howard B. Clarke. London: Weidenfeld & Nicolson. 292 pp.

Glantz, Michael (1978). Render unto weather ... – An editorial. *Climatic Change*, **4**, 305–6.

Gould, J. D. (1962). Agricultural fluctuations and the English economy in the eighteenth century. *Journal of Economic History*, **XXII**, 313–33.

Harrison, C. J. (1971). Grain price analysis and harvest qualities 1465–1634. *Agricultural History Review*, **19**, 135–55.

Helleiner, Karl F. (1967). The population of Europe from the Black Death to the eve of the Vital Revolution. In *The Cambridge Economic History of Europe*, ed. E. E. Rich & C. H. Wilson, vol. 4, pp. 1–95. Cambridge: Cambridge University Press.

John, A. H. (1976). English agricultural improvement and grain exports, 1660–1765. In *Trade, Government and Economy in Pre-Industrial England: Essays presented to F. J. Fisher*, ed. D. C. Coleman & A. H. John, pp. 45–67. London: Weidenfeld & Nicolson.

Jones, E. L. (1964). *Seasons and Prices. The Role of the Weather in English Agricultural History*. London: George Allen & Unwin. 193 pp.

Jones, Gwyn (1968). *A History of the Vikings*. London: Oxford University Press. 504 pp.

Lamb, H. H. (1966). *The Changing Climate*. London: Methuen. 236 pp.

Lamb, H. H. (1977). *Climate: Present, Past and Future* vol. 2, *Climatic History and the Future*. London: Methuen. 835 pp.

Le Roy Ladurie, Emmanuel (1971). *Times of Feast, Times of Famine: A History of Climate Since the Year 1000*. New York: Doubleday. 426 pp.

Lunden, Kåre (1974). Some causes of change in a peasant economy, interactions between cultivated area, farming population, climate, taxation and technology. A theoretical analysis of the Norwegian peasant economy *c.* 800–1600. *Scandinavian Economic History Review*, **XXII**, 117–35.

McClelland, Peter D. (1975). *Causal Explanation and Model Building in History, Economics, and the New Economic History*. Ithica: Cornell University Press, 290 pp.

Manley, Gordon (1958). The revival of climatic determinism. *The Geographical Review*, **XLVIII**, 1, 98–105.

Manley, Gordon (1962). *Climate and the British Scene*. London: Collins. 314 pp.

Miller, Edward & Hatcher, John (1978). *Medieval England – Rural Society and Economic Change 1086–1348*. London: Longman. 302 pp.

Mokyr, Joel (1977). Demand vs. supply in the Industrial Revolution. *Journal of Economic History*, **XXXVII**, 981–1008.

North, Douglass C. & Thomas, Robert Paul (1970). An economic theory of the growth of the Western World. *The Economic History Review*, 2nd Series, **XXIII**, 1–17.

North, Douglass C. & Thomas, Robert Paul (1973). *The Rise of the Western World. A New Economic History*. Cambridge: Cambridge University Press. 170 pp.

North, Douglass C. & Thomas, Robert Paul (1977). The first economic revolution. *The Economic History Review*, 2nd Series, **XXX**, 229–41.

Parry, M. L. (1978). *Climatic Change, Agriculture and Settlement*. Folkestone: Wm Dawson. 214 pp.

Phelps Brown, E. & Hopkins, Sheila V. (1956). Seven centuries of the price of consumables, compared with builders' wage-rates. *Economica*, **XXIII**, 296–314.

Post, John D. (1977). *The Last Great Subsistence Crisis in the Western World*. Baltimore: The Johns Hopkins University Press. 240 pp.

Post, John D. (1980). The impact of climate on political, social, and economic change: a comment. *Journal of Interdisciplinary History*, **X**, 4, 719–23.

Postan, M. M. (1966). Medieval agrarian society in its prime: England. In *The Cambridge Economic History of Europe*, ed. M. M. Postan & J. J. Habakkuk, vol. 1, pp. 548–632. Cambridge: Cambridge University Press.

Postan, M. M. & Hatcher, John (1978). Population and class relations in feudal society. *Past & Present*, **78**, 24–37.

Reutlinger, Shlomo (1976). A simulation model for evaluating worldwide buffer stocks of wheat. *American Journal of Agricultural Economics*, **58**, 1–12.

Russell, J. C. (1969). Population in Europe 500–1500. In *The Fontana Economic History of Europe: The Middle Ages*, ed. C. M. Cipolla, pp. 25–70. London: Fontana/Collins.

Saltmarsh, John (1952). Economic prehistory of Europe. *The Economic History Review*, 2nd series, **V**, 266–76.

Slicher van Bath, B. H. (1963). *The Agrarian History of Western Europe AD 500–1850*, transl. Olive Ordish London: Edward Arnold. 364 pp.

Slicher van Bath, B. H. (1977). Agriculture in the Vital Revolution. In *The Cambridge Economic History of Europe*, ed. E. E. Rich & C. H. Wilson, vol. 5, pp. 42–132. Cambridge: Cambridge University Press.

Smith, Philip E. L. & Young, Jr, T. Cuyler (1972). The evolution of early agriculture and culture in Greater Mesopotamia: a trial model. In *Population Growth: Anthropological Implications*, ed. Brian Spooner, pp. 1–59. Cambridge, Mass.: MIT Press.

Steensgaard, Niels (1978). The seventeenth century crisis. In *The General Crisis of the Seventeenth Century*, ed. Geoffrey Parker & Lesley M. Smith, pp. 26–56. London: Routledge & Kegan Paul.

Titow, J. Z. (1972). *Winchester Yields. A Study in Medieval Agricultural Productivity*. Cambridge: Cambridge University Press. 151 pp.

Topolski, Jerzy (1974). Economic decline in Poland from the sixteenth to the eighteenth centuries, transl. Richard Morris. In *Essays in European Economic History 1500–1800*, ed. Peter Earle, pp. 127–42. Oxford: Clarendon Press.

Utterström, Gustaf (1955). Climatic fluctuations and population problems in early modern history. *Scandinavian Economic History Review*, **III**, 3–47.

Vives, Jaime Vicens (1970). The decline of Spain in the seventeenth century. In *The Economic Decline of Empires*, ed. C. M. Cipolla, pp. 121–67. London: Methuen.

Wilkinson, Richard G. (1973). *Poverty and Progress. An Ecological Model of Economic Development*. London: Methuen. 225 pp.

15

Climate and popular unrest in late medieval Castile

ANGUS MACKAY

Abstract

This chapter describes and uses some of the documentary and printed sources – both quantitative and descriptive in nature – which provide data for the reconstruction of climatic fluctuations in the late medieval kingdom of Castile. Aspects of weather factors in the two regions of Old Castile and Murcia are then explained in order to highlight the differences between the dry-farming lands (*secano*) and the *huertas* of the Levantine coast. The chapter also considers some aspects of the possible connections between weather fluctuations and late medieval episodes of popular unrest which, although frequently directed against Jews and *conversos*, also expressed grievances of a more general nature. It is argued that 'stimulus-response' interpretations are inadequate and that satisfactory explanations of popular unrest must take into account the kinds of evidence that are often discarded as being 'irrational'.

Introduction

Two objectives lie behind the writing of this chapter. The first is to demonstrate that, even in the initial stages of research, it is clear that untapped documentary sources do exist which can be combined to construct a picture, albeit 'primitive', of Castilian climate during the later medieval period; the second is to examine, in a rather cautionary way, some aspects of the impact of short-term climatic fluctuations on popular unrest. Attention is confined to some of the areas which constituted the late medieval kingdom of Castile. Even so, of course, evidence from any one locality would hardly be sufficient to allow for generalisations which would have to cover such disparate regions as the 'bread-carrying lands' (*tierras de pan llevar*) of Old Castile and the irrigated landscapes or *huertas* of the Levantine coast. The areas which provide evidence for the study are Burgos and Palencia (representing the *meseta* of Old Castile), Murcia, and southwest Andalusia. This leaves some gaps – in particular the peripheral, humid zones of the north (Galicia and the Basque Provinces), and the lands of New Castile.

Weather data

The data available for reconstructing the weather record of late medieval Castile include both indirect evidence of climate, some of which is quantitative, and direct but invariably qualitative descriptions of weather events. The first category contains information related to tithes. The account books (*Libros Redondos*) of the cathedral of Burgos, for example, provide figures for the annual income derived from the *préstamos de pan* (also known as *los tercios del cabildo que son a pan*). This was income related to tithes of grain, and each year the tithe-rights for the different localities involved were auctioned. It is unfortunate that a series of precise tithe auction dates cannot be constructed, but it is nevertheless clear that the tithe farmers had a good idea of the state of the harvest before the bidding started. A similar system was operated by the cathedral of Palencia. In both the Burgos and Palencia accounts the figures are in kind (wheat and barley); the Burgos data cover the period 1352–1450, the Palencian ones the period 1412–63.[1] From such evidence it is possible to study the grain harvest fluctuations for some thirty localities of Old Castile (examples in Fig. 15.1). There are, of course, gaps: the Burgos data are more plentiful than those of Palencia, and figures for the fourteenth century are 'patchy'. Similar material is available for southwest Andalusia and has recently been used for a study of cereal production for the period 1408–1503 (Fig. 15.2) (Ladero Quesada & González Jiménez, 1978). Finally, if appropriate caution is observed, it is also possible to enlist evidence for grain prices in Aragon, Valencia, Seville, and Carmona (Hamilton, 1936; MacKay, 1972; González Jiménez, 1976).

All this quantitative evidence can be used to calculate the years of 'good' or 'bad' harvests in the respective regions, but the fluctuations do not reveal why one year, or group of years, was better or worse than another. This is where the 'descriptive' evidence comes into its own because, when combined with the quantitative data, it frequently provides the explanation for the severest fluctuations. Cross-checking between the descriptive and quantitative information is a laborious task, and a wide variety of sources has to be brought into play.

The most picturesque evidence is provided by documentation on *rogativas* or *plegarias*. These were prayer processions of a type illustrated in the miniatures of the thirteenth-century *Cantigas de Santa María* (Fig. 15.3). In the cathedral archive of Burgos, for example, references to such *rogativas* are to be found in the *actas* of the canons and even in the very same *Libros Redondos* from which the quantitative evidence on tithes is taken. Such information is valuable not only because it makes cross-checking easier but also because, although the so-called *lecha* accounts provide useful information, the absence of a 'normal' demesne economy in Castile (MacKay, 1977:70–5) has meant that manorial accounts, which have proved valuable to the English climatologist, are not available.[2]

References to *rogativas* are invariably succinct. The *actas* of the municipal archives on the other hand tend to give lengthy entries relating

Fig. 15.1. Receipts from grain tithes in areas of Burgos and Palencia 1350–1460. 100 = 1439–42.

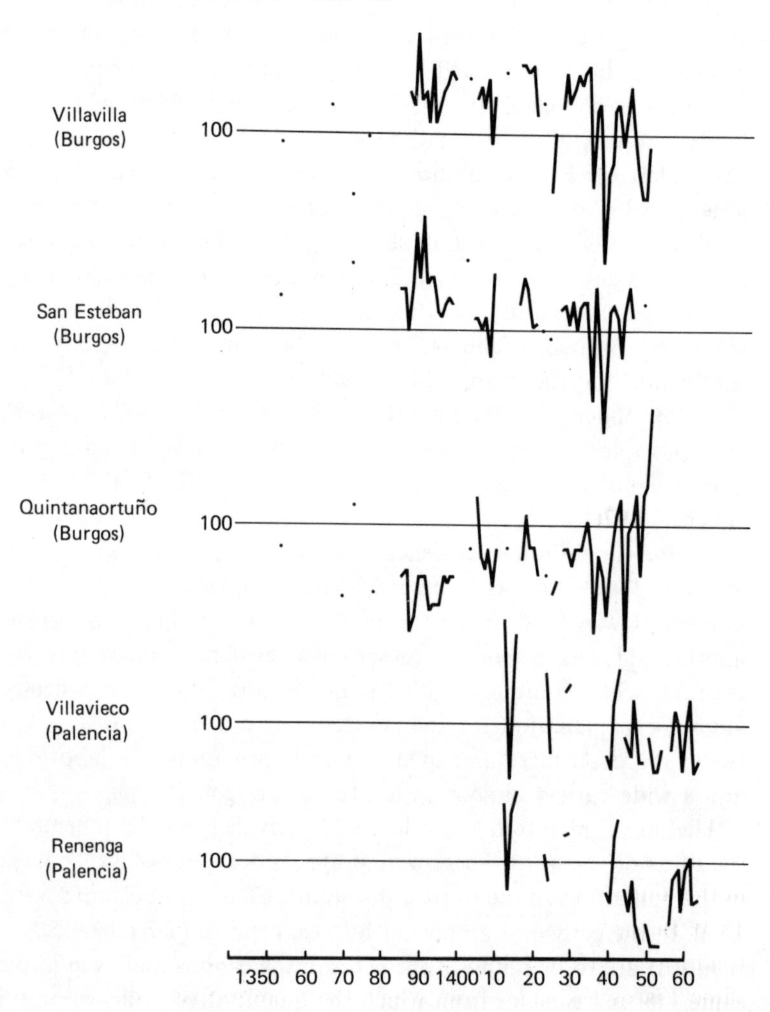

to the problem of the *saca* of food supplies – that is, the various mechanisms of control which regulated exports and imports into given localities, such as Murcia, Burgos, or Seville, during periods of scarcity. Bearing in mind the nature of the local or regional economy in question, the documentation on the frequency and intensity of such control mechanisms often provides information on the climatic 'triggers' of a crisis, as well as its relative seriousness and the adaptive strategies put into operation. For example, the deliberations of the oligarchs of the Murcian council usually reveal whether a crisis was induced by drought, floods, or locusts, and the complex attempts to secure imports of food from other areas afford valuable clues as to the geographical extent of the crisis.

Finally, there is the more problematical information provided by such sources as chronicles and the deliberations of *cortes* or parliaments. The difficulties are familiar to historians: each item of information has to be critically assessed in historical terms, and the variations in the 'quality' of the data can be enormous. There is, for example, a detailed and excellent account of the rain storms and floods in Seville and elsewhere during the winter of 1434–5 (Carrillo de Huete, 1946:182–95); on the other hand complaints in the 'national' *cortes*, which are often reported as being geographically general in nature, may in most cases have reflected nothing more than the views which unnamed *procuradores* from unspecified localities presented to a highly unrepresentative assembly. One peculiarity about the Castilian case, however, should be stressed. Apart from the rather slight *Anales de Garci Sánchez*, this period produced none of those urban chronicles which are so invaluable a source for the climatic data of other countries in western Europe.

Fig. 15.2. Cereal production in Western Andalusia: 1408–1503 (according to Ladero Queseda & González Jiménez, 1978). 100 = 1408–1503.

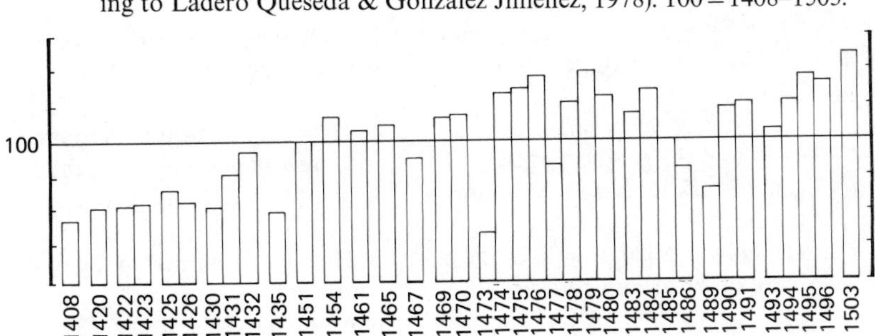

From all this evidence a pattern of 'good' and 'bad' years emerges, but it is one that has a regional rather than a kingdom-wide basis. For example, whereas the grain harvest of 1412 in Old Castile seems to have been relatively good, the opposite was the case in Andalusia. Moreover the sources tend to produce a biased picture because, while the quantitative data provide fluctuations 'impartially', the *actas* of Murcia only reveal the years of crisis. Still, bearing these factors in mind, it is easy enough to note the 'extremes' *a vista d'occhio* – to detect, for example, that the years 1413, 1435, and 1473 were particularly disastrous ones (Figs. 15.1 and 15.2).

Although the climatically-influenced extremes in the harvest record are easy to detect, the discovery of the precise weather conditions which determined these fluctuations presents a far more difficult problem. Moreover, not only must one keep in mind the possibility of regional differences in the weather patterns, but also the fact that the same factors of climate could have different effects because of variations in land and agrarian structures. A consideration of two regions – Old Castile and Murcia – will help to reveal the sharp contrasts that existed. This task is facilitated because, in addition to the historical records, there are good studies of a general nature (Linés Escardó, 1970; Cabo, 1973; Tamames, 1974) as well as monographs on both regions (Huetz de Lemps, 1962; García Fernández, 1963; Rosselló Verger & Cano García, 1975; Chacón Jiménez, 1979).

In the *secano* or dry-farming lands of Old Castile the *año y vez* system of cultivation prevailed over the regions to the north of the Duero from

Fig. 15.3. A *rogativa* for rain in Jerez de la Frontera, (From the thirteenth-century *Cantigas de Santa María*)

1. How in Jerez near the waters of the Alquevir there was no water.

2. How the people earnestly beseeched Jesus Christ for rain.

3. How a friar began to preach and told the people that they should ask the Virgin Mary for rain.

4. How they all beat the ground with their hands and asked the Virgin Mary to bring them rain.

5. How it began to rain at the behest of the glorious Virgin Mary.

6. How they returned to the town with plenty of water and greatly praising the Virgin Mary.

the Tierra de Campos to the *páramos* of Burgos. Although these lands do not give abundant yields, they have *traditionally* been thought of as providing consistent grain harvests which were relatively unaffected by considerable fluctuations in the total annual rainfall. In fact the disastrous grain harvests of this period do seem to have resulted from extremes in the weather conditions – yet it is important to remember the seasonal timing of such conditions (despite the fact that the sources are infuriatingly vague about this crucial factor). Thus, although drought could and did have disastrous effects on harvests, the evidence of the documents cannot be used to posit an automatic correlation. For example, the phrase *grant seca* may refer to a drought, but it is also possible that it may refer to the phenomenon known as the *asurado* or *escaldado* – that is, the abrupt transition to summer which in itself could have detrimental effects on the harvest if the June temperatures were too high (twentieth-century rises to about 30 °C in June endanger the wheat harvest because the optimum temperatures at this point of time are in the order of 20–25 °C) (García Fernández, 1963).

In the context of the *año y vez* system prevalent in Old Castile, a drought in the autumn (August to October) or in May could have disastrous consequences. The crucial factor was perhaps the October sowing for, given the short margin of time for sowing in these regions, an imperfect start entailed a defective harvest. This could happen when the first rains of autumn (usually towards the end of September or early October) arrived too late or were insufficient to facilitate the working of the land and the task of sowing by the second half of October. For these reasons the timing of the drought was as important as the drought itself. A drought in 1392, for example, led to a postponement of the Burgos vine harvest until the very end of August,[3] but the grain harvest does not appear to have been seriously affected. The droughts which do appear to have had serious consequences were those related to the grain harvests of 1413, 1440, 1443, 1451 and 1452.[4] It may well be that droughts recorded in other regions of the kingdom in other years also affected Old Castile, but in the absence of specific evidence it would be unwise to make this assumption.

The torrential rains and floods of the winter of 1434–5 caused havoc in the agrarian system of Old Castile and resulted in a uniformly bad harvest in 1435 (Galíndez de Carvajal, 1877:519; Carrillo de Huete, 1946:182–4). Yet here again an extreme case such as this tends to obscure the differential effects (according to region and soiltype) of the rains in less severe years. The earth of the Burgos *páramos* does not easily become waterlogged during the rainy season. This is not true of the *campiñas* of the Tierra de Campos which frequently become waterlogged, 'heavy', and unmanageable after the rains, and resist or seriously impede the task of ploughing. This factor may partially account for differences between the Palencian and Burgos fluctuations (Fig. 15.1) because, bearing in mind the short margin of time available for sowing, relatively heavy autumnal rains meant that those working 'difficult' land not only found their task harder but also managed to sow less land in the time at their disposal. On this point much also depends on how the terminology of the documentation is to be interpreted. When *rogativas* were organised to bring drought to an end, the vocabulary used was specific enough: an entry in a *Libro Redondo*, for example, might typically record payments incidental to arranging 'a procession to pray to God for rain'.[5] On the other hand, it is perhaps justifiable to assume that a *rogativa* for *buenos temporales* (that is, 'good weather') was intended to stop rain. In this respect a *rogativa* of 1444 is interesting.[6] Organised in Burgos, it does not seem to have been followed by a particularly good or bad grain harvest – yet the harvest in Palencia was poor. It may well be that the autumnal rains of 1444–5 were higher than average and that the Burgos *páramos* coped but the Palencian *campiñas* became waterlogged.

Although rain (either too much or too little depending on the time of the year) could seriously affect the harvest, the evidence suggests that, in the Burgos region at least, the greatest fear was aroused by the threat of hail-inflicted damage to crops. Until very recent times, indeed, this fear was responsible for the use of specially blessed church bells in Villadiego (the ringing noise helping to 'disperse' the danger as well as providing a warning), and the *Cantigas* inevitably depict a miracle in which the crops of a peasant devoted to the Virgin escape damage from hail (Guerrero Lovillo, 1949: *cantiga* clxi, *lámina* 176). The accounts of *lechas* for this period confirm that hail during the summer months constituted a weather hazard of considerable importance. Moreover, since hail-damage inflicted on crops after the tithe auctions had taken place was not normally discountable, the accounts of *lechas* can on occasion provide a valuable extra dimension to the tithe auction data. Looked at from this point of view the years 1394, 1398, 1426, 1438, 1439, 1442, and 1446 were ones during which hail caused considerable damage to crops. In 1438 the 'disappearance' of some tithe farmers and their guarantors, the need finally to make some realistic readjustment in the value of the farms, and the accounts of the *lechas* all point to this particular year as being one in which the harvests fell victim to disastrous hailstorms.[7]

More than in Old Castile, the agrarian crises of the Levantine *huertas* were the product of climatic extremes. The society of the irrigated lands of the late medieval Valencian *huerta* has been the subject of an excellent study by Glick (1970); in general his picture of a 'democratically' and 'densely' organised society, based on the all-important canals and divisors, is applicable to the Murcian region. In a sense, of course, the *huerta* economy emerged as a response to the clearly defined wet and dry seasons which characterise the region: in modern times only about 5 per cent of the annual rainfall occurs during the summer months, and of the two rainfall maxima the one in spring is frequently associated with thunderstorms (Linés Escardó, 1970). The unequal pattern of rainfall, with concentrated downpours within relatively short periods of time, created problems for the Murcian economy which were in turn aggravated by the nature of the rivers with their irregular courses and their abrupt seasonal changes from drying trickles to overflowing torrents. However, these were recurring problems with which the economy normally coped remarkably well. What the Murcian documentation reveals, therefore, are those extremes of floods and droughts with which the system could not cope (Torres Fontes & Calvo García-Tornel, 1975).

The well-documented floods in Murcia during the later medieval period all fell outside the dry season and occurred when rain could be expected – for example, the winter of 1423–4, the winter of 1452–3, and during March and April of 1465. The term 'flood', however, is rather ambiguous because it can refer to two rather different, if connected, phenomena – concentrated downpours of rain which were 'heavier' than usual or lasted a longer period of time, or the overflows of the river systems. In this latter sense, the floods of the Turia river in Valencia were notorious: for example, the flood of September 1328 destroyed the bridges and part of the walls of Valencia, and that of August 1358 came after a prolonged drought and destroyed 1000 houses (Cabo, 1973:92–3). The *actas* of Murcia not only make it clear that the fear of floods was associated with the rivers Guadalentín and Segura, but that active measures were taken to reduce the danger, for example by operating diversion dams (such as that of the Luçra in the Sangonera area) and by continually cleaning out debris from the river and canal beds during the dry season (Torres Fontes & Calvo García-Tornel, 1975). Despite such measures, the flood of 1424 destroyed part of the town walls and over 600 houses.[8] In general terms it seems clear that the recorded floods of the period caused considerable havoc to the agrarian economy, although frequently the severest effects tended to be rather indirect. For example, in 1465 one of the *acequías* or canals had to be broken in order to avert a major disaster, with the result that one sector of the *huerta* had its vineyards and cereal-lands destroyed. Similarly, in 1452, the greatest hardship seems to have arisen from the fact that flooding destroyed stores of grain in some quarters of the city (Torres Fontes & Calvo García-Tornel, 1975). In general, the Murcian floods usually caused extensive damage in three ways: the canal systems were disrupted, large areas of the *huerta* reverted to a livestock economy for a considerable period of time, and in the short run many of the local mills ceased to function.

The *actas* of Murcia reveal a greater preoccupation with droughts than with floods. Almost all the periods when emergency measures were taken to avoid the export of grain and stimulate imports from other areas were preceded by lengthy droughts. Combining the evidence for Valencia and Murcia, moreover, it would appear that the droughts affected the *huerta* regions in clusters of years, with some decades naturally suffering more than others: thus the 1350s, perhaps the 1370s, the first fifteen years of the fifteenth century, and the 1440s seem to have had more than their fair share of droughts.

Since the *huerta* economy depended on irrigation, a prolonged drought could be disastrous. Emergency measures, such as water diversion plans and the arranging of 'turns' to share out the dwindling supply, could stave off the worst effects, but if the drought disrupted sowing then there was little that could be done except arrange for the import of large supplies of grain and other food. A detailed description of the effects of the drought of 1412–13 on the Valencian *huerta* has been given by Glick (1970:135–45),[9] and although this was probably the worst episode of its kind in the late medieval period, the droughts of the 1440s were not far behind in their severity.

Also associated in some way with climatic factors were the not infrequent plagues of locusts in the Murcian region. A seventeenth-century chronicler's account of locust plagues in nearby Orihuela in 1358, 1407, and 1459 cannot be accepted without verification (Bellot, 1954:vol. 2, 96–7), but the *actas* of Murcia provide detailed information on the practical and magico-religious measures which were adopted to deal with the locusts in February and March 1438, April 1440, the autumn of 1463, and the summer of 1464.[10] Although we have a lengthy description of the Andalusian locust plagues of 1508–9 (Bernáldez, 1962:535–7), it is still difficult to ascertain the particular species involved (probably *Dociostaurus maroccanus*), and it is surely impossible to infer climatic changes from the occurrence of the plagues. Modern studies (for example, Pedgley & Symmons, 1968) suggest that initial locust upsurges may depend on unusual rains, but subsequent locust behaviour in other areas may be independent of this climatic 'trigger'.

Since there are no general studies on the Iberian climate of the periods preceding and succeeding the later Middle Ages, it is manifestly impossible to place the evidence of this paper within a specifically Hispanic secular trend. Even for the period under review a focussed picture will only emerge when more evidence from other regions and archives becomes available. Nevertheless, within a general European context some interesting patterns do emerge. Inevitably, and partly because of the nature of the evidence, the prevailing pattern is one which stresses climatic extremes: for example, the floods of 1402 and the drought of 1412–13 were calamitous enough to move contemporaries to pen descriptions of them (Hinojosa, 1893:108; García de Santa María, 1972:140–1). Such evidence is useful for two reasons. In the first place, it not only allows us to put flesh on the bare bones of the quantitative data, but it also makes it possible to check descriptions which may, at first sight, appear dubious. For example, an Englishman's description of the 'great hot summer' of 1473 'in south parts of the world, in Spain, Portugal, Granada' (Warkworth, 1839:23) acquires added significance when it is found to be corroborated by the Spanish evidence for this year (Ladero Quesada & González Jiménez, 1978:79–81). Secondly, although this cannot be attempted here, there are sufficient data for analyses of specific crises – a study, for example, of the regional variations in the phenomena associated with the great drought of 1412–13. Of greater interest is the fact that the Spanish evidence in general seems to fit in well with the view of the period in Europe given by Lamb (1977). In this sense, of course, the coincidences or similarities apply not to specific years but to runs of years, such as the dry summers of the second decade of the fifteenth century or the unsettled years of the 1430s. This latter case is particularly interesting because the 1430s 'seem to have brought a possibly unparalleled run of severe winters, or winters which included severe spells, in most parts of Europe' (Lamb, 1977:457). So far the Castilian evidence has yielded little that can be added to complement the map of the winter of 1431–2 (Lamb, 1977:33), although there is a brief reference to 'the great rains and upsets in the weather' (Galíndez de Carvajal, 1877:502). In February 1433, however, there were severe snowstorms – inevitably 'the worst in living memory' – which lasted on and off for forty days and were said to be heaviest in the northeastern regions of the kingdom (Galíndez de Carvajal, 1877:512). Later, from 28 October

1434 to the 7 January 1435, there occurred the rain- and snowstorms which, covering most of the kingdom, must have been among the very worst of the period and for which there exists an account of some 5000 words in length (Carrillo de Huete, 1946:182–95). Finally, January and February of 1437 witnessed further heavy snowstorms and extremely cold weather (Galíndez de Carvajal, 1877:532). This evidence corroborates the picture of the 1430s as being a decade of most unusual climate in Europe.

Climate and popular unrest

Aware of the great range of possible connections between different phenomena, most historians adopt a suitably cautious approach in discussing the impact of climatic fluctuations on the history of a given period. In the case of late medieval Castile several possible lines of investigation suggest themselves. Some may be dismissed immediately: I doubt whether anything useful would emerge from an analysis of the relationship between climatic changes and constitutional developments. At the other extreme it would be relatively easy to demonstrate that particular weather conditions affected the conduct of warfare and the organisation of Castilian frontier supplies and defences. Certain problems which fall between these extremes are of greatest interest. For example, the Castilian documentation makes one immediately aware of the 'inexact' or 'syncopated' coincidences in time between climate and outbreaks of plague. Thus *rogativas*, such as those in Burgos, were as much concerned with plague as with weather conditions (see, for example, López Martínez, 1961:230–3). Although in almost all cases the outbreaks of plague outlasted the adverse climatic conditions to which they seem to have been linked, the evidence does suggest a correlation in the intensities of both phenomena. For example, the disastrous drought year of 1413 was also described by contemporaries as one when 'the pestilence was particularly harsh', yet the plague easily outlasted the drought and one can sense the relief of the Palencian canon who wrote on the title page of the *actas* for 1415 'in this year, by the grace of God, the pestilence somewhat abated and ceased'.[11] This discussion could be extended further. However, for lack of space, I will turn to another problem and concentrate on some aspects of the relationship between climate and popular unrest.

It seems obvious that there was a fairly close connection between climatic extremes, which gave rise to bad harvests and famine, and the popular tumults known as *alborotos de hambre* or hunger riots; though other variable factors, such as demographic stress or monetary debasements, also have to be weighed in any explanation of popular unrest (MacKay, 1972). An apparently straightforward example of such a riot in Seville in 1462 is described by the urban chronicler Garci Sánchez (1953:45–7):

> In the year 1462 there were great floods during January in Seville and a *fanega* of wheat rose to 120 *mrs*. People could not find food to eat and so they ate rotten biscuits from the Crown [naval stores] during this month of February ...
>
> Monday, the 29th of March of this same year. At seven o'clock a hunger riot started in Seville. The mass of the people armed themselves and searched for bread. The gangs went from house to house searching for bread and saying that it had been stored away as a remedy for hard times. Everyone hid their valuables, thinking that they wished to loot their coffers. This lasted till mid-day.

Other accounts often tend to be more complex, involving such 'irrational' elements as the widespread tendency to explain disasters in terms of the sins of the people. In this respect Garci Sánchez' description appears almost clinical in comparison with the way in which Alvar García accounted for the famine and plague of 1412–13 (García de Santa María 1972:140–1):

> And all these things have afflicted and troubled Spain because of the sins of the people, and because of the bad way of life of those in Spain. Frequently they are bold enough to be drawn to tyranny, which is a great sin, in a way which does not happen in other lands, because in other lands each person puts himself in his proper station and accepts what God gives him. But it is the opposite in Castile where each man who takes the tonsure wants to be pope, and the man who obtains wealth wants to be of royal lineage, even though his lineage is lowly, so that there is no contentment from the lowest to the highest, and there is a great deal of discontent in all the estates of society ... And, therefore, God in his justice against Spain sends these plagues, starvation, and the loss of people, as a punishment ...

To the climatologist, such 'irrational' explanations may appear intrusive, requiring elimination after careful study and analysis of the text. But to the historian who seeks to assess the impact of climatic fluctuations on social behaviour, irrational explanations may be of great interest.

In a seminal article published in 1971, E. P. Thompson attacked the 'spasmodic view of popular history'. This view, according to Thompson, is one in which 'the common people can scarcely be taken as historical agents before the French Revolution. Before this period they intrude occasionally and spasmodically upon the historical canvas, in periods of sudden social disturbance. These intrusions are compulsive, rather than self-conscious or self-activating: they are simple responses to economic stimuli. It is sufficient to mention a bad harvest or a down-turn in trade, and all requirements of historical explanation are satisfied.' (Thompson, 1971:76) Thompson's words were directed for the most part against 'quantitative' or 'growth' historians who 'are guilty of crass economic reductionism, obliterating the complexities of motive, behaviour, and function' which affected popular behaviour. There is clearly a danger that historians and scientists interested in the connections between climate and popular movements may commit similar acts of 'crass reductionism'.

It follows from Thompson's arguments that it is essential to give attention to the participants' view of human behaviour, and to elucidate the logical structures which underlie 'primitive' or pre-industrial belief systems and the behavioural patterns to which they give rise. On this view, the fact that people believed that sins explained the occurrence of a natural disaster is of crucial significance, and it follows that the reactions of contemporaries to disaster can only be properly explained within this context. In other words, whereas the climatologist or scientist might remove 'irrational' or 'religious' elements in order to obtain 'objective' data on weather, the historian, if he is interested in the wider issues of social impact, must put 'sin' and other similar elements back into the centre of the picture. Unless we wish to reduce men to the role of mere puppets who react to climatic fluctuations without thinking about it, the effort to understand the participants' view is well worth making.

Certainly the late medieval Castilians found that such matters as the orderly progression of the seasons or extreme fluctuations in climate were powerful aids to those processes of contemplation which explained the order – or, more important, the disorder – of the world they lived in. Indeed, like other medieval communities, they went further and imposed their own patterns of popular and religious beliefs on the yearly cycle with the result that progress through time was frequently envisaged as a series of ritual rites of passage from one 'block' of time to another. In addition, therefore, to the crises arising from wars, plagues, and climatological phenomena, there were patterns of self-imposed times of feast and times of famine. The famous 'battle' between Don Carnal and Doña Cuaresma in the *Libro de Buen Amor* (Ruiz, 1972:295–333), for example, is an allegorical episode which Juan Ruiz based on the realities of medieval social life: from a period of festivity, Carnival, Christians entered into the long austerity and 'famine' of Lent, and then re-emerged into the celebrations of Easter when 'Flesh' and 'Love' once more triumphed (Laurence, 1970). In other words the 'battle' between Carnival and Lent on Shrove Tuesday, depicted by Bosch and Brueghel in art and Juan Ruiz in poetry, was a yearly event which ushered in the 'reign' of 'Old Woman Lent', and a period of fasting, repentance, and confession. Of course in the examples just cited it is possible that the artistic purpose has changed or deformed social reality. Yet, if we turn to the much more pedestrian narrative of the *Hechos del Condestable Don Miguel Lucas de Iranzo* (Escavias, 1940:152–83), we find the chronicler describing the year 1464 in terms of the ritual *fiestas*, fastings, and religious observances which structured the successive 'blocks' of time making up the year in the frontier town of Jaén. Moreover the chronicler was not inventing his own version of structured time: on the contrary, he was describing the collective vision of the participants. Thus when, in 1468, a conspiracy was hatched in Jaén to assassinate the constable and kill those of Jewish extraction, the conspirators chose a specific evening during Easter when they knew exactly where the participants would be acting out their various roles according to the dictates of the collective vision (Escavias, 1940:372–3). The example should be kept in mind because, although the conspirators did not in fact strike in 1468, a similar plot did succeed in the famine year of 1473.

Feasts and famines stemming from the collective religious or cultural vision, on the one hand, and those arising from climatological phenomena or other 'impersonal' forces, on the other, constituted the opposite sides of the same coin: for, if voluntary abstinence and confession expiated sins and averted the divine wrath, plagues and famines confirmed that unrepentant and stubborn sinners were being punished by God. The *Cantigas* miniatures illustrating a *rogativa* in Jerez demonstrate this rather obvious point that extremes of weather conditions stimulated the preaching of sermons, the organisation of processions, and the belief that society's sins were responsible for the disasters (Fig. 15.3). The preaching element was without doubt of great importance. In Spain it usually involved the friars, who acted as intermediaries between society and the deity who imposed – or at least failed to prevent – floods or droughts. Sermon material from the most bizarre sources, therefore, is likely to throw up climatological data – from, for example, the thirteenth-century *Cantigas* to Fray Francisco Ortiz' famous sermon in 1529 which led to an Inquisition trial about a *beata* or holy woman (Selke, 1968:31).

The connection between natural disasters and contemporary accounts of sin is complex. At a simple level, of course, the call on each individual sinner to repent is easy to understand. Similarly, prophesying disaster and using such prophecies as a form of moral control appears straightforward: when, in Murcia in 1411, St Vincent Ferrer preached a sermon in which he foretold that 'the wolf will devour the lamb', he both remembered the great Valencian flood of 1358 and anticipated that of 1424 in Murcia (Rosselló Verger & Cano García, 1975:14). But sin was also thought of not in terms of the individual but in terms of generalised evils which might affect the social structure as a whole. In this sense it might well be worth investigating whether natural disasters provoked Castilians into apportioning blame on those deliberate evildoers whom they held to be responsible for the crucifixion, because there does seem to be some correlation between weather 'triggers' and Iberian pogroms of Jews and *conversos*.

This correlation, of course, was not inevitable or monocausal. The ambivalent social and religious status of the *conversos*, for example, was of crucial importance. The *conversos* or 'New Christians' were those Jews who, having converted to Christianity in the aftermath of the massacres of 1391, found that the religious impediments to their social advancement were no longer operative. Jealousy of the worldly success of *conversos*, combined with the suspicion that they were crypto-Jews, therefore, helps to explain why they became the chief victims of fifteenth-century mob violence. This violence was also undoubtedly linked with royal debasements of the coinage, severe price inflation, increases in taxation, and monetary reductions (MacKay, 1972). It should also be remembered that the religiously imposed periods of feast and famine may have been more significant than those arising out of climatic fluctuations. The study by Roth (1933), which draws attention to the frequent clash in France between the Jewish carnival-feast of Purim and the Christian seasons of Lent and Easter, is one which deserves to be extended into an Iberian context. Such a study would have to take into account the fact that movable festivals as well as different calendars were involved (Purim falls in the Hebrew month of Adar). Inevitably, however, the suspicion is aroused that Christians emerging from the 'famine' of Lent would view the 'feast' of Purim with considerable animosity. Would they not suspect that the burning of the effigy of Hannam was in fact a mockery of the crucifixion? Some such clash of ritual may have sparked off the pogrom of 1473. Nor is there any evidence as yet to suggest that climatic factors played any part in the great pogrom of 1391 (Wolff, 1971).

Nevertheless it is surely no coincidence that the disastrous years of 1411–13, which brought drought, bad harvests, and plague, also witnessed the acceptance and popularity of Vincent Ferrer's anti-semitic preaching tours and the formulation of oppressive anti-Jewish legislation (Baer, 1966:166–9). It may also prove to be the case that the uprising and massacres in Toledo in 1449 were not unconnected with harvest failures in that particular region in 1448. The most striking example, however, relates to the massacres in Andalusian towns in 1473. From September 1472 until May 1473 there was a prolonged drought; sowing, which had been delayed as long as possible, produced a deficient harvest, and the hopes of the spring sowing were dashed by yet another drought during the remainder of 1473 (Ladero Quesada & González Jiménez, 1978:79). These grim conditions, however, seem to have been at their worst in Andalusia, and it was there that there occurred the ferocious killings of *conversos* which, beginning in March 1473 in Cordoba, quickly enveloped other southern towns (MacKay, 1972). In the case of Cordoba the trouble began when a procession, organised from a Franciscan house, erupted into violence and resulted in three long days of looting and massacre (Nieto Cumplido, 1977). In Jaén the constable on this occasion did not escape the machinations of conspirators – he was hacked to death during mass at the cathedral and the mob went on the rampage against the *conversos* (MacKay, 1972:62). Were these massacres, then, induced by the drought and related famine? Perhaps they were, but it should also be remembered that the killings coincided both with the emotionally charged period leading up to Easter and with the most anarchic point in the monetary history of late medieval Castile.

Were the town mobs or friars, moreover, so naive as to lay all the blame on Jews or *conversos*? A remarkable study of one of the two surviving versions of the Castilian *Dance of Death*, the *Dança de la Muerte*, shows that popular unrest was fascinatingly complex and not merely anti-semitic in nature (McKendrick, 1979). The *Dance* begins and ends with the preacher in the pulpit and with calls to repent from sin. A close textual analysis of this version, however, reveals that the twenty-five new characters which it contains were not 'abstract' people accused of vague 'theological' sins, such as avarice or pride. On the contrary, they were exclusively urban characters drawn from the streets of Seville who were charged with specific 'social' crimes – for example, selling dear bread, shortweight fish, putrid meat, and adulterated wine. In other words the catalogue of these 'sins' expressed consumer grievances about practices which outraged popular moral assumptions regarding legitimate market-place norms. This version of the *Dance* was printed in 1520 and was linked to the revolt of the *comuneros*, but it was evidently composed in the fifteenth century and its text relates most closely to the communal disturbances of 1462–3 which, as can be seen from the passage cited from Garci Sánchez, were associated with the floods of January and February 1462 and with one of the worst harvest failures of the period in Andalusia. For this reason the most reprehensible 'sins' in the Dance were those committed by the small shopkeepers and breadsellers of the Gradas district of Seville who were held to be responsible for the shortages and high prices of food. People, in short, did not just pray for good weather and confess their own sins – they identified 'social' sins, transferred guilt, and went on the rampage. Thus, although in this complex process climatic factors may have played a minor – but not unimportant – role, care must be taken to avoid the 'spasmodic' or 'stimulus-response' view of popular behaviour as identified by Thompson (1971).

Conclusion

On the basis of the evidence presented in this chapter it should be clear that sources of weather data exist for late medieval Castile. Further study of such sources should enable scholars to modify and improve the weather record which has been presented here. But, in studying the connections with popular unrest, it is evident that a useful pattern of regional or kingdom-wide weather fluctuations already exists. The problem, therefore, is not that of detecting the primary weather 'stimuli' or the 'responses' in terms of popular behaviour. The real problem is that of assessing the ways in which such behaviour was conditioned or modified by cultural and religious factors. What to modern scholars may appear to be totally irrational and irrelevant explanations of weather phenomena, can turn out to be essential elements of the belief systems of those who had to account for the events of the world in which they lived. Moreover, although we may label such belief systems 'primitive' or 'medieval', they were just as complex and, given the initial premises, just as logical as those prevailing in subsequent centuries. It could reasonably be objected that evidence such as the 'comic-strip' from the *Cantigas* miniatures (Fig. 15.3) presents us with 'crass reductionism' of another variety, but it is precisely evidence of this nature which enables us to glimpse the complexities of culturally mediated behaviour.

Acknowledgements

Research in Spanish archives was facilitated by a generous grant from the Twenty-Seven Foundation. Professors Torres Fontes, Ladero Quesada and González Jiménez kindly advised me about certain aspects relating to Murcia and Andalusia. Finally, I would like to thank Ms Geraldine McKendrick and Mr Anthony Goodman for drawing my attention to weather observations from the period in question.

Notes

1. In addition to the *L[ibros] R[edondos]*, the following registers provided quantitative data: *A[rchivo de la] C[atedral]*, Burgos, *libros* 66–8, *registro* 2, and a miscellaneous collection marked *Quentas de marabedises*. The Palencia figures are given in the canonical deliberations: AC, Palencia, *actas capitulares*.
2. The *lechas* form a sub-section of the *Libros Redondos* and, in listing discounts against income owed to the cathedral, data on weather conditions are sometimes recorded.
3. AC, Burgos, *actas capitulares*, *registro* 2, fo. 9.
4. In addition to the chronicle evidence for 1413, data on *rogativas* to end droughts are in AC, Burgos, LR 1441, fo. 22; LR 1443, fo. 23; LR 1451, fo. 24; LR 1452, fo. 22. Related to the droughts are the shortages of grain and increases in bread prices, such as those mentioned in A[rchivo] M[unicipal], Burgos, *actas capitulares*, 1453, fos. 21, 31.
5. For example, AC, Burgos, LR 1443, fo. 23.
6. Processions, in fact, were organised 'for peace, health, and good weather' ('*por la paz e por la salud e por buenos temporales*'): AC, Burgos, LR 1444, fo. 26. The prayers for '*salud*' were undoubtedly linked to an outbreak of *pestilencia*: AC, Burgos, *actas capitulares*, *registro* 5, fo. 134.
7. Discounts for damage from hail are in the *lechas* of the *Libros Redondos* for the years cited. The reference to absconding tithe farmers and guarantors is in AC, Burgos, LR 1439, fo. 57.
8. Discussions on the flood and its consequences are recorded in AM, Murcia, *actas capitulares*, 1423, fos. 13, 16–17, 21, 31–2. Subsequently the damage was briefly described in a petition presented to the *cortes* of Palenzuela of 1425.
9. Data on drought-related phenomena in AM, Murcia, *actas capitulares*, 1412–13, fos. 58, 63, 92, 98–9, 101, 128–9, and 134–5, complement the picture given by Glick (1970) for Valencia. For the famine of 1374–5 see Valdeón Baruque (1969:245–8).
10. For example, discussions recorded in AM, Murcia, *actas capitulares*, 1463, fos. 29, 43, 66, 67, 71, and 74–5, give information on the *cuadrilla* organisation for dealing with the locust menace.
11. The documents for Burgos, Palencia, Seville, and Murcia all confirm that these were years of *pestilencia*. The examples cited are from AC, Palencia, *actas capitulares*, 1413, fo. 34 ('la pestilencia de mortandat andaba muy de rezio'), and the title page of the *actas* for 1415 ('in quo anno per dei gratiam pestis mortalitatis aliquanticulum placata cessavit').

References

Baer, Y. (1966). *A History of the Jews in Christian Spain*, vol. 2, *From the Fourteenth Century to the Expulsion.* Philadelphia: The Jewish Publication Society of America.

Bellot, P. (1954). *Anales de Orihuela*, ed. Juan Torres Fontes, 2 vols. Orihuela: Casino Orcelitano.

Bernáldez, A. (1962). *Memorias del reinado de los Reyes Católicos*, ed. M. Gómez-Moreno & Juan de M. Carriazo. Madrid: Real Academia de la Historia.

Cabo, A. (1973). Condicionamentos geográficos de la historia de España. In *Historia de España Alfaguara*, vol. 1, ed. Miguel Artola, pp. 1–183. Madrid: Alianza Editorial.

Carrillo de Huete, Pedro (1946). *Crónica del halconero de Juan II*, ed. Juan de Mata Carriazo. Madrid: Espasa Calpe, S.A.

Chacón Jiménez, F. (1979). *Murcia en la centuria del quinientos.* Murcia: Universidad de Murcia.

Escavias, Pedro de (1940). *Hechos del Condestable Don Miguel Lucas de Iranzo*, ed. Juan de Mata Carriazo. Madrid: Espasa Calpe, S.A.

Galíndez de Carvajal, Lorenzo (1877). *Crónica de Don Juan II de Castilla.* In *Biblioteca de Autores Espanoles*, vol. 68, ed. Don Cayetano Rosell. Madrid: Real Academia Española.

García de Santa María, Alvar (1972). *Le parti inedite della 'Crónica de Juan II' di Alvar García de Santa María*, ed. Donatella Ferro. Venice: Consiglio Nazionale delle Richerche.

García Fernández, J. (1963). *Aspectos del paisaje agrario de Castilla la Vieja.* Valladolid: Universidad de Valladolid.

Glick, T. F. (1970). *Irrigation and Society in Medieval Valencia.* Cambridge, Mass.: Harvard University Press.

González Jiménez, M. (1976). Las crisis cerealistas en Carmona a fines de la Edad Media. *Historia, Instituciones, Documentos*, **3**, 283–307.

Guerrero Lovillo, J. (1949). *Las Cantigas. Estudio arqueólogico de sus miniaturas.* Madrid: Consejo Superior de Investigaciones Científicas.

Hamilton, E. J. (1936). *Money, Prices, and Wages in Valencia, Aragon, and Navarre, 1351–1500.* Cambridge, Mass.: Harvard University Press.

Hinojosa, Gonzalo de la (1893). *Continuación de la Crónica de España del Arzobispo Don Rodrigo Jiménez de Rada por el Obispo Don Gonzalo de la Hinojosa.* In *Colección de Documentos Inéditos para la Historia de España*, vol. 106, ed. El Marqués de la Fuensanta del Valle. Madrid: José Perales y Martı ´nez.

Huetz de Lemps, A. (1962). Les terroirs en Vieille Castille et León: Un type de structure agraire. *Annales: Economies, Sociétés, Civilisations*, **17**, 239–51.

Ladero Quesada, M. A. & González Jiménez, M. (1978). *Diezmo eclesiástico y producción de cereales en el reino de Sevilla* (*1408–1503*). Seville: Universidad de Sevilla.

Lamb, H. H. (1977). *Climate: Present, Past and Future*, vol. 2, *Climatic History and the Future.* London: Methuen & Co. Ltd.

Laurence, Kemlin M. (1970). The battle between Don Carnal and Doña

Cuaresma in the light of medieval tradition. In *Libro de Buen Amor Studies*, ed. G. B. Gybbon-Monypenny, pp. 159–76. London: Tamesis Books Ltd.

Linés Escardó, A. (1970). The climate of the Iberian Peninsula. In *World Survey of Climatology*, vol. 5, ed. H. E. Landsberg, pp. 195–239. Amsterdam: Elsevier Publishing Company.

López Martínez, N. (1961). Don Luis de Acuña, el cabildo de Burgos y la reforma (1456–1495). *Burgense*, **2**, 185–317.

MacKay, Angus (1972). Popular movements and pogroms in fifteenth-century Castile. *Past and Present*, **55**, 33–67.

MacKay, Angus (1977). *Spain in the Middle Ages. From Frontier to Empire, 1000–1500.* London: The Macmillan Press.

McKendrick, G. (1979). The *Dança de la Muerte* of 1520 and social unrest in Seville. *Journal of Hispanic Philology*, **3**, 239–59.

Nieto Cumplido, M. (1977). La revuelta contra los conversos de Córdoba en 1473. In *Homenaje a Antón de Montoro en el V centenario de su muerte*, pp. 29–49. Montoro: Monte de Piedad y Caja de Ahorros de Córdoba.

Pedgley, D. E. & Symmons, P. M. (1968). Weather and the locust upsurge. *Weather*, **23**, 484–92.

Rosselló Verger, V. M. & Cano García, G. M. (1975). *Evolución urbana de Murcia (831–1973)*. Murcia: Ayuntamiento de Murcia.

Roth, C. (1933). The Feast of Purim and the origins of the Blood Accusation. *Speculum*, **8**, 520–6.

Ruiz, Juan (1972). *Libro de Buen Amor*, ed. R. S. Willis. Princeton: Princeton University Press.

Sánchez, Garci (1953). *Los anales de Garci Sánchez, jurado de Sevilla*, ed. Juan de Mata Carriazo. Seville: Universidad de Sevilla.

Selke, A. (1968). *El Santo Oficio de la Inquisición. Proceso de Fr. Francisco Ortiz.* Madrid: Ediciones Guadarrama.

Tamames, R. (1974). *Estructura económica de España*, 7th edn. Madrid: Sociedad de Estudios y Publicaciones.

Thompson, E. P. (1971). The moral economy of the English crowd in the eighteenth century. *Past and Present*, **50**, 76–136.

Torres Fontes, J. & Calvo García-Tornel, F. (1975). Inundaciones en Murcia (Siglo XV). *Papeles del Departamento de Geografía de la Universidad de Murcia*, **6**, 29–49.

Valdeón Baruque, J. (1969). Una ciudad castellana en la segunda mitad del siglo XIV: El ejemplo de Murcia. In *Cuadernos de Historia. Anexos de la revista Hispania*, **3**, 211–54.

Warkworth, John (1839). *A Chronicle of the First Thirteen Years of the Reign of King Edward the Fourth*, ed. J. O. Halliwell. London: Camden Society.

Wolff, P. (1971). The 1391 pogrom in Spain. Social crisis or not? *Past and Present*, **50**, 4–18.

PART IV

CLIMATE–HISTORY INTERACTIONS: SOME CASE STUDIES

16

Climate, environment, and history: the case of Roman North Africa*

BRENT D. SHAW

Abstract

North Africa has long been renowned not only in academic literature, but also in popular imagination as the 'Granary of Rome'. This myth has been the principal basis upon which theories of a climatic change that afflicted the Maghrib since the period of Roman rule have been founded. When colonial archaeologists and historians began their reconstruction of North Africa's Roman past in the decades after the annexation of the western Ottoman Empire by European powers, beginning with Algeria in 1830, they laid stress on what seemed to be a lost but glorious episode of development and wealth at the time of the Roman Empire. Remains of large Roman cities, such as Lepcis Magna and Timgad, discovered in the midst of desolate landscapes, as well as villa and farmstead complexes in the deserted countryside, appeared to confirm their suspicions. The problem of the putative *decline* of North Africa since the Roman period was then solved by the elaboration of two aetiological myths. The first attributed the destruction of the Roman achievement to armed invaders, principally the Vandals and the Arabs. Buttressing this myth was the contention that there had existed more favourable climatic conditions in North Africa, marked above all by higher levels of precipitation, which had aided and permitted the economic development of the region as part of the Roman Empire.

These arguments, developed on the basis of false assumptions about the past, dubious literary evidence, and misunderstood archaeological data, took on an impetus of their own and continue to affect many investigations into climatic change in the ancient Mediterranean event to the present day. To prevent further abuse of the historical data, this chapter investigates the historiographical origins of the question of climatic change in this central region of the Roman Empire. It attempts to demonstrate that the question has been determined by factors quite extraneous to a proper scientific study (e.g. colonial biases, historical misconceptions about the Roman Empire, mistaken ideas about past agricultural productivity etc.). Most of the literary and archaeological arguments that have been deployed hitherto are largely

* For the *prehistoric* period the reader is referred to B. D. Shaw, Climate, environment and prehistory in the Sahara, *World Archaeology* **8** (1976), 133–49.

mythical and should be discarded from any serious discussion in the future. The question must be formulated afresh, free from its historical roots.

Three aspects of the data that have traditionally been employed in the debate (fauna, water resources and control systems, and the production of grain surpluses) are investigated, in order to reject them in favour of the quantifiable evidence relating to the geological and biological changes in the environment of the Maghrib. A brief survey of our present state of knowledge based on the latter type of data would lead one to believe that the *environmental* conditions of the Roman period probably were more favourable than at present (e.g. in terms of vegetal ground cover), but that these more favourable conditions may easily fall within the putative limits of a 'fossil' environment remnant in the Maghrib and Sahara since the last African 'pluvial'. Hence, the agricultural developments of the Roman period (in any event much exaggerated by moderns) probably took place under climatic conditions analogous to those of the present. There seems to be an arguable case for general climatic conditions shifting to a substantially more humid regime between *c.* AD 1000 and 1750, though even this hypothesis faces serious unanswered questions about the relationship between ground data and atmospheric fluctuations.

Introduction

Climate is one of those ever-present factors in the historical process that most historians prefer to ignore and yet cannot. In their reconstruction of the past they find themselves unable to assume without question that climate is an invariable element that can safely be disregarded. Certainly the debate over the influence of changing climatic regimes on the course of the history of the Roman Empire, and indeed of most other early circum-Mediterranean civilisations, is as old as modern historiography. But it seems that the factor of climatic change most often enters historical argument whenever an explanation is sought for catastrophic fluctuations in ancient civilisations. As a causal factor, deteriorating climate has been invoked most readily if the total collapse of a civilisation seems otherwise inexplicable. In addition to Imperial Rome, the Hittite Empire, New Kingdom Egypt, Mycenae, and the Indus Valley civilisation of Mohenjo-Daro have all been the subject of theories of climatic change that sought to explain their downfall. (For Rome, see Huntington, 1917; Demougeot, 1965, 1969: 25, 43–4; Baynes, 1943: 30–1. Egypt – Sandars, 1978: 20, 183. Mycenae – Carpenter, 1966; Wright, 1968; Bryson, Lamb & Donley, 1974; Bryson & Murray, 1977; Lamb, 1977: 420–22; and the rejoinder by Dickinson, 1974. Indus Valley – Singh, 1971; Hare, Kates & Warren, 1977: 333. See also McGhee, this volume.)

Amongst the numerous factors adduced in modern scholarship in connection with the 'Decline and Fall' of the Roman Empire, that of climatic change can be traced as far back as Gibbon, and beyond. Gibbon felt certain that the climate of Europe had been 'much colder formerly than at present', and gave what he called two 'less equivocal' reasons for his belief. First, he claimed that ancient literary sources showed that the Rhine and Danube rivers were 'frequently' frozen over and were thus capable of bearing the enormous weight of barbarian armies that were able, as he put it, to invade the Empire 'over a vast and solid bridge of ice'. Second, Gibbon presented a somewhat more sophisticated argument based on fauna. He centred this proof on the reindeer, 'that useful animal, from whom the savage of the north derives the best comforts of his dreary life'. Gibbon claimed that the reindeer needed intense cold, and that in his day the animal was not to be found south of the Baltic, whereas in Caesar's time reindeer and elk were commonly found in the forests of Poland and Germany. He also offered the reader what he believed to be a precise parallel to the climate and environment of Europe at the time of the Roman Empire. 'Canada', he opined somewhat phlegmatically, 'at this day is an exact picture of ancient Germany. Although situated in the same parallel as the finest provinces of France and England, that country experiences the most rigorous cold. The rein-deer are very numerous, the ground is covered with deep and lasting snow, and the great river of St Lawrence is regularly frozen, in a season when the waters of the Seine and the Thames are usually free from ice.' I shall omit further comment on Gibbon's forbiddingly chilling view of my homeland (too harsh a prospect, I fear, for a cultivated eighteenth-century English gentleman), but attention should be drawn to Gibbon's acceptance of the determinate effects of changing climatic regimes 'over the *minds* and bodies' of men (Gibbon, 1966: 209–11). And, in historical writing on the subject, climatic influences, along with aristocratic sloth, sexual excess, barbaric corruption, and the evils of Christianity, became one of the factors normally included in any consideration of the causes of the Empire's decline and fall. Recently, however, as modern-day Roman historians have become more sophisticated, the question of climatic change seems to have receded quietly from view. In their greater sense of professionalism the historians have gradually abandoned the consideration of what are perceived to be 'exotic' causal forces, and have developed a sort of disciplinary agnosticism on the subject. Thus a functional amnesia has been born, more, one must suspect, from a hesitation to grapple with a protean problem concerning which historians have little expertise than from any explicit denial of the importance of the question (e.g. two of the largest and most influential books on the subject, Rostovtzeff, 1957, and Jones, 1964, do not consider the factor of climate at all).

Within the context of the Roman Empire, the question of climatic change in historical times has probably been broached with the greatest persistence in scholarly works concerned with the six provinces that constituted Rome's empire in North Africa west of Egypt. These provinces covered the land area of the northern zone of the present-day state of Libya and the three countries of the Maghrib: Tunisia, Algeria, and Morocco. Even to the untrained observer of the present day who views the environment of the semi-arid regions of these countries, there is an apparent contrast between the extensive and impressive ruins of the Roman period, both cities and large rural villas, and the desolation of the countryside about them. Although this contrast is plainly visible only in certain places, the striking antithesis between past and present has been considered testimony of a conspicuous general ecological decline that has affected the whole of the Maghrib during historical times. Indeed, the contrast was regarded by early investigators as *prima facie* proof of a decisive climatic change marked primarily by an overall fall in the level of precipitation (Toutain, 1903:339; Bovill, 1929:416–17; Boudy, 1949:112; Brooks, 1949:18ff; Robert, 1950; Murphey, 1951:120ff; Capot-Rey, 1953: 90; Butzer, 1958:5ff). Once the fact of a general climatic deterioration was accepted, historians and archaeologists then searched deliberately for more evidence in support of the received view. They believed that a comparison of the apparent agricultural wealth of North Africa under Roman rule with what seemed to be a relatively economically backward agrarian economy in early colonial times was one such piece of supportative evidence. To them, it did not seem possible that present climatic conditions would permit the same level and extent of agricultural exploitation once practised by the Africans who lived under the Roman administration; so they claimed that 'more favourable' climatic factors (principally higher levels of precipitation) had permitted the greater agricultural achievement of Roman times (Carton, 1894, 1896; Goodchild & Ward-Perkins, 1949:95; Trousset, 1974:15–21, 162).

The problem with the question of 'climate and history' in ancient North Africa, as it has been posed hitherto, is therefore twofold. First, the question was created and developed on the basis of false hypotheses about the African past that were more ideological than empirical in nature. Secondly, it was assumed that ecological deterioration must be due to climatic change *alone*, and hence all the evidence that related to the problem was marshalled accordingly. There were further problems with the data themselves. Broadly speaking, the evidence employed in the debate fell into two categories: the brief, often tangential, references to weather conditions in classical literary texts, and the modern archaeological data. For the most part, however, the archaeological evidence consisted merely of the presence of 'Roman' ruins in regions that had experienced a degree of desertification since Roman times. It was simply the apparent contrast between past development and present desiccation that was regarded as sufficient proof of climatic change. The ancient literary sources were also plundered systematically in a search for signs that the environment had once been radically different from that of the present. But the literary sources suffered from the same basic flaw as the archaeological evidence deployed in the argument: they were simply not the *type* of data required to solve the problem of long-term climatic variation. Whereas it is true that some ancient historians, such as Livy, occasionally included reports of unusual atmospheric phenomena amongst the portents with which they began their account of the events of a given year, they were neither consistent in the criteria they used for these notices, nor are the data themselves in any way quantifiable. Constraints of space forbid any detailed criticism here of the priestly records, such as the *Annales Maximi*, upon which many Roman historians of the Republican period depended for their 'weather reports'.[1] As for other sources, we learn from casual references, such as those in Seneca (*Epistles* 23.1; 77.1) that the weather could be reported in prose writing, but that it was usually regarded as a banality not to be mentioned except when the writer was at a complete loss for anything else to say. Even though they are obviously part of the ominous *portenta* determined by the literary and ideological demands under which historical authors wrote till the end of antiquity (e.g. Iulius Obsequens, *Prodigiorum Liber*, and Gregory of Tours, *Historia Francorum*, late fourth and fifth centuries) haphazard literary notices of weather conditions continue to be used by modern analysts to decide questions of climatic change. One can only be mildly surprised at the persistence and continued acceptance of a futile and indefensible methodology (for but one example see McDougall, 1965). In fact, both the classical literary sources and much of the archaeological evidence hitherto used are so imprecise and lacking in quantitative measurement that quite contrary conclusions have often been drawn from the same set of data.[2]

But there is a further problem that specifically bedevils North African studies and which affects the investigation of climate. Put briefly, and not too elegantly, it is this: there was a hardened preconception in the minds of most Europeans, especially the *colons* and colonial administrators who went to North Africa, that they were the latter-day counterparts *or* heirs of a glorious Roman (that is to say, European) past. In encountering the material remains of 'Roman' towns, villas, and irrigation projects, they were certain that the past had been much happier than the present, and that it was their duty and privilege to restore the lost productivity. For them the problem was how to explain the apparent economic deterioration of Africa since the putative Roman achievement. In response there arose two fundamental 'myths' that have haunted the writing of North African history. The first could be labelled the 'Invasion Hypothesis'. Founded on the apparently unimpeachable authority of Ibn Khaldûn (*viv.* AD 1332–1406) who wrote fearfully of the nomadic tribesmen who destroyed the flower of Maghribî civilisation in the eleventh century 'like a cloud of locusts', colonial historians postulated the destructive forces of the Vandal invasion of the fifth century, and the Arab invasions of the seventh and eleventh centuries, as the critical causal factors in the collapse of the *Wirtschaftswunder* forged by the Romans. Only gradually was it demonstrated that the period of Vandal rule and the first Arab raids did not perceptibly alter the fabric of ancient North African society (Marçais, 1913, 1946; Courtois, 1955; Djaït, 1973; Vanacker, 1973). And only recently has the myth of the utter destruction wrought by the Banû Solaïm and Banû Hillâl been dismantled in favour of a more credible interpretation of the internal collapse of Fatimid society in Ifriqiyya and by a new understanding of Ibn Khaldûn (Poncet, 1967, 1968; Idris 1968a, b; Cahen, 1968; Lacoste, 1978; though de Planhol, 1968, still accepts both this myth and that of climatic change).

The other useful aetiological myth was, of course, that of 'climatic change'. This myth was also tied to the then current colonial debate over the viability of attempts to restore the lost agrarian prosperity of the Roman past. On the one side were ranged those who accepted the damage to the Roman achievement as reparable and due to secular causes (e.g. the Arab invasions). On the other side were those who questioned the effort that would be needed to combat what they perceived as the near-irreversible trend of ecological decline due to climatic deterioration. The latter men, no less than the former, were trapped by their excessively optimistic view of the past. So, for example, Coudray de la Blanchère, who prepared a survey of Roman hydraulic works at the behest of the *Resident Général* Massicault, stated flatly that, '... les textes anciens formels nous montrent le bord du Sahara comme une espèce de grande jungle', ['... the ancient literary texts show us that the periphery of the Sahara was a sort of great jungle'] (de la Blanchère, 1897:2). Of course the 'textes formels' could demonstrate no such thing, though it was firmly believed that they did.

The essential conclusion of this brief survey is that the question of climatic change as it was originally created and perpetuated was embedded in a series of responses to unproved assumptions about the past. When theories of climatic change and determinism were in vogue, the tenuous body of information from North Africa was made to fit that mould. But when a hypercritical rationalistic reaction set in, the new historians were able to point, quite correctly, to the 'unscientific' nature of the argument as propounded by those who favoured climatic change. The latter reaction began around the turn of the century and culminated in the near exhaustive analysis of the available evidence by Leiter (1909). Leiter's sceptical views were then canonised by Stephane Gsell, the doyen of Maghribî historians, who surveyed the question in similar depth in 1911 in preparation for the publication of the first volume of his monumental *Histoire ancienne de l'Afrique du Nord.* He arrived at the following general conclusion (Gsell, 1911:410): 'en somme, si le climat de la Berbérie s'est modifié depuis l'époque romaine, ce n'a été que dans une très faible mesure.' ['... in conclusion, if the climate of North Africa *has* changed since Roman times, it is only in some very small measure.'] This opinion might be right, it might be wrong. The point is that it is based on a type of data that could not provide an answer, and so reflected Gsell's own sane and reasonable judgement more than it did any reliable empirical analysis. It is no wonder then that the two schools of opinion on climatic change, of which Gsell's position was the sober middle ground, hardened into sheer dogmatism. The indecisive nature of the data meant that the prior theoretical disposition of the writer was the critical factor in the decision reached.[3] If the more questionable literary data relating directly to weather conditions are set aside, there are still three persistent arguments for climatic change in North Africa that deserve scrutiny because of their tendency to resurface repeatedly in historical works on the subject.

Extinct faunal species and the argument for climatic change

In arguments in favour of climatic change, as with Gibbon's reindeer, frequent recourse has been made to the disappearance from the range of fauna in the Maghrib and Sahara of certain animal species that require fairly humid environmental conditions for their continued survival. The number of such species is small and is usually limited to large mammals. References to the presence of lions, panthers, leopards, elephants, hippopotami, giraffe, and other such species in North Africa during Roman times, although striking at first, do not seem to form a reliable index of climatic variation or even, in a narrower sense, of endogenous changes in the environment. Indeed, most of these species have disappeared only within recent memory. According to the accounts of numerous eighteenth- and nineteenth-century travellers in the Maghrib, packs of lions roamed the Qabaliyya littoral of Algeria and the Moroccan Atlas, while other such reports indicate that lions were still present in these same regions until the early decades of this century. Panthers and lions were also reported frequenting the Constantine Plains of north-central Algeria in the last decades of the nineteenth century (Carcopino, 1943:20; Morizot, 1962:37–9; Brahimi-Chapuis, 1976). As for other subtropical species, the ostrich was found in such numbers early in this century that colonial officials were formulating serious programmes to have immigrant farmers exploit them on ranches (Sainte-Hilaire, 1919:336–60; Capot-Rey, 1953:91ff; Camps-Fabrer, 1962–63).

As for the other large mammalian species like the hippopotamus, rhinoceros, and elephant, the evidence consists mainly of rock art depictions and osteological finds of the prehistoric period from the Maghrib and Sahara. This body of evidence would seem to indicate that both the hippopotamus and rhinoceros had become extinct in the region by the beginning of the Neolithic – a fact which might show that both species were only highland 'fossil' fauna from the last pluvial.[4] But there are problems. This conclusion depends on the veracity of the repertoire of prehistoric artists as an accurate reflection of the whole range of animal life in their environment, and on a truly exiguous number of bone finds. But there are at least two grounds on which these data bases are suspect as reliable guides to the existence of the animal species in question. First, they are flatly contradicted by the literary record which shows that hippopotami survived, in however small a number, in parts of the Maghrib well into historical times. Secondly, the whole pattern of the pictorial and osteological data would lead us to believe, contrary to every expectation, that these large mammalian species disappeared *first* from the Maghrib and only *later* from the Saharan highlands. On these bases alone, any direct correlation between these data, the extinction of the large mammals, and climate must be severely questioned. A more plausible, though equally unprovable, reconstruction would hold that many of the large mammalian species like the hippopotamus and the elephant were relict species of the last African pluvial and that they did survive into the historical period though in such small numbers that they were very vulnerable to human depredation (Gsell, 1911:381–8; Balout, 1955:97–107, 163, 262; Vaufrey, 1955:371–8, 384–5, 398–9; Mauny, 1956; Boulière, 1963).

The central problem is: how are we to specify the relative effects of human action and climatic factors on the gradual disappearance of these large mammalian species during historical times? Of climate we cannot yet be certain, but the partial responsibility of Man cannot be doubted. To take but one of the more spectacular aspects of the human side of the equation, it is well known that Africa was considered in Roman times to be a type of reservoir of exotic wild animals that was systematically exploited to provide animal fodder for human entertainment in the arenas of the Empire, including North Africa itself (see Friedländer, 1968: vol. 4, p. 66 for some figures, along with 1968: vol 4, append. 36 for a survey of imperial arenas). But the tens of thousands of such animals purposefully hunted down for the arena were, of course, a small proportion of the total that must have yielded to more mundane processes such as the systematic destruction of their habitat by the expansion of agricultural settlements. Even so, only the least numerous and most vulnerable of the larger species did succumb: the rhinoceros, the hippopotamus, and the elephant. All the rest, like the panther, lion, leopard, and ostrich, survived. Their final disappearance can be placed in the early decades of this century and is probably directly attributable to the introduction of firearms in the mid-decades of the nineteenth century (Mauny, 1956:270).

Thus the insuperable difficulty with arguments based on the disappearance of large mammalian fauna is that the residual effects of a deteriorating climate cannot be sieved out neatly from the one major discernible cause, the hand of Man. What is required is a thorough study of less dramatic faunal species, such as ground snails, Mollusca, and insects, whose reaction to changing climatic regimes is more sensitive and less affected by intrusive human factors. To my knowledge, such evidence is not yet available for any archaeological site of the *historical* period.

Ground water resources and the question of climate

Another argument that appears with some frequency in the literature concerning climatic change is the repeated assertion that the surface and ground water resources of the Maghrib have declined severely since antiquity, and are insufficient to support agriculture on the scale once practised in North Africa. A corollary holds that the subsurface and runoff water is insufficient to permit the full functioning of wells, cisterns, aqueducts, and irrigation systems of the Roman period. Of course, such claims are restricted to the semi-arid regions of North Africa since there has never been any reasonable doubt that the present rainfall regime and ground water resources of the tellian zone are more than sufficient to support agriculture on any scale in the past.

I would argue that, to be persuasive, the number of testable cases that are adduced in support of these contentions must be large enough to indicate that it is the deficiency of precipitation or ground water that is at fault. Unfortunately, the evidence arraigned so far is inconclusive at best; even, it must be admitted, *against* the case for climatic change. A great number of wells and cisterns constructed during the Roman period are still in active use and show no sign of drying up or producing insufficient supplies of water (for some examples see Gauckler, 1897:43ff, 52, 59; Birebent, 1964:37, 51, 55, 59, 61; these could be multiplied many times over). Of those systems which no longer function properly, the three most commonly suggested reasons for failure are (Gauckler, 1897:27, 35–6; 1904:132–6; Birebent, 1964:33, 51, 54, 57, 65, 68, 71, 77, 81–5, etc.):

(*a*) a shift in the course of a wadi or stream bed, or a shift in ground water levels (i.e. not necessarily a lowering of the water table),

(*b*) the silting up of dams, cisterns, and aqueducts,

(*c*) actual physical damage to the system so that it no longer functions because of structural disrepair.

Although failure due to the first factor is not generally reparable, it should be possible to rectify the second and third situations in order to test the viability of ancient water systems under present conditions. In virtually every case in which an effort has been made to clean out cisterns, wells, aqueducts, and irrigation systems, and to repair the damage to them, the water has flowed again to the maximum capacity of the system concerned. The number of times that such repairs have failed, even allowing for biased reportage on the matter, is so small that it cannot be attributed directly to climatic change. The general viability of all these systems *taken as a whole* surely argues the reverse (see Gauckler, 1897:47–8; 1902:3–35; Gsell, 1911:377–8; and Birebent, 1964:38–9, 51, 65, 71, etc.).

Still, there can be little doubt that the desert periphery of the Maghrib, and some regions of the semi-arid zone, have suffered almost irreparable ecological damage in historical times. But we are only beginning to obtain an accurate measurement of the extent of this desertification. In a study of aerial photographs of irrigation networks on the southern flank of the Aurès Mountains (Algeria), Achenbach (1973) has been able to show that the modern indigenous irrigation systems have recovered from a marked post-Roman retreat, but in no case are they as extensive as the networks of the Roman period. Another analysis of aerial photographs, this time of the steppelands of southern Tunisia, has revealed evidence of widespread cultivation of lands in the Roman period which are now badly eroded and not as extensively exploited for agricultural purposes (Mensching & Ibrahim, 1975). Both studies are virtually unique. One would need dozens of studies of comparable rigour and excellence before any tentative judgements could be essayed on the relative roles of Man and climate in the process of desertification. For what it is worth, however, Achenbach, and Mensching and Ibrahim specify human activities, especially detrimental farming practices, as responsible for the retreat of arable land in the regions they studied.

Agricultural productivity and climatic conditions

The last argument that has traditionally formed the backbone of modern assumptions about climatic change in the Maghrib is what I call the 'Granary of Rome' myth. Indeed, the phrase 'Granary of Rome' is probably the one idea that is constantly associated with North Africa in the Roman Empire, by layman and scholar alike (e.g. Knight, 1929:51; Raven, 1969:ch. 5). What makes the myth so credible is that it *is* partially true; there can be no doubt that North Africa was one of the major grain exporting regions that supported the urban populace of Italy and, above all, of the city of Rome itself. But the grain exports of the Roman provinces were no new factor in north African history. The local 'kings' of pre-Roman times were already shipping substantial quantities of wheat and barley to other Mediterranean lands (Camps, 1960:197–200; Deman, 1975:19). The problem, as ever with the ancient world, is with *quantifying* the extent of these exports, since they are the mainstay of the argument for Africa's agricultural wealth. Conservative guesses would place Africa's shipments of grain to the city of Rome alone at *c.* 20 000 000 *modii* (somewhat less than 5 000 000 bushels) per annum (Charles-Picard, 1956; cf. Haywood, 1938:43 and Deman, 1975:45ff; the ancient references are Josephus, Bellum Iudaicum, 2.380–3, 386; Historia Augusta, *Vita Septimii Severi*, 27.3; and Aurelius Victor, *Epitome de Caesaribus*, I). The total grain exports to other places in the

Mediterranean region must have been well in excess of this figure, though by what approximate level we do not know.

But even if one accepts the fact of considerable amounts of grain being exported from North Africa in Roman antiquity, what does this reveal about climatic change? On the basis of the exports alone, very little. If the argument hinges, as it does, on the availability of large surpluses for export, the implication must be that modern North Africa has not had this same potential. But such an assertion is not tenable, for there is hardly a period when North Africa was *not* exporting considerable quantities of grain. Indeed, Africa appears to have been joined in a sort of economic 'yoke' with Sicily, the other historic breadbasket of the western Mediterranean (Schneider & Schneider, 1976:20–7). Catalan traders controlled much of the export trade from the Maghrib and Sicily as early as the late 1200s and early 1300s, their principal interest being grain (Dufourcq, 1966:20–92, 535–8, 565–9). The detailed records of correspondence between the Beys and French consuls in Tunis, and the French royal court between *c.* 1600 and 1750 attest a continuing large-scale grain trade between Tunis, Bône, La Calle, Algiers, and Marseilles during the whole period. Indeed, the grain trade appears to have been the sole *raison d'être* of the French Consul in Tunis in the seventeenth and eighteenth centuries (Plantet, 1893; cf. Masson, 1903, 1908). The same is true of Algeria during most of the period of Ottoman rule (1518–1830) when large amounts of tribute grain were shipped to the *Sublime Porte*, and a thriving trade with western Mediterranean countries, principally France, was also maintained (Danziger, 1977:22–5; Valensi, 1977: ch. 1). This considerable trade came to be dominated by a few Jewish families in Algiers engaged in export businesses, notably the Baqrî and the Bushnâq, who incurred massive debts of up to eight million francs to the Dey of Algiers for grain they had purchased to supply Napoleonic armies (Julien, 1964:18–23). In fact, these grain debts were the direct cause of the famous 'fly swatter' incident of April 1827 which led to the French military intervention of July 1830 and the formation of colonial Algeria (Abûn-Nasr, 1975: 176, 235–8).

There were at least three major non-climatic factors that have contributed historically to grain surpluses in North Africa and Sicily, ancient and modern:

1. Given the primitive technology of premodern agriculture, the very aridity and barrenness of the Sicilian and North African plains regions were, ironically, their greatest asset. Not requiring

laborious clearing and drainage, they were more naturally suited to widespread cultivation of cereal grains than were the more humid north Mediterranean lands.

2. The *relatively* low population levels in both regions in antiquity and the medieval period meant that there were genuine 'surpluses' available for export.
3. The colonial situation in which both regions have found themselves, with a dominance of external ownership of the best land, meant that marginal or artificially created 'surpluses' could be exported, *even in the face of local starvation* (Haywood, 1938:44; Charles-Picard, 1956:172–3; Staerman, 1964:200–2). An analogous case is provided by the history of modern Ireland, where Anglo–Irish landowners continued to export large quantities of grain out of Ireland in the 1840s when nearly a million Irish were starving to death (Beckett, 1971:144ff).

Given these factors, it is dangerous to argue that a high level of grain exports necessarily indicates favourable climatic conditions or even high levels of local productivity. The principal identifiable causes that ended North Africa's role as an exporter of cereal grains to the Mediterranean seem not to have been climatic, but rather the changes in the abovementioned factors: the rise of a competitive area of grain production in lands north of the Mediterranean, the dramatic rise of population in the Maghrib within the last century, and the decreasing power of external political control (Valensi, 1977:4–5). In spite of these changes, including the negative demographic factor, all the countries of the Maghrib, Libya included, continued to export substantial portions of their grain crops (i.e. 15–20 per cent of wheat; 10–15 per cent of barley) until the years just before the Second World War (Knight, 1929:54–5; Admiralty, 1945: 289–92; Guernier, 1947, 1948; and the *Statistique Générale de l'Algérie*). Hence, *Roman* North Africa was not unusual in terms of its export of grains, and the myth of superior productivity cannot contribute substantially to an understanding of climatic change in the historical period.

The problem of ecological change and the climate

One of the obvious results of an unfavourable change in the general climatic regime is a concomitant deterioration of environmental conditions. For the sake of brevity, I shall consider only one dramatic aspect of environmental decline in the Maghrib during the historical period, namely, the loss of vast areas of forest cover. The questions that must be posed concerning this phenomenon are: 'when did the greatest retreat in forest cover take place?' and 'due to what main factors?' For the Roman period, one cannot ignore the impact of the great expansion of an agrarian economy and, even though we have no statistics, the undoubted general rise in the total population reflected in the greater number and size of urban centres (Lassère, 1977). The lines of Roman centuriation (land survey) show that more and more marginal land (e.g. poor soils in the semi-arid zone, mountain slopes) were being brought under cultivation. Heavy deforestation of the mountain regions is indicated by the Elder Pliny who states that by his time (*c.* AD 70) the *Mons Ancorarius* (the Warsenis-Babors range in Algeria) had been so exploited for citrus wood that there was none left to be had (Pliny, *Historia naturalis*, 13.95; see Carton, 1895; Lavauden, 1927:85; Jodin, 1970:414, Poncet, 1973; and Lawless, 1973, on the exploitation of forests in antiquity). Classical literary sources like the Elder Pliny, however, do not give us any measurement of ecological decline; but they do inform us of many wooded regions in North Africa as late as the sixth century AD (when Roman rule had become a thing of the past), which are now barren (e.g. numerous passages in the *Johannid* of Corippus; see Diehl, 1896:405–6; Toutain, 1896:39–41; Gsell, 1913:151–2; Lavauden, 1927:85). How and when did they become denuded? In most cases there is simply no reliable information on the rate of change in the climax vegetation. Certainly the Arab geographers of the Middle Ages often report vast areas of forested shade for travellers in the North African interior as late as the thirteenth and fourteenth centuries, but it is rightly suspected that their fanciful accounts are no more to be trusted than those of their more imaginative classical counterparts (Tissot, 1884:278, 286–7; Bourde, 1889:6–7, 21; Gsell, 1913:151; Lavauden, 1927:87).

We can at least be relatively certain that the mountains of the *Tell* remained deeply forested; dependable evidence reveals that the Qabaliyya of Algiers was heavily treed and was an important source of timber for both Ifriqiyya and Egypt in the high Middle Ages (Lawless, 1973:230; Lombard, 1959). But we also know that most of the present forest cover of this same region was lost during the intensive warfare of 1871–72. That single fact gives an indication of the principal general cause. One of the most detailed and learned studies of forests ever made, Boudy's massive four volume *Economie forestière nord-africaine*, reveals that by far the greatest proportion of the loss of North Africa's forests took place within the last century, primarily in the half-century between 1890 and 1940 (Boudy, vol. 4, 1955:241–2). The process was twofold. A large amount of the cutting was done by the *colons* themselves, and by the forest companies which had been granted official concessions. To this must be added the damage done by an insidious process whereby the indigenous peoples, forced off their lands by confiscation, retreated upwards onto the marginal lands on mountain slopes which they then proceeded to clear of trees. Boudy estimated that at least 15 000 hectares per annum were cleared 'legally' and another 10 000 hectares per annum by the latter process. The total forest cover lost in Algeria alone during this period was well over 1 000 000 hectares; there are no comparable statistics for Morocco and Tunisia, though much the same processes must have been at work in these regions as well (Lefebvre, 1900; Guernier, 1947:246–52; 1948:367–87; Boudy, vol. 4, 1955:557–60).

The occasional accounts of colonial officials and travellers of the seventeenth to the nineteenth centuries tend to substantiate Boudy's gloomy picture. Reboud and Payen could report extensive forest cover and even good fishing in the mountains and streams of the Hodna Basin (south-central Algeria) in the 1860s where today the landscape is almost totally desolate. Similarly forests around Sebkha Naïl, reported in 1853, are now gone, and there were brush forests filled with wild boar around Wed Jedî and Biskra in a region where it is difficult to find many trees today (Payen, 1864; Reboud, 1865; Lavauden, 1927:87). These reports could be multiplied greatly, but they all point to the same general conclusion reached by the *Service forestière*: there was extensive brush and forest cover in North Africa only a century ago that does not exist today, and the principal agent in its destruction was Man not climate. The same causes are discernible in the deforestation of Sicily where ground cover has alternately advanced or retreated whenever men were incited to clear it in response to the fluctuation of the demand for grain in Mediterranean markets (Mack-Smith, 1965:109ff; 1968:183ff). So too, in his classic study of the Languedoc, Le Roy Ladurie concluded that the oscillations in the forest and brush cover of the region were due primarily to the rise and fall of population density and the concomitant intensity of agricultural exploitation of the countryside (Le Roy Ladurie, 1976:11–23).

The evidence of historical botany and geology

None of the above analysis is meant to suggest that the question of climatic change is not one of great importance to the historian, or that it is a desperately intractable problem, but only that the evidential basis on which it has been approached is at best ambiguous and at worst simply not the type of data that could provide reliable answers. It is here that the record of historical botany and geology might hope to provide a remedy. The irony is that, at present, there is so little to report. I know of no studies on the palaeobotany of *historical* archaeological sites in North Africa save for a few preliminary reports for the site of Roman Carthage that have not, as yet, produced anything of note (Ford & Miller, 1978). In an interesting and suggestive study of carbonised plant remains from a late prehistoric site published in 1948, Le Dû and Saccardy emphasised that the science of analysing botanical specimens preserved in archaeological sites in North Africa was still in its infancy, which is where it has remained for the last three decades (Le Dû & Saccardy, 1948:115). Their study of late Neolithic sites around Tébessa (southeast Algeria) led them to believe that there was no evidence for climatic change, though they found charcoal cinders in such profusion that they did recognise the factor of human cutting of forests was already significant in prehistoric times. More recent studies by Couvert on the late Neolithic site of Khanguet Si Mohammed Tahar (Aurès) seems to indicate the opposite. The tendency of xerophytic species to invade the lower mountain slopes and of the humid species to retreat to higher altitudes, together with tree rings indicating stronger spring rains, would seem to suggest a damper climatic regime than present at the end of the Neolithic, *c*. 4000–2500 BC (Couvert, 1969:215ff; 1972; 1976). Couvert's suspicions might be proved to be correct by other corroborative studies, such as those by Lubell, though we still have only the most tenuous bases for believing that the climate of *historical* times was markedly different from the present (Lubell, 1978; 1979:172–4). Insofar as botanical evidence for the latter period is concerned, we are still largely dependent on literary evidence, and although this does contain some useful information such as the extinction of the enigmantic *silphium* plant due to overgrazing and exploitation for markets, it is insufficient for our task (Pliny, *Historia naturalis*, 1915; Theophrastus, *Historia plantarum*, 6.3; Arrian 3.28; Strabo 17.3.22).

With historical geology we enter a more promising field. In his massive study of the Atakor massif (Hoggar, Sahara), Rognon noted the presence of a distinct phase of deposition and formation of sedimentary terraces that formed one of the few arable soils of the region. Rognon confessed his surprise that such sedimentary depositions in the wadi beds of the Hoggar existed at all, given the extreme aridity of the region. And yet, based on pottery finds, Rognon showed that this terracing must have been formed some time between 1000 and 1500 BP. He also voiced his suspicion that many other soil terraces of similar nature found throughout the Sahara might have been assigned arbitrarily to the supposed

'Neolithic subpluvial' whereas in reality they belonged to the more recent, historic phase of sedimentary deposition (Rognon, 1960: 37–43; 1967: 214ff, 514ff, 522–6). What is more striking is that precisely the same type of sedimentary deposits have been identified in most of the coastal wadis of the Maghrib itself. These fills are definitely post-Roman with ^{14}C dates extending from AD 1150 to 1750 (Vita-Finzi, 1969: chs. 1–5). Indeed, the sedimentary terraces studied by Vita-Finzi in North Africa appear to be connected to a general phase of alluviation that affected the entire periphery of the Mediterranean (cf. Potter, 1976, 1979, with Judson 1963a,b). But there are problems of interpretation of the significance of these sedimentary deposits for climatic history. As Judson, Butzer and others have pointed out, the sediments do not seem to conform chronologically to a well-defined 'Mediaeval Humid Phase', and even less so to the so-called European 'Little Ice Age'. Moreover, there would appear to be a number of anomalous phases of alluviation in the Mediterranean that would argue for a more general, long-term erosional history of the circum-Mediterranean lands not tied in any direct manner to climatic trends (Judson, 1963a,b; Butzer, 1974:67–8). Bintliff, however, has demonstrated that most of the objections are due to a misunderstanding of the cycle of erosion and deposition entailed by Vita-Finzi's model, which includes an important early phase of deltaic sedimentation that must not be confused with the later true valley fills (e.g. as at Ephesus, see Bintliff, 1975; 1977:35–50). The core of the numerous Mediterranean valley sedimentations that remain seem to fall within a fairly well-defined chronological horizon of *c.* AD 1000/1100 to 1750/1800, and it is fairly certain that no comparable general deposition is taking place today. On balance, a climatic explanation would seem to be warranted, especially since the synchronous sedimentation took place both after and before periods of heavy land-use by Man.

If this schema is accepted, it would suggest that the Roman achievement, not only in North Africa, took place under climatic conditions that were more like those of the present than those of medieval times. But once the employment of historical myths and questionable proxy data has been discarded, there are still questions that remain. For example, is it not possible that large-scale land-use by Man in the circum-Mediterranean region may actually have served, on the whole, to impede massive erosion rather than to encourage it? And is it not possible that in the centuries after *c.* 1000, in a period during which intensive agricultural exploitation was at its lowest ebb, 'natural' forces might have taken their toll of the environment when the artificial restraints of human cultivation were not at their most effective? It is only a question, but one that deserves some consideration before we follow, too readily, the sort of answers already proferred by Gibbon over two centuries ago.

Acknowledgements

I would like to thank the following for their assistance, encouragement, and criticism while I was preparing materials and writing earlier drafts of this chapter: Dr K. W. Butzer (Department of Geography, University of Chicago), Mr A. T. Grove (Downing College, Cambridge), Joyce M. Reynolds (Newnham College, Cambridge), and Dr C. Vita-Finzi (University College, London).

Notes

1. The *Annales Maximi* were the annual records kept by the *Pontifex Maximus* at Rome. The high priest, so the Elder Cato informs us, recorded agrarian events of note during the past year, including good or bad harvests, the price of grain and, quite naturally, significant meteorological phenomena (see Crake, 1940; Cato is quoted by Aulus Gellius, *Attic Nights*, 2.28.6). But these data, amongst the most specific alluded to consistently by ancient historians, are so fraught with pitfalls as to make them virtually useless for our purposes. Only *unusual* weather events were recorded, and we do not know the criteria for the choices made by the high priests, nor how consistently such weather reports were made. We cannot be certain what areas, if any, outside the immediate environs of Rome were reported in a given year. Further, we do not possess a complete set of even second-hand reports from the *Annales Maximi*.
2. The literary passages are collected by Gsell, 1911:393–402, essentially dependent on Leiter, 1909:84–7, 101–23. Some examples will suffice to demonstrate the vagueness and subjectivity of the reports. Augustine (Enarrationes in Psalmos, 236.9) reports that olives went to the press in November; Gsell remarks that this is the same as the present day, hence 'no climatic change'; no rain in 'Africa' for five years when the emperor Hadrian visited the country in AD 128 (Historia Augusta, *Vita Hadriani*, 22.14); apart from the questionability of the source and the influence of imperial 'favour', what is meant by 'Africa'?; Arnobius (*Adversus nationes*, 1.16) reports a long period of dryness in two of the Roman provinces, *Tingitana* (modern NW Morocco) and *Caesariensis* (modern NW Algeria), but says that there were still good harvests; Cyprian, writing in the mid-third century AD (*Ad Demetrianum*, 2.3.7–10 and 8 *init.*) cites amongst other proofs of the decline of the world, a growing diminution in rainfall; and other 'dry' spells reported by Victor of Vita (*De persecutione Vandelorum*, 3.55, AD 457) and Corippus (*Iohannide*, 6.247, AD 547).
3. The evidence for and against climatic change, employed by the various protagonists between the 1860s and the 1960s, remained virtually the same set of archaeological and literary data as used by Leiter (1909) and Gsell (1913), and yet diametrically opposed camps arose on the subject. In favour of climatic change were Carton (1894, 1896), Kilian (1934), and Carcopino

(1943:18ff); opposed to it were Negro (1915), Rivière (1920) – one of the best surveys – Gribaudi (1928), Fantoli (1931), and Murphey (1951:117), although both Gribaudi and Fantoli could be said to have had vested colonial interests in their conclusions.

4. I use the term 'pluvial' with some hesitancy since there is some doubt as to whether the traditional concept of synglacial pluvials is valid for Saharan and sub-Saharan Africa. There is now considerable evidence to show that those regions were arid rather than humid during the last phase of large-scale glacial activity in the northern latitudes (Street & Grove, 1976). This revisionist interpretation, however, does not appear to hold true for the Mediterranean latitudes with which I am primarily concerned. For these regions, even the revisionists still postulate a synglacial pluvial, the end of which I place *c.* 12 500 BP (Street & Grove, 1976; Sarnthein, 1978:43).

References

Abun-Nasr, J. M. (1975). *A History of the Maghrib*, 2nd edn. Cambridge: Cambridge University Press.

Achenbach, H. (1973). Römische und gegenwärtige Formen der Wassernutzung im Sahara-Vorland des Aures (Algerien). *Die Erde*, 104, 157–75.

Admiralty (1945). *Tunisia*, Geographical Handbook Series, BR 523. London: Naval Intelligence Division.

Balout, L. (1955). *Préhistoire de l'Afrique du Nord: Essai de chronologie*. Paris: Arts et métiers graphiques.

Baynes, N. H. (1943). The decline of the Roman Empire in western Europe. Some modern explanations. *Journal of Roman Studies*, **33**, 29–35 = D. Kagan, ed., *The Decline and Fall of the Roman Empire: Why did it Collapse?*, 1962, pp. 79–80. Boston: D.C. Heath.

Beckett, J. C. (1971). *A Short History of Ireland*, 4th edn. London: Hutchinson University Library.

Bintliff, J. L. (1975). Mediterranean alluviation: new evidence from archaeology. *Proceedings of the Prehistoric Society*, **41**, 78–84.

Bintliff, J. L. (1977). *Natural Environment and Human Settlement in Prehistoric Greece*, 2 vol. Oxford: BAR Supplementary Series no. 28.

Birebent, J. (1964). *Aquae Romanae: recherches d'hydraulique romaine dans l'est Algérien*. Algiers: Baconnier Frères.

Boudy, P. (1949). Considérations sur l'évolution du climat en Afrique du Nord et en particulier en Maroc depuis la périod préhistorique. *Bulletin de la Société des sciences naturelles au Maroc*, **25–7**, 112–18.

Boudy, P. (1948 & 1955). *Economie forestière nord-africaine*, vol. 1: *Milieu physique et milieu humain*; vol. 4: *Description forestière de l'Algérie et de la Tunisie*. Paris: Editions Larose.

Boulière, F. (1963). Observations on the ecology of some large African mammals. *Viking Fund Publications in Anthropology*, **36**, 43–54.

Bourde, P. (1889). *Rapport sur les cultures fruitières*. Tunis: Imprimerie générale.

Bovill, E. W. (1929). The Sahara. *Antiquity*, **2**, 414–23.

Brahimi-Chapuis, D. (1976). *Voyageurs français du XVIIIe siècle en Barbarie*. Paris: Diffusion H. Champion.

Brooks, C. E. P. (1949). *Climate through the ages*, 2nd edn. London: Faber & Faber. (Reprint, New York: Dover, 1970.)

Bryson, R. A., Lamb, H. H. & Donley, D. L. (1974). Drought and the decline of Mycenae. *Antiquity*, **48**, 46–50.

Bryson, R. A. & Murray, T. J. (1977). A drought in Ancient Greece. In *Climates of Hunger: Mankind and the World's Changing Weather*, pp. 3–17. Madison-London: University of Wisconsin Press.

Butzer, K. W. (1958). *Studien zum vor- und frügeschichtlichen Landwirtschaftswandel der Sahara*, Abhandlungen der Akademie der Wissenschaft und Literature (Mathematische-Naturwissenschaftliche Klasse), Mainz, 1, pt. 1: Die Ursachen des Landwirtschaftswandels der Sahara und Levant seit dem klassischen Altertums, 5–19.

Butzer, K. W. (1974). Accelerated soil erosion: a problem of man–land relationships. In *Perspectives on Environment*, ed. I. R. Manners & M. W. Mikesell, pp. 57–77. Washington, DC: Association of American Geographers.

Cahen, C. (1968). Quelques mots sur les Hilâliens et le nomadisme. *Journal of the Economic and Social History of the Orient*, **11**, 130–3.

Camps, G. (1960). *Aux origines de la Berbérie: Massinissa ou les débuts de l'histoire. Libyca* (*Arch.-épigr.*), 8.

Camps-Fabrer, H. (1962–63). La disparition de l'autruche en Afrique du Nord. *Revue africaine*, **106**, 33–74 = 1963, Algiers: Travaux du Centre de Recherche d'Anthropologie, Préhistoire, et Ethnologie.

Capot-Rey, R. (1953). *Le Sahara français*. Paris: Presses Universitaires de France.

Carcopino, J. (1943). *Le Maroc antique*. Paris: Gallimard.

Carpenter, R. (1966). *Discontinuity in Greek Civilization*. Cambridge: Cambridge University Press.

Carton, L. (1894). Climatologie et agriculture de l'Afrique ancienne. *Bulletin de l'Académie d'Hippone*, **27**, 1–45.

Carton, L. (1895). L'occupation forestière dans l'Afrique ancienne. *Bulletin archéologique du Comité des travaux historiques*, 338–42.

Carton, L. (1896). Note sur la diminution des pluies en Afrique. *Revue tunisienne*, **3**, 87–94.

Charles-Picard, G. (1956). Néron et le blé d'Afrique. *Cahiers de Tunisie*, **4**, 163–73.

Courtois, C. (1955). *Les Vandales et l'Afrique*. Paris: Arts et métiers graphiques.

Couvert, M. (1969). Etude de quelques charbons préhistoriques de la grotte Capéletti (Aurès-Algérie). *Libyca* (*APE*), **17**, 213–16.

Couvert, M. (1972). Variations paléoclimatiques en Algérie: traduction climatique des informations paléobotaniques fournies par les charbons des gisements préhistoriques. *Libyca* (*APE*), **20**, 45–8.

Couvert, M. (1976). Traduction des éléments de la flore préhistorique en facteurs climatiques. *Libyca* (*APE*), **24**, 9–20.

Crake, J. E. A. (1940). The Annals of the *Pontifex Maximus*. *Classical Philology*, **35**, 375–86.

Danziger, R. (1977). *Abd al-Qadir and the Algerians*. New York: Homes & Meier.

de la Blanchère, M. du Coudray (1897). L'aménagement de l'eau et l'installation rurale dans l'Afrique ancienne. *Nouvelles Archives des Missions scientifiques*, **7**, 1–109.

Deman, A. (1975). Matériaux et réflexions pour servir à une étude de développement et du sous-développement dans les provinces de l'Empire romain. In *Aufstieg und Niedergang der römischen Welt*, ed. H. Temporini, 2.3, 3–97. Berlin: Walter de Gruyter.

Demougeot, E. (1965). Variations climatiques et invasions. *Revue historique*, **233**, 1–22.

Demougeot, E. (1969). *La formation de l'Europe et les invasions barbares*. Paris: Aubier.

de Planhol, X. (1968). La ruine de Maghrib, 1: Le problème des variations climatiques; 2: Les invasions hilâliennes et leurs conséquences. In *Les fondements géographiques de l'histoire de l'Islam*, pp. 134–55. Paris: Flammarion.

Dickinson, O. (1974). Drought and the decline of Mycenae: some comments. *Antiquity*, **48**, 228–30.

Diehl, C. (1896). *L'Afrique byzantine: histoire de la domination byzantine en Afrique (533–709)*. Paris: Imprimerie nationale.

Djaït, H. (1973). L'Afrique arabe au VIIIe siècle (86–184/705–800). *Annales (ESC)*, **28**, 601–21.

Dufourcq, C. E. (1966). *Espagne Catalane et le Maghrib au XIIIe et XIVe siècles*. Paris: Presses Universitaires de France.

Fantoli, A. (1931). Le notizie meteorologiche sulla Tripolitania e Cirenaica nell'Antichità. *Rivista delle Colonie Italiane*, **5**.2, 638–46.

Ford, R. I. & Miller, N. (1978). Paleoethnobotany I. In *Excavations at Carthage 1976 conducted by the University of Michigan*, vol. 4, ed. J. H. Humphrey, pp. 181–7. Ann Arbor: University of Michigan Press.

Friedländer, L. (1968). *Roman Life and Manners under the Early Empire*, 4 vols., transl. L. A. Magnus of the 7th edn. of *Sittengeschichte Roms*. New York: Barnes & Noble.

Gauckler, P. (1897–1911). *Enquête sur les installations hydrauliques romaines en Tunisie*. Tunis: Imprimerie nationale.

Gibbon, E. (1966). *The History of the Decline and Fall of the Roman Empire*, vol. 1. London: Dent. (1st edn: 1776).

Goodchild, R. G. & Ward-Perkins, J. B. (1949). The *Limes Tripolitanus* in the light of recent discoveries. *Journal of Roman Studies*, **39**, 81–95 = *Libyan Studies*, ed. J. Reynolds, pp. 17–34. London: Elek.

Gribaudi, D. (1928). Sono mutate in epoca storica le condizioni climatiche della Libia? *Bollettino della Società geografica Italiana*, series 6, **5**, 171–213.

Gsell, S. (1911). Le climat de l'Afrique du Nord dans l'Antiquité. *Revue africaine*, **55**, 343–410 = chapter 3 in *Histoire ancienne de l'Afrique du Nord*, 1913, pp. 40–99. Paris: Librairie Hachette.

Gsell, S. (1913). Faune et flore de l'Afrique du Nord dans l'Antiquité. chapter 4 in *Histoire ancienne de l'Afrique du Nord*, vol. 1, pp. 100–58. Paris: Librairie Hachette.

Guernier, E., ed. (1947). *Tunisie*. In *L'Encyclopédie coloniale et maritime*. Paris: Editions coloniales.

Guernier, E., ed. (1948). *Algérie et Sahara*. In *L'Encyclopédie coloniale et maritime*. Paris: Editions coloniales.

Hare, F. K., Kates, R. W. & Warren, A. (1977). The making of deserts: climate, ecology, and society. *Economic Geography*, **53**, 332–46.

Haywood, R. M. (1938). Roman Africa. In *An Economic Survey of Ancient Rome*, vol. 4, ed. T. Frank, pp. 332–46. Baltimore: Johns Hopkins Press.

Huntington, E. (1917). Climatic changes and agricultural decline as factors in the fall of Rome. *The Quarterly Journal of Economics*, **31**, 173–208.

Idris, H. R. (1968a). L'invasion hilâlienne et ses conséquences. *Cahiers de Civilisation médiévale*, **11**, 353–69.

Idris, H. R. (1968b). De la réalité de la catastrophe hilâlienne. *Annales* (*ESC*), **23**, 390–6.

Jodin, A. (1970). L'exploitation forestière au Maroc antique. *93e Congrès national des Sociétés savantes, Tours, 1968* (section archéologique), 413–22.

Jones, A. H. M. (1964). *The Later Roman Empire, 284–602: A Social, Economic and Administrative Survey*, 2 vols. Oxford: Blackwell.

Judson, S. (1963a). Erosion and deposition of Italian stream valleys during historic times. *Science*, **140**, 898–9.

Judson, S. (1963b). Stream change during historic times in east-central Sicily. *American Journal of Archaeology*, **67**, 287–9.

Julien, C. A. (1964). *Histoire de l'Algérie contemporaine, I: La conquête et les débuts de la colonisation* (*1827–1871*). Paris: Presses Universitaires de France.

Kilian, C. (1934). Une variation du climat dans la période historique: dessèchement progressif du Sahara depuis l'époque précameline et les Garamantes. *Comptes rendus sommaire des séances de la Société géologique de France*, mai, 110–11.

Knight, M. M. (1929). Water and the course of empire in North Africa. *The Quarterly Journal of Economics*, **43**, 44–93.

Lacoste, Y. (1978). Le mythe de l''invasion arabe'. In *Ibn Khaldoun: naissance de l'histoire passé du tiers monde*, 4th edn., pp. 87–105. Paris: F. Maspero.

Lamb, H. H. (1977). *Climate: Present, Past and Future*, vol. 2, *Climatic History and the Future*. London: Methuen.

Lassère, J. M. (1977). *Ubique Populus: peuplement et movements de population dans l'Afrique romaine de la chute de Carthage à la fin de la dynastie des Sévères* (*146 a.C. – 235 p.C.*). Paris: Editions du CNRS.

Lavauden, L. (1927). Les forêts du Sahara. *Revue tunisienne*, **33**, 67–94.

Lawless, R. I. (1973). Population, resource appraisal and environment in the pre-Saharan zone of the Maghrib. In *Maghrib et Sahara: études géographiques offerts à Jean Despois*, ed. X. de Planhol, pp. 229–37. Paris: Société géographique.

Le Dû, R. & Saccardy, L. (1948). Etude de quelques charbons préhistoriques de la région de Tébessa. *Revue africaine*, **92**, 111–19.

Lefebvre, H. (1900). *Les forêts de l'Algérie*. Algiers: Girault.

Leiter, H. (1909). Die Frage der Klimaänderung während geschichtlicher Zeit in Nordafrika. *Abhandlungen der Kaiserlich-Königliche Gesellschaft in Wien*, **8**.1, 1–143.

Le Roy Ladurie, E. (1976). *The Peasants of Languedoc*, transl. & intro. J. Day, Urbana-London: University of Illinois Press = 1969, *Les Paysans de Languedoc*, Paris: Flammarion.

Lombard, M. (1959). Une problème cartographié: le bois dans la Mediterranée musulmane (VIIe–XIe siècles). *Annales* (*ESC*), **14**, 234–41.

Lubell, D. (1978). Holocene prehistory and environment in Eastern Algeria. *Current Anthropology*, **19**, 391–2.
Lubell, D. (1979). Holocene environment and Capsian subsistence in Algeria. In *Palaeoecology of Africa and the Surrounding Islands*, ed. E. M. von Zinderen Bakker & J. A. Coetzee, vol. 2, pp. 171–8. Rotterdam: A. A. Balkema.
McDougall, T. (1965). *Climate in Roman Times.* PhD thesis, University of London.
Mack-Smith, D. (1965). The *Latifundia* in modern Sicilian history. *Proceedings of the British Academy*, **51**, 85–124.
Mack-Smith, D. (1968). *A History of Sicily: Mediaeval Sicily (800–1713).* London: Chatto & Windus.
Marçais, G. (1913). *Les Arabes en Berbérie du XIe au XIVe siècle.* Paris: E. Leroux.
Marçais, G. (1946). *La Berbérie musulmane et l'Orient au Moyen Age*, Aubier: Editions Montaigne.
Masson, P. (1903). *Histoire des établissements et du commerce française dans l'Afrique barbaresque (1590–1793).* Paris: Librairie Hachette.
Masson, P. (1908). *Les compagnies du corail: étude sur le commerce de Marseille au XVIe siècle et les origines de la colonisation française en Algérie-Tunisie.* Paris: Fontemoing.
Mauny, R. (1956). Préhistoire et zoologie: la grande 'faune éthiopienne' du Nord-Quest africain du paléolithique à nos jours. *Bulletin de l'Institut français de l'Afrique noire*, sér. A, **18**, 246–79.
Mensching, H. & Ibrahim, F. N. (1975). Problems of desertification in the northern Saharan boundary zone: the steppe region of the Maghrib and the Sahelian zone. *Cambridge Meeting on Desertification*, International Geographical Union, Working Group on Desertification.
Morizot, J. (1962). *L'Algérie kabylisée*, Cahiers de l'Afrique et l'Asie no. 6. Paris: Peyronnet.
Murphey, R. (1951). The decline of North Africa since the Roman occupation: climatic or human? *Association of American Geographers, Annals*, **41**, 116–32.
Negro, C. (1915). Sul clima della Libia attraverso i tempi storici. *Memorie della Pontifica Accademia, Nuovi Lincei*, series 2, no. 1, Rome.
Payen, Capt. (1864). Les travaux hydrauliques anciens ... dans la partie du Hodna (dépendant de la province de Constantine). *Recueils et mémoires de la Société archéologique de Constantine*, **8**, 1–14.
Plantet, E. (1893). *Correspondance des Beys de Tunis et des Consuls de France avec le Cour, 1577–1830*, vol. 1. Paris: F. Alcan.
Poncet, J. (1967). Le mythe de la 'Catastrophe Hillâlienne'. *Annales (ESC)*, **22**, 1099–120.
Poncet, J. (1968). Encore à propos des Hilâliens: la 'mise au point' de R. Idris. *Annales (ESC)*, **23**, 660–2.
Poncet, J. (1973). De Platon au problème de l''environnement' méditerranéen. In *Maghrib et Sahara: études géographiques offerts à Jean Despois*, ed. X. de Planhol, pp. 325–30. Paris: Société géographique.
Potter, T. W. (1976). Valleys and settlement: some new evidence. *World Archaeology*, **8**, 207–19.

Potter, T. W. (1979). *The Changing Landscape of South Etruria*. London: Elek.
Raven, S. (1969). *Rome in Africa*. London: Evans.
Reboud, J. (1865). Notice sur les ruines romaines de l'Oued Chair. *Revue africaine*, **9**, 131–44.
Rivière, C. (1920). L'invariabilité du climat en Afrique du Nord depuis les temps historiques. *Revue historique et naturelle appliquée*, 71, 136, 145, 163, 197, 263, 302.
Robert, J. B. (1950). A propos de l'évolution du climat en Afrique du Nord depuis le début de la périod historique. *Etudes Rhodaniennes = Revue de géographie de Lyon*, **25**, 53–6.
Rognon, P. (1960). L'évolution morphologique des vallées de l'Atakor. *Travaux de l'Institut des recherches sahariens*, **19**, 25–49.
Rognon, P. (1967). *Le Massif de l'Atakor et ses bordures (Sahara Central): étude géomorphologique*. Paris: CNRS.
Rostovtzeff, M. (1957). *The Social and Economic History of the Roman Empire*, 2nd edn, rev. P. M. Fraser. Oxford: Clarendon Press.
Sainte-Hilaire, H. G. (1919). *L'Elevage dans l'Afrique du Nord (Maroc-Algérie-Tunisie)*. Paris: Challamel.
Sandars, N. K. (1978). *The Sea Peoples: Warriors of the Ancient Mediterranean*. London: Thames & Hudson.
Sarnthein, M. (1978). Sand deserts during glacial maximum and climatic optimum. *Nature*, **272**, 43–6.
Schneider, J. & Schneider, P. (1976). *Culture and Political Economy in Western Sicily*. New York-London: Academic Press.
Singh, G. (1971). The Indus Valley culture seen in the context of post-glacial climatic and ecological studies in north-west India. *Archaeology and Physical Anthropology in Oceania*, **6**, 177–89.
Štaerman, E. M. (1964). *Die Krise der Sklavenhalterordnung im Westen des römischen Reiches*, transl. W. Seyfarth. Berlin: Academie Verlag.
Street, F. A. & Grove, A. T. (1976). Environmental and climatic implications of the Late Quaternary lake-level fluctuations in Africa. *Nature*, **261**, 384–90.
Tissot, C. (1884). *Géographie comparée de la province romaine d'Afrique, t. I: Géographie physique, géographie historique, chorographie*. Paris: Imprimerie nationale.
Toutain, J. (1896). *Les cités romaines de la Tunisie: essai sur l'histoire de la colonisation romaine dans l'Afrique du Nord*. Paris: Fontemoing.
Toutain, J. (1903). Notes et documents sur les voies stratégiques et sur l'occupation militaire du Sud tunisien à l'époque romaine. *Bulletin archéologique du Comité des travaux historiques*, 272–409.
Trousset, P. (1974). *Recherches sur le Limes Tripolitanus du Chott el-Djerid à la frontière Tuniso-Libyenne*. Paris: CNRS.
Valensi, L. (1977). *On the Eve of Colonialism: North Africa before the French Conquest, 1790–1830*. New York & London: Holmes & Meier = K. J. Perkins transl., 1969, *Le Maghreb avant la prise de l'Alger*. Paris: Flammarion.
Vanacker, C. (1973). Géographie économique de l'Afrique du Nord selon les auteurs arabes du XIe siècle au milieu du XIIe siècle. *Annales (ESC)*, **28**, 659–80.

Vaufrey, R. (1955). *Préhistoire de l'Afrique, t.I: Maghreb*. Paris: Librairie Masson.

Vita-Finzi, C. (1969). *The Mediterranean Valleys: Geological Changes in Historical Times*. Cambridge: Cambridge University Press.

Wright, H. (1968). Climatic change in Mycenaean Greece. *Antiquity*, **42**, 123–7.

17

The economics of extinction in Norse Greenland

THOMAS H. McGOVERN

Abstract

The extinction of the Norse Settlements in Greenland (*c.* AD 985–1500) has often been cited as a demonstration of the impact of climatic deterioration on a human society. Archaeological data, especially that collected by the 1976–77 Inuit–Norse Project, has allowed a working reconstruction of the Norse economy. Combined with existing paleoenvironmental data, this paleoeconomic model indicates that the Norse would have been particularly vulnerable to fluctuations in important marine and terrestrial resources during the fourteenth and fifteenth centuries. However, archaeological and ecological data also indicate that the Norse never made use of alternative resources and ignored efficient Inuit technology. The Norse extinction thus may be seen as a failure of human managers to select effective responses to climatic stresses. A speculative organisational model is presented to account for this failure of decision-making.

Introduction

The Norse colony of west Greenland was, for nearly five hundred years (*c.* AD 985–1500), the westernmost outpost of Atlantic Europe. West Greenland was also the scene of the first prolonged contact between North American hunters and European farmers. This sub-arctic contact period lasted at least one hundred years longer than the span between the founding of Jamestown and the battle of Wounded Knee. In contrast to such later temperate-zone contacts between Native Americans and Europeans, Greenland saw the extinction of the Norse farmers and the survival and spread of the Inuit hunters.

Many theories have been proposed to account for the death of the Scandinavian colony. Inuit invasions, Basque pirates, declining trans-atlantic trade, plague, and climatic change have all provided speculative explanations for the extinction of Norse Greenland. Such largely mono-causal explanations have never been entirely successful (cf. Jansen, 1972, for a critical review). Recently collected data on Norse paleoeconomy (Meldgaard, 1977; McGovern, 1979a) and on medieval Greenlandic paleoecology (Vibe, 1967, 1970, 1978; Fredskild, 1973; Dansgaard *et al.*, 1975) combined with the theoretical approach of anthropological cultural ecology (Jochim, 1979) may allow a more systemic approach to this complex problem of economic–climatic interaction. This chapter will thus present yet another speculative model for Norse Greenland's extinction. Such first-approximation models are necessarily crude and should be as short-lived as research can make them. However, such integrative 'fairy tales' employing different disciplines' data in a common interpretive framework may serve to crystallise debate around potentially falsifiable hypotheses and to emphasise current gaps in data and analysis. Before we begin this attempt at integration, it may be well to consider briefly our archaeological, documentary, and paleoecological data base.

The evidence

Extensive archaeological research in Norse Greenland, dating from the end of the last century (Bruun, 1896; Nørlund, 1924; Roussell, 1941; Vebaek, 1958, 1962, 1965; Krogh, 1967) has provided a fund of locational, artifactual, and architectural evidence for the lost settlements. Generally excellent conditions of organic preservation, combined with an enlightened appreciation of the value of unmodified animal bone on the part of the early excavators, further enhances the value of this record for paleoeconomic analysis. In 1976 and 1977 the Inuit–Norse Project (jointly sponsored by Grønlands Landsmuseum and the Danish National Museum) attempted to improve this resource through an international, interdisciplinary approach to the problem of Inuit contact and Norse collapse. Employing probabilistic midden sampling strategies, systematic 4 mm mesh sieving of deposits, and computer registration of faunal remains, the Inuit–Norse Project investigated three Norse ruins and four Historic Inuit sites. We were extremely fortunate in finding conditions of excellent organic preservation and in the discovery of a deep, well-stratified Norse midden at V48 Niaqussat on the Ameragdla fjord. Over 33 000 identifiable animal bone fragments were recovered from the Norse sites, and their identification and analysis has recently been completed.

As space is limited and the details of excavation and analysis are to be published shortly in monograph form, we must cut short further discussion of specific techniques of archaeological excavation and analysis and turn instead to complementary sources of data. The written evidence for Norse Greenland is comparatively scanty and of very uneven character: ranging from sagas to Icelandic annals to papal bulls and Norwegian court records. Fortunately for our purposes, the critical review of Jansen (1972) and the thorough and scholarly synthesis of Gad (1970) have greatly clarified the compilation of Rafn (1876) and provide an invaluable guide to the non-specialist in assessing and employing the primary sources' evidence. Apart from the sagas (see Jones, 1964) perhaps our best written sources for Norse Greenland are the *Historia Norvegiae*, composed *c.* 1170 (Jansen, 1972:14), the *King's Mirror*, composed *c.* 1250 (Larsen, 1917; Gad, 1970:27), and the garbled report of episcopal steward Ivar Bardsson (stewardship *c.* 1341–68). Bardsson's account is the only surviving administrative document to describe the settlements' political and economic geography in any detail, and it survives only in a late (probably mid-sixteenth century) copy much embellished with fantastic interpolations (Gad, 1970:141). Though severely handled by transcription, Bardsson's list of church property is coherent and topographically consistent. It has proved of great practical value in the location and identification of actual farm and church sites (Vebaek, 1958), so we will follow precedent and make cautious use of this source here. Thus while our written evidence is hardly comparable to that available for medieval Iceland or the continent, we are spared much of the preliminary cultural–historical modelling of the prehistorian.

Our evidence for climatic fluctuation and ecological impact during the medieval period in west Greenland is also imperfect, but is rapidly expanding. Most useful in providing a long-term measure of shifts in paleotemperature are the results of the Greenland Ice Sheet Project, whose record of variations on the intervals of 10, 30, and 60 years is particularly valuable for paleoeconomic analysis (Dansgaard *et al.*, 1975). The extensive palynological work of Fredskild (1973) in the immediate vicinity of the Norse farms and the zoological projections of Vibe (1967, 1970, 1978) based upon population cycles observable in catch statistics, together provide the possibility of relating such variations in temperature to ecological variables directly affecting the faunal and floral resources exploited by the Norse Greenlanders.

Though this discussion of our evidence is by no means as complete or as critical as might be desired (cf. McGovern, 1979a), we must now turn to an equally brief outline of the history of the Norse settlements' interaction with their fluctuating environment before attempting a more theoretical explanation of the forces shaping that interaction.

Ecology and economy

Greenland is the world's largest island, but its habitable area might be better thought of as a kind of archipelago of ecological islands bordered by sea and continental ice sheet and divided by deep fjord systems and rugged mountain ranges. Greenland's west coast may be broadly divided into two major botanical and climatic zones: a very wet, oceanic coastal zone and a drier, but florally lusher, continental zone at the heads of a few of the deep fjords of the southwest (Bocher, 1954). Like most island chains, the climate of Greenland's ice-free coast is dominated by the ocean currents that wash it. As Vibe (1967) has demonstrated, fluctuations in the relative amount of warm North Atlantic Drift water and cold East Greenland water that reaches the west coast have profound effects upon both marine and terrestrial ecosystems.

When the Norse colonists arrived from Iceland *c.* AD 985, they settled in the ecological pockets formed by the restricted inner-fjord continental biome (Fig. 17.1). An extensive Eastern Settlement around modern Narsaq and Julianehaab districts with perhaps 4000–5000 inhabitants and a more northerly Western Settlement around modern Godthaab district with around 1000–1500 inhabitants made up the colony. While these estimates (based on a sampling of floor-space) are hardly precise, they are of the right order of magnitude. As a map of the smaller Western Settlement reveals (Fig. 17.2), the Norse farms were closely associated with the lowland areas of the inner fjords and the comparatively lush floral communities these biomes supported. Large, well-built byres and hay barns (Roussell, 1941), animal bones (McGovern & Bigelow, 1977; McGovern, 1979a), and the few written sources (Jansen, 1972) all indicate that the pasturage of cattle, sheep, and goats was the determining factor in the location of the Norse farms. Pollen studies (Fredskild, 1973) and the *King's Mirror* (Larsen, 1917) both suggest strongly that grain agriculture was not successful in Norse Greenland. The regular, dispersed pattern of Norse farms and lack of village-like concentrations may suggest the approximate minimum pasturage necessary for individual family economies.

While settlement pattern suggests a strongly terrestrial economy tied to the inner fjords, the animal bone remains reveal a major maritime component to Norse subsistence in Greenland (Table 17.1). Note that even inland farms (IF) some several hours' walk from the fjordside were heavily dependent upon seals (see McGovern, 1979a, for explanation of statistical presentation). The migratory harp seal (*Pagophilus groenlandicus*) makes up the bulk of this material. Harp seals arrive in west Greenland in the spring and migrate up the west coast in huge numbers.

While some do enter the fjord systems in search of capelin (*Mallotus vilosus*) the largest concentration of seals remains in the outer-fjord zone.

Hypothetical maritime catchment areas (McGovern, 1980b), based on historically recorded sailing rates (Brøgger & Shetelig, 1971) and performance of similar early modern craft (Morrison, 1973) and experimental reconstructions, suggest that the outer-fjord sealing grounds would be beyond the range of the commonest class of boat available to the majority of Western Settlement colonists. The recent discovery of a seasonal base in the outer fjords thus need come as no surprise (Berglund, 1973) (Fig. 17.3).

Our reconstruction of the Norse seasonal round may help to explain

Fig. 17.1. Distribution of Norse Settlement in Greenland. (From McGovern, 1980a)

the evident importance of harp seals to the Norse subsistence economy (Fig. 17.4). During the long Greenlandic winter, the cattle and probably the sheep and goats were kept shut in heavily insulated byres, standing in their own dung and living off stored fodder. Their milk production must have dropped off rapidly, and by mid-winter the Norse must have been subsisting mainly on stored dairy products and dried meat. The arrival of the harp seals in May–June would thus be of vital importance to the whole community.

As no harpoons or barbed spears of any kind have ever been found on any Norse site in Greenland, it is likely that seals were taken in communal boat drives that forced seal pods (i.e. small herds or schools of seals) onto beaches or into nets. This was the pattern of early modern sealing and small whale hunting in Scotland (Clark, 1948), Iceland

Fig. 17.2. Known Norse sites in the Western Settlement area. Note concentration of farms in the inner-fjord zone and in the lower elevations. (From McGovern, 1980a)

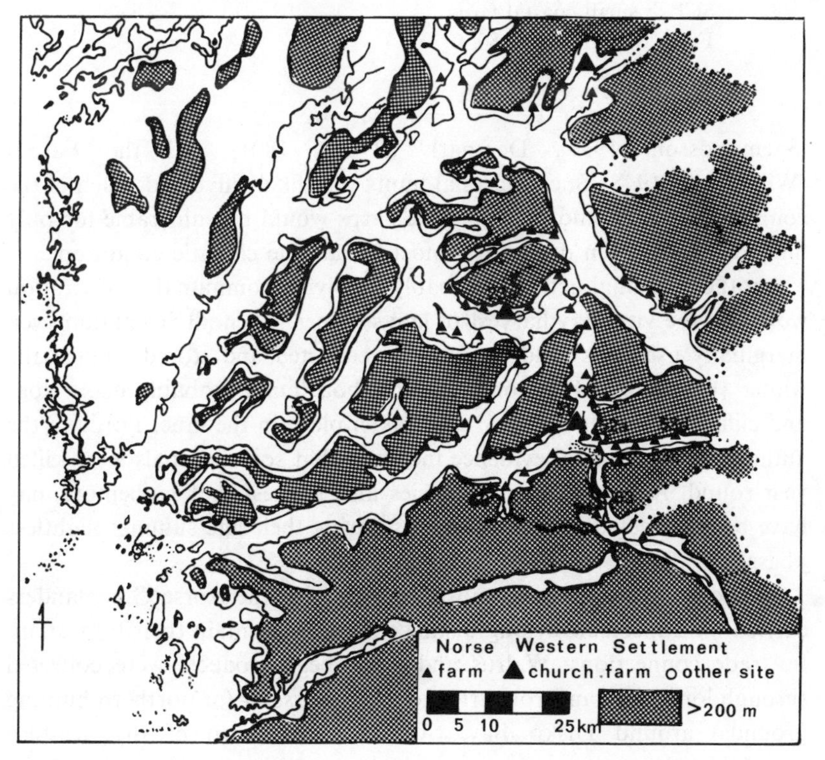

Table 17.1. *Western Settlement animal bone collections (over 300 identifiable fragments). Adjusted relative frequencies expressed as percentages (see McGovern 1979a for discussion of statistics and data sources)*

Site	Type	Cattle	Sheep/goats	Caribou	Seals
V51	LCFC	17.22	17.43	47.40	17.95
V35	IF	9.53	27.52	29.13	33.82
V54	IF	9.91	33.51	24.23	32.34
V53d	IF	9.57	20.93	33.07	36.43
V53c	IF	9.19	23.35	30.27	37.18
V52a	IF	8.97	24.11	41.63	25.28
V48 III	SCF	2.56	12.63	7.32	77.48
II	SCF	4.62	18.55	7.49	69.34
I	SCF	5.41	18.10	9.66	66.26

LCFC: large coastal farm with church.
IF : inland farm.
SCF : small coastal farm.
From McGovern (1980a).

(Saemundsson, 1939), Denmark (Bynch, 1801) and the Faroes (Williamson, 1948). Such seasonal hunts very likely involved much of the communities' men and boats, as the harps would be vulnerable for only limited periods even during the most favourable climatic conditions.

Another seasonal activity probably involving communal co-operation would be the vital hay harvest of late summer, when all vegetation even marginally useful as fodder would be collected and stored against the winter (Roussell, 1941). Communal caribou hunts probably using dogs and cliffside jump drives may have taken place in the inner fjords in the autumn, though faunal evidence indicates that some animals were killed year-round. Animal bone frequencies also suggest that guillemots may have been collected in some numbers during their late-summer flightless phase.

In addition to such basic subsistence activity, the Norse Greenlanders carried on a spectacular long-distance hunt to maintain their transatlantic trade connections. Walrus and polar bear products were collected through long and dangerous trips to the *Nordrsetur* (or northern hunting grounds) around Disko Bay, over 800 km north of the Western Settlement. Faunal remains strongly indicate that most of the Norse farms participated in this hunt at one time or another. This long-range hunting must have been costly in men, boats, and summer days for a community chronically short of all these. However, the Nordrsetur hunting seems to have been the price paid for continued contact with Europe. As the *King's Mirror* (Larsen, 1917) makes clear, traders risking the long Greenland voyage were interested in low-bulk, high-value cargoes like tusk and hide. Even in the best of times, transatlantic contact was irregular and subject to long delays. Unlike Shetland and the Faroes, Greenland was beyond the range of emergency famine relief from the mainland of Europe.

As Gad's quantitative analysis has demonstrated (1970:122–5) the irregular timing of transoceanic contact and the small cargo capacity of

Fig. 17.3. Estimated Norse maritime catchment areas. The 8 km and 16 km radius circles suggest the probable working range of the smallest and commonest class of Norse boat. Note the movement costs between the inner and outer fjords suggested by the catchments. (From McGovern, 1980a)

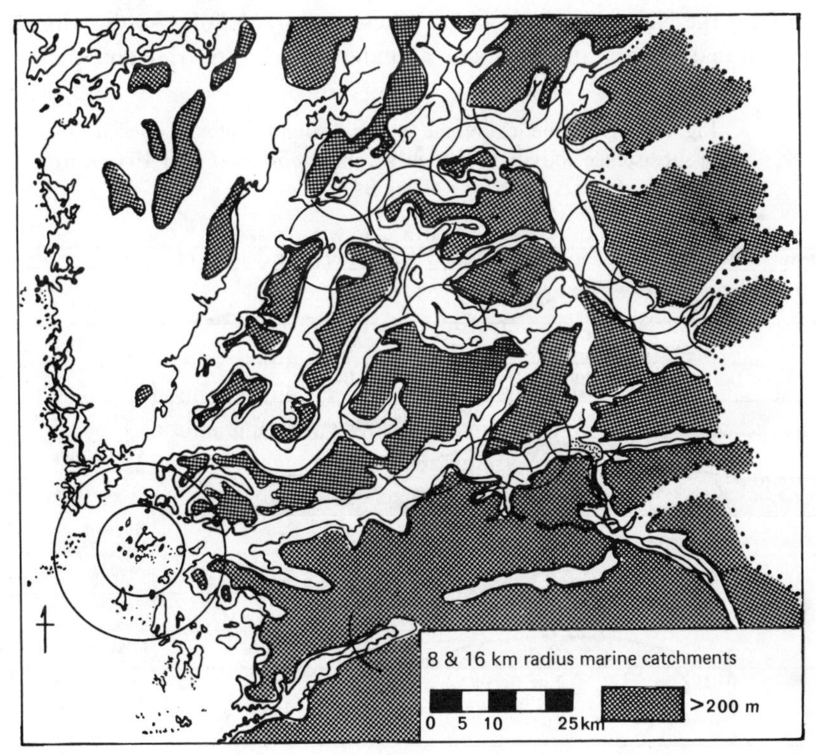

Norse vessels meant that the Greenlanders could hardly have been dependent on overseas trade for day-to-day subsistence. The crown trade monopoly that followed Greenland's submission to Norway *c.* 1262 and subsequent decline in sailings (as much the result of fourteenth-century political developments in the Baltic-oriented Danish–Norwegian state as of increasing thirteenth–fourteenth century drift ice) thus cannot bear the full blame for the colonies' end. However, the effect of even sporadic contact on the organisation of the Norse economy and politics in Greenland should not be undervalued either. The durable goods imported were clearly highly desired by Greenlandic consumers, and the Nordrsetur hunt seems to have played an important role in local mythology, as evidenced by the finds of walrus and polar bear carvings and ceremonial burials of walrus and narwhal skulls in church precincts (McGovern, 1979a). The scale of the Nordrsetur hunt is suggested by the crusade tithe of 635 kg of walrus ivory paid in 1327 (Gad, 1970). In exchange, the Greenlanders imported iron and wood vital to subsistence tasks as well as churchbells, stained glass, and rich church vestments which played their role in administrative and religious ceremony.

Thus the Norse economy was characterised by skilful co-ordination of communal labour and seasonal abundances of terrestrial and marine

Fig. 17.4. Hypothetical Norse seasonal round. Note the concentration of subsistence activity in the few summer months. (From McGovern, 1980a)

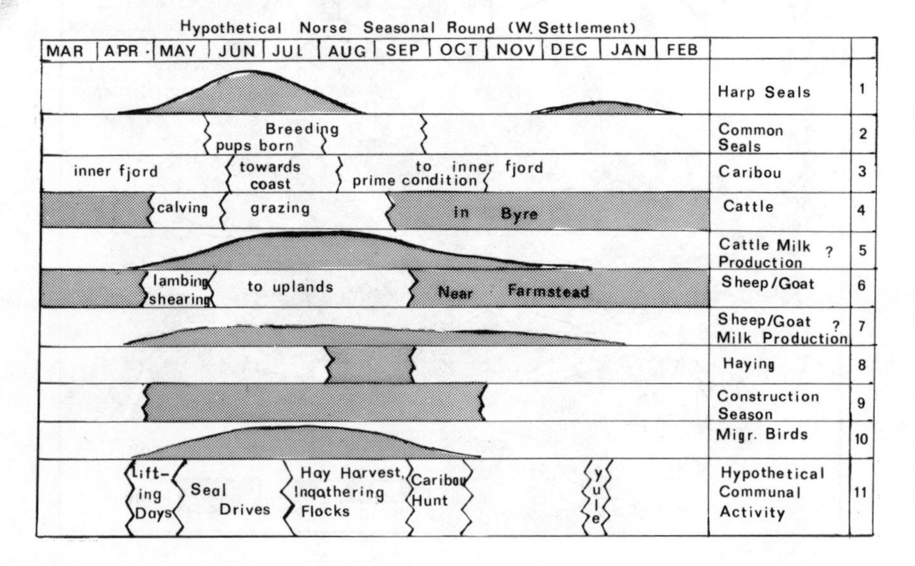

resources. The economy always operated on a thin edge, without the buffering effects of storable grain or much unused inner-fjord resource space. The economy's tight scheduling (particularly in summer months) would thus produce its best results in a context of high predictability of seasonal and year-to-year resource fluctuation. Contact with temperate Europe was always tenuous, but important enough to affect seriously the pattern of the whole economy. This economic system seems to have functioned well enough during the first 150 years of the settlement.

Social and political structure

Along with its population and basic economy, Norse Greenland's political system derived originally from Iceland. Like Iceland, Norse Greenland was never a community of equals. Both island communities were settled by *Landnamsmem* (first settlers) who took the best farm sites for themselves and parcelled out the remaining locations to followers. In Iceland the relationship between the chieftains (*gothar*) and their retainers (*thingmen*) became progressively less patriarchical and redistributive and more authoritarian and extractive as the Icelandic republic drew towards its bloody end in the 1260s (McGrew, 1970). The already hierarchically stratified Icelandic church played a major role in promoting the hierarchical tendencies of the later republic, and profited from Iceland's incorporation into the Norwegian state in 1262–64 (Sveinsson, 1953). Following the systems analysis of Flannery's classic paper on state origins (1972), we might say that the political–managerial institutions of Iceland were undergoing horizontal segregation and specialisation and vertical centralisation as the republic passed from chiefly oligarchy to royal control. Associated with this transition, and probably necessary to it, was an elaboration of status differentiation and class-consciousness that went in step with economic stresses that depressed the fortunes and independence of the Icelandic smallholder to the benefit of a few great landholding families, church magnates, and royal agents. Can we trace a similar process in Norse Greenland?

The patchiness of west Greenland's plant communities may supply part of the answer. While the site territories of the Western Settlement may be roughly regular in their spacing (Fig. 17.5), the contents of these territories are far from equal. It is possible to rank Greenlandic plant communities according to their pasturage value, on the basis of medieval husbandries as well as commonsense assessment. Table 17.2 presents a preliminary assessment of the relative richness of 1 km radius site territories of three Norse farms. While there are very few farms in the Western Settlement as rich as V51 Sandnes, there are many as poor as V48 Niaqussat. Large coastal farms with broad, level pastures commonly have churches or chapels, and no other pastures in Greenland are as rich as those of the episcopal manor at Gardar.

The relative scale of byres, store-houses, and dwelling rooms (Table 17.3) supports the picture provided by pasturage assessments. Note especially the large numbers of cattle that could be housed in the church-farm byres. Animal bone frequencies (Table 17.1) likewise seem to confirm this pattern, with most cattle (and least seal) bones being found on large farms with capacious byres and rich pastures.

The scale of the byres and meat stores (*skemmas*) of the large farms and the small size of byres and store-houses of the small farms suggests that some sort of redistributive mechanism may have operated in the

Fig. 17.5. Estimated Norse terrestrial catchment areas.Note the regular spacing of farms, and how fully the settlement pattern covers the inner-fjord lowlands. (From McGovern, 1980a)

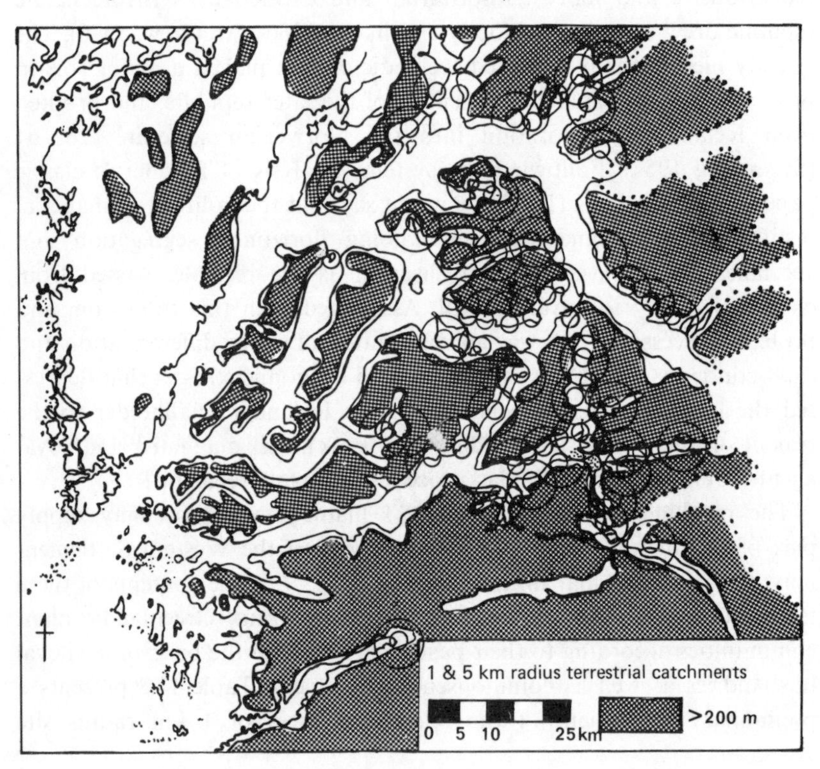

Table 17.2. *Preliminary quantification of pasturage potential of modern plant communities of three Norse farms in the same part of the Western Settlement. Note the dramatic differences in the fodder and grazing potentially available to the different farms*

Site	V51 Sandnes	V54	V48
Type	LCFC	IF	SCF
Landing/harbour	Excellent	—	Poor–fair
average slope (%)	0–10	0–30	35–40
Below 200 metres (%)	100	75	50
Pasture assessment (rating)			
% Marsh/wet meadow (+20)	35	20	0
% Grass-slope (+15)	40	20	10
% Modified copse (+15)	15	35	70
% Herbslope (+10)	5	0	0
% Steppe/heath (+5)	5	15	5
% Rock/fellfield (0)	0	5	5
% Screeslope (−10)	0	5	10
Raw pasture score:	1600	1250	1125
Adjusted pasture score:	320	83.3	30

Notes:
LCFC = large coastal farm with church.
IF = inland farm.
SCF = small coastal farm.
Raw pasture score = percentage of plant community × community rating (+20 through −10).
Adjusted pasture score = raw pasture score/average slope.
From McGovern (1980a).

Norse economy. The tiny three–four-animal cattle herds of the small farms would be given needed security from catastrophic losses or random infertility by access to the larger herds of the centrally located church farms for replacement stock. Alms of stored dairy produce or meat might become increasingly vital to the smallholders' domestic economies if conditions for stock-raising worsened. Such largess must have had its price in eventually reducing small holders to tenant status.

In 1125 the Greenlandic *thingmen* felt prosperous enough to trade a live polar bear for a Norwegian bishop, Arnald (*Einar Sokkason's Story*; in Jones, 1964). A large episcopal manor farm and cathedral were built

Table 17.3. *Estimates of store-house (skemma) capacity, cattle byre capacity, and human living space for selected Norse farms. Note that '?' denotes absent data or estimate based on analogy to better-documented structures of the same form and type*

Site	Type	Storage (m^2)	Byre est. cattle	Living space (m^2)
Ø 47 Gardar	EF	52.5	150–175	388
V7 Anavik	LCFC	3823	20–30	225–250
V51 Sandnes	LCFC	?	20–30	225
V54	IF	?	5–10	67
V35	IF	24	?	?
V52a	IF	33	12–15	?
V53c	IF	?	10–13	71
V53d	IF	30	5–10	69
V63	IF	35	?	?
V48	SCF	25?	?2–4?	?
V8	SCF	?	2–4	41
V16	SCF	?	4–5	?

EF : episcopal farm.
LCFC: large coastal farm with church.
IF : inland farm.
SCF : small coastal farm.
From McGovern (1980a).

on some of Greenland's best pasturage at Gardar, modern Igaliko. The growing spiritual and material power of the medieval church in Greenland is suggested not only by imported churchbells and stained glass, but also by the account of episcopal steward Ivar Bardsson of *c.* 1350. Some two hundred years after the accession of Arnald, the church seems to have owned some two-thirds of the best grazing in Greenland.

Perhaps the most telling evidence of the power of the clerical and secular élite in this arctic outpost is the spectacular program of church construction carried out between 1125 and 1300 (Roussell, 1941). The later churches of Norse Greenland are among the largest stone structures in the Atlantic Islands, and they were built by one of the smallest communities. These structures were formally planned after contemporary European models and were constructed of carefully fitted blocks of stone (a few of which weigh several tonnes). These later ecclesiastical structures contrast dramatically with both the informal and relatively impermanent semi-subterranean Norse secular architecture and also with the earlier, small, turf and stone chapels they replaced. Considering the tiny size of the colony and the multiple scheduling conflicts with summer subsistence activities imposed by construction (Fig. 17.4, line 9), the managerial skill and authority needed to command and co-ordinate such quantities of labour in an apparently marginal environment is particularly impressive.

Thus it seems highly probable that Greenland became at least as hierarchically organised as Iceland during the thirteenth–fourteenth centuries. Economic, political, religious, and ideological authority seem to have become concentrated in the hands of a lay and clerical élite probably arranged in a two- or three-tiered management hierarchy. How did this hierarchical organisation of decision-making and enforcement affect the response of Norse society to changing climate and southwards migrating Thule Inuit?

Climatic and cultural stresses

In answering this question, we must first consider which aspects of climatic change and culture contact were likely to have significant impact on Norse economy and society in Greenland. A wide range of climatic indicators (Lamb, 1977) suggest that the increasing cold of the later twelfth–fifteenth centuries may have resulted in hard times for the communities of the North Atlantic. Our brief outline of Norse economy in Greenland (as we now understand it) may suggest several 'pressure-points' vulnerable to probable Little Ice Age conditions. As so often, we must reason backwards from modern floral, faunal, and meteorological data, using the ice cores' long-term paleotemperature record to extend recent patterns into the medieval past. Thanks to the patient efforts of the Zoological Museum of the University of Copenhagen and several generations of local informants, an invaluable record of changing seal and caribou catches extends back to the early years of the last century in some localities. While it must be stressed that these are catch records and not population censuses, the cyclically recurring patterns observable in the data seem mainly to reflect climatically controlled variation in animal numbers and accessibility to human hunters (Vibe, 1967). Data on the most recent fluctuation in the harp seal catch in Godthaab District (former Norse Western Settlement) may illustrate the sorts of resource fluctuations involved (Table 17.4). Note that all stations show decreasing mean catch accompanying increasing interannual variability, and that the inner-fjord station of Kapisigdlit (in the main Norse settlement area)

Table 17.4. *Harp seal catches reported from four stations in modern Godthaab District. Preliminary figures based on the Danish Ministry for Greenland's annual* 'Summary of Hunting Statistics', *Copenhagen. Mean catches, relative percentage of harp seals in total seal catch, and inter-annual coefficient of variation show dramatic changes associated with the slight cooling of the early 1960s to 1970s*

	1954–58	1959–74
Kangek		
(outer fjord)		
Mean	38	1.7
% of total seal catch	58%	8%
Coeff. of var.	56%	159%
Godthaab		
(fjord mouth)		
Mean	139	52.5
% of total seal catch	79%	32%
Coeff. of var.	14%	97%
Kornok		
(mid-fjords)		
Mean	96.5	14.2
% of total seal catch	70%	27%
Coeff. of var.	22%	113%
Kapisigdlit		
(inner fjord)		
Mean	30.6	3.91
% of total seal catch	30%	4%
Coeff. of var.	4%	228%

shows the second largest decline in mean catch and the greatest increase in variability of catch between the periods 1954–58 and 1959–74. While these data are incomplete and in need of further analysis, the preliminary results serve to illustrate the sort of disruption that relatively small-scale cooling and increased variability of migratory pattern (on an annual or decadal scale) could inflict upon the vital, tightly scheduled maritime component of the Norse subsistence economy.

The terrestrial component of the Norse economy would be most adversely affected by increased precipitation in the continental inner fjords in either winter or summer. Deeper winter snow, especially when associated with ice crusting, has repeatedly triggered dramatic 'crashes' and local extinctions of caribou herds in Southwest Greenland (Vibe, 1967:163). Persistent winter snows and sudden spring storms could disastrously prolong byring time of the vulnerable Norse cattle, sheep, and goats. A series of long, wet winters would thus threaten the survival, first of new-born calves, lambs and kids, and eventually of the adults as well. Poorer farms with scantier pasture and flimsier byres would be hardest hit by such conditions, and their small herds least able to survive infection or malnutrition.

Increased summer precipitation could prove almost as threatening to the inner-fjord farmers. Wet weather during the haying season seems to have been a major cause of stock loss in medieval Iceland (Thorarinsson, 1956). A wet summer combined with a wet, early winter would probably be the worst combination for Norse stock-raising and caribou hunting.

Present evidence suggests that these kinds of stresses may in fact have affected the dual maritime–terrestrial economy of Norse Greenland. Vibe has for some time been working with the problem of projecting such recurring environmental conditions back beyond the limits of the observational evidence. Vibe originally made use of Koch's (1945) unfortunately unreliable sea ice compilations (cf. Ogilvie, 1978; Vibe, 1967). More recently he has turned to a combination of tidal cycles and the ice core data (Vibe, 1978). While research on these lines is not complete, the closeness of fit between the most recent of the faunal cycles and the 10- and 30-year oxygen-isotope curves (Vibe, 1967; Dansgaard in Lamb, 1977: 99; Dansgaard *et al.*, 1975) strongly suggest that this research is on the right track.

Additional support for Vibe's model is provided by the relative frequencies of seal bones excavated from Norse sites in both settlement areas. According to the model, common seals (*Phoca vitulina*) should have been fairly numerous in the Western Settlement area under most climatic conditions, while forming breeding colonies in the Eastern Settlement area only during periods of prolonged warmth with little East Greenland drift ice. Migratory hooded seals (*Cystophora cristata*) are not now accessible in any numbers from the Western Settlement area. They should tend to become more accessible in the Eastern Settlement during periods with heavy East Greenland ice. Thus the warmth of the first century or so of Norse occupation should have produced different sealing conditions from the later phases. At the site of Ø 17a near modern Narssaq in the Eastern Settlement area, common seal relative frequencies drop from 59 per cent in the lower layers (dated to the

Table 17.5. *Radiocarbon dates (excluding those on sea-mammal bone) currently available for Western Settlement Norse sites V48 and V54. Dates confirm evidence of stratigraphy and associated artifacts, and indicate the early and prolonged occupation of these two rather marginal farm sites. Data courtesy of Dr H. Tauber*

No.	Source	Radiocarbon yr	Calendar yr (AD)
K 3058	Willow charcoal V54 midden	1000±70	1030±70
K 3059	Sheep/goat dung V54 midden	1010±70	1040±70
K 3060	Willow charcoal V54 midden	1200±70	1255±70
K 3061	Structural turf /willow charcoal, V54	1410±50	1405±50
K 3062	Structural turf V54	1500±65	1440±65
K 3063	Willow twigs V48 midden (lowest layers)	990±75	1020±75
K 3197	Terrestrial mammal bone V48 midden (lowest layers)	990±50	1020±50
K 3199	Terrestrial mammal bone V48 midden (lowest layers)	960±40	1000±40
K 3201	Terrestrial mammal bone V48 midden (upper layers)	1310±50	1355±50
K 3203	Terrestrial mammal bone V48 midden (upper layers)	1390±50	1395±50

From McGovern (1980a).

eleventh century on runic and artifactual grounds (Vebaek, 1965) to 13 per cent in the upper layers (*c.* twelfth–fourteenth centuries). Hooded seal relative frequencies rise from 8 per cent to 20 per cent in the same collection (McGovern & Bigelow, 1977). As the model would predict, the Western Settlement stratified site of V48 shows no such dramatic fluctuation in common seal relative frequencies, and hooded seals are virtually absent from this and other Western Settlement collections (see Table 17.5 for radiocarbon dates for the V48 collection).

These partial confirmations of the Vibe model's usefulness in predicting common and hooded seal population dynamics suggest that we may be justified in provisionally accepting its predictions of harp seal accessibility as well. Thus we may hypothesise that the irregular cooling of *c.* 1250–1300 and the extreme inter-decadal fluctuations of the fourteenth century (Dansgaard in Lamb, 1977:99) produced harp seal migrations that were less regular and significantly less accessible to Norse communal hunting parties. While Eastern Settlement hunters may have been able to exploit the increasingly accessible hooded seals, this second option seems to have been denied Western Settlement hunters.

The inner-fjord precipitation that would have threatened the terrestrial component of the Norse economy may also have been a feature of this period. The increased fourteenth-century inter-decadal variability in the ice core paleotemperature record noted above suggests the sort of instability (in the sense of Vibe, 1967) on a shorter interval that has historically been associated with increased penetration of the continental inner fjords by oceanic storm systems. Fredskild's (1973) pollen profiles from both settlement areas generally agree in their indication of a shift to moister conditions in the inner fjords some time well after the initial Norse *Landnam.* While some radiocarbon dating problems remain, there seems no good reason to reject Fredskild's estimate of *c.* AD 1300 (1973:151) for this transition.

Thus we have some good grounds for suspecting that the Norse subsistence economy faced some significant challenges from the climate of the later thirteenth–fifteenth centuries. Present data and analysis suggest that the periods *c.* 1270–1300, *c.* 1320–60, *c.* 1430–60, and *c.* 1480 may have been particularly difficult.

This same general period saw the southward migration of Thule–Inugsuk Inuit along the west coast (Jordan, 1979; McGovern, 1979b; Tauber, 1973); from Thule in the north to Disko Bay to the extreme southwest. Our written evidence and some Inuit legends (Krogh, 1967; Gad, 1970; McGovern, 1979b) indicate that the relations between cultures were at least sporadically hostile. The large, unusually nucleated Inuit site of Sermermiut in the middle of the Nordrsetur has produced a preferred radiocarbon dating of AD 1240±120 (Mathiassen, 1958:48). Such settlements would have competed directly with Norse Nordrsetur hunting, and could have posed a serious threat to the hunters themselves. During the fourteenth century, Inuit winter houses seem to have spread all along the oceanic outer-fjord zone of southwest Greenland, often in nucleated, apparently defensive clusters (Mathiassen, 1936). The presence of permanent Inuit settlements in these prime sealing areas would greatly complicate Norse seasonal exploitation of the vital migratory seals. Even small-scale ambushes or occasional plundering of caches would dramatically raise the costs, hazards, and scheduling uncertainties of Norse subsistence sealing – just as the predictability and accessibility of the seal migrations themselves may well have been declining. Norse raids on Inuit settlements would surely invite blood-feud retaliation by survivors and their kin. As the security of upland flocks and lonely cots declined, so the costs of all aspects of Norse subsistence farming would rise. Thus the Inuit expansion of the thirteenth–fourteenth centuries probably reduced the viability of the Norse settlements. Certainly both steward Bardsson and King Magnus Eriksson (Gad, 1970:145) seem to have believed that the heathen *skraeling* had something to do with the collapse and extinction of the Western Settlement *c.* 1350.

As we have noted above, the fourteenth century also saw a decline in transatlantic contacts. When Bishop Alf died in 1378 he was never replaced, and official contact seems to have ended entirely after 1408. While increasing drift ice in Denmark Strait (Ogilvie, 1978) undoubtedly played a role in this decline, the precise mixture of political, economic, and climatic factors that resulted in the waning of Danish–Norwegian commerce in the western Atlantic remains to be determined. Declining overseas trade may not have immediately threatened Norse Greenlanders' survival, but it must have steadily worsened their bargaining position (Thorlaksson, 1978) and left them open to harsh exploitation or open piracy (Ingstad, 1966).

Thus we have some cause to model increasingly hard times for Norse Greenland in the later thirteenth–fourteenth centuries. A society whose economy is based on close co-ordination of communal labour and predictably abundant (but spatially distant) seasonal resources; whose critical resources become increasingly unpredictable in their occurrence, reduced in quality, quantity, and aggregation, and significantly more costly to exploit; is a society under dangerous stress. It might be argued that the death of Norse Greenland was caused by the synergistic interaction of the stresses of climatic fluctuation, culture contact, and shifting orientation of European markets. Had these factors not coincided when they did, Norse Greenland very probably would have survived the fifteenth century and might well have endured to the present in some form. However, this sort of explanation treats human response to climatic stress as a minor and dependent variable. Such an approach ignores the growing historical and anthropological evidence for the wide ranges of responses to declining carrying capacity (Glasgow, 1978) available to human populations in highly marginal environments. Much recent research (Flannery, 1972; Jochim, 1976, 1979; Johnson, 1978) has focused upon the processes by which individual societies select specific adaptive responses from a potentially wide range of alternatives. We must consider not only the nature of the external stresses that seem to have killed Norse Greenland, but also the reasons for that society's selection of ultimately unsuccessful responses to such stress.

Responses to stress

The success of the Inugsuk Inuit in medieval West Greenland demonstrates that the Little Ice Age fluctuations in temperature, sea ice conditions and faunal resources did not make the region totally uninhabitable. Much of the stress such fluctuations placed on the Norse economy was due to the strength of its linkage to the few ecological pockets suitable for stock-raising. The Norse settlement pattern, firmly based upon inner-fjord pasture, imposed considerable movement costs on both the maritime component of the subsistence economy and the Nordrsetur hunt that fuelled overseas trade. Had the Norse de-emphasised pasture and cattle-keeping and shifted a portion of their population permanently to the broader oceanic zone, they would have been far better able to respond effectively to fluctuating concentrations of seals and other marine resources. Marginal farms like V48 could have been abandoned or converted to herding stations and the pastures of the major farms could have been turned over to sheep and goat flocks less demanding of human labour and winter forage. A still more radical restructuring of the Norse settlement pattern might have produced the sort of whaling, fishing, and sealing villages that characterise modern Greenland.

However, our present documentary, locational, faunal, and artifactual data strongly indicate that the Norse Greenlanders took up none of these possible economic options. Cattle bones as well as sheep and goat remains are found in the topmost layers of V48. No radical change in settlement pattern or seasonal round is detectable in any of our other data. The Norse seem to have simply intensified their existing subsistence strategy, probably enlarging the byres and store-houses of the larger church farms (Nørlund & Stenberger, 1934; Roussell, 1941) in an attempt to increase the homeostatic buffering effects of redistribution.

Part of the cause of the persistently terrestrial orientation of the Norse settlement may have been a growing shortage of seaworthy wooden boats. While boat-fragments, steatite models, and scattered written references (Roussell, 1941; Gad, 1970) testify to the persistence in Greenland of traditional Scandinavian clinker-built wooden boat tech-

Table 17.6. *Comparison of modern seal catch data from Kapisigdlit (in the middle of the Norse Western Settlement area) and quantifiable Norse Western Settlement excavated bone collections. Note the contrasting patterns of seal exploitation and the insignificance of ringed seals in the Norse economy. Modern data from the Danish Ministry for Greenland's annual* 'Summary of Hunting Statistics', *Copenhagen.*

Seal	Kapisigdlit mean catch (%)[a]		Norse faunal collections total identified fragment (%)									
species	1954–58	1959–74	V48I	II	III	54	59	51	35	52a	53d	53c
Harp	30	5	58	58	55	71	38	67	56	64	68	61
Common	11	4	41	39	39	29	55	27	35	27	17	26
Ringed	60	91	0.5	0	0	0	4	2	3	2	2	2
Bearded	Trace	Trace	0.5	2	6	0	2	2	6	7	12	11
Hooded	0	0	0	0	0	0	0.7	0	0	0	0	0

[a] Preliminary figures.

nology, there is no evidence at all that the Norse adopted techniques of skin boat construction from the Inuit. *Umiaq*-like skin boats could have dramatically enhanced Norse coastal navigation and might have promoted the sort of dispersal of population and exchange of products involved in either of the coastal options. Efficient Inuit sea mammal hunting gear would have even more significantly improved the economics of Norse subsistence. Toggling harpoons and an associated complex of ice-hunting equipment allow the Inuit to exploit effectively the ringed seal (*Phoca hispida*) during the winter months. The value of this adaptation may be illustrated by reference to modern catch data (Table 17.6). Note that when harp seal catches declined at Kapisigdlit (in the middle of the Western Settlement area) in the late 1950s, the local Inuit hunters shifted their efforts to ringed seals (and caribou hunting) and the settlement remained occupied. The excavated Norse faunal collections show no significant exploitation of ringed seal, though these animals must have been present in the immediate neighbourhood of the Norse farms. They could have been hunted in the winter, without conflicting with other Norse subsistence tasks, *if* the Norse had acquired or developed the necessary hunting skills and technology.

Nor did the well-designed and skilfully tailored Inuit skin clothing make much impression on the Norse. Frozen woollen gowns, hoods, and caps from the late site of Herjolfsnes in the Eastern Settlement copy the latest European fashions (Nørlund, 1924). During the 300 years of contact, the Inuit seem to have had no such reluctance to acquire Norse equipment and materials (Mathiassen, 1930).

Rather than exploring the possibilities of new technology and searching out alternative resources, Norse society in Greenland seems to have stuck resolutely to its established pattern; elaborating its churches rather than its hunting skills. This single-minded conservatism and loss of adaptive resilience (in the sense of Holling, 1973) in the face of rising economic costs and declining returns contributed more directly to the Norse extinction than any one of the multiple stresses discussed above.

When Bardsson took up his stewardship *c.* 1342 he administered some 90 farms and 4 churches in the Western Settlement. By *c.* 1350 the Western Settlement was vacant, and Bardsson's account seems to indicate that its collapse was sudden rather than gradual.

The decline of the large Eastern Settlement may have been less catastrophic, but by AD 1500 Norse society in Greenland was dead. While this extinction is often cited as evidence of the determining impact of climate on human society, we might better consider it a cautionary example of the consequences of adaptive inflexibility in a changeable world.

The Norse Greenlanders of the initial *Landnam* period showed remarkable flexibility and ingenuity in modifying their Icelandic economy to exploit Greenland's very different resources. Why did their descendants show such fatal inflexibility in adjusting to later fluctuation in those resources?

This question's answer seems firmly bound to our earlier question about the role of élite management in Greenland's collapse. If Norse Greenland perished largely because of maladaptive responses to climatic and economic stresses, then we should concentrate our attention on the problem of how social decisions were made and enforced in Norse Greenland. As argued above, Norse Greenland seems to have supported a partly ecclesiastical élite that exercised increasing control over decisions affecting daily subsistence, transoceanic trade, and ceremonial architecture. Rather than enhancing the adaptability of Norse society in Greenland, this hierarchical management organisation seems to have fatally reduced its resilience.

To explain this failure of management, we may first consider the basic nature of decision-making in a stratified society (Flannery, 1972; Wright & Johnson, 1975; Johnson, 1978). Decision-making may be divided into three phases:

(*a*) collection and interpretation of data;
(*b*) choosing responses (making decisions);
(*c*) implementing and enforcing decisions.

As Moore (1980) has demonstrated, basic economic data (resource location, aggregation, and abundance) is not free. Its collection and interpretation can be very costly in time, energy, and foregone subsistence activity. When the margin between a human population and minimum accessible resources is comfortably broad (Fig. 17.6, point A) small-scale fluctuations (low in both amplitude and frequency) in accessible resources are relatively unimportant from the human standpoint. Under these conditions, we should expect investment in environmental data collection and interpretation to be relatively modest. As population increases, or the amplitude and frequency of resource fluctuation in-

Fig. 17.6. Schematic illustration of possible interaction of resource fluctuation (increasing in both amplitude and frequency over time), human population size and investment in management elaboration and specialisation.

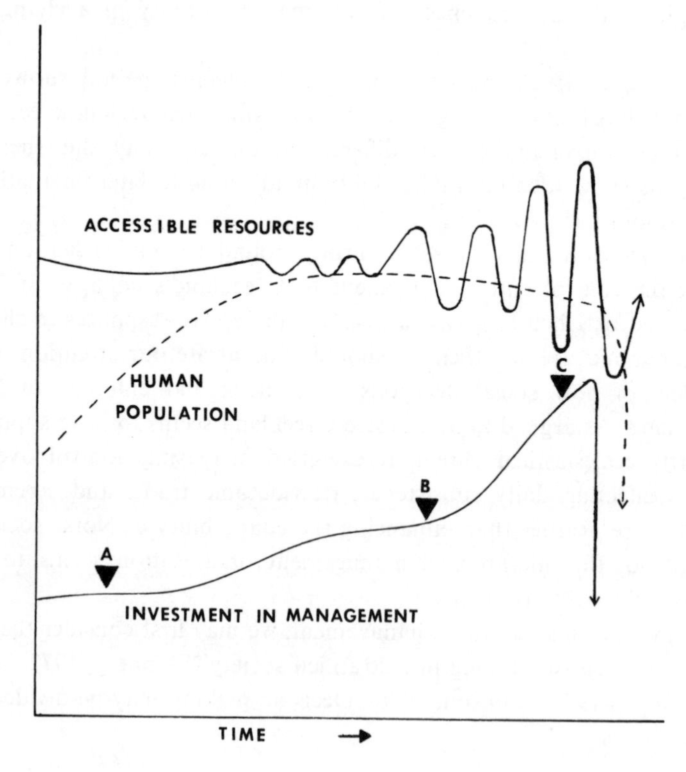

creases, this margin narrows (Fig. 17.6, point B). Under these circumstances, we might expect an increasing interest in good and bad seasons and a corresponding increase in investment in the collection and interpretation of environmental data of all sorts. This interest and investment will in turn lead to a dramatic increase in the amount of information being received by decision-makers and probably in the number of information sources that require monitoring. This increase is likely to overload the sort of low-cost, low-efficiency information management network suitable to our luxurious point A conditions. Human managers are capable of effectively monitoring a very limited number of information sources producing a strictly limited volume of data. If the number and detail of field reports reaching a single administrator multiply rapidly, a fall-off point will soon be reached in the administrator's processing efficiency. The administrator's 'span of control' will swiftly become so wide that he will be incapable of using the incoming data to administer effectively (Johnson, 1978). A common structural response to this problem is to engage in Flannery's (1972) horizontal specialisation (local administrators restrict their attention to topical or areal concerns) and vertical elaboration (a second or third tier of specialists dealing only with a restricted number of lower level administrators emerges). As Johnson (1978) has elegantly demonstrated, such horizontal and vertical specialisation of management structure significantly reduces the costs of information management and allows for the integration of a much greater number of information sources. Reduction of span of control may thus be achieved by the kind of increase in hierarchical stratification that we can document in medieval Iceland and have tried to demonstrate archaeologically in medieval Greenland.

The emergence of such a multi-tiered management hierarchy is not without its costs. One such social cost is the intensification of class distinctions and the divergence of class interests. The elaboration and maintenance of data-gathering institutions and administrative hierarchy represents an 'overhead' (Flannery, 1972) that must be paid for by successful resource management, profitable external trade, or intensification of effort and production by peasant farmers. Such intensification, while necessary to the maintenance of an élite lifestyle and administrative structure, may not be directly in the best interests of the individual farmer. Thus administrators are likely to be faced with local-level opposition to enforcing and implementing their decisions. This opposition may be crushed through an élite monopoly of military force (Fried, 1967). However, such direct military action is inevitably expensive and economically disruptive, and may ultimately prove counter-productive if excessively employed.

Much more effective is the spread and indoctrination of an ideology of hierarchy that provides religious justification for social inequities and stresses collective, class-specific duties and rewards over individual interests. If such an ideology finds widespread acceptance, it can greatly improve the effectiveness of the managerial élite. If the individual peasant carries an internalised social conscience that reminds him of his subordinate role in earthly hierarchy and divine plan, then the costs of decision implementation and enforcement can be considerably reduced and the élite can concentrate on the other components of decision-making.

This ideological conditioning may be achieved and reinforced through élite-sponsored ritual and impressive ceremony, emphasising a hierarchi cal social order and community solidarity behind ordained leaders. This process of conditioning, that can lead individuals to ignore an unfavourable balance between effort and reward, is what Harris (1979) has termed 'mystification'.

Therefore, it is probably not so surprising that the thirteenth–fourteenth centuries saw such expensive elaboration of ceremonial architecture and ritual paraphernalia in Norse Greenland. The degree of communal co-ordination required by the Norse subsistence economy, the unequal distribution of pasture and wealth, and the dispersed settlement pattern all suggest the need for a degree of social mystification to ease the enforcement of élite decisions.

If élite managers had performed well, searching out additional resources and alternative technologies to compensate for fluctuating and declining traditional resources, this separation of class interests and social mystification could have worked to the benefit of the society as a whole. However, élite managers in Greenland chose to ignore the rich fund of Inuit technology and expertise, to maintain an inner-fjord settlement pattern that favoured a few large farms at the expense of the many marginal ones, and to cling fatally to existing economic patterns. In short, they managed badly and their society died as a result.

So far, we have concentrated upon administrative structure rather than ideological content. However, *what* administrators believe about causation profoundly affects how they react to the data they collect and interpret. Baldly put, a society whose administrators (as well as its peasants) believe that lighting more candles to St Nicholas will have as much (or more) impact on the spring seal hunt as more and better boats is a society in serious trouble. To Flannery's (1972) list of hierarchical pathologies we may thus add 'auto-mystification'. This ideological malady will affect all three components of decision-making. Is the correct calculation of the date of Easter more important than accurate hay-harvest returns? Should communal labour be allocated to pasture drainage or church construction? How severely should we punish those who fraternise with the heathen?

As ceremony and ideology take up more of the necessarily limited attention of élite managers, there may be a pervasive tendency to simplify the threefold process of decision-making by mechanically repeating past choices, reducing the need for data gathering and interpretation. This simplification also effectively eliminates the possibility of significant innovation. Thus the ideological assessment as to which aspects of management responsibility are most vital and which can be de-emphasised and put on the 'automatic pilot' of tradition is critically important to the entire cultural system. Decisions in the realm of religious dogma or élite taste are thus intimately linked to subsistence economy.

As the society comes under greater stress, the élite may tend to elaborate the hierarchical structure of its organisation to process environmental and economic data more effectively (Fig. 17.6, point C). This process will tend to magnify the social distance between classes and thus call forth greater investments in mystification to smooth the enforcement of élite decisions. Such elaboration of mystification would have an immediate impact on administrative overheads (churches now 'need' stained glass and stone walls). It might also have the more insidious effect of gradually detaching élite managers from the phenomenological world and obscuring for them (as for their peasantry) the actual relationship between effort expended and material reward collected. As the (materialist) rationality of élite decision-making wanes and management efforts are increasingly diverted to calendrical ritual, ceremonial architecture, and the like, the survival of the society as a whole may be threatened unless external stresses are relaxed. Should management fail at this stage, the whole society (dependent upon co-ordinated communal labour and effective redistribution) may fall victim to Flannery's (1972) pathology of hypercoherence, bursting like a soap bubble rather than shrinking gradually like a deflating balloon. Auto-mystified élite managers under increasing stress may become particularly obsessed with conformity in both ritual and daily life and may ferociously repress attempts by lower-ranking members to find alternative solutions. It may not be insignificant that our last written source for Norse Greenland records both a proper Christian marriage and a burning for witchcraft at Hvalsey church in the Eastern Settlement (Gad, 1970).

Did Greenland's managerial élite succeed in enforcing ideological purity, only to fail in maintaining economic security? We may never be able satisfactorily to sort out and weigh the variables that killed Norse Greenland, and it is certainly not my intention to spawn another monocausal explanation of its end. However, it may be hoped that this preferably ephemeral integrative model may serve to direct future research towards a fuller investigation of Norse society in Greenland, and away from simple deterministic explanations of climate change, culture contact, and growing isolation.

Acknowledgements

I would like to thank Tove Hatting, Jeppe and Ulrik Møhl, Knud Rosenlund, and Christian Vibe of the Universitets Zoologiske Museum, C. L. Vebaek, K. J. Krogh, and Jørgen Meldgaard of the Danish National Museum, Jens Rosing and Claus Andreasen of the Grønlands Landsmuseum, Gregory Johnson of Hunter College, and Gerald Bigelow of Cambridge University for their advice and assistance in laboratory and field. The research reported here was made possible by the generous support of the National Science Foundation and the American–Scandinavian Foundation. Any errors of fact or interpretation are wholly my responsibility.

References

Berglund, J. (1973). Paa den Yderste Nøgne Ø. *Skalk*, no. 4.

Bocher, T. W. (1954). Oceanic and continental vegetational complexes in SW Greenland. *Meddelelser om Grønland*, **148**, no. 1.

Brøgger, A. W. & Shetelig, H. (1971). *The Viking Ships*, 2nd edn. New York: Twayne.

Bruun, D. (1896). Arkaeologiske Undersogelser i Julianehaabs Distrikt. *Meddelelser om Grønland*, **16**.

Bynch, L. (1801). Om saelhundefangsten paa Anholt, Iris, og Hobe *Maanedskrift* April 1801, 1–23, Copenhagen.

Clark, J. G. D. (1948). Seal hunting in the stone age of Northwest Europe: a study in economic prehistory. *Proceedings of the Prehistoric Society*, **12**, 12–48.

Dansgaard, W., Johnsen, S. J., Clausen, H. B., Reen, N., Gundestrup, N. & Hammer, C. U. (1975). Climatic changes, Norsemen, and modern man. *Nature*, **255**, 24–28.

Flannery, K. V. (1972). The cultural evolution of civilizations. *Annual Review of Ecology and Systematics*, **3**, 399–426.

Fredskild, B. (1973). Studies in the vegetational history of Greenland. *Meddelelser om Grønland*, **198**, no. 4.

Fried, M. (1967). *The Evolution of Political Society: an Essay in Political Anthropology.* New York: Random House.

Gad, F. (1970). *A History of Greenland*, vol. 1. London: Hurst & Co.

Glasgow, M. (1978). The concept of carrying capacity in the study of culture process. In *Advances in Archaeological Method and Theory*, vol. 1, ed. M. Schiffer. New York: Academic Press.

Harris, M. (1979). *Cultural Materialism: the Struggle for a Science of Culture.* New York: Random House.

Holling, C. W. (1973). Resilience and stability of ecological systems. *Annual Review of Ecology and Systematics*, **4**, 1–23.

Ingstad, H. (1966). *Land Under the Pole Star.* New York: St Martin's Press.

Jansen, H. M. (1972). A critical account of the written and archaeological sources' evidence concerning the Norse settlements in Greenland. *Meddelelser om Grønland*, **182**, no. 4.

Jochim, M. (1976). *Hunter-Gatherer Subsistence and Settlement: a Predictive Model.* New York: Academic Press.

Jochim, M. (1979). Breaking down the system: recent ecological approaches in archaeology. In *Advances in Archaeological Method and Theory*, vol. 2, ed. M. Schiffer pp. 77–117. New York: Academic Press.

Johnson, G. A. (1978). Information sources and the development of decision-making organizations. In *Social Archaeology: Beyond Subsistence and Dating*, ed. C. Redman *et al.*, pp. 87–112. New York: Academic Press.

Jones, G. (1964). *The Norse Atlantic Saga.* New York: Oxford University Press.

Jordan, R. (1979). Inugsuk revisited: an alternative view of neo-eskimo chronology and culture change in Greenland. In *Thule Eskimo Culture: An Anthropological Retrospective*, ed. A. P. McCartney. National Museum of Man, Mercury Series no. 88, Ottawa.

Koch, L. (1945). The east Greenland ice. *Meddelelser om Grønland*, **130**, no. 2.

Krogh, K. J. (1967). *Viking Greenland.* Copenhagen: National Museum.

Lamb, H. H. (1977). *Climate: Present, Past, and Future*, London: Methuen.

Larsen, L., transl. (1917). *The King's Mirror.* New York: American–Scandinavian Foundation.

McGovern, T. H. (1979a). *The Paleoeconomy of Norse Greenland: Adaptation and Extinction in a Tightly Bounded Ecosystem.* Ann Arbor Michigan, University Microfilms. 395 pp.

McGovern, T. H. (1979b). Thule–Norse interaction in Southwest Greenland: a speculative model. In *Thule Eskimo Culture: an Anthropological Retrospective*, ed. A. P. McCartney, National Museum of Man, Mercury Series no. 88, pp. 171–89, Ottawa.

McGovern, T. H. (1980a). Cows, harp seals, and churchbells: adaptation and extinction in Norse Greenland. *Human Ecology*, **8** (3), 245–75.

McGovern, T. H. (1980b). Site catchment and maritime adaptation in Norse Greenland. In *Site Catchment Analysis*, ed. F. Findlow. Los Angeles: University of California Press.

McGovern, T. H. & Bigelow, G. F. (1977). Animal Bones from the Norse Eastern and Middle Settlements, MS on file, Universitetets Zoologiske Museum, Copenhagen.

McGrew, J. H., transl. (1970). *Sturlunga Saga*. New York: Twayne.

Mathiassen, T. (1930). Inugsuk, a medieval Eskimo settlement in Upernavik District, West Greenland. *Meddelelser om Grønland*, **77**, no. 4.

Mathiassen, T. (1936). The Eskimo archaeology of Julianehaab District. *Meddelelser om Grønland*, **118**, no. 1.

Mathiassen, T. (1958). The Sermermiut excavations 1955. *Meddelelser om Grønland*, **161**, no. 3.

Meldgaard, J. (1977). Inuit–Nordbo Projektet 1976–77. *Nationalmuseets Arbiedsmark*, 1977, Copenhagen.

Moore, J. (1980). Effects of information networks on hunter-gatherer societies. In *Hunter-Gatherer Foraging Strategies*, ed. B. Winterholder & E. A. Smith. University of Chicago Press.

Morrison, I. (1973). *The North Sea Earls*. London: Gentry Books.

Nørlund, P. (1924). Buried Norsemen at Herjolfsnes. *Meddelelser om Grønland*, **67**.

Nørlund, P. & Stenberger, M. (1934). Brattahlid. *Meddelelser om Grønland* **88**, no. 1.

Ogilvie, A. E. J. (1978). Reconstructions of climate from weather compilations: a case study. Paper presented at the Nordic Symposium on Climatic Change and Related Problems, April 1978, Copenhagen.

Rafn, O. C. (1876). *Grønlands Historiske Mindesmaeker*. Copenhagen.

Roussell, A. (1941). Farms and churches of the medieval Norse settlement in Greenland, *Meddelelser om Grønland*, **89**, no. 1.

Saemundsson, H. (1939). *The Zoology of Iceland*. Copenhagen: Munksgaard.

Sveinsson, E. O. (1953). *The age of the Sturlungs*. Islandica, XXXVI. New York: Ithaca.

Tauber, H. (1973). Copenhagen radiocarbon dates X. *Radiocarbon*, **15**(1), 86–112.

Thorarinsson, S. (1956). *The Thousand Years Years Struggle Against Ice and Fire*. Reykjavik.

Thorlaksson, H. (1978). Comments on ports of trade in early medieval Europe. *Norwegian Archaeological Review*, **11**(2), 112–14.

Vebaek, C. L. (1958). Topographical and archaeological investigations in the medieval Norse settlements in Greenland: a survey of the work of the last ten years. *Third Viking Congress*, Reykjavik.

Vebaek, C. L. (1962). The climate of Greenland in the 11th and 16th centuries. Paper presented at the Conference on the Climate of the 11th and 16th centuries, Aspen, Colorado.

Vebaek, C. L. (1965). An 11th century farmhouse in the Norse colonies in Greenland. *Fourth Viking Congress*, York.

Vibe, C. (1967). Arctic animals in relation to climatic fluctuations. *Meddelelser om Grønland*, **170**, no. 5.

Vibe, C. (1970). The Arctic ecosystem influenced by fluctuations in sun-spots and drift-ice movements. *Production and Conservation in Northern Circumpolar Lands* ICUN, NS 16, 115–20.

Vibe, C. (1978). Cyclic fluctuations in tide related to season as key to some important short and long term fluctuations in climate and ecology of the North Atlantic and Arctic regions. In *Proceedings of the Nordic Symposium on*

Climatic Changes And Related Problems, ed. K. Frydendahl. Danish Meteorological Institute Climatological Papers no. 4, Copenhagen.

Williamson, K. (1948). *The Atlantic Islands.* London: Collins.

Wright, H. & Johnson, G. (1975). Population, exchange, and early state formation in southwestern Iran. *American Anthropologist*, **77**(2), 267–89.

18

Weather and the peasantry of Upper Brittany, 1780–1789

DONALD SUTHERLAND

Abstract

Historians of Old Régime France have long been aware of some of the relationships between meteorological change and peasant communities. The disastrous harvests and famines of the seventeenth century and the effects of the poor harvest of 1788 on the insurrections of the following year have all been thoroughly studied. From these a model of weather conditions and rural societies has developed. Thus a poor harvest is generally held to produce greater death rates, migration, indebtedness, unemployment, sales of peasant property and falling wage rates. Yet the model may only hold good for the effects of spectacular weather conditions. Thus in the region around the town of Vitré in Britanny in the 1780s, peasants were able to withstand the droughts of 1782, 1785–86 and the poor harvest of 1788 surprisingly well. The major victims were infants and children who could be replaced easily, while the poor may have held their own because secondary sources of income were only partially dependent upon the grain harvest. Other signs of a crisis in rural society are missing. The well-off peasantry actually benefited from a poor harvest. Employment for farmhands and wages remained more or less steady. The local economy was able to support the same number of households despite poor harvests, indebtedness did not increase and peasants were not forced to sell land to supposedly rapacious urban dwellers.

There are a number of reasons why rural society proved so resilient. The return on landed property was only 5 per cent so there was no point in buying land when it was easier to raise rents or gain windfall income from seigneurial dues and tithes paid in kind. Peasants also had their own expedients for dealing with hard times. Unlike many other regions, the parish priests controlled most of the considerable resources of the tithe, much of which must have gone to charity. Priests were also responsible for raising and distributing alms. In hard times, landlords were lenient in collecting rent or in accepting substitute payments in labour or produce.

Thus for all the mesmerising poverty, and despite the greater suffering of the poor and the defenceless, many peasant communities were not as vulnerable to the weather as we generally think.

Thanks to the work of specialists of the seventeenth century, historians of the Old Régime have adopted a model of the relationship of weather and peasant communities (Goubert, 1960; Deyon, 1967; Meuvret, 1971; Jacquart, 1974). Thus a harvest failure, as in 1660–61, 1693–94 or 1709–10, is generally supposed to have produced a dramatic rise in the incidence of deaths, mendicity and migration. The excessively cold and long winters of these years also produced more long-lasting effects. Peasants were quickly stripped of their resources, fell into debt and were forced to sell their land. Thus, meteorological conditions played a role in giving France its peculiar agrarian physiognomy of a country where small peasant property co-existed alongside medium sized and large noble and bourgeois estates. The assumption behind this line of argument is that rural communities were extremely vulnerable to bad harvests. In this chapter, I propose to examine the notion of vulnerability more closely and to show two things: that bad weather and poor harvests did not affect peasant society evenly, that certain groups were practically immune while others were indeed vulnerable; and that poor harvests did not set off an inevitable chain reaction reverberating through the entire structure of rural society. That is, poor harvests are better conceived as acting on a range rather than as a single, crushing blow. An examination of the case of one small region (the area around the town of Vitré in Upper Brittany in the 1780s) shows that bad weather and poor harvests did not always produce all the effects involved in the usual model. Rather, peasant society here and no doubt in many other regions was resilient enough to be spared some of the worst effects of bad weather.

Meteorological information about Britanny is skimpy but there is enough to indicate that the human consequences of bad weather were serious. The 1780s were certainly considerably cooler than the previous decade; and in this respect the figures for Brittany coincide well with those published for central England in the same period (Meyer, 1966; Lamb, 1977). Thus a comparison of the mean seasonal temperatures of the period 1778–81 with those of 1782–88 shows a marked cooling. Only winter temperatures remained constant while spring temperatures dropped one degree (11.1–9.7 °C), summer over three degrees (16.8–13.3 °C) and autumn two (8.1–5.7 °C). This cooling was not necessarily disastrous, however. Rye and buckwheat, the two principal food crops, can flourish at such temperatures and there does not appear to be any relationship between the falling temperatures and what production figures we have. In any case, the problem which struck contemporaries was that of rainfall (AD, I–V C1653–55, C1737, C3912). In 1781, for instance, there was little rain until the end of June, while July witnessed excessive rain and high winds and the subsequent late summer heat completed the ruin of many grain and garden crops. In 1784, a cold winter with hail at the end of April, spring floods and then lack of rain until the end of June produced poor forage crops and therefore a considerable blow to cattle-raising. 1785 was the worst year of the decade, ranking with 1740–41 and 1770 as among the most disastrous of the century. In some parishes, there was no rain between January and August while local hailstorms in early August and heavy thunderstorms in September produced such a poor crop that there were fears there would not even be enough seed grain. There was a less severe drought in 1786 but buckwheat and forage crops were harmed nonetheless, while 1788, so famous elsewhere for its spring hailstorms, suffered from excessive winter rains which adversely affected the export trade in wheat and rye although there was enough for local needs. Excessive rains in 1789 brought the buckwheat crop below normal.

Many years in the 1780s were poor. Although we have no actual production figures, the harvest reports indicate that no year was abundant. On the whole, harvests were moderately less than normal in the 1780s (AD, I–V C1653–55). Out of ten years (1780–89) in the Subdelegation of Rennes, wheat harvests were above normal in four years, less than normal in six. In the Subdelegation of Vitré, wheat was average in six years, below average in four. The success rate for the other grains was comparable. Though these figures may not seem unduly alarming, there is certainly evidence to suggest that the run of poor harvests which they represent could entail extensive human misery. Insufficient hay forced many peasants to sell off their horses and cattle at low prices before they starved, with the consequent loss of income from dairying, meat and work animals. Similarly, it was feared that the failure of the flax crop would put as many as 200 weavers and spinners out of work in the little parish of Vergéal (AD, I–V C3912). At the same time, poor harvests meant less work for labourers while tenant farmers and share-croppers were in difficulty. Worse still, Breton peasants lived in such shocking sanitary conditions and were so ignorant of the principles of elementary hygiene that poor harvests were usually followed by epidemics of typhoid and typhus, epidemics made all the more murderous by the constant movement of troops throughout the province during the War of American Independence (Goubert, 1974). While contemporaries undoubtedly exaggerated some of the worst effects of these years, independent evidence indicates that the situation was definitely serious. The number of foundlings cared for by the general hospital in Rennes increased and, in the province as a whole, the death rate rose after every poor harvest (Goubert, 1974). While we should be extremely careful about asserting a correlation between weather, food supply and death and while the relative contributions of food supply and disease to the number of deaths is unknown for the moment, it would appear that contemporaries had some basis for their fears.

Historians should nevertheless be more wary of these sources than many of them have been, especially with regard to the longer-term effects of adverse weather conditions and in terms of their impact on the society as a whole. Lurid descriptions of a worsening pre-revolutionary crisis often derive from individuals who could only speak impressionistically, naturally enough in a pre-statistical age, and who felt that the problems of charity and police were getting beyond their control, for reasons that related to declining charitable donations and government cutbacks rather than to the growing magnitude of the problem itself (Vovelle, 1973; Hufton, 1974; Fairchilds, 1976). To assess the extent of misery involves going beyond descriptive sources to others scattered throughout the archival system.

One way of approaching this question of misery is to construct a budget for a hypothetical family. The tax rolls which survive provide a great deal of information about land-use and grain yields and the household inventories provide information about livestock and employment. Finally, various governmental bodies drew up harvest reports and kept track of prices. Along with other sources of information, it is possible to reconstruct a reasonably accurate model of the budget of a family which had enough land to be self-sufficient. Such a family would have worked ten *journaux* (roughly five hectares) of land, have had two children, hired a male and female farmhand and indulged in small-scale textile manufacture and cattle-raising as business side-lines. If we run their model budget through the 1780s, it is evident that their income did not suffer from the droughts. Indeed, since shortages pushed up prices, a family of owner–occupiers increased its average annual income by a little over 2 per cent over the decade, and their income reached its highest peaks in the drought years (Sutherland, 1981). Thus, the increase in grain prices which followed a poor harvest, rising cattle prices and the relative decline of labour costs, taxes and seigneurial dues in relation to gross income, allowed them to hold their own. People like this were not enriching themselves, by any means. A family of tenants working ten *journaux* still could not produce enough to live from the grain economy alone. Any family, tenants or owners, would have had to scrimp on food and other essentials in hard times. Yet it would have taken a major meteorological disaster and harvest failure and eventual exhaustion of credit to push people like this off the land. Thus there existed within rural society a group of substantial peasants whose tenure was almost immovable.

This group was not very large, however. Ten *journaux* was the approximate cut-off between subsistence and market agriculture and, according to the land tax records, well over half the holdings in Upper Brittany did not produce enough to support their occupants. In areas where lease-holding predominated, close to three-quarters of the tenants must have had insufficient land (Sée, 1906). For the poorer peasantry, therefore, survival depended on two things: production rather than price-levels and secondary economic activities. Estimating the contribution of these two factors to the poorer peasantry is much more difficult because of the paucity of the sources but it would appear that, all things considered, their position did not deteriorate as much as might have been expected. The greatest difficulty lay in the realm of grain production. As we have seen, the harvests of most grains in the 1780s were less than average and the inadequacy of the food supply must have imposed great hardship. It was reported, for instance, that some peasants in particularly deprived areas had to rely on a gruel made of chestnuts for up to six weeks of the year, and some claimed that others drank so much cider in order to satiate a nearly perpetual hunger (AD, I–V C1653–55, L516; Goubert, 1974). Yet the higher-yield and tougher buckwheat crop, which provided basic nourishment, fared better than other grains and some rural consumers must have benefited from its falling price.

It is likely, however, that some of the poor's secondary sources of income, which were of greater importance to them than to the well-off, did undergo some perturbations. The overall contribution of textiles is ambiguous. It is not possible to reconstruct a price series for hemp, the main textile crop, from the household inventories, because they list textiles in so many different states of preparation. If we assume, however, that the sail-cloth market reflected the demand for hemp, we can make some reasonable guesses. Since the Breton ports were major staging posts during the War of American Independence, the demand for sail-cloth boomed, only to fall off dramatically with the conclusion of peace. Thereafter, there was some recovery, followed by a sharp fall after 1788, perhaps a reflection of the pre-revolutionary industrial crisis. These fluctuations were no doubt felt by most hemp producers in one way or another, yet just as demand fell off after the war, the succession of good seasons brought the harvest of flax and hemp back to normal (AD, I–V C1547–52, C1653–55). Thus, the poor had a little more to sell, so that the more abundant production may have compensated for the decline in demand.

Cattle-raising was another important source of secondary income which underwent a crisis, more serious perhaps than that of any other. If animals did not starve in the droughts, they were sold in considerable numbers because of the shortage of forage. In some parishes in 1785, where there had been no rain from January to July, some claimed that between one-half and three-quarters of the cattle either died or were sold at low prices. This must have been an exaggeration but, where figures do exist, it is clear the losses were extensive. In St-Aubin-des-Landes, one-tenth of the horses and nearly one-fifth of the cattle were lost. By September, fifty-four farm animals had died in Erbrée and there were fears that half the rest would have to be sold before the following Easter. Everyone suffered in a situation like this. There was not enough fertiliser and substitutes like ashes and nightsoil suddenly became expensive; weakened oxen and horses could not plow effectively and the price of milk and butter shot up. But the poor, for whom cattle-raising was a more important resource, suffered most of all. And unlike a mediocre grain harvest, an animal shortage had more long-lasting effects. One could still see the effects of the 1785 and 1786 droughts on cattle prices in 1787 and many were still being replaced in 1789 (AD, I–V C1655, C1735, C3912).

In Brittany as elsewhere, real wages declined over the century but it is likely that in the short-term perspective of the 1780s, real wages rose slightly. Information on daily wages is impossible to collect. Indications in the household inventories are too scattered and infrequent to be of use and, in any case, it is impossible to know how many days in a year a labourer worked. Furthermore, casual labourers, particularly harvesters, were often paid in cash and kind so that a wage index would be particularly hard to construct (Sée, 1906; AD, I–V L1345). It is possible, however, to infer something about the labour market generally from the experience of permanent farm servants. For this class of labourer, the inventories and the *capitation* (head tax) rolls provide abundant and complementary information.

On the whole, both from the point of view of annual wages and employment demand, the situation for permanent farm help was more or less favourable. From 1783 to 1787, the years for which the *capitation* rolls survive, employment demand remained fairly constant. Thus, the mean number of male farmhands employed in the two parishes of Argentré and Balazé during these six years was 193 (standard deviation = 7.3) and the mean number of female farmhands was 216 (standard deviation = 10.7). Employment opportunities scarcely varied from year to year and appear to have been independent of harvest conditions, a situation which suggests that permanent hands were paid irregularly. In any case, annual money wages remained more or less stable. Men averaged 47.3 *livres* (standard deviation = 4.2), women 30.0 (standard deviation = 4.5) AD, I–V, C4501–64, series 4Bz). Thus for at least one class of labourer, poor harvest conditions did not produce marked unemployment or wage cuts.

If we can extend our conclusions about the labour market of permanent farm help to the labour market generally, it seems likely that those of the poor who were wholly or partly dependent on a wage for a living did not suffer. At least from this angle, their position did not deteriorate as the Revolution approached. Indeed, this was the case with most poor peasants. Worse off as grain and cattle producers, holding their own in textiles, they were reasonably secure as employees. Wages held up and the decline in the cost of the principal food, buckwheat, improved their position as consumers. Thus, even for the most vulnerable members of rural society, there were limits to the deterioration which poor harvests inflicted on their capacity to survive.

There was, however, a more vulnerable element still in rural society: children. It was they above all who were most sensitive to poor economic conditions, a fact tragically evident in death rates. The population of Brittany grew by about 10–15 per cent in the eighteenth century, about half the national average and, where the nation's growth merely slowed down after 1770, Brittany's actually fell by about 4 per cent (Goubert, 1974). If the experience of two parishes in the 1780s was typical, this was due to an excessive child mortality rate (Table 18.1(*a*), (*b*)). Less than half the children reached the age of twenty. And this frightful infant and child mortality rate was probably related partially to the food supply since deaths in this age group rose after years of poor harvests. But Malthusian relationships do not appear to have affected adult deaths as much, if at all. Since the reproductive age cohorts survived better, the lost children could be quickly replaced, particularly in the case of Brittany which had one of the highest fecundity rates in Old Régime France. Thus the resiliency of adults to poor harvests showed something of the strength of rural society.

Table 18.1. (a) *Deaths in Balazé and Noyal-sur-Vilaine, 1780–89*

	1780–81	1781–82	1782–83	1783–84	1784–85	1785–86	1786–87	1787–88	1788–89	Totals	Percentage change per year
Infants (0–11 months)	48	40	53	66	48	41	49	51	48	444	0
Children (1–19 years)	52	57	57	93	71	71	69	89	40	599	1.17
Adult working population (20–49 years)	29	38	85	54	48	25	48	38	41	406	−1.67
Old (50–89 years)	42	40	56	53	65	33	36	26	36	387	−3.63
Stillborn	6	17	4	8	2	6	11	13	3	70	−2.07
Totals	171	175	251	266	232	170	202	204	165		−0.91

Table 18.1. (*b*) *Summary: Births, deaths and marriages in Balazé and Noyal-sur-Vilaine*

	Births	Deaths	Marriages
1780–81	162	171	44
1781–82	147	175	53
1782–83	174	251	37
1783–84	148	266	78
1784–85	278	232	57
1785–86	170	170	53
1786–87	140	202	41
1787–88	172	204	38
1788–89	197	165	48
Totals	1588	1836	449
Percentage change per year	1.52	−0.91	−1.29

Sources: AD, I–V 3E, parish registers of Balazé and Noyal-sur-Vilaine.
Notes: Still-births have not been included in either the birth or death columns. Those who are from Vitré and are recorded as having died 'en nourrice' and those born in Vitré and dying in Balazé under aged two have been supposed to have died 'en nourrice'. Neither category is included in the deaths columns. All annual percentage changes in this chapter have been computed on the basis of trend values.

If mediocre harvests could affect the poor and the young, they could not make many further inroads into the structures of rural society. This is evident from an examination of indebtedness, migrations and land sales.

This is particularly clear in the case of indebtedness. The sources for an examination of this factor are the household inventories. The most serious limitation of the inventories are that they were taken at any time during a particular year. Measuring indebtedness is therefore difficult because the finances of a peasant family would have looked very different before or after a harvest or before or after rents were paid. Taking a large number of inventories goes some way towards dealing with this problem and thus allows us to grasp the essentials of indebtedness: who owed what to whom, why and in what proportion.

Table 18.2. *Evolution of debts*

Year	Number of inventories	No declared debts (%)	Debts per those in debt	Rents per those in debt	Other debts per those in debt
1765–72	101	19.80	158.05	84.84	73.21
1779–81	11	0	268.25	164.48	103.77
1782	46	23.91	347.76	129.44	218.32
1783	49	12.24	258.55	136.50	122.05
1784	31	19.35	330.99	250.48	80.51
1785	52	17.31	223.70	137.87	85.83
1786	57	19.30	232.32	125.48	106.84
1787	29	62.07	240.23	162.50	77.73
1788	60	36.67	351.87	162.55	186.32
1789	28	35.71	266.07	76.46	189.61
Annual change (1779–89)		64.38 or 24.27 eliminating 1779–81	−0.68%	−2.82%	2.33%

Most people were in debt (Table 18.2). Of the 460 odd inventories chosen from the Vitré area, about three-quarters of the households had debts to declare. Almost everyone owed money to his farmhands or to local labourers. This was only natural and, on average, the sums were quite small. In Upper Brittany, unlike other regions, a comparatively small proportion of debts was due to loans from merchants and to services from artisans. Either loans or unpaid merchandise could have been used by the local bourgeoisie and well-off peasants to turn the country people into a debtor class. Yet this did not happen to any significant extent. Outstanding rents and arrears were the most important component of debts yet the weight of rents appears to have declined slightly. Furthermore, the proportion of families in debt in the 1780s declined appreciably. Those who remained in debt saw a slight decline in the weight of their obligations. Both of these trends were probably a reflection of the small general rise in incomes. In broad terms, indebtedness became less of a problem in rural society as the decade wore on.

The same was generally true of the rate of migration. The procedure here has been to compare the names of the heads of households on the tax rolls of two parishes, Argentré and Balazé, for every year between 1783 and 1788 and then to check names against the parish registers to make sure that a death was not recorded as a migration. Those who disappeared only to return a year or so later (never a very large number anyway) have been eliminated from the calculations. The results of this tedious procedure show a remarkable stability in the number of households. The number of households in these two parishes averaged 695 with a standard deviation of only 7.5 and the net migration over the period was exactly zero (AD, I–V C4501–64). Despite economic problems in individual years, people still found it possible to continue to make a living. The local economy was well able to support its population.

There were, nonetheless, some extraordinary changes in migration movements. Although the number of households remained more or less fixed over the six years, there was a very high rate of emigration and immigration. In Balazé, 46.9 per cent of the households disappeared between 1783 and 1788 through migration, in Argentré 48.6 per cent. At this rate, there would have been a complete change in personnel in less than a generation, a turnover rate which was double that of many other French villages. Yet it is unlikely that much of this migration was related to misery which produced the desperate long-range migration of the hordes of beggars which plagued the northern plains. There was plenty of migration of this sort in Brittany too but it appears to have been limited to travel between cities where the opportunities for charity, work, crime and prostitution were greater (AD, I–V 8B 57–60, 76–77; Sée, 1930). The turnover in parishes was more related to taking advantage of kin and friendship ties for jobs, tenancies and marriages, as people moved about within an invisible but small circumference in a community as real as that of the parish. To move outside this frontier was a sign of genuine deprivation and, on the whole, people stayed within it.

Another sign of a weakened peasantry was the loss of land. During the great subsistence crises of the seventeenth century, the urban bourgeoisie was able to take advantage of economic dislocations of the peasantry to acquire more land. But if the spoliation of one class by another is a sign of rural crisis, then there was no such crisis in Upper Brittany. Within the sixty odd parishes within the jurisdictions of the land registries of St-Aubin-du-Cormier and Vitré, land came increasingly under local control as former owners from well outside the region put their land on the market (Table 18.3). Furthermore, land sales did not pit country against town. There was no urban onslaught on rural property. On the whole, the urban bourgeoisie tended to lose land or, in the case of Vitré after 1770, marginally to increase it. They were weak participants in a competition which largely took place between the rural bourgeoisie and

Table 18.3. *Evolution of land sales in the parishes of Balazé, Etrelles and Erbrée, 1750–70*

	Noble	Priest	Urban bourgeois	Rural bourgeois	Peasant	Outside bourgeois
Livres	−1950	+22 690	−30 385	+15 099	+26 942	−32 396
%	−3.01	35.05	−46.94	23.33	41.62	−50.05

Evolution of land sales within the bureaux of St-Aubin-du-Cormier and Vitré, 1770–90

	Noble	Priest	Urban bourgeois	Rural bourgeois	Peasant	Outside bourgeois
St Aubin-du-Cormier						
Livres	46 660	1910	−43 924	144 548	3521	−152 715
%	23.73	0.97	−22.34	73.51	1.79	−77.66
Vitré						
Livres	59 978	25 029	31 566	132 681	103 550	−352 804
%	17.00	7.09	8.95	37.66	29.35	−100.00

Sources: AD, I–V $2C^{38}$ 233–41, bureau de St Aubin-du-Cormier and AD, I–V $2C^{45}$ 245–69, bureau de Vitré.
Notes: ‘Urban bourgeois’ means anyone who lived in the *chef-lieu* of the jurisdiction, ‘Outside bourgeois’, anyone who lived outside it, even though the distinction between rural and urban bourgeois is probably artificial in the case of St Aubin-du-Cormier which was only a large bourg. The large gain registered by priests in the first section is misleading since it includes one massive acquisition of 16 000 *livres* in 1767.

the peasantry. Yet, even here, the number of participants was fairly small. On the whole most peasants did not have the resources to enter the competition. A sizeable majority operated upon an extremely small margin and it is doubtful many were in a position to save the 300 *livres* or so required to buy even one *journal.* Inevitably, land purchases were restricted to the well-off. The major beneficiaries then were the rural bourgeoisie and wealthy peasantry but they did not achieve their relative success at the expense of the poorer peasants.

Moreover the land which was sold off was fairly well distributed among local buyers in spite of their relatively small number. No one apparently used his wealth to monopolise acquisitions. Of the eighty-nine land-buyers in a half-dozen parishes around Vitré between 1785 and

1790, only two bought four pieces of property, three bought three pieces and thirteen bought two pieces. The other seventy-one bought one piece of land each. Thus, there was no hoarding of land within the elite which could afford to buy. In fact the hierarchy of land holdings and the social distribution of property remained remarkably stable. If one compares the value of all the land which could have been sold around Vitré between 1770 and 1790 with the value of all the land that was sold, less than 2 per cent of the land was exchanged between social groups through land sales. Put another way, the peasantry gained only 0.50 per cent of the arable in the generation before 1790. Practically nothing disturbed the patterns of land holding.

Viewed from a broader perspective, peasant society was extremely stable, so much so that despite the extraordinary turnover of the population, it appears to have reached a state of structural equilibrium. The basic demographic stability, the success in keeping debts within manageable limits and the stability in incomes meant that peasant society simply reproduced a photographic image of itself over the decade. It would be risky to extrapolate the situation of the 1780s forward or back in time but everything known about demographic movements, land sales and price changes, suggests a fundamental stability at least since the 1760s.

Such a conclusion poses an interesting paradox. If unfavourable weather conditions, particularly the droughts of 1782 and 1785–86, produced poor harvests, how was it that peasant society felt few long-lasting effects? In a society whose productive capacity was so low that it was impossible to stock much grain between harvests and in which a majority of the population had insufficient land, why was a poor harvest not more devastating? The demographic situation was admittedly serious. The population did not replace itself in the 1780s but even here the number of births exceeded the numbers of infant and child deaths and the trend of adult deaths was downward, so that a basis was being laid for future demographic growth. Beyond this, poor harvests had nothing like the effect of a seventeenth-century *crise de subsistances*. There were no mass migrations, mounting indebtedness or enforced land sales.

The explanation for this resiliency lies in the economic structures and cultural patterns of rural society. The rich, for instance, did not have the opportunity to exploit peasants' financial distress. Peasants were not a debtor class in this period, so that they still had a great deal of credit to exhaust before being forced to put their land up for sale. In any case, with land bringing a return of only 5 per cent and therefore a twenty-year wait before an acquisition could make a profit, the rentier class did not have much incentive to buy (Young, 1969). Peasants also had resources of their own to withstand such pressures. The well-off among them invariably profited from high grain prices while the poorer peasants had secondary sources of income which were only indirectly tied to the grain economy: cattle-raising, textiles and, in Upper Brittany, poaching and salt-smuggling. Nor did the well-off always exploit the hardships of the weak. In hard times, for instance, farmhands were often not paid although they continued to be fed and housed by their employers. Landlords proved willing to accept substitute forms of rents, particularly labour rather than cash or produce. Sometimes they allowed arrears to build up. The Marquis de Chateaugiron was owed far more in arrears than the total annual value of his rents while the Marquis de Piré often retained tenants and their descendants despite a history of faulty payments (Meyer, 1966; AD, I–V 2 Er 342).

The attitudes of paternalistic nobles had their counterparts in the First and Third Estates. Rural society was one in which wealth in itself brought no prestige and in which economic man had not yet emerged. This was most evident in attitudes to charity. The parish clergy was at the centre of charitable activity in Upper Brittany. The parish priests were the only tithers in one-third of the parishes and the principal ones in another third. This meant that in many cases one man controlled close to 10 per cent of all the production of his parish and a great deal of it was redistributed to the poor, old and sick in the form of food or medical supplies. Priests were also the conduit for much of lay charity, because they were held to be responsible for stimulating the consciences of the rich and organising the distribution of the derisory sums set aside by the government. Such exhortations did not always work; an examination of charitable expenditures in account books shows that the volume of complaints about the niggardliness of seigneurs, urban bourgeois and large ecclesiastical tithe-owners was often justified (Sutherland, 1981). But the clergy also worked in a milieu in which the laity still saw the poor as somehow tinged with saintliness and as being worthy objects of a sometimes ambivalent compassion. Thus the poor still received alms at weddings and the thefts of a poor widow, for instance, were tolerated even if rarely forgiven, far more than those of unknown wandering beggars (Le Goff & Sutherland, 1974). An examination too of the criminal records of the Old Régime shows that an enormous number of disputes centered around defending insults to the family. The inference is that the circle of relatives was a major co-operative unit in which the more fortunate members were expected to aid the lesser, a fact which can be easily verified in guardianship cases where orphans or semi-orphans usually found the required twelve members to fill the family council. In short, it was at least partially true that hard times could be overcome by co-operation. A society so poor had to limit the capacity of the strong to exploit the weak in order that the whole should survive.

Such efforts were not always successful and the economic effects of such values cannot be calculated. The extent of the misery in the countryside which so horrified such differently placed observers as Arthur Young and the royal intendant, Bertrand de Molleville, should not be underestimated. Nor was everyone equally protected. The death rates among infants and children and the propensity of the poor to migrate in greater numbers than the well-off, among other examples, show there were severe limits to good intentions. But the existence of brakes on ruthless exploitation, whether they existed in the form of deferring financial obligations or of charity did mean that rural society did have some defences against poor harvests and bad weather. Short of a meteorological disaster, peasant communities were not entirely passive victims of their environment.

References: printed

Deyon, P. (1967). *Amiens, capitale provinciale. Etude sur la société urbaine au 17e siècle.* Paris: Mouton.

Fairchilds, C. C. (1976). *Poverty and Charity in Aix-en-Provence, 1689–1789.* Baltimore: The Johns Hopkins University Press.

Goubert, J.-P. (1974). *Malades et médecins en Bretagne, 1770–90.* Paris: Librairie Klincksieck.

Goubert, P. (1960). *Beauvais et le Beauvaisis de 1600 à 1730.* Paris: SEVPEN.

Hufton, O. (1974). *The Poor of Eighteenth Century France.* Oxford: Oxford University Press.

Jacquart, J. (1974). *La crise rurale en Ile-de-France, 1550–1670.* Paris: Armand Colin.

Lamb, H. H. (1977). *Climate: Present, Past and Future*, vol. 2, *Climate History and the Future.* London: Methuen.

Le Goff, T. J. A. & Sutherland, D. M. G. (1974). The Revolution and the rural community in eighteenth century Brittany. *Past and Present*, no. 62, 96–119.

Meuvret, J. (1971). Les crises de subsistances et la démographie de la France d'Ancien Régime. In his *Etudes d'histoire économique*, pp. 271–8. Paris: Armand Colin.

Meyer, J. (1966). *La noblesse bretonne au XVIIIe siècle.* Paris: SEVPEN.

Sée, H. (1906). *Les classes rurales en Bretagne du XVIe siècle à la Révolution.* Paris: V. Giard & E. Brière.

Sée, H. (1930). Les essais de statistiques démographiques en Bretagne à la find de

l'Ancien Régime. In his *Etudes sur la vie économique.* Paris: Imprimerie nationale.
Sutherland, D. M. G. (1981). *The Chouans. A Study of the Social Origins of Popular Counterrevolution in Upper Brittany, 1770–1796.* Oxford: Oxford University Press (in press).
Vovelle, M. (1973). *Piété baroque et dechristianization en Provence au XVIIIe siècle.* Paris: Plon.
Young, A. (1969). *Travels in France During the Years 1787, 1788, 1789,* ed. J. Kaplow. New York: Anchor Books.

References: manuscript

AD = Archives Départementales d'Ille-et-Vilaine
4Bz: Juridictions seigneuriales. Inventaires après décès.
8B 57–60, 76–77: maréchaussée de Bretagne. Interrogatoires.
C 1547–52: toiles présentées aux bureaux de visite.
C 1653–55: états des produits des récoltes.
C 1735, 1737, 3912: subsistances.
C 4501–64: capitation. Rôles des paroisses.
$2C^{38}$ 233–41, $2C^{45}$ 245–69: registres d'insinuation du centième denier, bureaux de Saint-Aubin-du-Cormier et de Vitré.
2 Er 342: livre de marque ... du marquis de Piré.
3E: parish registers of Balazé and Noyal-sur-Vilaine.
L 516: statistique agricole, produit des récoltes.
L 1345: subsistances, district de Rennes.

19

Climatic stress and Maine agriculture, 1785–1885

DAVID C. SMITH, HAROLD W. BORNS, W. R. BARON AND ANNE E. BRIDGES

Abstract

The purpose of this study is to document the relationship between climatic stress and Maine agriculture during the period 1785–1885. The research focuses on Maine's prime agricultural area of the early and middle nineteenth century; present-day Kennebec County. To locate evidence of possible climatic stress a climate record for the region was reconstructed. Instrumental records for Maine locales during the nineteenth century were kept by amateur meteorologists and by various agents of federal and local government. The earliest records begin in the 1810s. For the period prior to 1820 there exist a number of qualitative weather diaries. These have been analysed by means of a content analysis technique to develop a quantitative record which is generally comparable to the later instrumental climate records. To supplement this climate information a record of changes in Maine agriculture has been compiled, utilising state and local agricultural society reports as well as federal records. To round out these records, local newspapers and diary materials have been surveyed. These data are used to examine the interrelationships between Maine climate and Maine agriculture in the century after 1785. During the early part of this period a slight amelioration in climate occurred and large population movements into the region took place. Fairly large scale commercial agriculture began. However, this initial development was never consolidated, and out-migration became an issue of considerable concern. The possible importance of climate as a factor which helps to explain out-migration has not previously been considered by historians. The preliminary results of our analysis suggest that variations in climate and contemporary perceptions of them did, in fact, play a significant role.

Introduction

Nineteenth-century Maine agricultural history is a story of failed potential and missed opportunities. The period might be described as one of social and political uncertainty as this area of somewhat marginal agriculture felt many pressures. Contemporary observers (and historians since have followed them) dealt extensively with the perceived problem of 'keeping the boys at home'. After an initial period of settlement, Maine suffered constantly from out-migration to the west and to the cities of the east, and the state never achieved its anticipated population. This shortfall is usually explained in terms of poor lands at home, the decline of lumbering, the availability of rich and fairly cheap lands in the west, and a seductive propaganda from other areas. In response, Maine leaders disseminated propaganda of their own, advocating better education, state support for agricultural colleges, subsidies on crops, and the use of new and exotic crops to provide the necessary profit margins to dissuade people from leaving the state. But still the flight persisted.

This chapter will suggest that, although much of this traditional explanation of Maine's difficulties was and remains correct, it omits one factor of possible major importance – the role of climatic variations and of perceptions of them. The climate of Maine throughout the period 1785–1885 was cool with short intervals of warmer weather. No major climatic shift was at issue. But the variation in the climate was prominent enough to hamper the rational selection of potentially profitable cash crops, because the warm episodes raised expectations of an agricultural boom which were dashed when the climate reverted to cool conditions. In this sense, climatic factors may have contributed significantly to the agricultural and demographic problems of Maine.

Climate reconstruction

Over the last two years members of the Maine climate unit have reconstructed a climate record for Maine, with emphasis on the prime agricultural area of Kennebec County, still the finest agricultural valley in the state. This research involved compilation of a number of instrumental records of both temperature and precipitation. Information was gathered from many different sources, including local newspapers, agricultural society reports, and reports kept by military establishments and scientific observers. The two longest records were kept at the town of Gardiner on the Kennebec River just south of present-day Augusta and at the city of Portland. These records began in 1836 and 1816 respectively and comprise monthly mean temperature and precipitation readings. For the period prior to available instrumental records (in this case before 1810), several long-term daily weather diaries were subjected to a content analysis technique (see Source Note, p. 463). They include diaries from North Yarmouth, Maine (Rev. Tristam Gilman), 1770–1807; Lebanon, Maine (Rev. Isaac Hasey), 1764–1809; and Hallowell-Augusta, Maine (Henry Sewall), 1783–1842. To supplement the temperature and

Fig. 19.1. (*a*) Annual temperature (in °F) and (*b*) precipitation (in inches) for selected Maine stations based on instrumental records.

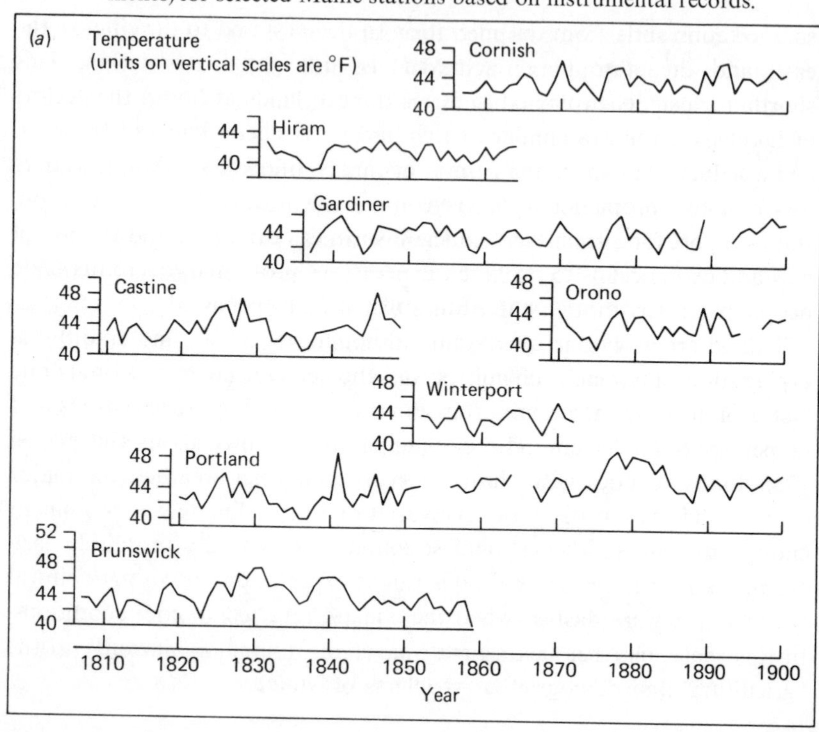

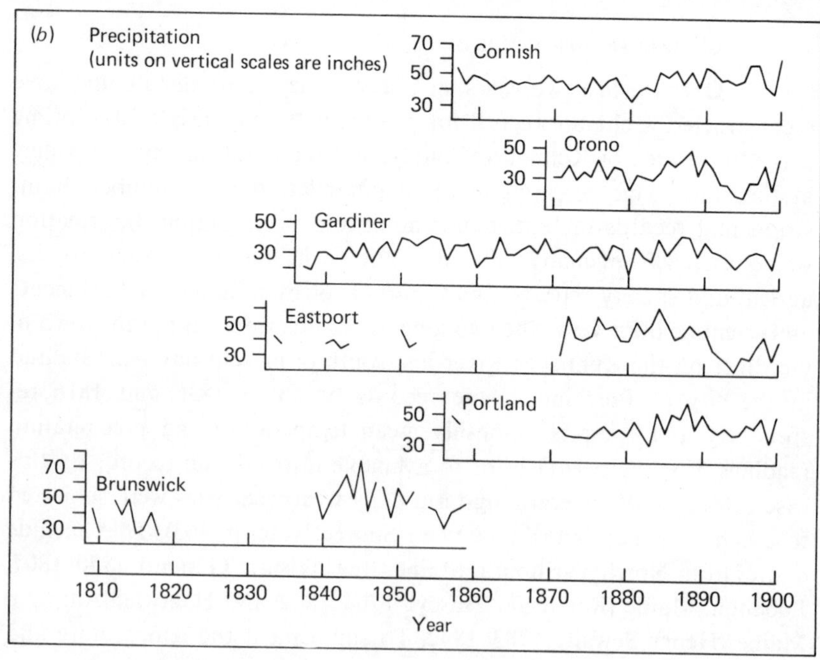

precipitation series, we compiled time series of ice-out and freeze-up data for major rivers, and analysed local records of the times when trees began to leaf, when corn began to show its spindles, and other such indirect indicators of climate.

Figs. 19.1–19.3 show the results of our compilation for southwestern Maine including the Kennebec region (i.e. the area from $68\frac{1}{2}°$–71° W and $43\frac{1}{2}°$–45° N). The series indicate temperature and rainfall peaks in the 1760s. The period 1775–1820, by contrast, was in general very cool and somewhat drier, though there were warm and dry years in 1784, 1786, 1792 and the early part of the first decade of the nineteenth century. In the 1820s an important change occurred when the temperature began to rise significantly. However, a decline set in in the early 1830s, with substantially cooler annual mean temperatures (down perhaps 3 deg F). There is also evidence, not shown here, of somewhat higher precipitation,

Fig. 19.2. Temperatures and precipitation estimates derived by content analysis of the diaries of Rev. Tristam Gilman and Rev. Isaac Hasey. Solid lines indicate estimates from Hasey, dashed lines estimates from Hasey based on incomplete data, dotted lines estimates from Gilman.

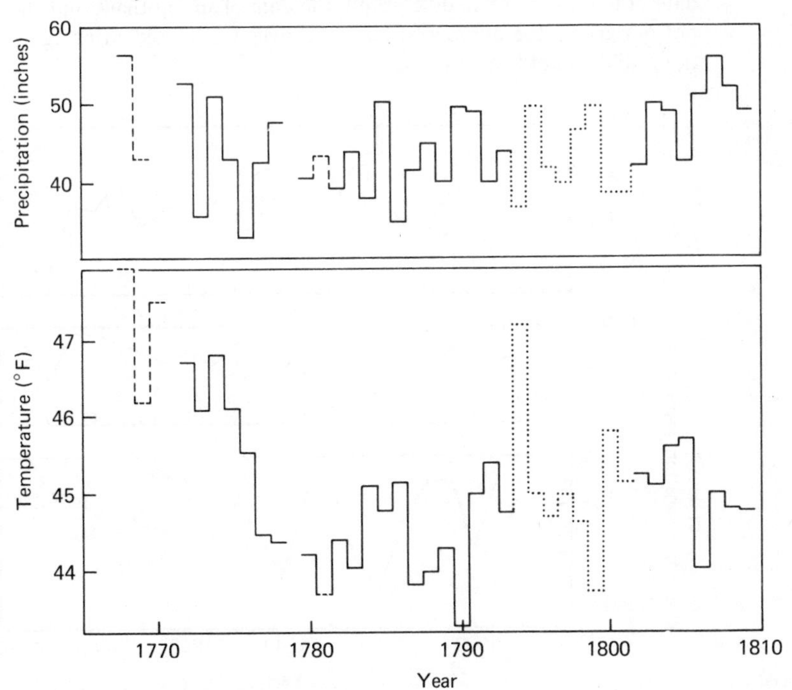

often occurring in the autumn harvest months. Precipitation continued to increase until around 1850, and conditions were relatively cool and damp until 1885. Thereafter, annual mean temperatures rose until the end of the century, while annual precipitation declined until around 1915.

Economic and social impact

One of the most prominent features of the economic and social history of Maine in the nineteenth century is the fact that the state, after the period of initial settlement, suffered constant out-migration. Population figures from 1850 to 1880 show an actual decline, and even before that the growth was less than would have been expected on the basis of earlier trends (Table 19.1). These demographic facts were symptomatic of severe economic problems. It is difficult to assess rigorously the contribution which climatic stress made to these problems, because many other factors operated to complicate the picture. However, analysis of newspapers and other records indicates that contemporaries

Fig. 19.3. Kennebec River freeze-up/ice-out. The top graph shows the date of freeze-up; the middle graph, the date of spring-thaw, and the bottom graph, the number of days the river was closed to navigation. All data are 5-year means.

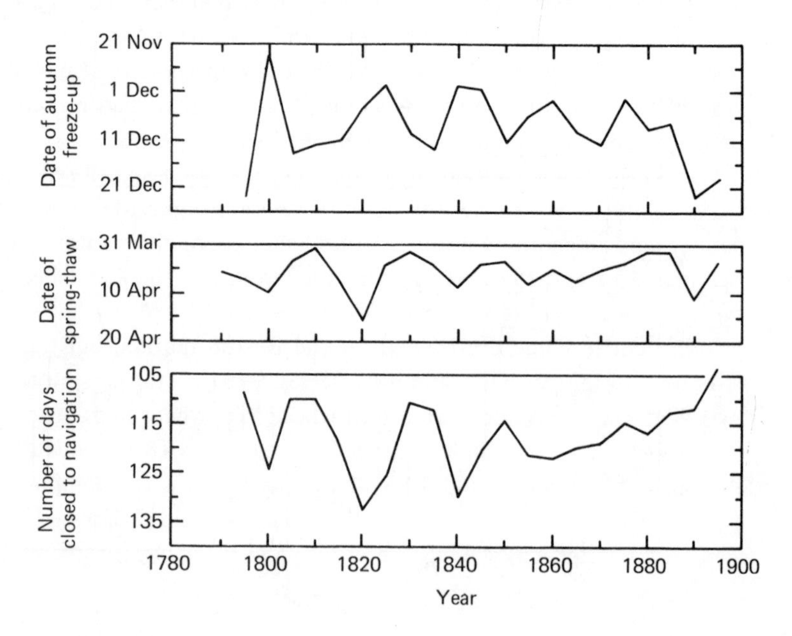

Table 19.1. *Kennebec County population 1830–80*

Year	1830	1840	1850	1860	1870	1880
Population	38 929	51 384	57 908	55 655	53 223	53 601

Source: The Maine Register, 1882.
Note: (1) Some small changes in the area of Kennebec County occurred in this period.
(2) Unfortunately there are at present no comparable figures available for the population of Kennebec County before 1830.

definitely perceived climate as a significant factor, a point ignored by previous historians. In the following pages we shall sketch the agricultural history of Maine in the nineteenth century in a way which emphasises this neglected element, but without attempting to gauge accurately just how important the climatic factor was relative to others.

It seems clear that throughout most of the period 1785–1885, climatic conditions for agriculture were significantly less favourable than in the recent past. Thus, at the beginning of the period of this study the growing season was at least 22 days shorter than the average for the period since 1940. Severe frosts were not unusual in May, and they occasionally happened in June (1778, 1794, 1797, 1800, 1807, 1816, 1817, 1832, 1833). In the autumn severe frost often occurred in September, and occasionally in late summer, August (1796, 1816, 1835, 1836). In these conditions farmers constantly worried about their crops. In the spring many seeds were planted indoors, and transplanted after the frosts had passed, while early frosts in late summer often damaged the harvest.

For the men who settled in Maine after the American Revolution (a heavy influx of Massachusetts military veterans came to Maine on war bounty lands, beginning about 1780), such cool conditions were 'normal' (see Figs. 19.1 and 19.2). At this time the farming of the Kennebec region was primarily for subsistence. Some farmers, especially in southern sections, cut hay and cord wood for the Boston market, but most crop sales were to lumbermen (often themselves part-time farmers) in the local area. The staple products were buckwheat, rye, wheat and barley.

This pattern of economic activity changed in the 1820s, which (as our climate reconstruction indicates) were marked by the temporary incidence of higher temperatures. As the warming trend progressed, diarists and others noted changes in the crops. The most usual remark during this period was that Indian corn (maize) was an assured crop, although it had been a marginal one at this latitude earlier. But more important than these actual changes was the fact that they raised *expectations* of even warmer conditions than were in reality forthcoming. The amount of comment from observers on climatic conditions at this time was substantial. Many felt that the warming presaged a very large increase in temperatures, and a number of 'explanations' of the changed conditions were developed to justify this sanguine view. For example, some argued that the cutting of the original forest cover was the reason for the temperature amelioration: in cleared land runoff was quicker and, because the sun's rays penetrated better, duration of snow cover was reduced. Farm newspapers, reports of the federal agricultural department, and books written about the area all discussed the new theories. Most felt that new crops were needed to meet these new conditions, as well as new techniques. Indeed the warmer temperatures of the late 1820s persuaded most farmers that the opportunity to grow more and better cash crops was at hand and large-scale commercial agriculture began to develop. A notable feature of this period was a silkworm craze, a fact that gives some indication of the extent of the hopes. Because expectations were raised so high, a deteriorating climate (or, more accurately, a return to a more 'normal' regimen) was bound to be felt more savagely when it did come.

Thus the period of cooler, more difficult weather from roughly 1830 to 1845 came as a great shock. Frosts occurred in low-lying areas in July 1831. Later, cool summers generally prevailed, and very heavy rains often fell in the autumn. These were followed by severe snow storms, high winds, and very cold winter weather. This cold weather, combined with increased snowfall, hampered transportation and severely damaged fruit trees. Springs were late and planting was often delayed over this period.

It must be recognised, however, that climatic stress was certainly not the only problem which faced Maine agriculture at this juncture. The difficulties posed by the weather were exacerbated by animal, human and plant diseases: in particular, sheep were affected, humans suffered from scarlet fever and cholera, and there were outbreaks of wheat blast or rust and early manifestations of potato blight. Some of these diseases may, in fact, have been partly conditioned by the climate; but other problems were entirely independent of the weather. Particularly important was the competition resulting from the construction of the Erie Canal (begun 1817, completed 1825) which permitted the transportation of produce from the rich agricultural lands of Ohio, Indiana and Illinois. Clearly, these non-climatic factors have to be taken into account when assessing the reasons for the chief symptom of economic stress, out-migration, which increased throughout the 1830s (despite being handicapped by excessive speculation in lands, the high cost of prairie farming, and the panic of 1837). But it may be significant that the steady stream of newspaper articles and leaders on the migration, which began to appear from 1836, focused considerable attention on the weather. Newspapers printed large numbers of summaries of weather diaries, lists of snow storms, cold temperatures, depths of snow, and other such melancholy data. Over and again the seasons were described as 'uncommonly cold and wet'. Some farmers began to discover cycles in the weather – the most prominent theories centering on a 17- or 19-year sequence. By 1838 several editors observed that for the past 7 years the Indian corn crop of most farmers had been severely curtailed, if not destroyed. More analysis of cyclic theories occurred throughout the 1840s, some of it based on references to diaries from more than a century before. Eventually, however, less attention was paid to the detail of the weather and climate and the cyclic theories were not so much discussed. Farmers were now apparently agreed that the warm weather of the 1820s was the unusual weather, and they devoted their efforts to attempting to stem the tide of migration, mostly by creating an agriculture that would deal with this reality of colder, wetter weather with its less secure crop returns. Even when the warm weather came at the end of the period, Maine farmers were still seeking solutions to their problems, and it was the new crops, coupled with developments in agricultural technology, that occupied their minds and their newspapers.

The process of adaptation was a painful one. We have observed that the warmer period of the 1820s had seen considerable agricultural development, cash crops succeeding the subsistence farming of the earlier period. Even in 1841 Maine raised substantial amounts of wheat (ranking fifteenth in the Union), barley (second), potatoes (second), hay (sixth), and wool (seventh in 1839 figures). This level of production was significant for a relatively small cultivated area. But, in the face of climatic stress and other problems, Maine agriculture suffered badly. By 1870 the leading crops were potatoes and hay, and the land was rapidly becoming an area of abandoned farms, already well into its period of rural decline.

The details of the decline make stark reading. In the early 1830s it was not unusual for farms to raise an impressive range of products. For example, in 1834 one particular farm produced hay, potatoes, apples, pears, parsnips, beets, carrots, onions, corn, beans, grapes, cabbage, cucumbers, cider, currants, and turnips, and the proprietor offered grape vines, mulberry trees, and other fruit trees for sale. But significantly, this farm was on the market by 1841. As far as the older (pre-1820s) crops were concerned, wheat production persisted longest, artificially encouraged by a massive newspaper campaign urging farmers to raise their own wheat as a measure of their independence from 'foreign' elements, and by a state-paid subsidy to wheat and corn growers in 1830s. However, by the end of the decade the legislature was no longer willing to subsidise this venture in a falling market now controlled by midwestern wheat producers.

The economic and demographic problems experienced in Maine stimulated a variety of responses. Some were irrational, idealistic, defeatist, or impractical. Agricultural societies in the area, seeking a scapegoat for the perceived agricultural decline and for out-migration in the 1830s, blamed mothers and daughters for not making a soft enough home, and young women for enticing their young men into lives of 'vice and extravagance'. Others called for the establishment of German and Swedish migrants on the abandoned and cutover lands, apparently feeling that these hardier people from northern latitudes might be able to make Maine pay.

A farmers' newspaper was begun in the heart of the region in 1833, with the object, according to the first number, of 'the mutual improvement of the Farmer and the Mechanic ...,' especially in the 'Siberian portion (as it has been churlishly called) of the Union'. The editor remarked that it was 'a time for patriotism as the Union was tottering', and the foundation must be 'an active, intelligent, enlightened and enterprising yeomanry'. (*Maine Farmer*, 31 January, 1833).

Others, less sanguine of the possibilities than these observers, were blunt. One newspaper report (*Franklin Register*, 4 April, 1844) remarked strongly that:

> No sophistry can prevent our arriving sooner or later at the bold position, that our agriculture is ruined, unless we can turn our attention within ourselves and learn to compete successfully on our own soil with the farmer in the fertile valley of the West!!

When the Maine State Board of Agriculture was created in 1856 in response to these problems, the secretary sent out a questionnaire of 30 questions dealing with farming methods, crops, and related matters. Question 30 was the key: 'How many in agriculture have emigrated since 1850? Why? and Where To?' According to the secretary, the results were unpleasant; and he said that, if they were true, Maine had better submit in good grace.

A more positive, but impractical, response was to seek exotic substitutes for products which could no longer be grown profitably. Among crops which were grown on an experimental basis were hemp (on which the state paid a subsidy for a brief period), sugar beet, sweet corn, flax, grapes, sunflowers, chestnuts, broom corn, maple sugar, winter wheat, silkworms (and their necessary mulberry trees of which more than 40 000 were planted), sweet potatoes, hops, blueberries, and cranberries. Cheese and butter making were also attempted.

Other methods used to alleviate economic and climatic stress included deeper ploughing, ditching, draining, liming, reclaiming swamp lands, and, of course, the universal panaceas, public education, school books, mandatory courses in soil chemistry, co-operative efforts, farmers' clubs, and finally the establishment of the University of Maine in the mid-1860s, with a state-supported experiment station by the mid-1880s.

Thus, many farmers attempted to save their lifestyle without perceiving or dealing with the real problems, i.e. the competition from elsewhere and the limitations imposed by the varying but generally harsher climate. However, some Maine farmers proved more adept at identifying and dealing with the challenges which faced them. The obvious solution was to choose a cash producer for which prevailing climatic conditions were not marginal. Eventually, when the incidence of potato blight diminished after the period of the Civil War, potatoes were to be the most successful crop. In the earlier period, the raising of beef cattle for market provided a major source of income. Cattle-raising was buffered from the impact of severe seasons because, if a poor season damaged the hay crop, farmers simply sold their cattle earlier.

A regular market was held each week outside Boston, at Brighton, where beef cattle and numerous other items of livestock and agricultural produce were sold at auction. This market, which dated from the late Revolutionary War days, provided most of the meat for the growing city of Boston. In earlier days, New England farmers drove their cattle overland to this market. By around 1820 regular groups of drovers began to purchase the cattle, drive them to a central location, sometimes fatten them, and then market the animals in the normal way. The autumn was an especially busy time as farmers disposed of the cattle they did not expect to be able to feed over the winter. The centre of the droving trade in Maine was in Kennebec county.

Fig. 19.4 shows the numbers of cattle sold at the Brighton auction each year from 1833 to 1852. It is difficult to tell exactly how many of these cattle came from Maine, but, until the railroad allowed economically viable competition from the west in the 1850s, there is good reason to believe that a considerable proportion came from this state. Maine products were often large and extra fat, commanding high prices. In 1836 nearly 10 000 cattle passing in droves were counted by one individual in southern Maine. In 1838 the local agricultural society reported that 'immense herds of stock and sheep have been sent to Brighton ...' In addition, some cattle from the western half of the area were driven to Quebec.

As potato blight became more prevalent in Maine, and rust damaged

Fig. 19.4. Brighton Market. Number of animals auctioned in thousands, 1833–52.

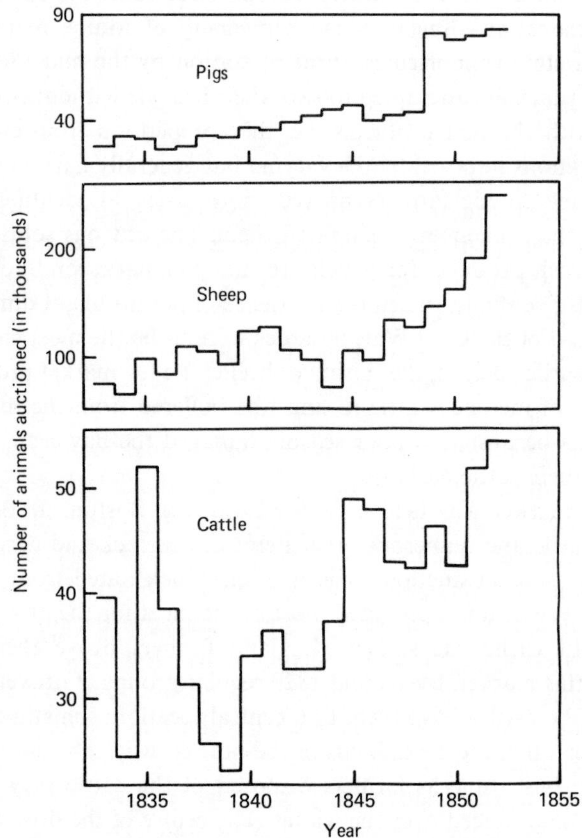

grain crops, farmers increasingly began to rely on hay and cattle as cash crops. At the Kennebec County Agricultural Fair in 1845, cattle were so prominent that Massachusetts observers travelled to see the exhibition. By the 1850s, however, the trade had diminished, probably largely as a result of competition from rail-freighted cattle from the west, but perhaps also influenced by warmer temperatures. Once the railway reached central Maine, in the 1850s, the trade improved slightly. Although the trade was quite active in the 1860s and 1870s, especially as farmers in northern Maine sold their cattle to itinerant drovers who moved the cattle to the central railroad depots for trans-shipment, it suffered a decline relative to its competition. The cattle trade had helped farmers in a time of need, but was not sufficient. As this trade declined absolutely, Maine drovers, especially in the west and north of the state, continued to do their work; but in central Maine, including Kennebec County, farmers changed to crops with specialty markets, such as apples, cranberries, butter, and milk, while farmers further to the north and east concentrated on potato growing (see above).

Conclusion

Farmers in Maine found themselves under pressure throughout the period in review. At first the climate was very severe, and farming was primarily of a subsistence nature. Later, as the climate ameliorated somewhat, population increased, and fairly large scale commercial agriculture began. Soon, however, the production of many crops began to decline. Traditionally, historians have ascribed this to the impact of western migration, better, cheaper land elsewhere, and the result of the transportation revolution begun by the Erie Canal and hastened by railroad communication with the west. Undoubtedly these were all important factors. But in Maine more farmers might have survived the competition if the climate had not varied markedly. In fact, the climate deteriorated at the very time the competition began to be strong. Recourse to new crops, and other efforts to improve the profit margin stayed the state's agricultural and demographic decline somewhat, but only temporarily. By the 1880s the state found itself suffering its worst farming period. Climatic conditions during the bulk of the nineteenth century were not suitable for the types of agriculture initially implemented in Maine, and the climate deserves consideration in the story of Maine agriculture in the period from 1785 to 1885.

Acknowledgements

Partial funding was received from Maine Agriculture Experiment Station, Quaternary Institute, History Department, and the Vice-President for Research and Public Service, University of Maine at Orono.

Source note

Maine was a part of Massachusetts until 1820, and remained a frontier zone throughout most of the period covered by this chapter. Accurate quantitative data for the area are very limited, though we hope to discover more as our work continues. In cases where quantitative material is lacking, it is necessary to fall back on qualitative sources. However, we recognise the limitations of the data derived from such sources, and to take account of them have presented the conclusions of the chapter in tentative form. It is hoped that further research will furnish more concrete data to substantiate the ideas suggested here.

The most important sources used in the preparation of this chapter are as follows.

The reconstruction of the climatic record

Instrumental series

Blake, J. F. (1873–4). Survey of Waldo County. *Annual Report of the Secretary of the Maine Board of Agriculture*, p. 195 (Table 1, Maine Temperature from 1852 to 1873 for Winterport).

Cleaveland, Parker (1687). Results of meteorological observations made at Brunswick, Maine, between 1807 and 1859. In *Smithsonian Contributions to Knowledge*, no. 204, ed. Charles A. Schott, pp. 2–25, 44, 253–4. Washington DC: Smithsonian Institute.

Gardiner, R. H. & Gardiner, F. (1865). Climate and season – meteorological tables. *Annual Report of the Secretary of the Maine Board of Agriculture*, p. 183 (Table 4, Mean temperature from 1837 to 1865 from records kept at Gardiner).

Goodnough, X. H. (1930). Rainfall in New England. *New England Waterworks Association Journal*, **44**, no. 2, 164–5, 168–70, 176–7.

Stevens, James S. (1907). *Meteorological Conditions at Orono, Maine*. Orono: University of Maine Press, Studies, 1st series, **7**.

United States Department of Agriculture (1866). *Report*, pp. 611–40.

United States Department of Agriculture (1867). *Report*, pp. 429–62.

United States Department of Agriculture (1868). *Report*, pp. 612–14.

United States Department of Agriculture (1869). *Report*, pp. 638–79.

United States Department of Agriculture (1870). *Report*, pp. 621–65.

United States Department of Agriculture (1890). *Report*, p. 188.

United States National Archives of the United States, Climatological Records on Microfilm. T-907, rolls 200, 201 (1856–92); roll 205 (1869–89).

United States Weather Bureau (1896–1900). *Climatological Data, New England Section*. Washington DC: US Government Printing Office.

United States Weather Bureau (1930). Section 85 – Maine. *Climatic Summary of the United States*, p. 7. Washington DC: US Government Printing Office.

Willis, Henry (n.d.). MS. The Weather Register, 1856–1864 (for Portland, Maine). Orono, Maine: University of Maine Climate Project Office.

Temperature and precipitation series based on diary observations

The series are based on the analysis of Rev. Tristam Gilman, MS diary, 1770–1809 (North Yarmouth, Maine), Maine Historical Society, Portland, Maine; Rev. Isaac Hasey, MS diary, 1764–1809 (Lebanon, Maine), Maine Historical Society, Portland, Maine. The content analysis techniques applied to these diaries are discussed in W. R. Baron, Tempests, Freshets and Mackerel Skies; Climatological Data from Diaries using Content Analysis, Ph.D. thesis, University of Maine, Orono, 1980.

Kennebec River freeze-up/ice-out data

Gardiner, R. H. (1859). Observations on the opening and closing of Kennebec River, Maine. *Annual Report of the Smithsonian*, 35th Congress, 2nd Session, Senate Miscellaneous Document, No. 49, pp. 434–6.

Maine Board of Agriculture (1865). Climate and seasons; meteorological tables. *Annual Report of the Secretary of the Maine Board of Agriculture*, pp. 181–2 (Table 3, Opening and closing of the Kennebec River at Gardiner, Maine, from 1785 to 1865).

Maine Board of Agriculture (1892). Climate and seasons; meteorological tables. *Annual Report of the Secretary of the Maine Board of Agriculture*, pp. 213–16.

Economic and social impact

The principal sources available for studying the impact of climatic stress on the society and economy of Maine are local newspapers and journals. The most valuable is the *Maine Farmer* (1832–1920). This newspaper was published in the geographical centre of the area discussed and was edited from its founding until 1865 by the foremost student of Maine agriculture in the nineteenth century, Ezekiel Holmes. This newspaper and others from the area that focus on Maine agriculture, including *The Christian Intelligencer* (Gardiner, 1831–36) *The Sandy River Yeoman* (Farmington, 1831–32), *The Yankee Farmer* (Cornish and Portland, 1838–40), *Drew's Rural Intelligencer* (1854–58), *Farmington Chronicle* (1845–75), *Franklin Register* (1840–42), *The Farmer and Artisan* (1852–53), *Aroostook Valley Sunrise* (1867–68) and *Eastern Argus* (Portland, 1860–90), furnish information on all aspects of Maine history. Rigorous analysis of their contents reveals that the central focus of Maine history throughout the period was farm abandonment, crop change and out-migration (for further discussion of this idea see D. C. Smith, Towards a theory of Maine history, in *Explorations in Maine History*, ed. Arthur Johnson, Orono, 1970, pp. 45–74). For those interested, we are willing to provide detailed references to information on climate, climate change, weather, crop selection and out-migration to be found in nineteenth-century newspapers and periodicals.

In addition the study utilises a number of documents both from the federal government and the state of Maine. Examples include Moses Greenleaf, *A Survey of the State of Maine*, Portland, 1829; J. C. Gray, Remarks on the climate of New

England, Commissioner of Patents, *Report*, Washington, 1854; United States Department of Agriculture, *Fluctuations in Crops and Weather, 1866–1948*, Washington, 1951; Maine State Board of Agriculture, *First Report*, Augusta, 1856. The reports of the Maine Board of Agriculture and the federal agencies have been analysed throughout the period.

Further information relevant to the theme of this chapter may be found in D. C. Smith, Maine and its public domain: land disposal on the Northeast Frontier, in *The Frontier in American Development*, ed. David M. Ellis, Ithaca, 1969, pp. 113–37; and D. C. Smith, *A History of Maine Lumbering, 1860–1960*, Orono, 1972. Eugene Graves, *A History of Maine's Nineteenth Century Beef Industry*, MA thesis, University of Maine, Orono, 1967, provides data on the activities of the Brighton market which supplement the information derived from newspapers. More details of the meteorological data are given in W. R. Baron, D. C. Smith, H. W. Borns, J. Fastook and A. E. Bridges, Long-time series temperature and precipitation records for Maine, 1808–1978, *LSA Experiment Station Bulletin 771*, University of Maine, Orono (1980). The data for Brunswick given here (Fig. 19.1(a)) are a revised version of those in *Bulletin 771* (p. 22).

20

Droughts in India over the last 200 years, their socio-economic impacts and remedial measures for them

D. A. MOOLEY AND G. B. PANT

Abstract

Data on the principal droughts which affected India during the period 1771–1977 have been collected from historical and other sources. The meteorological, political and economic background which has an important bearing on the suffering of the people during droughts has been reviewed. Application of the Mann–Kendall rank statistic test and Swed and Eisenhart's runs test to the interval between years of successive drought shows that droughts occur randomly in time. The stochastic occurrence of drought was examined to find out if it followed any probability law. The probability of occurrence of drought per year is low. In this situation, we expect that a Poisson probability model may fit the number of droughts in a 5- or 10-year period. A Poisson distribution was fitted to the data and the goodness-of-fit was tested by Chi-square. The fit was found to be very good. Thus the number of droughts in a 5- or 10-year period is a Poisson distributed variable. The probabilities of droughts on the basis of the Poisson model could be used in planning funds to mitigate the suffering resulting from droughts.

These droughts had serious effects on the people and economy of India. Food prices rose beyond the purchasing power of the rural population and led to reduced food-intake, resulting in low vitality and a large number of deaths from epidemics. Shortage of food resulted in exploitation by anti-social elements and in riots. Each drought pushed the farmer deeper in debt. The heavy cost of famine relief prevented the government from attending to such important large-scale problems as the illiteracy of the masses and the need for agricultural improvement.

Strengthening of agriculture and the rural economy together with special attention to the programmes for drought-prone areas, food reserves and planning of funds to meet drought situations would afford more protection against drought events.

Introduction

Droughts are natural calamities which in India are generally caused by failure of the summer monsoon. These put a severe strain on the Indian economy. By and large the behaviour of the summer monsoon is beneficial to the economy. Occasionally, however, the monsoons behave so erratically that the country suffers a substantial shortfall of rain water, the elixir of sound economy. The purpose of this study is both (*a*) to ascertain whether the Indian droughts which result from aberrations in monsoon behaviour exhibit randomness or non-randomness and whether they occur in accordance with any probability law, and (*b*) to examine the impact of drought on society and the remedial measures initiated by governmental and other agencies. The study covers former India (i.e. present India, Bangla Desh and Pakistan) up to 1947 and the present India thereafter.

Information on principal droughts over India during the period 1771–1977 was obtained from the *Report of the Indian Famine Commission* (1880, 1898, 1901), Bhatia (1967), Srivastava (1968), a number of independent accounts of Indian famines, and reports contained in year books, newspapers, and official gazetteers of the various provinces. Wherever available, rainfall reports were also taken into consideration. The major source of information is the report of the First Famine Commission. This commission was set up in 1878, immediately after the devastating drought of 1876–77, and covered all provinces of British India and many native states. It collected detailed information in respect of all famines and scarcities due to drought during the period 1769–1878. The commission prepared a carefully considered series of enquiries, and each local government was requested to entrust collection of the required information to the officials most competent to give valuable and trustworthy replies, and to invite qualified non-official people also to cooperate in this task. The commission supplemented and completed the enquiries by visiting several provinces in turn and by personal communication with some of the best-informed local officials or other people of importance. They also examined many official and private documents. The information collected included narrated and recorded accounts of the state of the monsoon, the area and population affected by famine, prices of essential commodities, crop conditions, mortality rates, severity of the famine, and relief measures including remission of land revenue, gratuitous relief, and other activities undertaken by official and private agencies. Though the commission could not derive any objective criteria, it arrived at its assessment of principal drought by considering the different factors in their totality and not by fixing any particular numerical value of any of the drought-sensitive factors.

Bearing in mind the information on the principal droughts during the period 1771–1878, as listed by the First Famine Commission, principal droughts during the period 1879–1977 were assessed on the basis of all the available information. Rainfall records were used as cross-checks on historical and other accounts. Since information on famines has been taken as a primary source for deriving the years of principal drought, due care has been taken to ensure that only those events are included where drought was the primary cause of famine. It may be mentioned that an element of subjectivity exists in the assessment of the principal droughts; however, the assessment is based on an overall consideration of the fairly detailed information relating to some of the drought-sensitive factors.

Meteorological, political and economic conditions in India

We shall consider these conditions in view of the fact that heavy concentration of rainfall in a particular season, lack of political freedom, low level of development and the economy of the country all have an important bearing on the suffering of the people in famines resulting from drought.

(*a*) The Indian rainfall is highly concentrated in the summer monsoon season, May–October (Fig. 20.1). More than 90 per cent of annual rainfall occurs in this season for 68 per cent of the area of the country, which contains about 70 per cent of the total population. This population is extremely vulnerable to any failure or highly irregular behaviour of the monsoon.

(*b*) The period 1760–1858 was characterised by marked political instability when the East India Company, a British trading company, progressively took over different areas of India. The British Government took over the administration of India in 1858 and the country only became independent in 1947. Under the alien rule there was little scope for native economic development of India.

(*c*) Extreme poverty afflicted the masses. According to Dada Bhai Naoroji (1900) the annual *per capita* income in British India towards the end of the last century was £2 (Rs. 30). A large section of the Indian population remained insufficiently fed.

(*d*) According to Baird Smith (1861) oppressive land revenue demand was the direct cause of famine in India. Dutt (1900) has also mentioned that the chronic poverty of the farmer was caused by over-assessment of land revenue. The peasant had often to sell or mortgage his land to pay the revenue (Srivastava, 1968:355).

(*e*) The composition of the population of the British Provinces was: agricultural 56 per cent, traders 18 per cent, labourers 16 per cent, professional and service 10 per cent (*Report of the Indian Famine Commission*, 1880: section 31). Labourers were mostly employed on land and many of the traders and artisans had smallholdings, so that a very large section of the British Indian population was dependent on agriculture for its livelihood.

(*f*) The Indian textile industry, which had done very well up to the first

Fig. 20.1. Normal monsoon (May–October) rainfall over India expressed as a percentage of the annual normal rainfall (based on data for 1901–60).

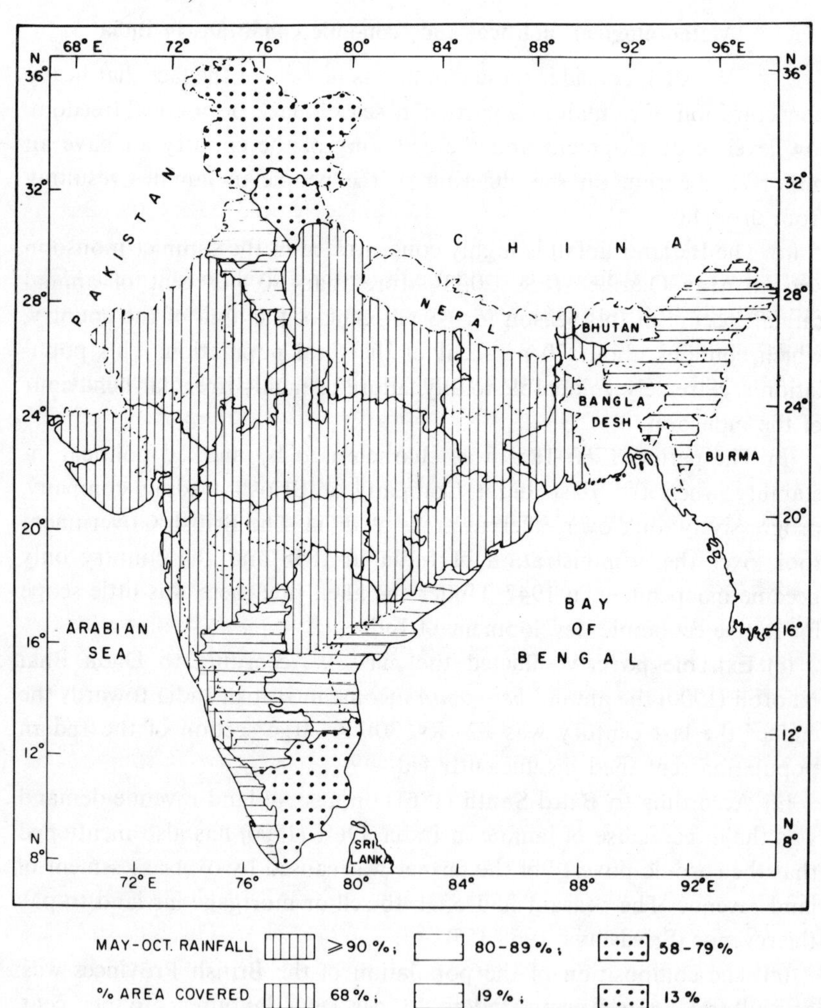

Table 20.1. *Indian exports and imports of cotton goods (millions of rupees)*

Year	Export	Import	Year	Import
1813–14	5.29	0.09	1840	38.6
1815–16	13.15	0.26	1850	52.2
1816–17	16.59	0.32	1856	63.0
1819–20	9.03	1.58	1859–60	96.0
1824–25	6.02	5.30	1864–65	110.0
1830–31	0.85	6.01	1869–70	135.0
			1874–75	163.0
			1879–80	169.0

Sources: Srivastava, (1968:11); Bhatia (1967:17, 37).

quarter of the nineteenth century, was crippled by 1833 as it could not stand the competition of the new techniques employed by the British textile industry (Bhatia, 1967:16). Table 20.1 illustrates this. These industrial workers then turned to agriculture for a livelihood, thus adding substantially to the already high percentage of the population dependent on agriculture.

(*g*) Despite the large proportion of the population dependent on agriculture, very little effort was made towards agrarian improvement until the end of the nineteenth century.

(*h*) The Government had no policy of procuring and storing food grains. However, up to the middle of the nineteenth century, farmers in northern India used to store 6 months' stock in pits.

(*i*) Bhatia (1967:23) has shown that agriculture had to bear the cost of India's administration since revenues from customs duty and income tax were very small. Moreover, interest had to be paid on the British investment in the Indian railways, which remained a losing concern till the end of the nineteenth century, and also on India's public debt held in England. Payment also had to be made on salaries, furloughs and pensions of the British civil and military personnel employed by the Indian Government. In order to make these payments export surplus had to be maintained and this was achieved by exporting agricultural produce, i.e. food grains, cotton, jute and opium (Table 20.2). The export of food grains continued up to the First World War. The movement of exports and imports was very much facilitated by the rapid expansion of railways from 1870 and the opening of the Suez Canal in 1869.

Table 20.2. *Total Indian exports and imports (millions of rupees)*

Year	Exports	Imports
1859–60	289 (188)	407 (116)
1864–65	695 (548)	495 (132)
1869–70	535 (359)	469 (162)
1874–75	580 (355)	443 (195)
1879–80	692 (396)	528 (196)

Source: Bhatia (1967:36–7).
Note: The bracketed figure under Exports gives the export of agricultural produce, i.e. food grains, raw cotton, raw jute and opium. The bracketed figure under Imports gives the import of cotton twist and yarn and cotton goods.

(*j*) Prices of food grains rose appreciably in the latter half of the nineteenth century (Table 20.3) but the wages of the agricultural labourer rose only slightly.

As a result of these conditions, the Indian rural population became very sensitive to price rises resulting from reduced food output even in a single isolated drought year, particularly in the nineteenth century and early part of this century.

Occurrence of principal droughts over India

There were 32 principal droughts (Table 20.4) resulting in famines in which a large number of people suffered hardship during the period 1771–1977. The table does not include 1974, since good monsoon activity continued in October of that year, and the seasonal rainfall deficit up to the end of September was removed, in a number of sub-divisions, by the end of October.

The question as to whether the occurrence of drought showed any significant non-randomness was examined by applying the Mann–Kendall rank statistic test for randomness against trend, and Swed and Eisenhart's runs test for runs above and below the median, as recommended in the World Meteorological Organisation's Technical Publications (1966a, 1966b), to the time interval between successive years

Table 20.3. *Mean prices of food grains in rupees per quintal (100 kg)*

Period	Wheat in NW provinces	Rice in Bengal	Bajra in Madrea	Paddy in South India
1821–30	3.88	2.10	4.25	2.33
1831–40	4.17	2.12	3.80	2.37
1841–50	3.45	1.76	3.47	1.81
1851–60	3.56	2.21	3.50	2.57
1861–65	4.88	4.27	4.65	4.38
1866–70	6.37	5.30	5.60	4.04
1871–75	5.85	5.18	4.15	3.12
1876–80	6.10	6.13	6.26	4.82
1881–85	5.55	5.37	3.99	3.35
1886–90	6.85	6.03	4.54	3.60
1891–95	7.40	7.40	5.62	4.31

Sources: Bhatia (1967:44, 141); Census of India (1951:170–1).

Table 20.4. *Years in which principal droughts occurred over India during the period 1771–1977*

1782	1783	1791	1802	1803	1806	1812
1823	1824	1832	1833	1837	1838	1844
1853	1860	1865	1868	1873	1876	1877
1896	1899	1907	1911	1918	1920	1941
1951	1965	1966	1972			

of drought. The latter test brings out whether the number of runs is too large or too small. If the number of runs is too large, it suggests oscillation and if it is too small, it suggests trend. If it is neither too large (i.e. not greater than the 95 per cent value), nor too small (i.e. not less than the 5 per cent value), the series can be taken to be random.

For the drought data, the value of the Mann–Kendall rank statistic is 0.103, whereas the significant value at the 5 per cent level is ± 0.245. Since the value of the statistic lies within these limits, it is not significant and the length of the interval between droughts can be taken to be random. Using the runs test, the number of runs above and below the median is 17. This lies between 11 (5 per cent value) and 20 (95 per cent

Table 20.5. *Goodness-of-fit of the Poisson probability model to the observed number of droughts in a 5-year period (1771–1977)*

No. of droughts in a 5-year period	Observed frequency	Theoretical frequency on Poisson hypothesis	Contribution to chi-square
0	17	18.41	0.108
1	16	14.72	0.111
2	8	5.90 }	0.002
$\geqq 3$	0	1.97 }	

χ^2 (chi-square) $=0.221$ (d.f.1).
$P(\chi^2>0.221)=0.67$ (d.f.1).
Note: d.f. = degrees of freedom.

value). Thus, neither significant trend nor significant oscillation is suggested, and the interval length can be taken to be random. The two tests suggest that the occurrence of drought is random in time.

Suitable probability model for the occurrence of drought

In order to aid the future planning of relief measures, we would like to know the probabilities of zero, one, two or three droughts in a 5- or 10-year period. We need, therefore, to find a probability model which shows a good fit to the observed data. The mean probability of occurrence of a drought year is low, so the Poisson probability model is expected to give a good fit to the number of droughts in a 5- or 10-year period. (Alternatively a binomial distribution model could be used, but we have chosen the Poisson model since the two give almost identical results.) The probability mass function of the Poisson model is given by

$$P(x)=\frac{e^{-m}m^x}{x!}, \text{ for } x=0,1,2\ldots,$$

where the parameter m is the mean obtained from the sample data, and x is the variable, i.e. the number of events in a specific period, say, 5 or 10 years.

The Poisson model was fitted to the data and the goodness-of-fit tested by chi-square (Tables 20.5 and 20.6). The fit was found to be very good. Thus the number of droughts in a 5- or 10-year period is a

Table 20.6. *Goodness-of-fit of the Poisson probability model to the observed number of droughts in a decade (1771–1977)*

No. of droughts in a decade	Observed frequency	Theoretical frequency on Poisson hypothesis	Contribution to chi-square
0	4	4.24	0.014
1	6	6.58	0.051
2	6	5.10	0.159
3	3	2.64 }	0.002
≧4	1	1.44 }	

χ^2 (chi-square)$=0.226$ (d.f.2).
$P(\chi^2 > 0.226) = 0.89$ (d.f.2).
Note: d.f. = degrees of freedom.

Poisson-distributed variable. On the basis of the model the probabilities of getting 0, 1, 2, 3 droughts in a 5-year period are respectively 0.449, 0.359, 0.144 and 0.038. The probabilities of 0, 1, 2, 3, 4 droughts in a 10-year period are, 0.212, 0.329, 0.255, 0.132, 0.046.

The theoretical probabilities on the basis of the Poisson distribution depend exclusively on the value of *m*, the mean number of droughts in a 5-year period. The mean for the 5-year period study has been obtained on the basis of 207 years (1771–1977) of data. There are 32 drought years in 207 years, so $m = 32 \times 5/207 \approx 0.77$. This value of *m* has been used to obtain the theoretical frequencies on the basis of the Poisson distribution for the 5-year periods, the first and the last 5-year periods commencing with 1771 and 1971. If the periods considered were 1772–1977, 1773–1977, 1774–1977 and 1775–1977, then *m* might change slightly and the theoretical probabilities of getting 0, 1, 2, ..., droughts in a 5-year period, on the basis of the Poisson distribution would change only slightly. The empirical probabilities may also change slightly, if the first 5-year period commenced some other year. These slight changes would not materially affect the result. The same argument applies to 10-year periods.

Socio-economic impacts of droughts

The droughts resulting from very deficient rains in some parts of the country led to reduced output in the affected areas. The reduced output, in the context of the prevailing Indian conditions, caused famine which affected the society in the following ways.

(*a*) The consequent rise in food prices in the affected area was 150–300 per cent above the prices prevailing in the area prior to the occurrence of drought. These high prices were beyond the purchasing power of the rural population. When we consider the All-India Index of food prices, the rise in the Index due to drought is relatively much less, being 20–60 per cent of that prevailing before the drought year (Fig. 20.2).

(*b*) A large section of the rural population had to subsist at a very low level of food-intake. This reduced substantially their resistance to epidemic diseases, notably influenza, cholera, plague, and smallpox, as was the case in 1877–78, 1899–1900 and 1918–19.

(*c*) Fodder shortage affected cattle-rearing, which had an important place in the rural economy.

(*d*) The policy of *laissez-faire*, i.e. non-interference in free trade, rigidly pursued by the government almost up to the end of the First World War, resulted in considerable exploitation by anti-social elements.

(*e*) Every drought and the resulting famine aided the moneylender by giving him a tighter grip over the rural population. According to the Famine Commission of 1880, one-third of the land-holding classes were deeply and inextricably in debt. Estimates of rural indebtedness in millions of rupees are 450 in 1895 (according to Sir F. Nicholson), 3000 in 1911 (Sir Edward Maclagan) and 6000 in 1928 (Mr Darling) (Jathar & Beri, 1939). The heavily indebted farmer was totally unable to improve the productivity of his land (Deccan Riots Commission Report, 1878).

(*f*) Those who could get neither food nor employment resorted to looting food grains to satisfy their hunger. This resulted in riots.

Fig. 20.2. Index of food prices in India 1861–1940 (base 1861 = 100). Years following principal droughts are marked.

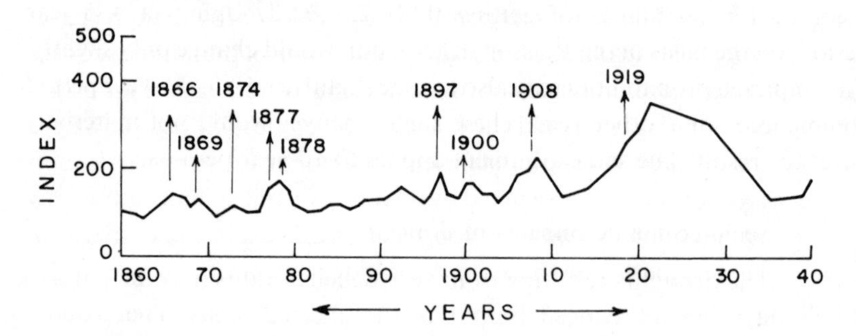

(*g*) Heavy costs were incurred by the government in starting and maintaining relief works to provide employment, poor-houses for disabled persons, and relief centres for those who deserved special consideration, and in giving tax remission to holders of agricultural land. The cost of relief became very high during the period 1870–1900, amounting to between Rs. 70 000 000 and Rs. 150 000 000 (*Report of the Indian Famine Commission*, 1880, 1898: Srivastava, 1968:221, 225). On account of the heavy cost of famine relief, no attention could be devoted to the important large-scale problems of illiteracy and agricultural improvement.

Remedial measures

Past and current remedial measures include the following:

(*a*) *Improvements in the communication network facilitating the movement of food grains to famine areas.* The railways, introduced in India in 1853, expanded most rapidly between 1879 and 1918 when the length of the rail network increased from 13 370 to 58 570 km (Sanyal, 1930). Thereafter, till Partition and Independence in 1947, the length increased slowly to 66 000 km. From 1947 to 1975, the length of the network in present India increased from 53 600 to 60 300 km (Rao, 1975). The metalled road length was 62 000 km in 1895, 80 000 in 1910 and 106 000 in 1947; thereafter, it improved considerably and was 493 000 km in 1975. (Bhatia, 1967:211; Hindustan Motors Ltd, 1968; Government of India, 1978).

(*b*) *Irrigation works to improve yields.* Even though irrigation works afforded greater protection from drought and famine conditions, and the Famine Commission 1880 had clearly pronounced in favour of higher priority for irrigation works than for the railways, the growth of irrigation in India was tardy until Independence. During the period 1878–1947 the irrigated area slowly increased from 14 per cent to 20 per cent of the area sown with crops. The factors responsible for the slow growth in irrigation were, a strong belief in official circles that the solution of the famine problem lay in putting more emphasis on the extension of railways than on the development of irrigation, and the financial difficulties of the government, which stood in the way of raising sufficient funds for investment in irrigation. While the railways were constructed by private investment under a system of government guarantee of interest, construction and extension of irrigation work had to rely mainly on the State. After Independence, the irrigated area rapidly increased to 31 per cent by 1975.

(*c*) *Improvements in agriculture.* Good progress has been made only during the last 20–25 years. Crop production is being increased by sowing high-yielding and drought-resistant varieties of crops and by using better fertilisers and more effective insecticides and pesticides. Since a major part of India has an annual rainfall of less than 100 cm, research on dry-land farming has been stepped up substantially by the Indian Council of Agricultural Research and useful research results are being implemented to boost production.

(*d*) *Strengthening of the rural economy.* Action is being taken to step up the development of village industries. Cattle have an important place in the rural economy and efforts to save them by transporting fodder to the famine area are being made by the State governments.

(*e*) *Greater attention to drought-prone areas.* Drought-prone areas constitute about 20 per cent of the total area and contain 12 per cent of the population. The Government of India initiated a Drought-Prone Areas Programme in 1970–71 (*Times of India*, 1977). It has been in operation in areas which have been identified as chronically drought-prone, and its emphasis is on the restoration of a proper ecological balance in these areas. The important components of this aspect of the programme are the development and management of irrigation resources, soil and moisture conservation and afforestation, restructuring of cropping pattern and pasture development, livestock development, popularisation of dry-land farming, and development of small and marginal farms and agricultural labour. The Fifth Five-Year-Plan made a provision of Rs. 1 870 000 000 for this programme.

(*f*) *Provision to meet a drought situation.* From 1877 to 1878, a Famine Relief Insurance Grant of Rs. 15 000 000, to be raised by extra taxation, was provided in the country's budget for about 38 years, but this did not function satisfactorily (Srivastava, 1968:369–74). More valuable was the Indian Peoples' Famine Trust, formed in 1900 with the generous donation of Rs. 2 000 000 by the Maharaja of Jaipur, and in operation until 1978 (Srivastava, 1968:268).

A Pilot Crop Insurance Scheme recently finalised by the General Insurance Corporation of India on behalf of the Government of India (*Times of India*, 1979) would provide cover against all climatic risks and pest attacks. Under this scheme, the State governments would be co-insurers, sharing claim as well as premium to the extent of 25 per cent. Premium indemnity tables have already been prepared for twelve States and one Union Territory. A crop insurance scheme for a particular variety of cotton has already been introduced in certain parts of Maharashtra and Gujarat.

(*g*) *Decrease in rural indebtedness.* The situation has improved since Independence. The network of nationalised and co-operative banks has been spreading to rural areas and it is possible for farmers to arrange loans from these banks for a number of agricultural purposes. This situation is expected to lead to an improvement in the rural economy, and a better life for the farmers.

(*h*) *Building up stocks of food grains.* Food grains are being procured and food reserves are being maintained to meet emergency situations. Stocks are held mostly in cities. However, a scheme to establish a network of rural centres for storing food grains has been under the active consideration of the Government of India (*Indian Express*, 1979). Under this scheme each centre would serve a group of villages and food grains would be rushed to famine areas.

Conclusion

The occurrence of drought over India has been shown to be random in time. The number of droughts in a 5- or 10-year period is found to follow the Poisson probability law closely. Probabilities on the basis of this model could be used for planning funds to mitigate the hardships resulting from drought.

Droughts resulted in acute suffering for a large number of people and death among the affected population. They also added significantly to rural indebtedness.

Strengthening of agriculture and the rural economy, effective implementation of the Drought-Prone Areas Programme, building up stocks of food reserves, suitable crop insurance schemes and planning requisite funds for mitigating the hardships resulting from drought would enable the country to face effectively the challenge posed by droughts resulting from monsoon failure.

Acknowledgements

The authors are grateful to the Director of the Indian Institute of Tropical Meteorology for the facilities to pursue this study. They would like to thank Miss C. P. Ghosh for typing the manuscript.

References

Bhatia, B. M. (1967). *Famines in India – A Study in Some Aspects of the Economic History of India* (1860–1965), 2nd edn. New Delhi: Asia Publishing House.

Census of India (1951). Vol. 1, part I–A – Report, pp. 170–1. New Delhi: Government of India Press.

Deccan Riots Commission Report (1878). Report of the Commission appointed in India to enquire into the riots of 1875 in Poona and Ahmadnagar districts of Bombay Presidency. London: Blue Book, C-2071.

Dutt, R. C. (1900). *Famine and Land Assessment in India*, pp. 14–17. London Oxford University Press.

Government of India (1978). *India: A Reference Manual 1977–78*, p. 551. New Delhi: Ministry of Information and Broadcasting.

Hindustan Motors Ltd. (1968). *Road Transport in India: A Study*, the Department of Economic and Market Research. Calcutta: Abinas Press.

Indian Express (1979). Newspapers of 17 April and 27 June 1979.

Jathar, G. B. & Beri, S. F. (1939). *Indian Economics*. London: Oxford University Press.

Naoroji, Dada Bhai (1900). *Poverty and Un-British Rule in India*. New Delhi: Publication Division Government of India, (Indian edition, 1962).

Rao, M. A. (1975). *Indian Railways*. New Delhi: National Book Trust, India.

Report of the Indian Famine Commission (1880). Part I, Famine Relief, London: Blue Book C-2591, Part II, Measures of Protection and Prevention. London: Blue Book, C-2735.

Report of the Indian Famine Commission (with-evidence volumes) (1898). Calcutta: Government of India Central Printing Office.

Report of the Indian Famine Commission (1901). Calcutta: Government of India Central Printing Office.

Sanyal, N. (1930). *Development of Indian Railways*. Calcutta: University of Calcutta Press.

Smith, Baird (1861). *Report on North Western Provinces Famine of 1860–61* (dated 14 August, 1861).

Srivastava, H. S. (1968). *The History of India Famines, 1858–1918*. Agra: Sri Ram Mehra & Co.

Times of India (1977). *Directory and Year Book*, ed. Sham Lal. Bombay: Times of India Press.

Times of India (1979). Newspaper of 18 April 1979.

World Meteorological Organisation (1966a). *Climatic Change*, Technical Note No. 79, WMO, No. 195, Geneva.

World Meteorological Organisation (1966b). *Some Methods of Climatological Analysis*, Technical Note No. 81, WMO 199. Geneva.

21

The effect of climate fluctuations on human populations: two hypotheses

MARTYN J. BOWDEN, ROBERT W. KATES,
PAUL A. KAY, WILLIAM E. RIEBSAME,
RICHARD A. WARRICK, DOUGLAS L. JOHNSON,
HARVEY A. GOULD, AND DANIEL WEINER

Abstract

The relationships between climatic fluctuations, population dynamics and social vulnerability are examined in the Tigris–Euphrates Valley 6000 BP to the present; the Sahel AD 1910–74; and the US Great Plains AD 1880–1979. Two major hypotheses are tested. The *lessening* hypothesis states that societies are able to lessen the impact of *minor* climatic stress (events with a return period of the order of less than 100 years) upon the resident population and indirectly on the entire society. The *catastrophe* hypothesis states that the increasingly elaborated technology and social organisation that insulate a livelihood system from recurrent minor climatic stress do little to reduce and may *increase* the system's vulnerability to *major* climatic stress (defined as events with a return period of the order of more than 100 years).

Comparison of the effects of the two similar droughts of 1910–15 and 1968–74 in the Sahel revealed that the Sahelian peoples, except for the nomadic minority, experienced a *lessening* in the toll of human mortality in the recent drought. Lessening resulted from the region's increased dependence upon external aid and consequent changes in internal structure. Exaggeration of these trends in a future drought could bring the region to the verge of a system collapse.

The long-term population declines that mark the two major population cycles in the Tigris–Euphrates Valley (6000–2500 BP and 2500–750 BP) seemingly result from a sequence or combination of environmental stress, political–social instability, and an absence of technological innovation.

In the first 60 years of Great Plains history, *lessening* is strongly evident in the impact of agricultural drought on population change and society at local and regional levels. The trend is less clear during the last 40 years, and drought–wheat-yield relationships demonstrate no strong evidence to support the lessening trend in the last half-century. The 1930s drought, a major climatic stress, provides a test of the catastrophe hypothesis. Adjustments to minor climatic stress in the Plains were insufficient to insulate the region from the major climatic stress of the 1930s drought. Recent and full integration into the national social system (made possible by recent changes in American attitudes to government involvement in the economy and in disaster relief) averted a system collapse in the Plains. 'Catastrophe' took the form of a devolution of impact into the nation. Thereafter, the effects of *all* subsequent droughts were diffused to the national system.

Calculations of the range of impact of Plains droughts at the global level suggest that a recurrence of the 1930s drought could cause declines of 7–19 per cent of total wheat export in 1985. Whether a future Plains catastrophe would be absorbed by the nation or ripple into other economies less able to adjust to sudden loss of staples is a question of global concern.

Introduction

Evidence for the effect of climate stress on human populations is meagre and consists primarily of a few detailed studies, many historical anecdotes, and much speculation. The dynamics of the relationship of climatic fluctuation to human well-being and social change are still elusive.

Our research tests the relationships between climatic stresses, population dynamics, and social vulnerability. Three contrasting cases were selected: the Sahel, the Tigris–Euphrates Lowland of Iraq, and the US Great Plains. In the Tigris–Euphrates River Valley, six millennia of historical and archaeological data may make it possible to capture the range of fluctuations and interactions existing in an irrigation-based society. The century-long history of settlement of the Great Plains is rich in detailed documentary material. This makes possible careful analysis of the changing vulnerability of the system over time. The Sahel is the setting for an examination of climate and society in the context of different theories of development in the Third World.

Two major hypotheses are examined. The first, the *lessening* hypothesis, states that all societies are able, through their technology and/or social organisation, to *lessen* the impact of *minor climatic stresses*, defined as events with a return period of the order of less than 100 years. By being adaptive they will experience over time diminished effects from such minor climatic stress.

The second hypothesis states, however, that success in insulating a livelihood system from minor climatic stress does little to reduce, and may increase, vulnerability to *major climatic stress*, defined as an event with a return period of more than 100 years. In a partially closed livelihood system, poorly integrated with the world economy, such extreme climatic stress may cause a collapse of the system, evidenced by major population decline (a loss of more than 10 per cent of pre-stress population). In more open and enlarging systems, the effects of extreme events will be devolved or shared ever more widely, rippling into previously non-dependent areas or societies that may experience ill-effects far from the locus of climatic impact. This is the 'catastrophe' hypothesis.

The Sahel

The Sahel–Sudan climate and vegetation zone is the home of 30 million people. Between 1968 and 1974 the area experienced low rainfall years that led to drought, famine, and worldwide interest. A major debate developed as to the drought's causes and effects. Most scholars concluded that the affected population had become more vulnerable to the impact of drought than it had been in the past (referred to below as the 'worsening' argument). The people were the victims of a colonial and neocolonial international economic and technical order that had increased their dependency and reduced their self-sufficiency by decreasing the area devoted to food crops, by draining off the agriculturally important labour supply by migration, by creating technical conditions that prompted rapid increase in population and livestock numbers, and by adopting policies that favoured the small urban élites.

A minority of scholars contend that the people, with all their difficulties, saw less vulnerability to drought than in the past because the Sahelian nations called upon the conscience of the world for assistance (the 'lessening' argument). Extended families were not entirely dependent on vulnerable crops or herds. Medical care restrained the childhood disease epidemics that previously accompanied famine, and a rudimentary infrastructure and national organisation were available with international aid to assist great numbers of people. Only where governments failed to act (Ethiopia) and in extremely remote areas was there great loss of life.

Meteorological drought

To test the assumptions of the contending scholars an analysis was made of the rainfall of the last great drought in 1910–15 and of the recent one (Fig. 21.1). The rainfall in the more extreme year of the pair 1912/13 or the pair 1972/73 is shown as a percentage of the average rainfall for the period 1931–60, to yield a rough comparison of the magnitude and areal extent of rainfall deficiencies. In both droughts deficiencies persisted over six years. In both there were at times average or even good rainfalls, marked variations from place to place and great seasonal variability within any given year. There is some suggestion of a greater total short-fall of precipitation in the more recent drought, but the indications are that the single most serious drought year this century (1913) occurred in the earlier drought (Sircoulon, 1976). We conclude that the two droughts were reasonably comparable except for their spatial pattern. Drought affected sedentary farmers in the southern Sahel and Sudan zones (east and south) more seriously in the earlier drought, whereas the recent drought more seriously affected the nomadic pastoralists of the sub-desert and northern Sahel (north and west).

Crops and animal losses

Except for Senegal and Nigeria, the Sahelian countries are among the poorest in the world, and their statistical reporting services for crop production are limited and improvised, other than for certain export crops. Similarly, the actual size and composition of the national herds is never really known. In the recent drought, there was some urgency in developing estimates of crop and animal losses, and many were made (Table 21.1). Analysis of these estimates suggests that in the worst years estimated losses of one-third to one-half of the food crops and herds are reasonable and conservative.

Against this we must compare occasional reports from 1910 to 1915. For example, in Senegal in 1912–13 it was reported that ‘grains are beginning to become scarce in certain *cercles*’; ‘bovine epidemic disease, mortality from 33–42%’ (Pourafzal, 1978)[1]; in Nigeria, 1914, ‘the Wodaabe pastoralist Fulani of Western Bornu Province were estimated

Fig. 21.1. Drought in the Sahel: rainfall deficiencies for 1912–13 and 1972–73.

	WEST			CENTRAL			EAST		
	Stations	1912-13	1972-73	Stations	1912-13	1972-73	Stations	1912-13	1972-73
SUBDESERT 100-300mm				10 Tombouctou-Kabara 220mm	58	47			
SAHEL 300-650mm	1 Podor 336mm	38	33	11 Niamey 576mm	58	64	16 Bol 341mm	13	18
	2 St. Louis 347mm	43	43				17 Zinder 508mm	42	58
	3 Kaedi 410mm	66	32				18 N'Djamena 639mm	48	49
	4 Dakar 578mm	53	20				19 Maidiguri 650mm	54	66
	5 Tivaouane 615mm	39	24						
	6 Rufisque 620mm	54	20						
SUDAN 650-900mm	7 Thies 694mm	34	32	12 Segou 730mm	76	69	20 Sokoto 719mm	57	54
	8 Kayes 743mm	89	66	13 Ouagadougou 872mm	46	81	21 Kano 843mm	57	49
WOODLAND >900mm	9 Sedhiou 1379mm	66	61	14 Bobo-Dioulasso 1170mm	69	75			
				15 Gaoua 1190mm	64	73			

☐ Percentage of Mean Annual Rainfall (1931-1960) in Worst Year Either of 1912 or 1913 and 1972 or 1973

Larger deficiency shown in grey

Table 21.1. *Estimated single year crop and cattle losses in the Sahel 1968–74*

	Crop losses[a] (%) (worst years:1968–74/1967–68)			Cattle losses (%)	
	Millet/ Sorghum	Cotton[b]	Peanuts	EDF[c] 1972–73	Other[d]
Mauritania	77	—	—	45	25, 30, 42, 60–80
Niger	42	86	76	36	33, 45, 50, 60–65
Mali	42	+2[e]	19	34	16, 32, 40, 50
Chad	38	36	15	34	34, 40
Senegal	51	+20	43	25	10, 15, 20
Upper Volta	11	25	14	15	2, 7–8, 13, 30

[a] Data taken from Berg (1975).
[b] (1968–69 base year).
[c] European Development Fund Study cited in Berg (1975).
[d] Various sources.
[e] Plus denotes a gain.

at 10 000 with 88 000 cattle in 1913 and 5500 with 36 000 cattle in 1914' (Watts, 1978)[1]; and in Niger, 1913, 'millet production [was] insufficient for subsistence needs ... some areas had only half of the average and others none at all' and 'animal mortality: cattle, 1/3; sheep and goats, 1/2; and camels, negligible' (Pourafzal, 1978)[1]. These qualitative assessments could well be interchanged with similar reports during the recent drought. In both of the drought periods the Sahel–Sudan inhabitants faced losses of their food supply on the order of one-third to one-half. What were the consequences of these comparable losses?

Human impacts

A review of seven studies (Center for Disease Control Village Surveys (CDC), 1973, 1975; Kloth, 1974; Caldwell, 1975; Garcia & Escudero, in draft; Imperato, 1976; and Faulkingham, 1977) leads us to two conclusions. First, in the recent drought the death rate rose in the drought-affected regions of the Sahel–Sudan zone in selected years by 25–100 per cent over the relatively high pre-drought death rates of 25–40 per thousand, and, second, the widely quoted figure of 100 000 deaths (Imperato, 1976; Ware, n.d.; Swift, 1977) is probably conservative. Premature deaths, particularly of young children, extended over a population at risk of 10–15 million for several years could easily number two or more times the widely cited estimate.

A review of malnutrition surveys (CDC, 1973, 1975; Seamon, Rivers, Holt & Murlis, 1973; Garcia & Escudero, in draft; and de Goyet & de Ville, 1976) suggests that the high rates of acute malnutrition endemic among children (5–10 per cent) doubled in the severe drought areas and throughout the Sahel during the worst years of the drought.

The difficulties in estimating drought-related mortality are grossly magnified in the period 1910–15. A handful of references appear in the colonial reports. In Niger in 1914 estimates of mortality note '1/4 of the population', '20 000 people', 'approximately 80 000'; and there are reports of intestinal disease from eating wild foods. Infant mortality was reported 'very high' in Segou, Mali, 1914, whereas in Senegal in the same year there were reports of plague, but of no famine deaths (Pourafzal, 1978)[1] Recent histories extend these figures. The Comité Information Sahel (1974) notes for Mali, 1913, that it was estimated that '10% of the population died, a number that is surely an underestimate', and Bonte (in draft) comments on Niger, 1914, that 'the 1914 famine (called *Kakalaba* by the Nigerian Hausa) was among the most murderous and is engraved in the collective memory down to this day', and taken to be severer than the 1973 famine. Writing of Nigeria, 1913–14,

> Hastings, in a flight of fancy, suggested that in Kana province alone the death toll was on the order of 50 000. Polly Hill states that 'many thousands of people died in Kano ... and at least 4000 in northern Katsina.' For the rest we have to be content with vague colonial references to 'very high death rates in Daura and Zongo', 'considerable mortality in Kozaure' and a 'rather heavy figure for Gunel' (Watts, 1978).

Many of these observations on mortality are 'unreliable, spatially inconsistent, and almost wholly non-quantitative' (Watts, 1978). Yet where more reliable quantitative data are available they show totals which are absolutely higher for a comparable area than similar totals for 1968–74. Given that the population in 1968 was almost three times the 7.5 million recorded in the area of French colonial administration in 1910, relative losses must have been far higher in the earlier drought than in 1968–74.

The case for lessening

The case for lessening in the recent drought rests on the lower mortality rate in the face of a similar meteorological drought, and on similar crop losses (perhaps higher animal losses). Watts' (1978) report, which carefully integrates what is known about the Sudan zone of Northern Nigeria, concludes that:

> Superficial parallels should not serve, however, to cloud other more important structural differences. Firstly, there are some simple but important differences in magnitudes; in 1913 human mortality was enormous but in 1973 minimal; price inflation of grains occurred in both famines but was proportionally much greater in 1913 ...
>
> Informants who had experienced both famines invariably stressed that the crucial distinction between 1913 and 1973 was food availability – and by extension the mechanics of its genesis. In Katsina and Daura elders emphasized that in 1913, for quite extended periods, grain was simply not available in the market-place irrespective of whether purchasing power was available. In 1973 conversely, only in a few isolated areas did grain disappear entirely; grain was available, said most informants, but purchasing power congruent with prevailing prices was not.

The case for worsening

The arguments for absolute worsening focus on the pastoral peoples and particularly on the nomads of the Sahel. It is the view of most observers, supported by the nutrition surveys and the census of the refugee camps, that nomads proportionately suffered the greatest burden in the recent drought. The data suggest greater climatic stress in 1968–74 than in 1910–15, a worsening of the livelihood systems of $2\frac{1}{2}$ million nomads, and an increase in the impact of drought. The sparseness of references to nomads made by colonial administrators 1910–15 is consistent with this conclusion, although this inference must be qualified by the knowledge of the administrators' hostility to the nomads over whom they had yet to extend power.

If absolute worsening of the nomads' situation did occur, it was probably the result of the severe decline in the nomads' access to land and resources and in their political power. The process of semi-sedentarisation and the lower population growth rate of the nomads compared to that of the sedentary peoples resulted in a cutting in half of the proportion of the nomads relative to the total population and in a cutting back of nomadic territories into the driest areas. The problem was compounded by the entrenchment in the current governments of most Sahelian countries of leaders from the sedentary agricultural regions, a permanent shift in the former balance of political power that resulted from the defeat of nomadic military power by the French. The problem was made worse by the tripling of the region's population between 1918 and 1968.

The case for relative worsening – a permanent and increasing dependence on outside aid – rests on the increasing inability of subsistence systems to provide a livelihood for those dependent on them, and on the gradual monetarisation of all exchanges. The development of trade in food crops, the increased sales of land, the emergence of agricultural wage earning, and even a rise in the sale of prepared food are symptoms of incipient failure of the systems (Raynaut, 1977).

None of the scholars who argue for worsening asserts that the people of the Sahel were better off when more died from drought in an earlier era. One must infer that it is the relative worsening, the gap between what is and what might be, that causes their concern. They envisage a society that could provide greater social and subsistence security in contrast to one that at best is dependent upon the slim reed of international aid received at a great cost to the social fabric.

Weighing the evidence

For the majority of the Sahelian peoples, particularly the 13–15 million whose equivalent numbers did not exist in 1910, the recent great drought comparable to that of 1910–15 saw a lessening in the grim toll of human mortality. This is less clear for the $2\frac{1}{2}$ million nomadic peoples whose political, social and subsistence positions worsened in the 60-year interim, and whose suffering probably reflected this worsening. But there should really be naught for our comfort in this relative improvement. For even with overall lessening, the toll of mortality was large and the morbidity great, stunting and wasting the youth of the region.

The processes which have, in the past, led to a lessening of the overall susceptibility to drought, could, in the future have opposite effects. Lessening where and when it occurred in the recent Sahelian drought resulted from the region's increased dependence upon external aid and consequent changes in internal structure. This provided for increases in the availability of imported food, external opportunities for work, ease of movement, and advances in preventing infectious disease. But these same factors are implicated in the loss of villages, food self-sufficiency, the growth of human and animal population, and the breakdown of traditional controls on land degradation. Exaggeration of these trends in a future drought could bring the region to the verge of a fundamental, irreversible change in the ability of either the region's fragile resource system to provide for its inhabitants' sustenance or of the social system to provide alternative livelihood.

A simulation of the agricultural carrying capacity of the Sahel–Sudan region, based on a considerably improved and ecologically sound agriculture, found the region capable of supporting only 33 million people (Matlock & Cockrum, 1974). In that model of improved agriculture, improvement came with the loss of jobs for 60 per cent of the work force. Moreover, while sustenance increased considerably in the model's projection, it would be insufficient to meet the needs of a population projected by the end of the century to be 42 million (Caldwell, 1975).

Tigris–Euphrates Lowland

The Tigris–Euphrates Lowland provides a research opportunity different from that of the Sahel and Great Plains where the case for short-term lessening can be tested and demonstrated, but where catastrophe, depending on its definition, can only be speculated upon. In the Tigris–Euphrates Lowland, the sketchy data base, extending back over 6000 years, does not permit consideration of climatic events with recurrence intervals of 20–50 years as in the Sahel and the Great Plains. Thus, we can say little about possible lessening of the impact of minor climatic stress.

Theories of social change

The Tigris–Euphrates Lowland does, however, provide an extended record of population growth made possible by Adams' (1965, 1981) 6000-year record of population change derived from archaeological data (mainly the surface concentration and areal extent of potsherds). This record (Fig. 21.2) clearly indicates $2\frac{1}{2}$ cycles of growth and decline and allows us to consider how these changes correlate with major changes in the natural environment, society, governmental structure, and technology. By allowing this long-term examination, the Tigris–Euphrates Lowland fills a particular hiatus in environment–society studies related to possible differences in the frequency and phasing of environmental, as opposed to social and demographic events. The last several centuries have seen unprecedented demographic, social and technological change. In contrast, there have been relatively few (and certainly geographically restricted) major environmental changes. Thus to consider the interaction of environment and society more completely, we sought a region of

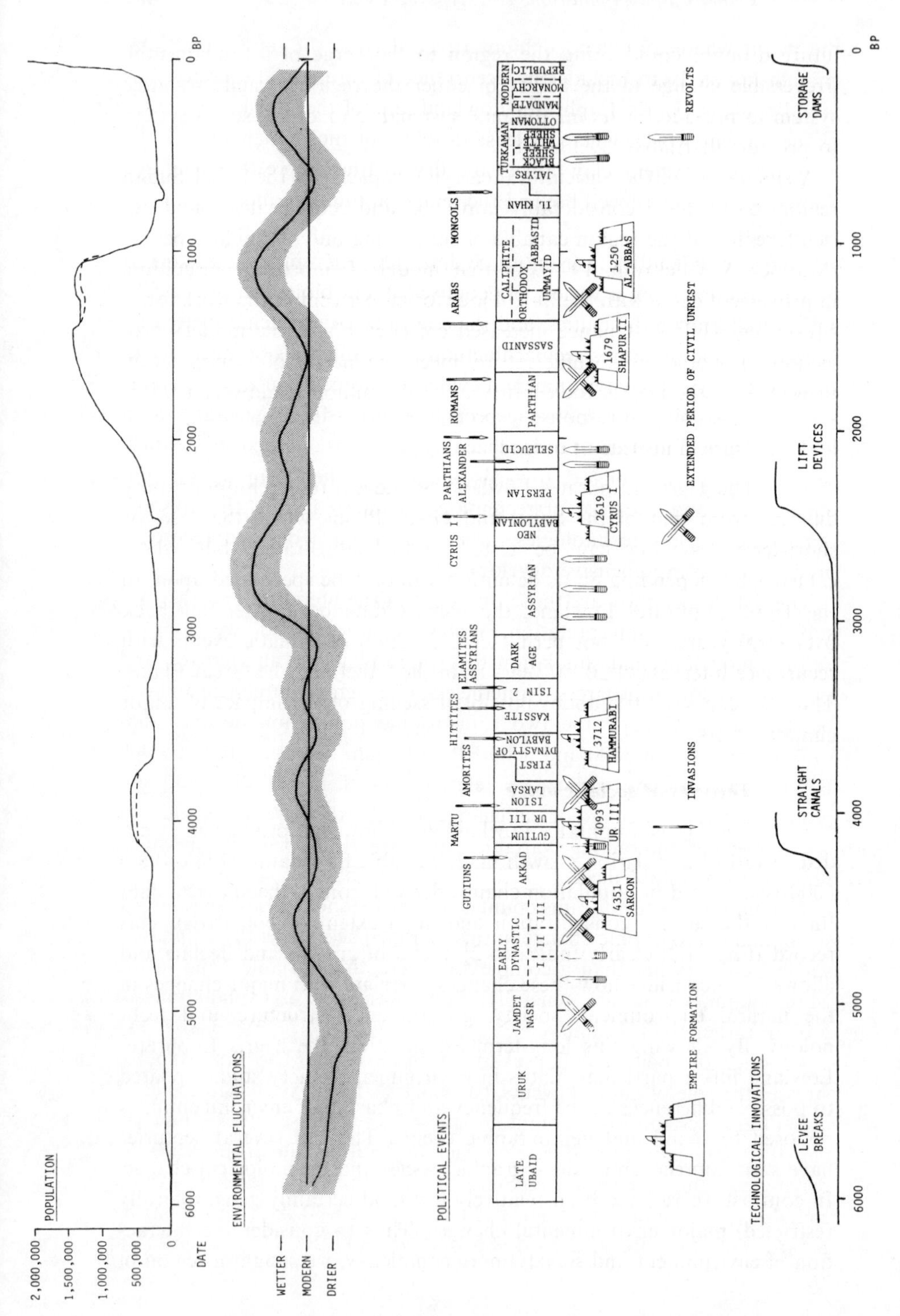

Fig. 21.2. Fluctuations in population, society, and environment in the Tigris–Euphrates Lowlands.

the world where an extended human record could be found, one long enough hopefully to include the occurrence of rare climatic events. We found it in the Tigris–Euphrates Lowland, one of the best known areas of the ancient world, where at least in terms of the human record, the coincidence of system, clay tablets and dry weather, serve to extend our knowledge of society well beyond the temporal limits of most regions of the world.

There are difficulties in using these data. The attempt to infer stream-flow fluctuation from a variety of geophysical and biological proxy data suffers well-known difficulties, and the contradictory evidence offered by such techniques captures secular fluctuations at intervals of a century at best. Not only is the environmental evidence imprecise on a human time scale, but also human response (especially if successful) is generally rapid and may pass unnoticed in the archaeological record. Moreover, political reconstructions are biased towards kingship rather than agricultural engineering, and the role of technology is inferred from the episodic appearance of key technologies rather than from a detailed record of gradual and comprehensive development.

The data we employ are those that exist in the literature and they are subject to varied interpretation. When we are puzzled, we hope to employ modelling and simulation to explore plausible alternatives, and this work is still in progress. Nonetheless, one significant conclusion can be demonstrated here: the two catastrophic population declines that took place between 3800 and 3000 BP and again between 1100 and 800 BP, do not correlate with any single recurrent factor of drought, war, political unrest, or technological change. Rather, they are the product of a complex of interacting variables, the exact importance of which remains to be established.

This initial conclusion runs counter to many standard explanations for social and demographic change in this region. These generally emphasise single factors rather than a complex of interacting variables, and in the Mesopotamian case, are generally better developed as explanations of growth rather than decline.

Theories of societal collapse are of three types. The first (type 1) stresses the role of natural events, particularly those external to the local agricultural system. Most commonly these theses utilise either long-term climate change or the adverse impact of natural catastrophes as the driving variable. Neumann & Sigrist (1978), for example, demonstrate that Mesopotamian barley harvest dates were 20–30 days earlier in the period 1800–1650 BC than for the period 600–400 BC. They suggest that the warmer, drier conditions in the first period encouraged over-irrigation and this resulted in soil salinisation that degraded the productive capacity of much of southern Mesopotamia. The more abrupt and catastrophic impact of violent climatic events, such as the floods produced by heavy precipitation in the mountainous headwaters of the Tigris and Euphrates, have been reviewed by Mallowan (1971) for the Early Dynastic period and by Sousa (1965) for the Islamic period.

Human causation external to the local system also plays a prominent role in some theories. Such hypotheses (type 2) often stress the adverse impact of invading nomadic hordes, or the depredations of neighbouring states, such as the Gutian invasions that ended in the final collapse of the Sargonid empire in 2159 BC (Bottero *et. al.*, 1967), or the *coup de grâce* administered to the Abbasid Caliphate and the population of Baghdad by Hulagu Khan in AD 1258 (Le Strange, 1900). Another related suggested explanation for population decline is as the result of disease often transmitted in the wake of military conflict (McNeill, 1976).

The best-developed theories of decline (type 3) examine the impact of factors internal to the Mesopotamian irrigation system. Most of these explanations emphasise the breakdown of managerial procedures. Thus, Redman (1978:235) points out the negative impact of excessive élite consumption practices and inept leadership among other factors. Development of these tendencies can lead to progressive deterioration of the irrigation system due to shorter fallow cycles (Gibson, 1974). This would compound the effects of siltation and salinisation (Jacobsen & Adams, 1958).

Except for the climate change/desiccation hypothesis articulated by Childe (1952) and Wright (1976), few growth theories give paramount place to environmental factors as promoters of social change. Rather, most theories identify societal developments as fundamental. Among them are Adams' emphasis on increasingly elaborate social organisation (1966), Wittfogel's (1957) insistence that the requirements of large-scale irrigation promoted the growth of centralising water-managing bureaucracies, Flannery's (1965) focus on trade and exchange mechanisms, Smith & Young's development of the population growth hypothesis (1972), and Redman's (1978) sophisticated attempt to construct an integrative systems–ecological hypothesis.

Of the factors involved in population growth and decline, four, natural environment, social and political events, and technology, are considered briefly.

Factors affecting population dynamics

For the agricultural societies of arid Mesopotamia, streamflow is a critical environmental variable. The temporal rhythm of the annual hydrograph is out-of-phase with the agricultural cycle. Peak flows occur in April–May, at the harvest season; minimum flows occur in August–October, when the second crop is growing. Unlike the Nile, there is no long historical record of environment in the Tigris–Euphrates Valley. Neumann & Sigrist (1978) derive growing season thermal conditions from barley harvest dates, but the data are few. If a history of streamflow is to be had, it must be inferred from the palaeoclimate of the basin as reconstructed from geological and biological proxy data. The Tigris and Euphrates rise in the mountains of eastern Anatolia and northwestern Iran. The melting of winter snowfall is the main source of water (al-Khashab, 1958). Information related to winter conditions in the head-waters must, therefore, be considered along with more direct proxy streamflow data.

Palaeoenvironmental data for the Middle East is of highly variable quality with respect to absolute dating, temporal resolution, and climatic interpretation. Fig. 21.3 exhibits some of the chronologies consulted in this study. The varve sequences (Kempe & Degens, 1978; Schoell, 1978; Lamb, 1977) offer temporal resolution on the scale of one year to one decade. The chronologies, however, are imperfectly fixed in time, and the climatic relationship is not well established. The pollen records (van Zeist & Woldring, 1978; van Zeist & Bottema, 1977; Niklewski & van Zeist, 1970; Beug, 1967; van Zeist *et al.*, 1975) are also poorly dated, and are not particularly sensitive environmental records. At best, temporal resolution is possible only to the scale of several centuries. Geological proxy evidence, such as marine sediments (Diester-Haass, 1973) and glacial deposits (Erinc, 1978) bear only inferred dates, and have temporal resolutions only of millennial scale. Fig. 21.3 shows a lack of synchrony in the timing of climatic episodes and in the direction of climate change (although a consideration of synoptic patterns indicates that these differences are not necessarily contradictory). With the data at hand, we can only derive the tentative and very generalised streamflow curve presented on the bottom of Fig. 21.3 and on Fig. 21.2.

Fig. 21.2 shows that there is no consistent relationship between the streamflow and population curves. Our hypotheses relate to the impact of extreme climatic events and so to climatic *variability* rather than to long-term means. A test of the hypotheses, using Mesopotamian popu-

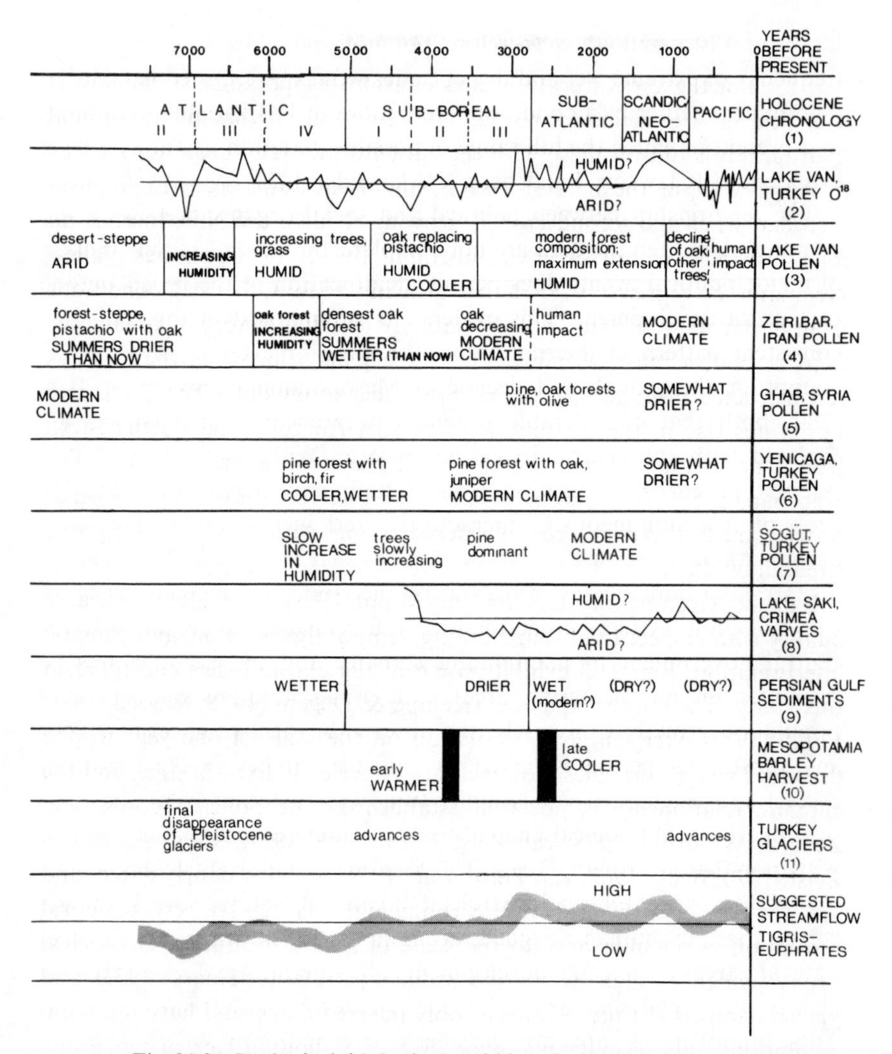

Fig. 21.3. Geological, biological and historical evidence for temperature and moisture conditions in the Tigris–Euphrates Basin, and adjacent regions, 7500 BP to the present. (Sources: (1) Wendland & Bryson, 1974; (2) Schoell, 1978; (3) van Zeist & Woldring, 1978; (4) van Zeist & Bottema, 1977; (5) Niklewski & van Zeist, 1970; (6) Beug, 1967; (7) van Zeist, Woldring & Stapert, 1975; (8) Lamb, 1977; (9) Diester-Haass, 1973; (10) Neumann & Sigrist, 1978; (11) Erinc, 1978.

lation and streamflow, is therefore not now possible: a record of the frequency of extreme wet and dry episodes would be required in order to determine whether past conditions were more or less variable than those of a present in which the full range of possible variation has not yet been experienced (Clawson, Landsberg & Alexander 1971).

The relationship between political and social organisation is equally complex. Although the data are not complete, the existing reconstruction of major political events does permit identification of the broad outline of societal development. This evidence is summarised in Fig. 21.2. No consistent pattern of internal empire building or external invasion accounts for the growth and decline of Mesopotamian population. It is clear, both that considerable private entrepreneurial activity was associated with the central irrigation bureau (Walters, 1970), and that population growth, irrigation construction, and bureaucratic activity took place simultaneously, interactively, and incrementally. This conclusion stands in contrast to those theories that stress the crucial role of a centralised bureaucracy in organising the labour required to construct the irrigation system. None-the-less, the role played by an efficient central government in maintaining stability and increasing access to resources on an increasing areal scale was an important aspect of population growth. Conversely, the absence of effective government for any substantial period is linked to reduced stability in the irrigation system and heightened vulnerability to adverse external events.

Similarly, technological change took place episodically, and in phase with population growth cycles. The four most important technological innovations were (*a*) small, artificial breaks in natural levees; (*b*) the substitution of human-designed, straight canals for the sinuosities of natural streams; (*c*) the development of waterlifting devices (*shaduf*, *na'ura*, later the *saqia* or Archimedean screw); and (*d*) storage dams. These innovations are not necessary preconditions for population growth. On occasion they serve to reinforce and stabilise other societal trends. However, they share two common characteristics. The first is their ability to bring into production new resources on a wider scale than those previously available. The second is the buffer they can provide for society against inopportune streamflows, particularly low flows.

Tigris–Euphrates conclusions

Given the reconstruction of the fluctuations in population, streamflow, society and technology at the scale of centuries rather than years, what might be said of lessening and catastrophe?

The environmental record does not allow us to test the hypothesis that societies develop in a way that allows the impact of minor climatic fluctuations to be lessened. In this region, population declines occur for reasons that as yet exhibit no consistent pattern. However, some sequential or concurrent combination of environmental stress, political and social instability, and absence of technological innovation seems required before a population's equilibrium is disturbed. Ultimately a new stability is achieved at a new, generally substantially lower population level.

Finally, we should note that, on this long time scale, catastrophe occurred (if at all) not with a bang but a whimper. Even the precipitous decline from the Abbasid period population of 1.5 million to the post-Mongol invasion low of one-tenth that number probably took three centuries. Measured on the human scale of generations, it was surely a time of troubles, possibly in Post's (1977) terms even a cascade of crises; but, by modern standards, it was a very long and drawn out collapse.

US Great Plains

The climate history of the North American Great Plains is punctuated by recurrent drought. Within the period of instrumental record major droughts have occurred in the region approximately every 20 years. Recent tree-ring analysis suggests a 22-year rhythm of widespread drought in the western US back to about 1600 (Mitchell, Stockton & Meko, 1978), although considerable controversy surrounds the notion of drought cycles. Within the period of agricultural settlement, droughts occurred during 1887–96, 1933–36 and 1952–56, and less important droughts occurred in the 1910s and 1970s. Dendroclimatological data indicate that the drought experienced in the Plains during the 1930s was the worst event in the entire 360-year record, a major climatic stress. The rarity of the 1930s drought is substantiated by simulation analysis of northern Great Plains moisture data (Eddy & Cooter, 1978) which assigns a 250–300-year return period to the 1930s drought in Kansas. Other droughts, minor climatic stresses, had a return interval of less than 100 years.

Ideally, the drought history of the Plains should be assessed in a consistent manner if the agricultural and social impacts are to be compared over time. In this vein, Fig. 21.4 presents a drought history for the US Great Plains as a whole. The plotted values represent growing-season accumulations of the number of climatic divisions exhibiting specified levels of the Palmer Index.[2] Thus, without specifying exact location within the area, these 'division-month' data represent one means of indicating relative drought severity by incorporating drought magnitude, areal extent, and duration. Peaks and troughs on the curve indicate maxima and minima in the overall extent of the drought: note in particular the extensive drought of the 1930s and 1950s.

Examining the lessening hypothesis

How did the Great Plains fare during each of the five major droughts? Our operating framework is to relate drought measures to agricultural production and these to societal impacts. We focus on annual yields of wheat, a crop which is sensitive to drought, important in world trade, and central to Great Plains agriculture.

Drought and wheat yields Average wheat yields per harvested acre for eight Great Plains states[3] were plotted over time, from 1890 to 1977 (Fig. 21.5). The number of abandoned acres was also plotted from 1929, when consistent data became available. A polynomial curve was fitted to observed yields in order to depict the trend over time. The departures from this trend provide a measure of the relative decline in yield during poor years.

Fig. 21.4. Great Plains drought area 1931–77; number of divisions exhibiting severe or extreme drought (≦ −3.00) conditions, based on Palmer Index values summed over 5 months, April through August. (Data source: NOAA)

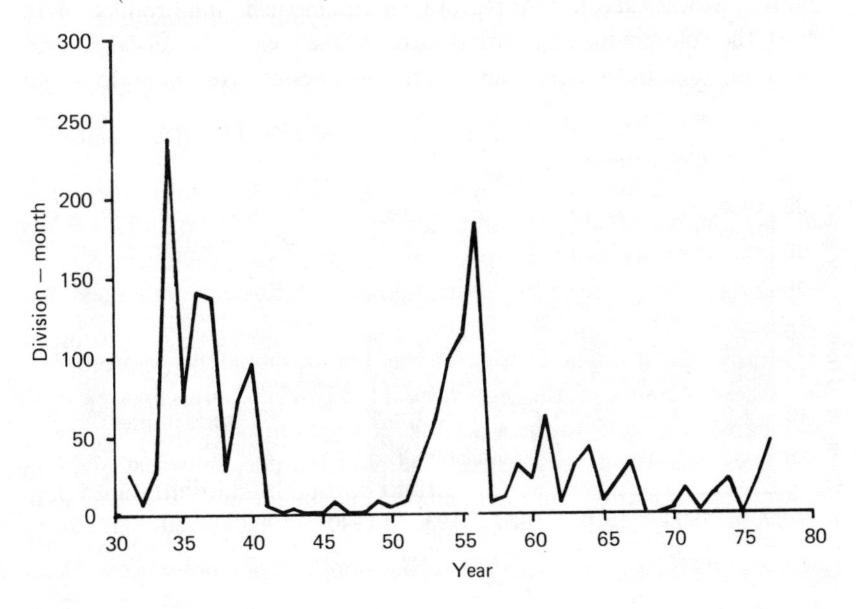

With each major drought occurrence since the 1890s, wheat yields were significantly reduced, reflecting the sensitivity of wheat to moisture conditions. There were other drops, such as that around 1950 due to rust. Since the 1930s, the relative impact of drought on yields, when defined according to the two worst years, has lessened over time (Table 21.2). However, after roughly normalising for severity by dividing the yield depressions by division months of drought for the years concerned, the relative impact of drought on yields appears approximately constant over time. On this basis there appears to have been no strong lessening of impact.

An extensive literature deals with climate–yield relationships based on regression-modelling using historical grain yield and concurrent weather data. The issues addressed relate to trends in expected yields: has there been a change, and, if so, which is responsible, technology or weather?

Newman (1978) concludes that relative variability in annual yields has been reduced, apparently due to technology. The United States Department of Agriculture (USDA) (1974) strongly concurs. McQuigg and others, writing for the National Oceanic and Atmospheric Administration (NOAA, 1973) argue forcefully that the apparent increases and reliability of yields in recent years have been due primarily to an unusual streak of very favourable weather, a point of view with which Schneider & Temkin (1978) and Haigh (1977) agree. The Institute of Ecology (1976) and Gasser (1976) claim that both technology and unusual weather account for recent increases in yields and production. In short, the voluminous literature is inconclusive and controversial regarding lessening. Both arguments have weaknesses. Newman does not

Fig. 21.5. Wheat yields in the Great Plains 1890–1972. (Date sources: USDA statistics)

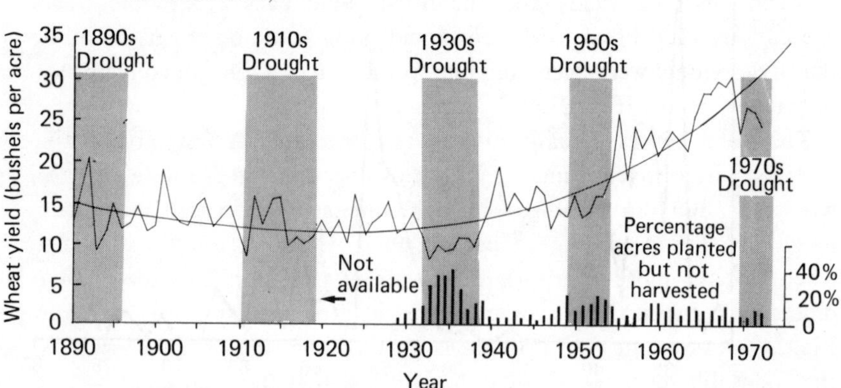

Table 21.2. *Relative wheat yield declines for historical droughts in the Great Plains*

	Drought decades				
	1890s	1910s	1930s	1950s	1970s
Percentage decline, from trend[a]	25	24	29	19	17
Percentage decline per division-month of drought (× 100)[b]	—	—	8	6	7

[a] Average of the two worst years.
[b] ≤ −1.00 Palmer Index summed over all climatic divisions during the March–July growing season.

account for persistent droughts. NOAA (1973) does not preclude the possibility that buffering technology also improved during the unusually favourable recent years, and both sides deal with a broad-scale analysis.

We dealt with these problems by employing smaller units of observation (state or crop reporting districts) and by including more recent drought events. Kansas was chosen for intensive study because it is the centre of winter wheat production in the Plains and because a number of droughts have occurred there since the 1950s. A continuous record of Kansas wheat yields from 1890 was assembled, and a third-degree polynomial curve was fitted to describe the trend. On the basis of the trend curve, years of 'good' yields and 'poor' yields were selected, being defined in terms of positive and negative residuals (i.e., the difference between observed yields and the fitted trend curve). Separate trend curves were then fitted to the good and poor years, on the assumption that if 'lessening' were occurring one would detect a relative convergence of the two curves (Fig. 21.6).

The trend curves actually diverge (see the table accompanying Fig. 21.6) because of overall increase in yields over time; in absolute terms, a worsening. But our notion of drought impact is concerned more with losses in relation to some expected 'good' yield, i.e., relative declines. Therefore, the more appropriate measure of yield-impact is the relative difference between the curves, expressed as a percentage (as indicated in Fig. 21.6). Assuming that drought is a major contributor to poor yields, the small differences in these values suggest that the relative impact of drought on yields has remained stable over time, i.e., neither lessening nor worsening.

Additional analysis employed a University of Wisconsin weather–wheat yield regression model (Michaels, 1978) constructed on the basis of climate and yield data obtained from 31 crop-reporting districts from 1932 to 1975. It incorporates a 'technology' trend, a habitat variable (a weighting to adjust baseline yields to differences in distinct growing environments) and an empirically derived weather component. The model explains 95 per cent of the variance in the wheat yields over 31×44 observations; the weather variables explain about 40 per cent of the variance remaining after the 'technology' and 'habitat' effects are factored out. As a way of testing for differences in sensitivity of yields to weather, a number of runs were made of the model, such that fewer recent years were included in successive runs (Table 21.3). If lessening had occurred then one might have expected the variance explained by climate to have decreased progressively from the 'older' years (1932–57). Clearly, lessening is not evident.

If, indeed, 'lessening' is not occurring, as our example indicates, the implications are far reaching. Generally, it is assumed that the arsenal of agricultural technologies, in particular the soil and water conservation practices so strongly encouraged since the 1930s, in addition to increas-

Fig. 21.6. Kansas wheat yield trends. (Data source: USDA)

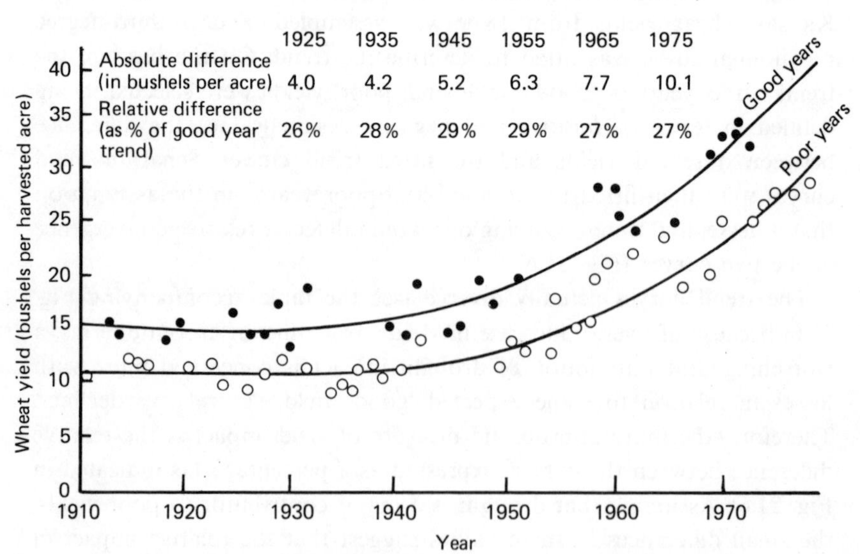

Table 21.3. *Wheat yield variance explained by climate by periods of increasing length*

Run years	Percentage variance reduced by climate
1932–57	46.1
1932–60	52.1
1932–63	48.5
1932–66	47.3
1932–69	45.7
1932–72	44.9
1932–75	45.2

ing yields and reducing production costs, have also been effective in buffering agricultural production from drought stress. Our data suggest that caution should be exercised in accepting such an assumption. This kind of knowledge of climate impact becomes critical as society faces the issue of which sets of management strategies to devise for future droughts on the Great Plains.

Impact of societal well-being Of the numerous indicators of societal well-being related to drought history in the Plains we first considered population migration. The alternation of wet and dry decades in the Great Plains presents a rare opportunity to estimate the relationship between climate and outmigration (displacement of population).

The severity of droughts within the dry decades was measured by summing growing-season division months with values of −3.0 and lower on the Palmer Index. The worst of the droughts was the 1930s, a major climatic stress. The other droughts were minor climatic stresses, in order of severity: the 1950s, the 1890s, the 1910s and the 1970s. Decennial census figures, the only population measures available Plainswide, reveal that the dry decades (1890s, 1910s, 1930s) experienced population losses and that the wet decades (1880s, 1900s, 1920s) saw population increases. However, the extent of population loss directly related to drought is consistently underestimated. This is because drought occurred in the middle of dry decades. Unaware that these were dry decades people flowed unabated into the Plains in the first three years before the onset of drought, and quickly resumed the inflow in the last three years of the decade after the drought was over. Thus, the population recorded by the census at the beginning of the decade *underestimated* the population level at the drought's onset, while census population figures recorded at the end of the decade *overestimated* the population level at the end of the drought. This is substantiated in those instances in which states (e.g., Kansas and Nebraska) collected or estimated intercensal population numbers. To achieve a more accurate estimate of population change during the drought we therefore used decennial census statistics to project the population trend of the preceding decade to the third year, and that of the succeeding decade back to the seventh year, and compared the difference, i.e.,

$$\Delta X(\%) = 100(X_B - X_A)/X_A$$

where

$$X_A = X_n + 0.3(X_n - X_{n-1})$$
$$X_B = X_{n+1} - 0.3(X_{n+2} - X_{n+1})$$

and where X_n means population at decadal year n (e.g., if $n = 1930$ then $n-1 = 1920$, $n+1 = 1940$, $n+2 = 1950$).

The declines in population between the third and seventh years are shaded for all climatic divisions in the decades of the 1890–1960 period (Fig. 21.7). Increases are unshaded.

The contrast between outmigration in the dry decade of the 1890s and the wet decade of the 1900s is remarkable. In the latter, only 3 of the 63 climatic divisions experienced any loss in population. Increase in population was general. Comparison of the minor drought decade of the 1910s and the wetter decade of the 1920s reveals a similar contrast, the more telling if it is noted that the population loss in the northwest Plains (Montana) in the 1920s is a delayed response to the late 1910s drought (1919–21) in that area. By the time of the 1950s drought, however, the pattern of population change is similar to that of the wet decades of the 1940s and 1960s.

Certainly the levels of loss in the 1950s are much lower than those common in the lesser droughts of the 1890s and 1910s. In the 1890s, the loss is intense in dry-farming margins of the Plains region, where many climatic divisions lost between one-half and three-quarters of their population and 6 of the 63 divisions experienced near total depopulation. The Plains were poorly integrated into the world economy and the people unprepared for drought. The pattern suggests complete failure of farming practice and a collapse of the system in the western Plains in the face of a minor climatic stress. There was also a major decline in population in the drought of the 1910s. This hit selectively in the western Dakotas/Montana and in the Dust Bowl sections of Texas, Oklahoma, Colorado, and Kansas, and produced system collapse and prompted population losses between 25 and 50 per cent over extensive areas. These were lands newly settled by inexperienced, under-capitalised farmers and losses were proportionately high (Bowden, 1977).

The impact of drought appears to be most immediate and, as measured by net migration, most severe in the wheat and cotton regions of the drier western margins. The stock-raising regions of the Flint Hills, Sandhills, and Black Hills respond to drought, but usually in a delayed fashion. Quite different are the diversified farming regions growing corn, wheat and other cereals, and fattening livestock in the Eastern Plains Transition Areas. Since the 1890s these wetter margins of the Plains have shown little response, in population numbers, to the presence or absence of drought.

The 1930s – major climatic stress The effects of lessening are well illustrated in a comparison of the impacts on population of the droughts of the 1890s and the 1930s. Although on meteorological grounds the drought of the 1890s was a minor climatic stress and that of the 1930s was a major climatic stress, the impact in the 1890s as measured by, for example, the area with above 25 per cent outmigration, was much greater. Nevertheless, it is as the one rare climatic event of the last century of Plains history that the major climatic stress of the 1930s interests us. Was the 1930s drought, with a return period variously estimated between 250 and 400 years, a 'catastrophe'?

Clearly, the Plains economy did not collapse in the 1930s even though a sizeable minority of writers in influential popular magazines and a majority of government spokesmen in 1934–36 seriously thought Plains agriculture to be permanently damaged and on the verge of collapse. In the Dust Bowl section of the southern Plains localised collapse did occur, as indicated by a major population decline – over 10 per cent and in some places over 25 per cent. This is supported in our analysis of Haskell County, Kansas, and Jackson County, Oklahoma, in which the scale of foreclosure of mortgages and of sales of tax-delinquent property for 1934–36 brought farming practically to a standstill that would have extended for four years or more, had not the state and federal governments enacted emergency legislation to extend credit to farmers and the length of the period of forgiveness for tax delinquency.

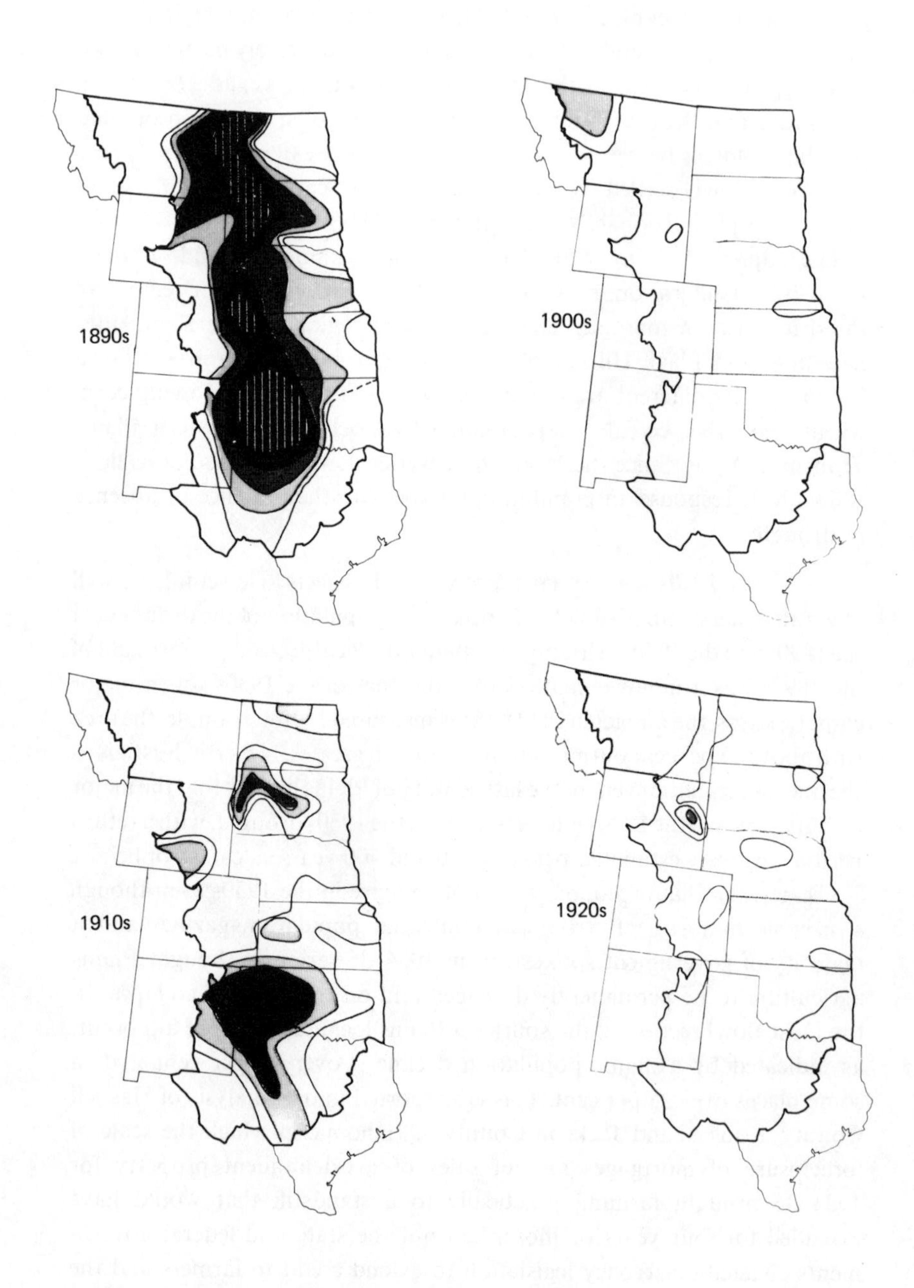

Fig. 21.7. Great Plains population declines.

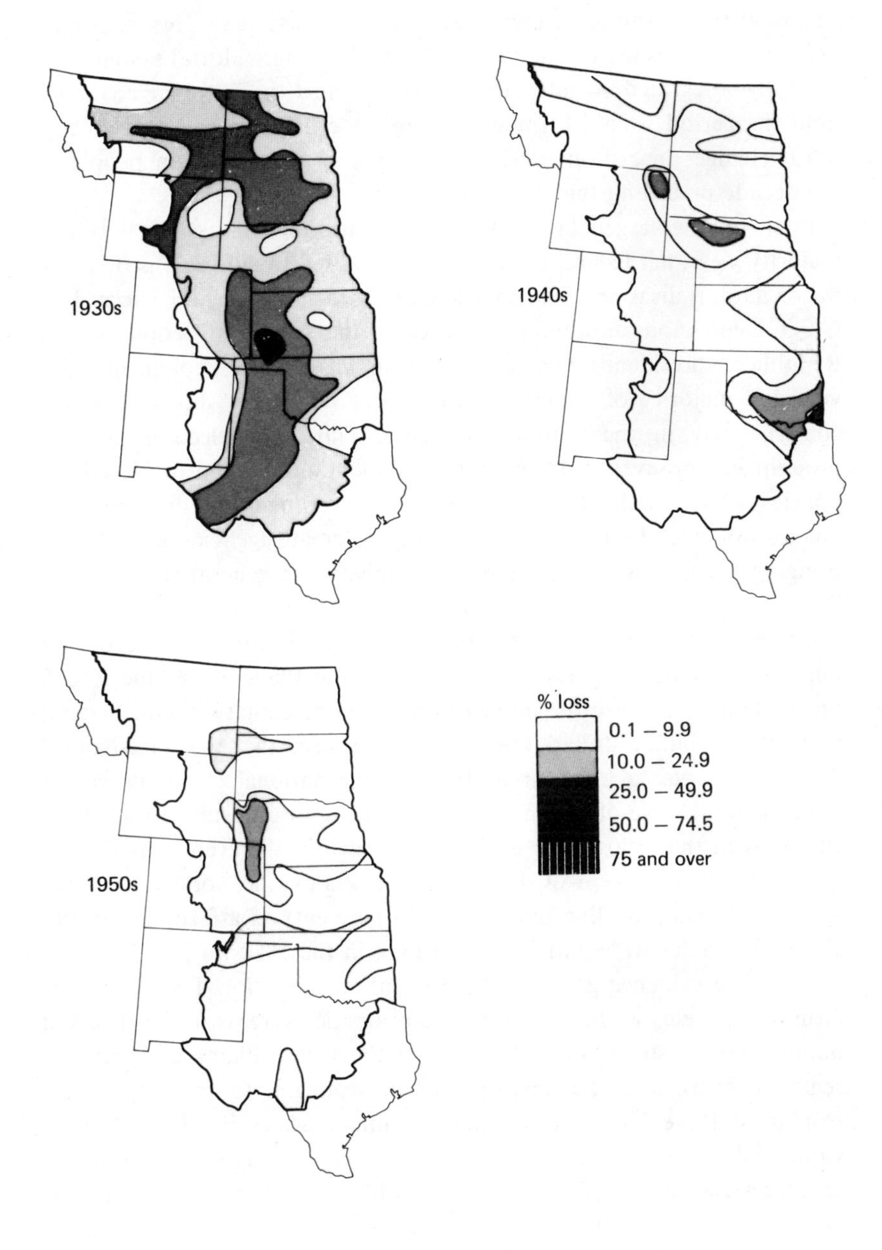

Fig. 21.7 (*continued*)

In the face of major climatic stress in the 1930s widespread system collapse like that in the western Plains in the 1890s and in the northwest Plains 1919–20 (minor climatic stresses) was averted. This apparent anomaly is explained by the fact that the Plains agricultural system was not fully integrated socially and politically into the national economy until the period 1925–34. Integration resulted from at least three significant changes in attitude and circumstance of the American people in the decade preceding the 1934–36 drought.

First, there emerged in the 1920s, following the Mississippi floods, a majority view that federal dollars should be used to aid regions beset by major natural disasters. Second, the Great Crash of 1929, the succeeding Great Depression, and the impatience of the American people with a Republican president's misplaced faith in *laissez-faire* capitalism, convinced a majority of Americans that the government should interfere constructively in the nation's economy. Third, the election of 1931 brought in Roosevelt, a Democratic president, committed to New Deal policies which made it possible for the government to disburse vast amounts of federal funds to the needy and to create agencies dedicated to bringing to the country's economic periphery the general standards of living enjoyed in the economic core.

These changes conspired to make massive federal aid to drought-impacted regions obligatory. The people of the Plains were able to call on the nation to share the impact of an extreme climatic event. Federal relief of 2.48 billion dollars (1957 dollars) ensured that the 'catastrophe' of the 1930s was a large ripple through the national economy, rather than the tidal wave of system collapse on the entire western front of the Plains as in the 1890s. This is clearly reflected in the evenly distributed loss of about 6 per cent of the population Plainswide, compared to the major population decline (more than 25 per cent) on a farming frontier 150 to 250 miles wide and 700 miles long in the 1890s.

From the evidence gathered on population migration, farm transfers, farm income, etc., we have witnessed a dramatic decrease in the effects of minor climatic stress on the society of the Great Plains at local and regional levels, and the development of strategies to mitigate 'catastrophe' at these levels during major climatic stress by devolution of vulnerability to the nation. A major issue is whether the impacts of successive major climatic stresses will be transferred outside the borders of the Great Plains to the international as well as the national level.

Examining the catastrophe hypothesis

In the 1978–79 crop year, 98 per cent of global exports came from five countries, the United States (45 per cent); Canada (20 per cent); Australia (14 per cent); France (14 per cent); and Argentina (4 per cent) (USDA, 1980). This represented 16 per cent of total wheat produced. Such a concentration of available wheat supplies has led to much speculation regarding the possible global impacts of a severe production decline in the United States. What would be the global repercussions if a severe drought were to cripple the Great Plains, a region where 60 per cent of the nation's wheat is grown? Could such an event precipitate an increase in global levels of starvation and malnutrition? To examine this question, we have investigated, in preliminary fashion, the possible range of impact that a recurrence of a drought of magnitude similar to the 1930s would have on US wheat production, and ultimately, on levels of international trade in 1985.

Two estimates of a 1930s-type drought impact on 1985 US wheat production were derived from existing studies: (*a*) a lower impact which assumes 'lessening' and (*b*) a higher impact which assumes effects similar to the 1930s. The lower impact estimate is based on Newman's 1978 study which argues that wheat production has become less variable over time. This he attributes to technology, and increased geographical diversification of the US wheat crop. If a major drought were to occur, so it is argued, not all wheat growing areas would be affected adversely. Based on Newman's data and reasoning, a recurrence of a 1930s drought would result in a 17 per cent decline in production for the Great Plains only.

The high impact estimate was derived from two separate studies, those of the Institute of Ecology (1976) and NOAA (1973). Both studies examine historical yield data and reach the conclusion that yield variability has been unaffected by technology over time. Based on their calculation of deviation from expected yields (fitted trend lines), a recurrence of the worst years of the 1930s could result in about a 21 per cent decline in US wheat production.

The two drought impact estimates were then applied to two 1985 production/export demand scenarios. These were derived from the United States Department of Agriculture (1978) model of the global grain, oilseed, and livestock (GOL) economy. The high-production/export scenario assumes a high level of US participation (63 per cent of total trade) in a high level of international wheat trade (3.2 billion bushels). The low-production/export scenario assumes a low US participation (30 per cent of total trade) in a low level of international trade (2.1 billion bushels). In all cases it is assumed that domestic consumption is constant at 0.89 billion bushels per year.

Amounts of wheat available for export under drought and no drought conditions were calculated for the two scenarios and are indicated in Table 21.4. Under the high-production/export scenario – which implies a high level of dependence on US supplies – the US is predicted to contribute 2.00 billions, or 63 per cent, to total trade. This assumes favourable weather. Under the low drought impact, US exports would be cut to 1.71 billion bushels, or 54 per cent of total trade, a 9 per cent decline. Under the high drought impact, with the same trade assumptions, US wheat available for export declines even further, by 19 per cent.[4]

An assessment of US export potential assuming a low-production/export demand reveals that the US contribution to total trade declines 7 per cent under the low drought impact and 15 per cent under the high impact case. This represents a decline of 0.16 and 0.32 billion bushels respectively.

In terms of unexpected declines in the US contribution to total wheat supply, a 19 per cent decline under a high dependency situation translates into 21 days of total wheat consumption. This assumes that total world wheat utilisation is equal to 0.027 billion bushels or 27 million bushels/day.[5] A 7 per cent reduction in available supplies, under the assumptions of low drought impact on a lower US involvement in world trade translates into 6.7 days of total wheat consumption. This assumes that total world wheat utilisation is equal to 0.024 billion bushels, or 24 million bushels/day.[6]

Whether a 19 per cent decline would have serious global consequences in terms of human impact is a question which we are presently addressing. Our first step has been to set some realistic ranges of potential Great Plains drought on the global system. The next step will be to translate these declines into effect on societal well-being.

Great Plains conclusions

We are not yet prepared to hand down final judgements about the lessening and catastrophe hypotheses as they apply to the Great Plains. But some of the pieces are coming together.

First, *with respect to drought–wheat-yield relationships we find no strong evidence to support the lessening effect.* Rather, the literature and our own analyses demonstrate considerable ambiguity.

Table 21.4. *Potential contribution and drought induced declines in US wheat to projected 1985 international exports, in bushels, and as a percentage of total demand*

Production/ Export demand scenarios	Drought impact		
	None	Low (17% decline in GP)	High (21% decline in US)
High dependency on US supplies	63% of a total trade of 3.2 bil. bu. =2.00 bil. bu.	1.71 bil. bu.= 54% of total trade 9% decline (0.29 bil. bu.)	1.39 bil. bu.= 44% of total trade 19% decline (0.61 bil. bu.)
Low dependency on US supplies	30% of a total trade of 2.1 bil. bu. =0.64 bil. bu.	0.48 bil. bu.= 23% of total trade 7% decline (0.16 bil. bu.)	0.32 bil. bu.= 15% of total trade 15% decline (0.32 bil. bu.)

Methodologically, our findings suggest that complementary analyses at smaller scales, such as counties or crop-reporting districts, would shed light on this question.

Second, *with respect to the relationship between agricultural drought and societal impact at local and regional levels, 'lessening' appears strongly evident.* Certainly, with the comparison of population migration and farm transfers through time, lessening is the case. Additionally, since the one major drought occurrence in the Great Plains in the 1930s and the massive involvement of the nation in (drought) relief, the effects of even the minor droughts, confined to local areas before the 1930s are being diffused through space to the greater social system of national and, possibly global scope.

Our preliminary efforts at defining the range of impact of a major Great Plains drought at the global level suggest that a *recurrence of a 1930s drought could cause declines of 7 per cent to 19 per cent of total wheat export in 1985*, under varying assumptions of production, export levels, and weather impact. We do not know whether a Plains 'catastrophe' would be mitigated by the nation and controlled within its confines as it was in the 1930s or whether 'catastrophe' would ripple into other economies less able to adjust to sudden loss of (food) staples.

Ultimately, it is the answer to this question that concerns us – and the world – the most.

Lessening and catastrophe

At the scale of decades in the Great Plains and in the Sahel, a *lessening* in the impact of climate fluctuation upon population is evident. Following the early agricultural occupation of the Great Plains by a people who were unaware that the region was prone to regular droughts, there was a disastrous loss of population during the first drought decade (1890s). The Great Plains (during the droughts of the 1890s and 1910s) and the Sahel in the two major droughts of this century (during which intensive agriculture was replacing extensive pastoral nomadism) were not fully integrated socially and politically into the national or global economy. They were economic colonies being brought into the 'periphery' (Frank, 1970). The wider (national or global) economies in which these regional economies were embedded were neither willing nor obligated to share the two regions' vulnerability to drought and to assume responsibility for aid. Both systems had limited technological ability and social–political organisation with which to adjust to minor climatic stress, and each experienced widespread losses of population through death by starvation and by outmigration.

Technological innovations, from within and outside the system, and the development of political and economic links with systems having different climatic regimes were matched by a relative lowering in the population loss during the drought periods that followed the first one. To this point the lessening hypothesis is confirmed. But the most recent climatic fluctuations in the Great Plains, when related to population changes and to finer-grained indicators of stress suggest that after the initial phases of lessening there develops a near immunity to minor climatic stress rather than a continued history of lessening.

In our three study cases the only clear test for the catastrophe hypothesis is the major climatic stress of the 1930s in the Great Plains. Sizeable population losses occurred uniformly across the Plains, and a major population decline and related temporary collapse of the social–economic system occurred in a few places in the heart of the Dust Bowl. But there was no system collapse in the 1930s, and nothing to approach the system collapse in the 1890s in which thousands of square miles on the dry margins of the Plains were effectively depopulated for at least four years. Changes in American political, economic, and disaster relief philosophies occurring in the ten years before the onset of drought (1934), and not prompted by the mid-1930s drought, ensured that the Plains were a full part of the open and enlarging national economy which directly shared the drought impact. Given the post-New Deal policy of the US Government toward its internal regions, system collapse within the Great Plains appears unlikely unless the soil and water resource is more fragile than is at present believed (Bowden, 1977). However, a serious setback in present levels of production and population could occur if major climatic stress Plainswide should coincide with a prohibitively sharp rise in the costs of energy now so necessary to maintain the increasingly irrigated Plains agriculture. This could prompt a widespread reversion to extensive ranching and dry-farming practices, and produce a ripple effect of economic/social stress into countries that depend increasingly on US grain.

There have been no '*major*' climatic stresses in the Sahel this century. Nevertheless, another '*minor*' climatic stress like that of 1910–15 or 1968–74, in conjunction with the population growth postulated in the Matlock & Cockrum (1974) model and with multifarious social, political, and economic changes could precipitate a system collapse in the Sahel, as could the continued and related exploitation of an already fragile resource base.

Of course, the climatic stresses of the last century in the Sahel and the Great Plains may prove to be events in long-term 'catastrophes' such as those that terminated the first and second cycles of population growth in the Tigris–Euphrates Valley. Attractive (and simple) as is the thesis of environmental stress as the cause of these long-term 'catastrophes', our research casts doubt on it. These two prolonged declines in population in the Tigris-Euphrates Valley constitute long-term 'catastrophes' of far greater magnitude than those experienced and foreseen in the Plains and the Sahel. They were the result, our studies maintain, of a combination of factors: political and social instability, an absence of technological innovation and, to a limited extent, climatic and hydrologic stress.

Acknowledgements

This chapter is based on research supported under grant ATM 77-15019 from the National Science Foundation and contract AID/otr 147–79006 with the Agency for International Development. The opinion expressed in this chapter are our own and do not necessarily reflect the views of the National Science Foundation or the Agency for International Development.

Notes

1 Pourafzal searched in Dakar the *Reports of the Governor-Generals* of the former French colonies of West Africa for the years 1910–15. Watts' unpublished report was based on field work and archival research. They are not responsible for our interpretation of their data and indeed we differ in interpretation in some instances.

2. Climatic divisions are sections of the country delimited by the US Weather Bureau for which climatological data are made available. The Palmer Index (PI) is a measure of soil moisture surplus or deficit in terms of deviation from a climatological and hydrological normal. Zero represents normal; negative deviation corresponds to moisture deficit, and positive deviation indicates surplus. PI values were obtained from the National Climatic Center, Asheville, North Carolina.

3. Colorado, Kansas, Montana, Nebraska, North Dakota, Oklahoma, South Dakota, and Texas.

4. The low drought impact figures are calculated by adding the adjusted production for Great Plains acreage with the acres not in the region, where no drought impact is assumed. Domestic consumption (0.89 billion bushels) is then subtracted from that sum to give wheat available for export. For the high-impact scenario, a yield reduction of 21 per cent was calculated for all US acres, and domestic consumption was then subtracted out.

5. Based on USDA (1978) GOL model assumption that 1985 total wheat utilisation is 9.8 billion bushels per year for the high-production/export demand scenario.

6. Based on USDA (1978) GOL model assumption that 1985 total wheat utilisation is 8.8 billion bushels per year.

References

Adams, R. McC. (1965). *Land Behind Baghdad: A History of Settlement on the Diyala Plains.* Chicago: University of Chicago Press.

Adams, R. McC. (1966). *The Evolution of Urban Society.* Chicago: Aldine.

Adams, R. McC. (1981). *Heartland of Cities: Surveys of Ancient Settlement and Land Use on the Central Flood Plain of the Euphrates.* Chicago: University of Chicago Press.

al-Khashab, W. H. (1958). *The Water Budget of the Tigris and Euphrates Basin.* University of Chicago, Department of Geography, Research Paper 54.

Berg, E. (1975). *The Recent Economic Evolution of the Sahel.* Ann Arbor: Center for Research on Economic Development, University of Michigan.

Beug, H. J. (1967). Contributions to the postglacial vegetational history of Northern Turkey. In *Quaternary Paleoecology*, ed. E. J. Cushing & H. E. Wright, Jr, pp. 349–56. New Haven: Yale University Press.

Bonte, P. (in draft). Drought in Sahel: transformation of the Sahelian pastoral and agricultural systems. In *The Roots of Catastrophe*, ed. R. Garcia, vol. 3, Chapter 2. (IFIAS Report on 'Drought and man: the 1972 case history').

Bottero, J., Cassin, E. & Vercoutter, J. (eds.) (1967). *The Near East: The Early Civilizations*, transl. R. F. Tannenbaum. New York: Delacorte Press.

Bowden, M. J. (1977). Desertification of the Great Plains: will it happen? *Economic Geography*, **63**, 397–406.
Caldwell, J. C. (1975). *The Sahelian Drought and its Demographic Implications* (Overseas Liaison Committee Paper No. 8). Washington: American Council on Education.
Center for Disease Control (CDC) (1973). Nutritional surveillance in drought affected areas of West Africa (Mali, Mauritania, Niger, Upper Volta), August–September 1973. Atlanta, Center for Disease Control. Unpublished Report.
Center for Disease Control (CDC) (1975). Protein/energy undernutrition surveys in the Sahel, 1974 & 1975. Atlanta: Center for Disease Control. Unpublished Report.
Childe, V. G. (1952). *New Light on the Most Ancient East*. New York: Praeger.
Clawson, M., Landsberg, H. H. & Alexander, L. T. (1971). *The Agricultural Potential of the Middle East*. Amsterdam: Elsevier.
Comité Information Sahel (1974). *Qui Se Nourrit de la Famine en Afrique*? Paris: Maspero.
Diester-Haass, L. (1973). Holocene climate in the Persian Gulf as deduced from grain-size and pteropod distribution. *Marine Geology*, **14**, 207–23.
Eddy, A. & Cooter, E. (1978). *A Drought Probability Model for the USA Northern Great Plains*. Norman: University of Oklahoma, Department of Meteorology.
Erinc, S. (1978). Changes in the physical environment in Turkey since the end of the last glacial. In *The Environmental History of the Near and Middle East Since the Last Ice Age*, ed. W. C. Brice, pp. 87–110. London: Academic Press.
Faulkingham, R. H. (1977). Ecologic constraints and subsistence strategies. The impact of drought in a Hausa village, a case study from Niger. In *Drought in Africa*, vol. 2, ed. D. Dalby, R. J. Harrison Church & F. Bezzaz, pp. 148–58. London: International African Institute.
Flannery, K. V. (1965). The ecology of early food production in Mesopotamia. *Science*, **147**, No. 3663, 1247–56.
Frank, A. G. (1970). The development of underdevelopment. In *Imperialism and Development*, ed. R. I. Rhodes, pp. 4–17. New York: Monthly Review Press.
Garcia, R. & Escudero, J. (in draft). *The Constant Catastrophe*, vol. 2 (IFIAS Report on 'Drought and Man: the 1972 Case History').
Gasser, W. R. (1976). *World Climatic Change and Agriculture: The Issues*. Reston, Va. Symposium on living with climatic change.
Gibson, M. (1974). Violation of fallow and engineered disaster in Mesopotamian civilization. In *Irrigation's Impact on Society*, ed. T. E. Downing & M. Gibson, pp. 7–19. Anthropological Papers of the University of Arizona, No. 25. Tucson: University of Arizona Press.
de Goyet, E. J. & de Ville, C. (1976). Preliminary report of an anthropometric survey during a nutritional relief program in Niger. *Social and Occupational Medicine*, **4**, 70.
Haigh, P. (1977). *Separating the Effects of Weather and Management on Crop Production*. Columbia, Mo.: James D. McQuigg Consulting.
Imperato, P. J. (1976). Health care systems in the Sahel before and after the drought. In *The Politics of Natural Disaster: The Case of the Sahel Drought*, ed. M. Glantz, pp. 282–302. New York: Praeger.

Institute of Ecology (1976). *Impact of Climatic Fluctuation on Major North American Food Crops.* Climate/Food Study Panel Report. July.

Jacobsen, T. & Adams, R. McC. (1958). Salt and silt in ancient Mesopotamian agriculture. *Science*, **128**, No. 3334, pp. 1251–8.

Kempe, S. & Degens, E. T. (1978). Lake Van varve record: the past 10 420 years. In *The Geology of Lake Van*, ed. E. T. Degens & F. Kurtman, pp. 56–63. Ankara: The Mineral Research and Exploration Institute of Turkey.

Kloth, T. I. (1974). Sahel nutrition survey, 1974. Atlanta: Center for Disease Control. Unpublished report.

Lamb, H. H. (1977). *Climate: Present, Past and Future*, vol. 2, *Climatic History and the Future*, London: Methuen.

Le Strange, G. (1900). *Baghdad during the Abbasid Caliphate.* Oxford: Oxford University Press.

McNeill, W. H. (1976). *Plagues and Peoples.* Garden City, New York: Anchor Press/Doubleday.

Mallowan, M. E. L. (1971). The early dynastic period in Mesopotamia. In *Cambridge Ancient History*, ed. I. E. S. Edwards, C. J. Gadd & N. G. L. Hammond, vol. 1, pp. 238–314. Cambridge: Cambridge University Press.

Matlock, G. W. & Cockrum, E. W. (1974). *A Framework for Agricultural Development Planning*, vol. 2. Cambridge, Mass: Center for Policy Alternatives.

Michaels, P. J. (1978). *A Predictive Model for Winter Wheat Yield in the United States Great Plains.* Report No. 94. Institute of Environmental Studies. Madison: University of Wisconsin.

Mitchell, J. M., Stockton, C. W. & Meko, D. M. (1978). *Evidence of a 22-year Rhythm of Drought in the Western United States Related to the Hale Solar Cycle Since the 17th Century. Proceedings, Symposium/Workshop on Solar-Terrestrial Influences on Weather and Climate* (July 24–28). Columbus: Ohio State University.

National Oceanic & Atmospheric Administration (NOAA). (1973). The Influence of Weather and Climate on United States Grain Yields: Bumper Crops or Droughts. A Report to the Associate Administrator for Environmental Monitoring and Prediction, NOAA. Washington, DC: United States Department of Commerce.

Neumann, J. & Sigrist, R. M. (1978). Harvest dates in ancient Mesopotamia as possible indicators of climate. *Climatic Change*, **1**, 239–52.

Newman, J. E. (1978). Drought impacts on American agricultural productivity. In *North American Droughts*, ed. N. J. Rosenberg, pp. 43–62. Boulder, Colorado: Westview Press.

Niklewski, J. & van Zeist, W. (1970). A late Quaternary pollen diagram from northwestern Syria. *Acta Botanica Neerlandica*, **19**, 737–54.

Post, J. (1977). *The Last Great Subsistence Crisis in the Western World.* Baltimore: The Johns Hopkins University Press.

Raynaut, C. (1977). Lessons of a crisis. In *Drought in Africa*, vol. 2, ed. D. Dalby, R. J. Harrison Church & F. Bezzaz, pp. 17–32. London: International African Institute.

Redman, C. L. (1978). *The Rise of Civilization.* San Francisco: W. H. Freeman.

Schneider, S. H. & Temkin, R. L. (1978). Climatic changes and human affairs. In

Climatic Change, ed. J. Gribben, pp. 228–46, Cambridge: Cambridge University Press.

Schoell, M. (1978). Oxygen isotope analysis on authigenic carbonates from Lake Van sediments and their possible bearing on the climate of the past 10 000 years. In *The Geology of Lake Van*, ed. E. T. Degens & F. Kurtman. Ankara: The Mineral Research and Exploration Institute of Turkey.

Seamon, J., Rivers, J., Holt, J. & Murlis, J. (1973). An inquiry in the drought situation in Upper Volta. *The Lancet*, **2**, 774–8.

Sircoulon, J. (1976). Les données hydropluviométriques de la sécheresse récente in Afrique intertropicale: comparaison avec les sécheresses '1913' et '1940'. *Cahiers ORSTOM*, séries hydrologiques, **13**, 75–174.

Smith, P. & Young, T. C. Jr, (1972). The evolution of early agriculture and culture in Greater Mesopotamia: a trial model. In *Population Growth: Anthropological Implications*, ed. B. J. Spooner, pp. 1–59. Cambridge, Mass: MIT Press.

Sousa, A. (1965). *Faydānāt Baghdad fi 't-Tarīkh* (in Arabic), vol. 2. Baghdad: Matbal'a al-adīb.

Swift, J. (1977). Sahelian pastoralists: underdevelopment, desertification and famine. *Annual Review of Anthropology*, **6**, 457–78.

United States Department of Agriculture (USDA). (1974). *The World Food Situation and Prospects to 1985.* Foreign Agricultural Economic Report No. 98.

United States Department of Agriculture (USDA). (1978). *Alternative Futures for World Food in 1985: World G.O.L. Model.* Foreign Agricultural Economic Report No. 146.

United States Department of Agriculture (USDA). (1980). *Agricultural Statistics.*

van Zeist, W. & Bottema, S. (1977). Palynological investigations in western Iran. *Palaeohistoria*, **17**, 19–85.

van Zeist, W. & Woldring, H. (1978). A postglacial pollen diagram from Lake Van in east Anatolia. *Review of Palaeobotany and Palynology*, **26**, 249–76.

van Zeist, W., Woldring, H. & Stapert, D. (1975). Late Quaternary vegetation and climate of southwestern Turkey. *Palaeohistoria*, **17**, 53–143.

Walters, S. G. (1970). *Water for Larsa: An Old Babylonian Archive Dealing with Irrigation.* New Haven: Yale University Press.

Ware, H. (n.d.). *The Sahelian Drought: Some Thoughts on the Future.* Mimeo. New York: United Nations, Special Sahelian Office.

Watts, M. (1978). Famine over northern Nigeria in 1913–14 and 1972–74. A report to the Climate and Society Project, Clark University. Unpublished Report.

Wendland, W. M. & Bryson, R. A. (1974). Dating climatic episodes of the Holocene. *Quaternary Research*, **4**, 9–24.

Wittfogel, K. A. (1957). *Oriental Despotism.* New Haven: Yale University Press.

Wright, H. E. Jr (1976). The environmental setting for plant domestication in the Near East. *Science*, **196**, No. 4263, 385–9.

Author index

Italics indicate pages on which references are listed

Perring, F. H., 117, *137*

Subject index